# 25th International Conference on Atomic Physics (ICAP2016)



Seoul, Korea
24 - 29 July 2016

ISBN: 978-1-5108-7737-5

**Printed from e-media with permission by:**

Curran Associates, Inc.
57 Morehouse Lane
Red Hook, NY 12571

**Some format issues inherent in the e-media version may also appear in this print version.**



Printed by Curran Associates, Inc. (2019)

For permission requests, please contact Institute of Applied Physics
at the address below.

Institute of Applied Physics
Seoul National University
Building 56-212
1 Gwanak-ro, Gwanak-gu
Seoul 08826
South Korea

Phone: 82-2-880-8515
Fax: 82-2-876-2590

http://phya.snu.ac.kr/IAP

**Additional copies of this publication are available from:**

Curran Associates, Inc.
57 Morehouse Lane
Red Hook, NY 12571 USA
Phone: 845-758-0400
Fax: 845-758-2633
Email: curran@proceedings.com
Web: www.proceedings.com

# ICAP2016

25th International Conference on
Atomic Physics

24 – 29 July, 2016
Seoul, Korea

Book of Abstracts

# Foreword

The 25th International Conference on Atomic Physics takes place in the COEX Seoul. Following the tradition of ICAP, the conference presents an outstanding program of invited speakers covering the most recent subjects in the field of atomic physics, such as:

- Ultracold gases and Bose Einstein condensates,
- Ultracold Fermi gases,
- Fundamental atomic tests and measurements,
- Precision measurements, atomic clocks and interferometers,
- Quantum information and simulations with atoms and ions,
- Quantum optics and cavity QED with atoms,
- Atoms and molecules in optical lattices,
- From two-body to many-body systems,
- Ultrafast phenomena and free electron lasers,
- Beyond atomic physics (optical imaging, optomechanics...).

The program includes 39 invited talks, 9 'hot topic' talks and one IUPAP young scientist award talk.

This book of abstracts gathers the contributions of these talks and of more than 400 posters, organized in three sessions. The proceedings of the conference will be available online in open access.

On behalf of the committees, we would like to welcome you in Seoul, and to wish you an exciting conference.

# Committees

## Program Committee

| | | | |
|---|---|---|---|
| Charles Adams | UK | Kyungwon An | Korea |
| Cristiane Morais-Smith | Netherland | Xinye Xu | China |
| Cappellaro Paola | USA | Xuzong Chen | China |
| Wolfgang Ertmer | Germany | Maciej Lewenstein | Spain |
| Massimo Inguscio | Italy | Victor Balykin | Russia |
| Peter Hannaford | Australia | Cindy Regal | USA |
| Anne L'Huillier | Sweden | John Doyle | USA |
| Philipp Treutlein | Switzerland | Gretchen Campbell | USA |
| Yoshiro Takahashi | Japan | Raymond Laflamme | Canada |
| Michele Leduc | France | | |

## Local Organizing Committee

| | |
|---|---|
| **Wonho Jhe** (Co-chair) | Seoul National University |
| **Donghyun Cho** (Co-chair) | Korea University |
| **Young Soon Kim** (Co-chair) | Myongji University |
| Jaewook Ahn | KAIST |
| Kyungwon An | Seoul National University |
| Jai-Min Choi | Chonbuk National University |
| Hyunseok Jeong | Seoul National University |
| Jaewan Kim | KIAS |
| Jin-Tae Kim | Chosun University |
| Jung Bog Kim | Korea National University of Education |
| Yoon-Ho Kim | POSTECH |
| Jinhyoung Lee | Hanyang University |
| Heung-Ryoul Noh | Chonnam University |
| Yong-il Shin | Seoul National University |
| Dae-Hyuk Yu | KRISS |
| Wonmin Son | Sogang University |
| Han Seb Moon | Pusan National University |
| So-young Shin | Secretariat |

## International Advisory Committee

| | |
|---|---|
| Ennio Arimondo | Italy |
| Hans Bachor | Australia |
| Vanderlei Bagnato | Brazil |
| Viktor Balykin | Russia |
| Rainer Blatt | Austria |
| Immanuel Bloch | Germany |
| Gretchen Campbell | USA |
| Claude Cohen-Tannoudji | France |
| Robin Côté | USA |
| Gordon Drake | Canada |
| Norval Fortson | USA |
| Philippe Grangier | France |
| Theodor Hänsch | Germany |
| Peter Hannaford | Australia |
| Serge Haroche | France |
| Ed Hinds | UK |
| Massimo Inguscio | Italy |
| Wolfgang Ketterle | USA |
| Daniel Kleppner | USA |
| Michèle Leduc | France |
| Hélène Perrin | France |
| William Phillips | USA |
| Lev Pitaevskii | Russia |
| Trey Porto | USA |
| Fujio Shimizu | Japan |
| Winthrop Smith | USA |
| Jook Walraven | Netherland |
| David Wineland | USA |
| Tsutomu Yabuzaki | Japan |
| Mingsheng Zhan | China |

# Sponsors

IUPAP

APCTP
Asia Pacific Center for Theoretical Physics

KOREA UNIVERSITY

KOFST
THE KOREAN FEDERATION OF SCIENCE AND TECHNOLOGY SOCIETIES

SEOUL NATIONAL UNIVERSITY

POSTECH
POHANG UNIVERSITY OF SCIENCE AND TECHNOLOGY

MenloSystems

TOPTICA PHOTONICS

SK telecom

ibs CoReLS
Center for Relativistic Laser Science

KPS 한국물리학회
The Korean Physical Society

KIAS KOREA INSTITUTE FOR ADVANCED STUDY

JOINT QUANTUM INSTITUTE

cua

iap INSTITUTE OF APPLIED PHYSICS
SEOUL NATIONAL UNIVERSITY

KAERI Korea Atomic Energy Research Institute

SEOUL METROPOLITAN GOVERNMENT

KOREA TOURISM ORGANIZATION

# Exhibitors

**Springer Nature**
Mapo-gu, Seoul, Korea

Smith Ahram Chae
Smith.chae@springer.com

**NUFERN**
East Granby, CT, United States

Robin Tonar
rtonar@nufern.com

**Jinsung Instruments, Inc.**
Seo-Gu, Deajeon, Korea

Jinsuk Kim
jskim@jinsunginst.com

**TOPTICA Photonics AG**
Lochhamer Schlag 19
82166 Gräfelfing, Munich
Germany
Dr. Stephan Falke
Stephan.falke@toptica.com

**OMA Company**
Yuseong-Gu, Daejeon, Korea

Dae-Gyu Kim
dgkim@omacom.co.kr

**American Physical Society**

Gordon W.F. Drake
pra@aps.org

**Menlo Systems**
D-82152 Martinsried, Germany

Christina Mariel
c.mariel@menlosystems.com

**Rhysics Photronics**
Yuseong-Gu, Daejeon, Korea

Senung-churl Lee
sclee@rphotronics.com

**RayVis INC.**
Seocho-dong, Seocho-gu, Seoul
Korea
Ha, Yoonmi
info@rayvis.co.kr

**M Squared Lasers Ltd**
Maryhill Rd, Glasgow
Scotland
Julie Phinn
Julie.Phinn@m2lasers.com

**Oxford University Press**
Great Clarendon Street, Oxford
UK
Katie Peaper
gab.exhibitions.uk@oup.com

**Quantaser Photonics Co. Ltd**
Roosevelt Rd. Da'an Dist. Taipei
Taiwan
Tina Hsing
tina@quantaser.com

# Table of Contents

## Invited Talks

Friday 29 July, 8:30 am - 10:15 am
**Cold molecules** Chair: J. Ye

**Direct laser cooling and trapping of diatomic molecules**
David DeMille (45min) ..77

**Laser cooling and trapping of molecules: experiments and models**
Michael Tarbutt (30min) ..78

**Coherent control of ultracold $^{87}Rb^{133}Cs$ molecules**
Simon Cornish (30min) ..79

Friday 29 July, 10:40 am - 12:20 pm
**Hot Topics** Chair: G. Jo

**Ultracold chemistry and asymptotic physics with diatomic strontium molecules**
Robert Moszynski (25min) ..80

**Experimental realization of a two-dimensional synthetic spin-orbit coupling in ultracold Fermi gases**
Jing Zhang (25min) ..81

**Cat states and quantum metrology with Rydberg atoms**
Jean-Michel Raimond (25min) ..82

**Einstein-Podolsky-Rosen entanglement in spinor Bose-Einstein condensates**
Carsten Klempt (25min) ..83

Friday 29 July, 12:20 pm - 12:35 am
**Closing** Chair: D. Cho

# Poster Session

## Monday 25 July, 12:25 pm – 14:50 pm

### Precision Measurements and Fundamental Tests

## Spectroscopy, Atomic and Molecular Structure

## Cooling and Trapping of Atoms and Ions

## Mesoscopic Quantum Systems



## Ultracold Mixtures and Molecules

## Intense Fields and Ultrafast Phenomena



## Few Body Interactions and Collisions

# Tuesday 26 July, 12:25 pm – 14:50 pm

## Atom Interferometry

## Bose Gases

## Fermi Gases

## Rydberg and Artificial Atoms

# Thursday 28 July, 12:25 pm – 14:50 pm

## Atomic Clocks



## Quantum Simulation

## Quantum Optics & Cavity QED

## Quantum Information

## Trapped Ions

## Postdeadline

# Invited Talks

# From evaporative cooling and BEC in hydrogen to microscopy innovation in biology

Harald Hess
HHMI, Janelia Research Campus
19700 Helix Drive
Ashburn, VA 20147 USA

The first search to observe BEC in spin-polarized hydrogen began in the early 1980's and employed confinement in cryogenic cells. Progress was halted abruptly with the recognition that three-body recombination blocked these approaches to BEC: at higher temperatures, volume recombination was prohibitive; at lower temperatures, recombination on the cryogenic surface was prohibitive. Switching to the high-field seeking states and employing a pure magnetic trap made surface-free confinement possible but required a new cooling method. We demonstrated that by lowering the depth of the trap, the more energetic atoms would escape and as the gas re-thermalized its temperature would drop [1]. Subsequently we switched to rf evaporation, which greatly improved the cooling efficiency. The process turned out to be extremely well suited to the alkali metal atoms where it enabled the first observation of BEC in a gas. Several years later, it enabled us to observe of BEC in hydrogen.

The lessons learned as a postdoc in the BEC challenge, both technical and philosophical, proved to be a good base for a career in numerous other physics topics. I will describe a personal path that lead to low temperature scanning tunneling electron microscopy of vortices and a series of other scanned probe microscopes for studying nanoscale physics at Bell Labs. A focus on the high throughput, large scale imaging challenges of semiconductor industry provided a new perspective to imaging and later on offered the liberating horizons of unemployment. Together with another jobless ex-Bell Labs colleague, we conceived of a new super-resolution optical microscope [2]. This led to employment and a variety of 3D volume microscopy challenges and new instrumentation for biology, some of which are inspired by research in atomic physics.

[1] H. Hess, Phys. Rev B **34**, 5, 3476 (1986).
[2] E. Betzig et al, Science **313**, 1642 (2006).

# Helium films at ultralow temperatures: from strongly correlated atomically layered films to topological mesoscopic superfluidity.

John Saunders

Department of Physics, Royal Holloway University of London, U.K.

Helium films provide model systems for strongly correlated quantum matter and topological superfluidity. The bottom-up approach is to grow atomically layered helium films on the surface of graphite. Here we find: the Mott-Hubbard transition of a $^3$He monolayer into a putative quantum spin liquid [1,2]; 2D solid $^3$He as a model for frustrated magnetism on a triangular lattice [3,4]; heavy fermion physics and quantum criticality in $^3$He bilayer films [5,6]; model 2D Fermi liquids with tuneable interactions in $^3$He atop a superfluid $^4$He film [7,8]; a novel condensate in a $^4$He monolayer with intertwined superfluid and density wave order [9]. The top-down approach involves the confinement of topological superfluid $^3$He in engineered nanoscale geometries. This controlled confinement has a profound effect on the p-wave superfluid order parameter [10], which can be determined by NMR [11]. New phases may be stabilized. In the quasi-2D limit, mesoscopic surface roughness can be used to introduce a quantifiable disorder potential [12]. Nano-fluidic confinement offers the prospect to study Majorana surface excitations (in time reversal invariant superfluid $^3$He-B) and Majorana-Weyl edge excitations (in chiral superfluid $^3$He-A), and to create hybrid structures for the study of topological mesoscopic superfluidity.

[1] A Casey, H Patel, J Nyéki, B P Cowan and J Saunders. Evidence for a Mott-Hubbard transition in a two dimensional $^3$He fluid monolayer. *Phys. Rev. Lett.* **90**, 115301 (2003)

[2] F Arnold, J Nyéki, B Cowan and J Saunders. *In preparation.*

[3] M Siqueira, J Nyéki, B Cowan and J Saunders. Frustration by Multiple Spin Exchange in 2D Solid $^3$He Films. *Phys. Rev. Lett.* **78**, 2600 (1997)

[4] A. Casey, M. Neumann, J. Saunders, N. Shannon. Two-Dimensional Ferromagnetism of a $^3$He Film: Influence of Weak Frustration. *Phys. Rev. Lett.* **111**, 125302 (2013)

[5] M Neumann, J Nyéki, B Cowan, J Saunders. Bilayer $^3$He: a simple two-dimensional heavy fermion system with quantum criticality. *Science* **317**, 1356 (2007)

[6] M Neumann, J Nyéki, L. Levitin, A. Casey, B Cowan, J Saunders. Kondo breakdown and frustrated magnetism in heavy fermion $^3$He bilayer films on graphite. *In preparation.*

[7] M Dann, J Nyéki, B Cowan and J Saunders. Quasiparticle interactions in two dimensional $^3$He on $^4$He films. *Phys. Rev. Lett.* **82**, 4030 (1999)

[8] A Waterworth, J Nyéki, B Cowan and J Saunders. *In preparation.*

[9] J Nyéki, A Phillis, A Ho, D Lee, P Coleman, J Parpia, B Cowan, J Saunders, Evidence for intertwined superfluid and density wave order in two dimensional $^4$He, *submitted.*

[10] L. Levitin, R. Bennett, A. Casey, B. Cowan, J. Saunders, D. Drung, T. Schurig, J. Parpia. Phase Diagram of the Topological Superfluid $^3$He Confined in a Nanoscale Slab Geometry. *Science* **340**, 841 (2013)

[11] L. Levitin, R. Bennett, E. Surovtsev, J. Parpia, B. Cowan, A. Casey, J. Saunders. Surface Induced Order Parameter Distortion in Superfluid $^3$He-B Measured by Non-Linear NMR. *Phys. Rev. Lett.* **111**, 235304 (2013)

[12] P. Sharma, A Corcoles, RG Bennett, JM Parpia, B Cowan, A Casey, J Saunders. Quantum transport in mesoscopic $^3$He films: experimental study of the interference of bulk and boundary scattering. *Phys. Rev. Lett.* **107**, 196805 (2011)

# Optical lattice clock and quantum many-body physics

Jun Ye[1]

[1]JILA, National Institute of Standards and Technology and University of Colorado, Boulder CO, USA

The progress of optical lattice clock to the accuracy of 2E-18 has benefited greatly from the understanding of atomic interactions. At the same time, the precision of clock spectroscopy has been applied to explore many-body spin interactions including SU(N) symmetry. Our recent work on this combined front of quantum metrology and many-body physics includes the probe of spin-orbital physics in the lattice clock and the investigation of a Fermi degenerate gas of tens of thousands Sr-87 atoms in a three-dimensional magic-wavelength optical lattice. We now demonstrate an atom-light coherent interaction time of 10 s, and show a clear path for improving the clock performance by another factor of 10.

# Frequency ratios of optical lattice clocks at the 17th decimal place

H. Katori[1,2,3,4], I. Ushijima[1,2,4], N. Nemitz[2,4], N. Ohmae[1,2,3], T. Takano[2,3], T. Akatsuka[1,2,3], S. Okaba[2,3,4], M. Das[1,2,4], T. Ohkubo[1,2,3], K. Yamanaka[1,2,3], T. Pruttivarasin[1], A. Yamaguchi[1,2,4] and M. Takamoto[1,2,4]

[1]Quantum Metrology Laboratory, RIKEN, Wako, Saitama 351-0198, Japan.

[2]Innovative Space-Time Project, ERATO, JST, Bunkyo-ku, Tokyo 113-8656, Japan.

[3]Department of Applied Physics, The University of Tokyo, Tokyo 113-8656, Japan.

[4]RIKEN Center for Advanced Photonics, Wako, Saitama 351-0198, Japan.

Optical lattice clocks benefit from a low quantum-projection noise by simultaneously interrogating a large number of atoms, which are trapped in an optical lattice tuned to the "magic wavelength" to largely cancel out light shift perturbation in the clock transition. About a thousand atoms enable the clocks to achieve $10^{-18}$ instability in a few hours of operation, allowing intensive investigation and control of systematic uncertainties. It is now the uncertainty of the SI second ($\sim 10^{-16}$) itself that restricts the measurement of the absolute frequencies of such optical clocks. Direct comparisons of optical clocks are, therefore, the only way to investigate and utilize their superb performance beyond the SI second.

In this presentation, we report on frequency comparisons of optical lattice clocks with neutral strontium ($^{87}$Sr), ytterbium ($^{171}$Yb) and mercury ($^{199}$Hg) atoms. By referencing cryogenic Sr clocks [1], we determine frequency ratios, $\nu_{Yb}/\nu_{Sr}$ [2] and $\nu_{Hg}/\nu_{Sr}$ [3], of a cryogenic Yb clock and a Hg clock with uncertainty at the mid $10^{-17}$ level. Such ratios provide an access to search for temporal variation of the fundamental constants. We also present remote comparisons between cryogenic Sr clocks located at RIKEN and the University of Tokyo over a 30-km-long phase-stabilized fiber link. The gravitational red shift $\Delta\nu/\nu_0 \approx 1.1\times10^{-18}\,\Delta h$ cm$^{-1}$ reads out the height difference of $\Delta h \sim 15$ m between the two clocks with uncertainty of 5 cm, which demonstrates a step towards relativistic geodesy. Finally, we report our ongoing experiments that reduce clock uncertainty to $10^{-19}$ by applying "operational magic frequency," [4], where light shifts due to dipole, multipolar, and hyper-polarizability effects effectively cancel out for a certain range of optical lattice intensity.

[1] I. Ushijima, *et al.*, Nature Photon. **9**, 185 (2015).
[2] N. Nemitz, *et al.*, Nature Photon. **10**, 258 (2016).
[3] K. Yamanaka, *et al.*, Phys. Rev. Lett. **114**, 230801 (2015).
[4] H. Katori, *et al.*, Phys. Rev. A **91**, 052503 (2015).

# Search for new physics with atomic clocks

Marianna Safronova
Department of Physics and Astronomy, University of Delaware, Newark, Delaware, USA and Joint Quantum Institute, NIST and the University of Maryland, College Park, Maryland, USA

The modern theories directed toward unifying gravitation with the three other fundamental interactions suggest variation of the fundamental constants and violation of Lorentz symmetry. While the energy scale of such physics beyond the Standard Model is much higher than that currently attainable by particle accelerators, both variation of the fundamental constants and Lorentz violation may nevertheless be detectable via precision measurements at low energies. I will give an overview of various searches for new physics with atomic clocks and focus on proposals for future experiments, in particularly with highly-charged ions.

Selected forbidden transitions in highly-charged ions (HCI) were shown to have very large sensitivities to $\alpha$-variation. Moreover, some highly-charged ions have level structure and other properties that are not present in any neutral and low-ionization state ions that may be advantageous for the development of atomic clocks and provide remarkable new opportunities for precision tests of fundamental science and quantum information. In 2015, sympathetic cooling of $Ar^{13+}$ with laser-cooled $Be^{+}$ ions has been demonstrated [1], elevating highly-charged ions to the realm of applications previously limited to singly-charged ions currently used for atomic clocks and quantum information. One of the main obstacles for the experimental work in this direction is the lack of experimental data for the relevant HCIs as well as difficulties in accurate theoretical predictions of the transition wavelengths and other properties. We carried out an exhaustive search of optical transitions in highly-charged ions that are particularly well suited for the experimental exploration [2]. I will present an overview of the HCI proposals highlighting the highly-charged ions that have all features of the best optical clock transitions leading to possibility of the frequency measurements with fractional accuracy at the level of $10^{-18}$ or better and have a factor of 100 extra enhancement of $\alpha$-variation in comparisons to experimental frequency ratio measurement accuracy [3].

I will also report a systematic study of sensitivities to Lorentz violation in atomic systems that identified ytterbium ion as an ideal system with high sensitivity as well as excellent experimental controllability [4]. By applying quantum information inspired technology to $Yb^{+}$, we expect tests of Lorentz violating physics in the electron-photon sector to reach levels of $10^{-23}$, five orders of magnitude more sensitive than the current best bounds. Similar sensitivities may be also reached with highly-charged ions.

[1] L. Schmöger *et al.*, Science **347**, 1233 (2015).
[2] M. S. Safronova, V. A. Dzuba, V. V. Flambaum, U. I. Safronova, S. G. Porsev, and M. G. Kozlov, Phys. Rev. Lett., **113**, 030801 (2014).
[3] V. A. Dzuba, M. S. Safronova, U. I. Safronova, and V. V. Flambaum, Phys. Rev. A **92**, 060502(R) (2015).
[4] V. A. Dzuba, V. V. Flambaum, M. S. Safronova, S. G. Porsev, T. Pruttivarasin, M. A. Hohensee, and H. Häffner, Nature Physics **12**, 465 (2016).

# Quantum magnetism with ultracold fermions

P. M. Duarte, R. A. Hart, T.-L. Yang, M. C. Revelle, B. A. Olsen, J. A. Fry, K. R. A. Hazzard, R. Telles, R. G. Hulet

Department of Physics and Astronomy, Rice University, Houston, Texas 77005, USA

Ultracold atomic gases confined to optical lattices have proven to be highly versatile and tunable systems for realizing novel quantum states of matter. We are using Fermi gases of $^6$Li atoms in our laboratory to explore several goals related to the strong correlations that arise in these systems. We have realized the Hubbard model, which has long been suspected of containing the essential ingredients of high temperature superconductivity. We measured the compressibility of the Mott insulating phase that occurs near half filling (1 atom/site), thus demonstrating the excitation gap of the Mott insulator [1]. Progress in this field, however, has been hampered by an inability to cool to low enough temperatures to achieve the most ambitious goals. To address this problem, we have developed the compensated optical lattice method to enable evaporative cooling in the lattice. With this method, we have cooled the Mott insulator sufficiently far to observe short-range antiferromagnetic correlations using Bragg scattering of light [2]. We are currently exploring new methods for entropy storage and redistribution to achieve even lower entropy in the antiferromagnetic phase.

Motivated by the enhancement of quantum correlations in low dimensions, we are also exploring Fermi gases in quasi-one-dimension (1D) [3]. A deep 2D optical lattice produces an array of 1D tubes which can be weakly coupled by reducing the lattice depth, thus increasing the lattice hopping $t$ between them. We observe a crossover from 1D-like to 3D-like behavior in the phase separation of a spin-imbalanced Fermi gas with increasing $t$. The crossover occurs at different $t$ for different interaction strengths, yet we find that the crossover location is universally dependent upon the scaled hopping $t/\varepsilon_b$, where $\varepsilon_b$ is the pair binding energy. The crossover regime is expected to be a fruitful regime for discovering the elusive polarized superfluid known as FFLO.

[1] P. M. Duarte, R. A. Hart, T.-L. Yang, X. Liu, T. Paiva, E. Khatami, R. T. Scalettar, N. Trivedi, and R. G. Hulet, Phys. Rev. Lett. **114**, 070403 (2015).

[2] R. A. Hart, P. M. Duarte, T.-L. Yang, X. Liu, T. Paiva, E. Khatami, R. T. Scalettar, N. Trivedi, D. A. Huse, and R. G. Hulet, Nature **519**, 211 (2015).

[3] M. C. Revelle, J. A. Fry, B. A. Olsen, and R. G. Hulet, arxiv:1605.06986.

# Many-body quantum phenomena in a dipolar Bose-Einstein condensate of erbium atoms

Francesca Ferlaino[1,2]
[1]Institut für Experimentalphysik, Universität Innsbruck
[2]Institut für Quantenoptik und Quanteninformation (IQOQI), Österreichische Akademie der Wissenschaften, 6020 Innsbruck, Austria

Given their strong magnetic moment and exotic electronic configuration, rare-earth atoms disclose a plethora of intriguing phenomena in ultracold quantum physics with dipole-dipole interaction. Here, we report on many-body phenomena in a Bose-Einstein condensate (BEC) with attractive dipolar interactions. When tuning the interactions, our measurements reveal the crossover from a dilute BEC to a high-density droplet-like state, which is stabilized by quantum fluctuations. Both the ground-state properties and the axial collective mode of the Bogoliubov spectrum agree with the Lee-Huang-Yang correction to the mean field approach. Finally, we present the first experimental study of an extended Bose-Hubbard model using bosonic Er atoms in a three-dimensional optical lattice and we report on the first observation of nearest-neighbor interactions.

# Quantum Computations and Quantum Simulations with Trapped $Ca^+$ Ions

Rainer Blatt

Institute for Experimental Physics,
University of Innsbruck, Technikerstrasse 25, A-6020 Innsbruck, Austria
and
Institute for Quantum Optics and Quantum Information,
Austrian Academy of Sciences, Otto-Hittmair-Platz 1, A-6020 Innsbruck, Austria

In this talk, the basic toolbox of the Innsbruck quantum information processor based on a string of trapped $Ca^+$ ions will be reviewed. For quantum information processing, the toolbox operations are employed for quantum computations [1] and for quantum simulations [2].

For quantum computation, a scalable Shor algorithm was realized [1] with a string of trapped $Ca^+$ ions. Towards scaling the trapped ion quantum computer, we encode one logical qubit in entangled states distributed over seven trapped-ion qubits. We demonstrate the capability of the code to detect one bit flip, phase flip or a combined error of both, regardless on which of the qubits they occur. Furthermore, we apply combinations of the entire set of logical single-qubit Clifford gates on the encoded qubit to explore its computational capabilities [3]. The quantum toolbox is further applied to carry out both analog and digital quantum simulations. The basic simulation procedure will be presented and its application will be discussed for a variety of spin Hamiltonians. In particular, it will be shown how strings of trapped ions can be used for quantum simulations of a lattice gauge field theory. As an example, we map the real-time evolution of the Schwinger mechanism to a string of trapped ions in a few-qubit quantum computer, simulating the spontaneous creation of electron-positron pairs [4].

[1] T. Monz et al., Science **351**, 1068 (2016).
[2] P. Jurcevic et al., Nature **511**, 202 (2014).
[3] D. Nigg et al., Science **345**, 302 (2014).
[4] E. Martinez et al., Nature **534**, (2016).

# Modular Quantum Computers and Quantum Simulators with Trapped Ions

C. Monroe, J. Amini, E. Birckelbaw, M. Cetina, K. Collins, C. Crocker, S. Debnath, C. Figgatt, A. Grewal, M. Hernandez, P. Hess, D. Hucul, K. Hudek, V. Inlek, K. Johnson, H. Kaplan, K. Landsman, A. Lee, M. Lichtman, N. Linke, J. Mizrahi, G. Pagano, P. Richerme, C. Senko, J. Smith, K. Sosnova, D. Wong-Campos, K. Wright, and J. Zhang.

Joint Quantum Institute and University of Maryland Department of Physics, College Park, MD 20742 USA

Laser-cooled and trapped atomic ions are standards for quantum information science, acting as qubits with unsurpassed levels of quantum coherence while also allowing near-perfect measurement. When qubit state-dependent optical dipole forces are applied to a collection of atomic ions, their Coulomb interaction is modulated in a way that allows the entanglement of the qubits through quantum gates that can form the basis of a quantum computer. We have demonstrated a fully programmable and reconfigurable quantum computer module composed of five fully-connected qubits [1], where the physical layer is buried behind a quantum compiler that translates textbook quantum gates into native gates between the ions. We have executed several algorithms that are blind to the underlying physical hardware and look forward to scaling this device to 30-100 qubits in the near future. In the long run, such ion trap quantum computer modules can be connected through photonic links, and we have recently shown the operation of both local (Coulomb) and remote (Photon) gates in a single system [2].

Trapped ions are also an ideal platform for the quantum simulation of interacting spin models. When the same optical dipole forces used for quantum gates are globally applied to an ion chain, an effective long-range quantum magnetic interaction arises, with reconfigurable and tunable graphs that are determined by the spectrum of the laser forces. We have have implemented transverse Ising or XY models with up to 25 trapped ions for studies of quantum dynamics [3,4], and added programmable disorder to the chain in order to display effects of manybody localization [5]. Recent experiments have investigated the implementation of interacting spin-1 models [6] and Floquet Hamiltonians in order to study topologically-ordered quantum states of the spin chain. Soon these experiments will be extended to >30 spins, where no classical computer can predict its behavior, particularly the many-body dynamics. Such results are expected to shed light on the behavior of spin-liquids and other interesting forms of magnetism that feature frustration and massive entanglement.

This work is supported by the ARO with funding from the IARPA LogiQ program, the AFOSR MURI on Quantum Transduction, the AFOSR MURI on Quantum Measurement, the ARL Center for Distributed Quantum Information, the ARO Atomic Physics Program, and the NSF PFC at JQI.

[1] S. Debnath, et al., *arXiv:1603.04512* (to appear in *Nature*, 2016).
[2] D. Hucul, et al., *Nature Physics* **11**, 37 (2015).
[3] P. Richerme, et al., *Nature* **511**, 198 (2014).
[4] C. Senko, et al., *Science* **345**, 430 (2014).
[5] J. Smith, et al., *arXiv: 1508.07026* (to appear in *Nature Physics*, 2016).
[6] C. Senko, et al., *Phys. Rev. X* **5**, 021026 (2015).

# Hybrid Quantum Information Processing

Akira Furusawa[1]

[1]Department of Applied Physics, School of Engineering, The University of Tokyo, 7-3-1 Hongo, Bunkyo-ku, Tokyo 113-8656, Japan

We are working on hybrid quantum information processing, which combines two methodologies of quantum information processing – qubits and continuous variables (CVs) [1]. More precisely, we encode quantum information onto single-photon-based qubits and utilize CV quantum processors to realize universal optical quantum computing. The advantage of this methodology is that we can have both high-fidelity nature of qubits and determinisity of CV quantum processors. In other words, we can enjoy both particle- and wave-nature of quantum mechanics. Towards this goal we performed various things, which include quantum error correction with nine-party CV entanglement [2], teleportation of Schrödinger's cat state [3], adaptive homodyne measurement with phase-squeezed states [4], deterministic teleportation of time-bin qubits [5], creation of ultra-large-scale CV cluster states [6], generation and measurement of CV entanglement on a chip [7], and synchronization of photons with cavity-based quantum memories [8].

[1] A. Furusawa and P. van Loock, *Quantum Teleportation and Entanglement: A Hybrid Approach to Optical Quantum Information Processing,* (Wiley-VCH, Weinheim, 2011).

[2] T. Aoki, G. Takahashi, T. Kajiya, J. Yoshikawa, S. L. Braunstein, P. van Loock, and A. Furusawa, Nature Physics **5**, 541 (2009).

[3] N. Lee, H. Benichi, Y. Takeno, S. Takeda, J. Webb, E. Huntington, and A. Furusawa, Science **332**, 330 (2011).

[4] H. Yonezawa, D. Nakane, T. A. Wheatley, K. Iwasawa, S. Takeda, H. Arao, K. Ohki, K. Tsumura, D. W. Berry, T. C. Ralph, H. M. Wiseman, E. H. Huntington, and A. Furusawa, Science **337**, 1514 (2012).

[5] S. Takeda, T. Mizuta, M. Fuwa, P. van Loock, and A. Furusawa, Nature **500**, 315 (2013).

[6] S. Yokoyama, R. Ukai, S. C. Armstrong, C. Sornphiphatphong, T. Kaji, S. Suzuki, J. Yoshikawa, H. Yonezawa, N. C. Menicucci, and A. Furusawa, Nature Photonics **7**, 982 (2013).

[7] G. Masada, K. Miyata, A. Politi, T. Hashimoto, J. L. O'Brien, and A. Furusawa, Nature Photonics **9**, 316 (2015).

[8] K. Makino, Y. Hashimoto, J. Yoshikawa, H. Ohdan, T. Toyama, P. van Loock, and A. Furusawa, Science Advances **2**, e1501772 (2016).

# Exploring frontiers of quantum optics with strongly coupled atoms and atom-like systems

Mikhail Lukin[1]

[1]Physics Department, Harvard University, Cambridge MA, USA

We will discuss recent developments at a new scientific interface between quantum optics, nanoscience and quantum information science. Examples include the use of quantum optical techniques for manipulation of individual atom-like impurities at a nanoscale and for realization of hybrid systems combining atoms and atom-like systems with novel nanophotonic devices. We will discuss how these techniques are used for exploring quantum nonlinear optics, scaling up quantum networks and probing non-equilibrium quantum dynamics in strongly interacting systems.

# Realizing quantum gates using cavity QED

Stephan Ritter, Bastian Hacker, Norbert Kalb, Andreas Reiserer, Stephan Welte, Gerhard Rempe

Max Planck Institute of Quantum Optics, Hans-Kopfermann-Str. 1, 85748 Garching, Germany

Single atoms are the smallest material system that can still be employed as ultra-sensitive probes, as detectors, and for the storage and processing of information. Their quantum-mechanical character even allows for applications beyond classical limits. Crucial prerequisites are full control of the atomic quantum states, such as its motion and spin, and deterministic interactions with other information carriers, such as optical photons, which are ideally suited to bridge long distances and connect distributed systems. To enhance light-matter interaction at the level of single quanta, a cavity can be employed. It was proposed more than a decade ago [1] that the robust, deterministic interaction mechanism provided by reflecting photons off a single-sided cavity containing a single atom can be used to realize quantum logic gates, the arguably strongest interaction between two quantum states. The talk will highlight how combination of a single-atom cavity QED system in the strong coupling regime with full control over photonic and atomic degrees of freedom has led to a series of intriguing experimental advances. Building on the implementation of a hybrid atom-photon quantum gate [2], the atom can act as a nondestructive detector for single photons [3], but also mediate interactions between two successively reflected photons. They are so strong that each photon can shift the other photon's phase by $\pi$ and so accurate that a quantum logic gate between the photons is realized [4]. These results provide a path to overcome the fundamental scalability problems of linear-optics and purely matter-based systems. Possible applications in photonic quantum computation and distributed quantum networks will be discussed.

[1] L.-M. Duan, H. J. Kimble, Phys. Rev. Lett. **92**, 127902 (2004).

[2] A. Reiserer, N. Kalb, G. Rempe, S. Ritter, Nature **508**, 237 (2014).

[3] A. Reiserer, S. Ritter, G. Rempe, Science **342**, 1349 (2013).

[4] B. Hacker, S. Welte, G. Rempe, S. Ritter, accepted for publication in Nature, arXiv:1605.05261 (2016).

# Spontaneous spin squeezing in a rubidium BEC

Théo Laudat[1], Vincent Dugrain[2], Carlos Garrido Alzar[1], Alice Sinatra[2], Peter Rosenbusch[1], Jakob Reichel[2]

[1]LNE-SYRTE, Observatoire de Paris-PSL Research University, CNRS, UPMC-Sorbonne Universités, 61 Avenue de l'Observatoire, 75014 Paris, France

[2]Laboratoire Kastler Brossel, ENS-PSL Research University, CNRS, UPMC-Sorbonne Universités, Collège de France, 24 rue Lhomond, 75005 Paris, France

Spin-squeezed states are the prime example of a highly entangled many-particle state having the potential to improve atomic clocks and interferometers beyond the standard quantum limit. The metrological prospect also applies to rubidium BECs, which, due to their minimum phase-space spread, are considered as precious source states for atom interferometry in spite of inherent phase fluctuations and losses. Therefore, spin-squeezed states of $^{87}$Rb BECs have met with wide interest. However, generating the required nonlinear interaction by collisional interaction, as originally envisaged in 2001 [1], is a challenge in $^{87}$Rb due to the almost identical scattering lengths of its ground state sublevels. Existing experiments have foregone this problem by using state-dependent potentials on an atom chip [2] or an optical lattice combined with a Feshbach resonance [3]. Both methods have led to spectacular results, but come at the price of considerable complications in the setup.

Using our Trapped-Atom Clock on a Chip (TACC) [4,5], we have performed an experiment where spin squeezing occurs spontaneously within a standard Ramsey sequence applied to a $^{87}$Rb BEC trapped on an atom chip. A single resonant $\pi/2$ pulse is applied to all the atoms in the condensate, which is then allowed to evolve freely. The trap is cigar-shaped and trap parameters are chosen such that demixing occurs, which then reduces the inter-state scattering length. We find a spin squeezing that improves with atom number $N_{at}$ and reaches 1.7dB for our maximum $N_{at}$ ~10000. 96% contrast is observed when the second Ramsey pulse is applied at the remixing time of ~1.1s. The squeezing is limited by the asymmetric losses occurring during the free evolution time. We are currently investigating the role of the aspect ratio of the trap.

[1] A. Sørensen, L.-M. Duan, J. I. Cirac & P. Zoller, Nature **409**, (2001).
[2] M. F. Riedel, P. Böhi, Y. Li, T. W. Hänsch, A. Sinatra & P. Treutlein, Nature **464**, 1170 (2010).
[3] C. Gross, T. Zibold, E. Nicklas, J. Estève & M. K. Oberthaler, Nature **464**, 1165 (2010).
[4] C. Deutsch, F. Ramirez-Martinez, C. Lacroûte, F. Reinhard, T. Schneider, J.-N. Fuchs, F. Piéchon, F. Laloë, J. Reichel & P. Rosenbusch, Phys. Rev. Lett. **105**, 020401 (2010).
[5] R. Szmuk, V. Dugrain, W. Maineult, J. Reichel & P. Rosenbusch, Phys. Rev. A **92**, 012106 (2015).

# Particle Physics with AMO Methods

Gerald Gabrielse[1]

[1]Leverett Professor of Physics, Harvard, USA

The most precise measurement of the property of an elementary particle was measured, using AMO methods in a university laboratory, in order to test the most precise prediction of the Standard Model of particle physics. The incredible precision of the Standard Model's prediction of what is measured is arguably the Standard Model's greatest triumph. At the same time, the Standard Model predicts that no matter universe should result from a big bang. The Standard Model is thus the great triumph and the great frustration of modern physics. Recent progress in several low energy tests of the Standard Model and its basic symmetries will be discussed.

# Searching for ultralight dark matter with atomic spectroscopy and magnetic resonance

Dmitry Budker[1,2,3]

[1]Helmholtz Institute, Johannes Gutenberg University, Mainz, Germany

[2]Department of Physics, University of California, Berkeley, California, USA

[3]Nuclear Science Division, E. O. Lawrence Berkeley National Laboratory, Berkeley, California, USA

Axions, axion-like particles (ALPs), dilatons, and other ultralight (masses from $10^{-4}$ down to $10^{-23}$ eV) particles have been discussed as possible candidates for dark matter. An interesting feature of these ideas is that they lead to predictions of potentially observable transient and oscillating effects. I will describe how we are looking for these as well as the relation of such experiments to tests of fundamental symmetries (P, CP, T, CPT ...). For up-to-date information on our various experiments in this area (CASPEr, GNOME, differential atomic-dysprosium clock, etc.), please refer to the web pages [1,2].

[1] https://budker.uni-mainz.de/
[2] http://budker.berkeley.edu/

# A Michelson-Morley test for electrons using trapped ions

T. Pruttivarasin[1,2], M. Ramm[1], S. G. Porsev[3,4], I. I. Tupitsyn[5], M. Safronova[3,6], M. A. Hohensee[1,7], H. Häffner[1]

[1]Department of Physics, University of California, Berkeley, California 94720, USA

[2]Quantum Metrology Laboratory, RIKEN, Wako, Saitama 351-0198, Japan

[3]Department of Physics and Astronomy, University of Delaware, Newark, Delaware 19716, USA

[4]Petersburg Nuclear Physics Institute, Gatchina, Leningrad District 188300, Russia

[5]Department of Physics, St. Petersburg State University, Ulianovskaya 1, Petrodvorets, St.Petersburg 198504, Russia

[6]Joint Quantum Institute, National Institute of Standards and Technology and the University of Maryland, College Park, Maryland, 20742, USA

[7]Lawrence Livermore National Laboratory, Livermore, California 94550, USA

Lorentz symmetry is one of the corner stones of modern physics. As such it should not only hold for photons, but also for other particles such as the electron. Here we search for violation of Lorentz symmetry by performing an analogue of a Michelson-Morley experiment for electrons. We split an electron-wavepacket bound inside a calcium ion into two parts with different orientations. As the Earth rotates, the absolute spatial orientation of the wavepackets changes and anisotropies in the electron dispersion would modify the phase of the interference signal. To remove noise, we prepare a pair of ions in a decoherence-free subspace, thereby rejecting magnetic field fluctuations common to both ions. After a 23 hour measurement, we limit the energy variations to 11 mHz, verifying the isotropy of the electron's motion at the 1E-18 level. Alternatively, we can interpret our result as testing the rotational invariance of the Coulomb potential. Assuming Lorentz symmetry holds for electrons and that the photon dispersion relation governs the Coulomb force, we obtain a new limit on anisotropies in the speed of light. We will also discuss how using Ytterbium instead of Calcium ions promises tighter bounds for Lorentz symmetry breaking in the electromagnetic sector.

# Ultrafast atomic and molecular dynamics explored with high harmonics

Chang Hee Nam[1,2], Hyeok Yun[1], Kyung Taec Kim[1,2], Hyung Taek Kim[1,3]
[1]Center for Relativistic Laser Science, Institute for Basic Science (IBS), Gwangju 61005, Korea
[2]Dept. of Physics and Photon Science, GIST, Gwangju 61005, Korea
[3]Advanced Photonics Research Institute, GIST, Gwangju 61005, Korea

High harmonics from atoms driven by intense femtosecond laser pulses are a unique light source with attosecond duration in the EUV and soft x-ray spectral range. The harmonic light source can be applied to probe ultrafast dynamics of atoms and molecules. By tuning the harmonic wavelength a specific state of an atom or molecule can be excited, which can be further excited or ionized by applying a time-delayed femtosecond laser pulse. This process of combining harmonic pulses and time-delayed femtosecond laser pulses was applied to investigate the dynamics of photoexcitation and photoionization processes of He. In addition high-harmonic radiation generated from molecules contains the information on the structure of molecules. When multiple molecular orbitals are exposed to a strong laser field, the highest-occupied molecular orbital (HOMO) is mostly ionized and thus emits strong high-harmonic radiation containing the characteristics of HOMO. In order to resolve the harmonic composition from multiple orbitals of $CO_2$ molecules we explored two-dimensional high-harmonic spectroscopy (HHS) by applying an orthogonally polarized two-color laser field consisting of the fundamental frequency and its second harmonic. In this case odd and even harmonics carried the characteristics of the HOMO and HOMO-1, respectively. The multi-orbital characteristics were thus revealed in the two-dimensional spectroscopy employing the two-color laser field, paving a new route to examine ultrafast molecular dynamics.

# Attosecond electron dynamics on surfaces and layered systems

Reinhard Kienberger[1]

[1]Fakultät für Physik, E11, Technische Universität München, James Franck Straße, 85748 Garching

Attosecond stereaking experiments performed on different solids, e.g. tungsten (110) and (100), rhenium, and magnesium (0001), leading to different delays – also depending on the excitation photon energy, will be disussed. We show results from time-resolved transport of different types of electrons through defined adlayers on a bulk material on attosecond timescale.

The attosecond streaking method [1] is the most established technique in attosecond science. Photoelectrons generated by attosecond extreme ultraviolet pulses (XUV), are exposed to a dressing electric field from well synchronized laser pulses. The energy shift experienced by the photoelectrons by the dressing field is dependent on the delay between the XUV pulse and the dressing field. Looking at photoelectrons at different kinetic energies makes it possible to measure the respective delay in photoemission between these electrons. Insight into the photoemission process is gained from this information.

With laser based sources (high-order harminc generation, HHG) first experiments have been carried out to measure sub-femtosecond behavior of matter. Recently, the dynamics of the photoionization process on solids has been studied [2]. Not only that attosecond metrology now enables clocking on surface dynamics, but also the individual behaviour of electrons of different type (core electrons vs. conduction band electrons) can be resolved. Here, we measured a time delay of about 100 as on the emission of the aforemention two types of electrons. The information gained in these experiments triggered many theoretical activities leading to different explanations on the physical reason of the delay. Attosecond stereaking experiments have been performed on different solids, e.g. tungsten (110) and (100), and rhenium, leading to different delays – also depending on the excitation photon energy.

In a more recent experiment by Neppl et al.[3], the time-delay in the emission of valence band (VB) electrons and core-level electrons has been measured for Mg(0001), in order to get a better understanding for these theoretical models. Mg is a well suited test material to discern the possible effects for the delay in photoemission since the band structure for this material is very "free-electron-like", so the influence from the group velocity of the electrons can be well calculated and accounted for. A simple estimation for the respective "travel-time" using free particle velocities and calculated penetration depths [4] shows that the vanishing time-delay can be well accounted for by simply considering transport effects – hinting at a negligible contribution from other phenomena like initial-state effects.

We show measurements of time-resolved transport of different types of electrons through defined adlayers on a bulk material on an attosecond timescale (Fig 1).

A systematic investigation of photon-energy-dependent (EPh = 95 eV…145 eV) delay times measured in tungsten giving insight into possible effects responsible for the delay is presented and will be discussed.

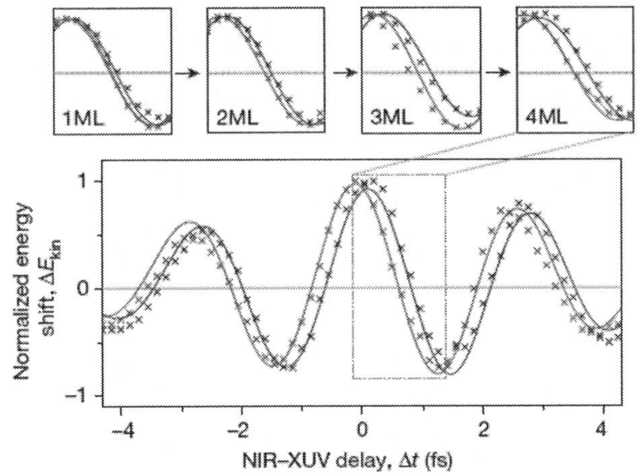


**Figure 1.** Delay of tungsten 4f electron with respect to MG electrons depending of the number of Mg adlayers (from [4]). The linear behaviour is explained by transport effects. The delay of conduction band electrons is more complex.

[1] R. Kienberger et al., Nature 427, 817 (2004).
[2] A. Cavalieri et al., Nature 449, 1029 (2009).
[3] S. Neppl et al., PRL 109 (8), 087401 (2012).
[4]S. Neppl et al., Nature 517, 342 (2015).

# Birth of a resonant photoelectron wavepacket

Luca Argenti
Dep. Physics & CREOL, 4111 Libra Dr., Univ. Central Florida, Orlando, FL32816, USA

Attosecond pulses can excite coherent superpositions of states across a wide energy range, thus providing a means to monitor the fast correlated dynamics of electrons in the continuum. For example, from simulated and experimental attosecond-transient-absorption spectra (ATAS), it has been possible to reconstruct the concerted motion of two excited electrons in the helium atom [1]. The shape and dynamics of the electron wave packet that emerges from a photoionization event is encoded in the phase of its spectral components across a continuous set of energies. In the presence of autoionizing states, the modulation of such phase bears information on the timing of the Auger decay. Using an analytical finite-pulse resonant two-photon model [2,3], we interpret two recent experiments in the region of the first autoionizing series of argon [4] and helium [5], based on two complementary implementation of the RABBIT technique (Reconstruction of Attosecond Beating by Interference of Two-photon Transitions). These two experiments provide a glimpse on the temporal structure of the resonant photoelectron wavepacket at the time of its birth and offer a new way to study the dynamics of structured continua with sub-femtosecond time resolution.

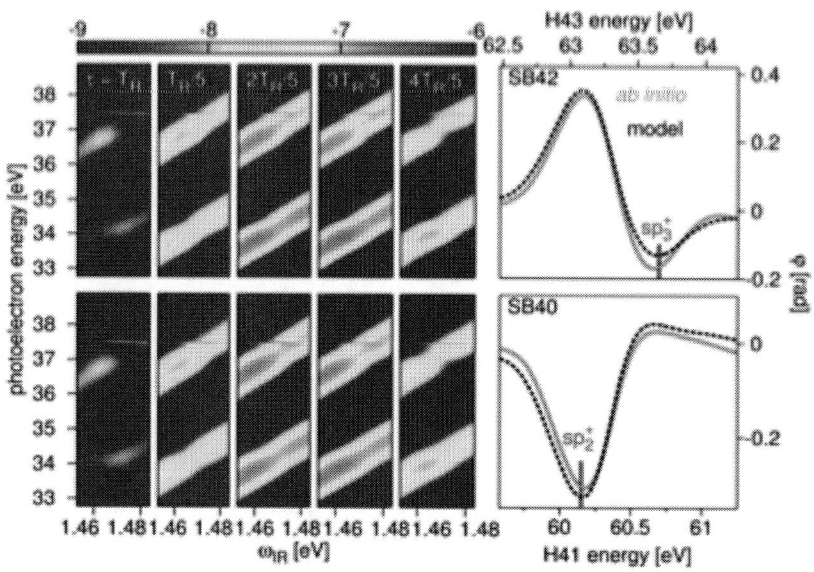


**Figure 1.** Sideband spectrum vs IR photon energy in RABITT ionisation of helium in the region of the N=2 autoionizing states of the atom, for selected pump-probe time delays. Upper left panels: ab initio results. Lower left panels: model. On the right panels, phase of the sideband beating vs energy of the H43 harmonic, which is resonant with the 2s2p doubly excited state..

[1] Ott C *et al.*, Nature **516** , 374 (2014).
[2] Jiménez Galán Á, Argenti L, and Martín F, Phys. Rev. Lett. **113**, 263001 (2014).
[3] Jiménez Galán Á, Martín F, and Argenti L, Phys. Rev. A **93**, 023429 (2016).
[4] Kotur M *et al.*, Nature Communications **7**, 10566 (2016).
[5] Gruson V *et al., submitted.*

# Multi-plate supercontinuum generation: New approach to single-cycle pulses and isolated attosecond pulses

Andy Kung[1,2]
[1]Institute of Atomic and Molecular Sciences, Academia Sinica, Taiwan
[2]Institute of Photonics Technologies, National Tsing Hua University, Taiwan

By strategically placing a number of thin fused-silica plates at the focused waist of an intense femtosecond beam, we have demonstrated the generation of an octave-spanning supercontinuum with high pulse energy without destroying its coherence properties. The approach has produced several orders-of-magnitude increase in output energy in an all-solid medium while maintaining excellent spatial and temporal quality [1,2]. This multi-plate approach is highlighted by the advantages of being relatively compact, easy to align, reliable and reproducible as compared to conventional methods for high intensity broad spectral generation using the gas phase approach.

The supercontinuum pulse from multi-plates has been shown to be fully coherent. The pulse can thus be phase-compensated and compressed effectively. Using a SLM shaper-assisted 4-f grating compensator we have successfully compensated and produced pulses with a duration of 2.8 fs that is very near a single-cycle width (2.6 fs) of the pulse envelope. Using custom chirped mirrors we achieved compression to 3.6 fs with a 90% efficiency in total energy transmission.

As a first application, we performed high harmonic generation experiments with these pulses and obtained isolated attosecond pulses as evidenced by a continuous high harmonic spectrum in the 45-60 eV region. The high-energy cut-off in HHG has been pushed to beyond 100 eV in Ne with 3.6 fs 25 GW peak-power pulses obtained with the multi-plate approach.

Our measurements show that polarization is preserved in the multi-plate system. We have thus also produced circularly-polarized near-single-cycle pulses. In this case, a broadband quarter-wave plate was used after compression to produce the pulse because the compressor requires linear polarized input for it to function. Nevertheless with the possibility of polarization control attosecond pulses of different polarization and duration may be synthesized using pulses generated from the multi-plate system [3].

[1] Chih-Hsuan Lu, Yu-Jung Tsou, Hong-Yu Chen, Bo-Han Chen, Yu-Chen Cheng, Shang-Da Yang, Ming-Chang Chen, Chia-Chen Hsu, and A. H. Kung, Optica **1**, 400-406 (2014).
[2] Yu-Chen Cheng, Chih-Hsuen Lu and A. H. Kung, Opt. Express **24**(7), 7224-7231 (2016).
[3] Pei-Chi Huang et.al. "Isolated, Circularly Polarized, Attosecond Pulse Generation," Postdeadline paper JTh4A.7 CLEO 2016.

# Spontaneous broadening of a driven dissipative Rydberg system

J.V. Porto

Joint Quantum Institute,
National Institute of Standards and Technology and the University of Maryland,
Gaithersburg, Maryland 20899, USA

The strong interactions in Rydberg atoms make them an ideal system for the study of correlated many-body physics, both in the presence and absence of dissipation. Using such highly excited atomic states requires addressing challenges posed by the dense spectrum of Rydberg levels, the detrimental effects of spontaneous emission, and strong interactions. A full understanding of the scope and limitations of many Rydberg-based proposals requires simultaneously including these effects, which typically cannot be described by a mean-field treatment due to correlations in the quantum coherent and dissipative processes. We study a driven, dissipative system of Rydberg atoms in a 3D optical lattice, and observe substantial deviation from single-particle excitation rates, both on and off resonance. The observed broadened spectra cannot be explained by van der Waals interactions or a mean-field treatment of the system. We attribute these effects to unavoidable blackbody-induced transitions to nearby Rydberg states of opposite parity, which have large, resonant dipole-dipole interactions with the state of interest. Even at low densities of Rydberg atoms, uncontrolled production of atoms in other states significantly modifies the energy levels of the remaining atoms. These off-diagonal exchange interactions result in complex many-body states of the system and have implications for off-resonant Rydberg dressing proposals.

[1] E. A. Goldschmidt, T. Boulier, R.C. Brown, S.B. Koller, J.T. Young, A.V. Gorshkov, S.L. Rolston, and J.V. Porto, Phys. Rev. Lett. **116** , 113001 (2016)

# Dynamics of strongly interacting atomic Fermi gases

Giacomo Valtolina[1,2], Francesco Scazza[1,2], Andrea Amico[1,2], Alessia Burchianti[1,2], Chiara Fort[2], Massimo Inguscio[1,2], Matteo Zaccanti[1,2] and Giacomo Roati[1,2]

[1]INO-CNR, via N. Carrara 1, Sesto Fiorentino, Italy

[2]LENS, University of Florence, via N. Carrara 1, Sesto Fiorentino, Italy

We will present two experimental studies, exploiting strongly interacting ultracold Fermi gases of $^6$Li atoms confined in optical potentials. In a first experiment, we create the analogous of a Josephson junction by bisecting BEC-BCS crossover superfluids with a thin optical barrier. We observe coherent dynamics in both the population and in the relative phase between the two superfluid reservoirs, extracting the Josephson coupling energy, which appears to be maximum for unitary superfluids. For critical parameters, we see how the Josephson dynamics is affected by the presence of topological defects entering the superfluid bulk [1]. In a second experiment, we create an artificial ferromagnetic state by segregating degenerate spin mixtures into two initially disconnected reservoirs [2]. We study the spin dynamics for different interaction strengths and temperatures. For sufficiently high values of the inter-spin repulsive interactions and sufficiently low temperatures, we observe a softening of the spin dipole mode connected to a time window during which spin diffusion is zeroed. We interpret our data as the evidence of the Stoner instability in our system for critical values of repulsive interactions. Our measurements provide exciting new insights into the physics of attractive and repulsive Fermi gases.

[1] G. Valtolina et al., Science **350,** 1505 (2015)

[2] G. Valtolina et al., arXiv:1605.07850v1 (2015)

# Combining Yb and Li: Feshbach Resonance and dual-species Bose-Fermi Superfluid

Subhadeep Gupta[1]

[1]Department of Physics, University of Washington, Seattle, WA, USA

I will present recent experimental work in an ultracold dual-species mixture of ytterbium ($^{174}$Yb) and lithium ($^{6}$Li) atoms. Using a combination of ground state Li atoms and excited (metastable) state Yb atoms, we observe an interspecies Feshbach resonance in an alkali+alkaline-earth-like system [1, 2]. By dynamically controlling the shape of a time-averaged optical trapping potential, we achieve efficient evaporative cooling, allowing for rapid Yb BEC production with 1.6 seconds cycle time [3]. Extending our cooling techniques to Yb and Li, we realize a dual-species Bose-Fermi superfluid mixture with large mass-mismatch ($m_{\mathrm{Yb}}/m_{\mathrm{Li}} = 29$) and investigate interactions in this system.

[1] A. Khramov et al, Phys. Rev. Lett. **112**, 033201 (2014)
[2] W. Dowd et al, New J. Phys. **17**, 055007 (2015)
[3] R. Roy, A. Green, R. Bowler, and S. Gupta, Phys. Rev. A **93**, 043403 (2016)

# Atomic vapor spectroscopy in integrated photonic structures

Tilman Pfau[1,2], Ralf Ritter[1,2], Nico Gruhler[3,4], Wolfram Pernice[3,4], Harald Kübler[1,2], Robert Löw[1,2]

[1]Center for Integrated Quantum Science and Technology, IQ$^{ST}$

[2]5. Physikalisches Institut, Universität Stuttgart, Germany

[3]Institute of Nanotechnology, Karlsruhe Institute of Technology, Germany

[4]Institute of Physics, University of Münster, Germany

We investigate an integrated optical chip immersed in atomic vapor providing several waveguide geometries for spectroscopy applications [1]. This includes integrated ring resonators [2], Mach Zehnder interferometers, slot waveguides and counterpropagating coupling schemes. The narrow-band transmission through a silicon nitride waveguide and interferometer is altered when the guided light is coupled to a vapor of rubidium atoms via the evanescent tail of the waveguide mode. We use grating couplers to couple between the waveguide mode and the radiating wave, which allow for addressing arbitrary coupling positions on the chip surface. The evanescent atom-light interaction can be numerically simulated and shows excellent agreement with our experimental data. This work demonstrates a next step towards miniaturization and integration of alkali atom spectroscopy and provides a platform for further fundamental studies of strong atom light coupling. Cooperativities on the order of 1 are within reach. In the future integrated optical and electronic circuits in atomic vapor cells will enable applications in quantum sensing and quantum networks.

[1] R. Ritter, et al., Appl. Phys. Lett. **107**, 041101 (2015)
[2] R. Ritter, et al., arXiv:1605.05961

# Oscillators for quantum information: the cat code, error correction, and two mode entanglement

Robert Schoelkopf[1] and the Yale circuit QED team
[1]Department of Applied Physics, Yale University, United States

The creation, manipulation, and measurement of non-classical states of light in oscillator modes has been an area of interest in atomic physics and cavity QED for many years. In quantum information, the canonical building blocks are usually qubits realized with two-level atoms or spins, which are then coupled together and arrayed to make more complex, higher-dimensional systems. An oscillator or cavity, however, can store multiple photons and thus serve as more complex building block, or even a logical qubit. Superconducting microwave cavities can also have long coherence times, and offer potential simplifications in the hardware requirements for error correction and other tasks. An example that our team at Yale has been pursuing is the "cat code," [1] which utilizes superpositions of coherent states to create a logical quantum bit that can be efficiently corrected for photon loss. I will present recent experimental progress on using cavities for quantum information, using the coupling to an ancillary superconducting "transmon" qubit. The first experiments [2] demonstrate the redundant encoding of quantum information using multi-photon states. The second set of experiments [3] perform rapid, QND measurements of the photon number parity, which is used to monitor and correct the naturally occurring decoherence, thereby extending the lifetime of quantum information by error correction for the first time. By using a transmon to couple to two cavity modes, we demonstrate the ability to create two-mode cat states [4], or the entanglement of two logical qubits in a Hilbert space of about 150 dimensions. Finally, using techniques from optimal control, we find efficient pulse sequences that can manipulate the encoded information, and we realize [5] high fidelity (>99%) universal gates on the cat code states in a cavity. These experiments show the potential of hardware-efficient approaches to quantum information processing, and open new capabilities for fundamental experiments in quantum optics.

[1] Z. Leghtas et al., Phys. Rev. Lett. **111**, 120501 (2013).
[2] B. Vlastakis, G. Kirchmair, et al., Science **342**, 607 (2013).
[3] N. Ofek, A. Petrenko, et al., arxiv/1602.04768 (2016).
[4] C. Wang, Y. Gao, et al., Science **352**, 1087 (2016).
[5] R. Heeres, P. Reinhold, et al., in preparation (2016).

# Rydberg atoms in a cavity

Wu-Ming Liu
(Institute of Physics, Chinese Academy of Sciences, Beijing 100190)
(Email: wmliu@iphy.ac.cn)

Laser-excited Rydberg atoms stored in large-spacing optical lattices or magnetic trap arrays offer unique possibilities for implementing scalable quantum information processors, for example, an efficient implementation of a universal quantum simulator for spin models involving *n*-body interactions, including such of higher order. We study an optical cavity coupled to a lattice of Rydberg atoms, which can be represented by a generalized Dicke model. Our results show that the competition between the atom-atom interaction and atom-light coupling induces a rich phase diagram. A novel superradiant solid phase is found, where both the superradiance and crystalline orders coexist. Different from the normal second order superradiance transition, here both the solid-1/2 and superradiant solid phase to superradiant phase transitions are first order. These results are confirmed by large scale quantum Monte Carlo simulations [1-4].

[1] Z. M. Wang, J. L. Lian, J. Q. Liang, Y. M. Yu, W. M. Liu, Collapse of superradiant phase and multiple quantum phase transitions for Bose-Einstein condensate in optomechanical cavity, Physical Review A 93, 033630 (2016).

[2] K. A. Yasir, W. M. Liu, Controlled electromagnetically induced transparency and Fano resonances in hybrid Bose-Einstein condensate optomechanics, Scientific Reports 6, 22651 (2016).

[3] K. A. Yasir, W. M. Liu, Tunable bistability in hybrid Bose-Einstein condensate optomechanics, Scientific Reports 5, 10612 (2015).

[4] X. F. Zhang, Q. Sun, Y. C. Wen, W. M. Liu, S. Eggert, A. C. Ji, Rydberg polaritons in a cavity: A superradiant solid, Physical Review Letters 110, 090402 (2013).

# Quantum spin dynamics, coherences and entanglement in trapped ion arrays

Ana Maria Rey

JILA, NIST and University of Colorado, Boulder, Colorado 80309, USA

One of the most important goals of modern quantum sciences is to learn how to control and entangle many-body systems and use them to make powerful and improved quantum devices, materials and technologies. However, as the size of quantum systems grow, it becomes extremely challenging to identify, and quantify the buildup of entanglement. In this talk I will discuss our analysis and benchmarking of recent Penning trap ion experiments at NIST-Boulder that use planar arrays of ions as a platform for quantum simulation of spin models. In those experiments strong spin-spin interactions can be engineered through optical dipole forces that excite phonons of the crystals. The number of ions can be varied from tens to hundreds with high fidelity control. I will report on recent efforts to validate the potential of this system as a quantum simulator by investigating the impact of coherent Ising interactions, spin-motion entanglement and decoherence in the quantum dynamics [1]. We performed measurements of the spin depolarization, spin squeezing, the Fisher information and the so called multi-particle quantum spectrum (first developed for NMR systems) and compared the data to a full solution of the master equation. This study opens a path for future simulations of more complex spin models which are intractable with current numerical techniques.

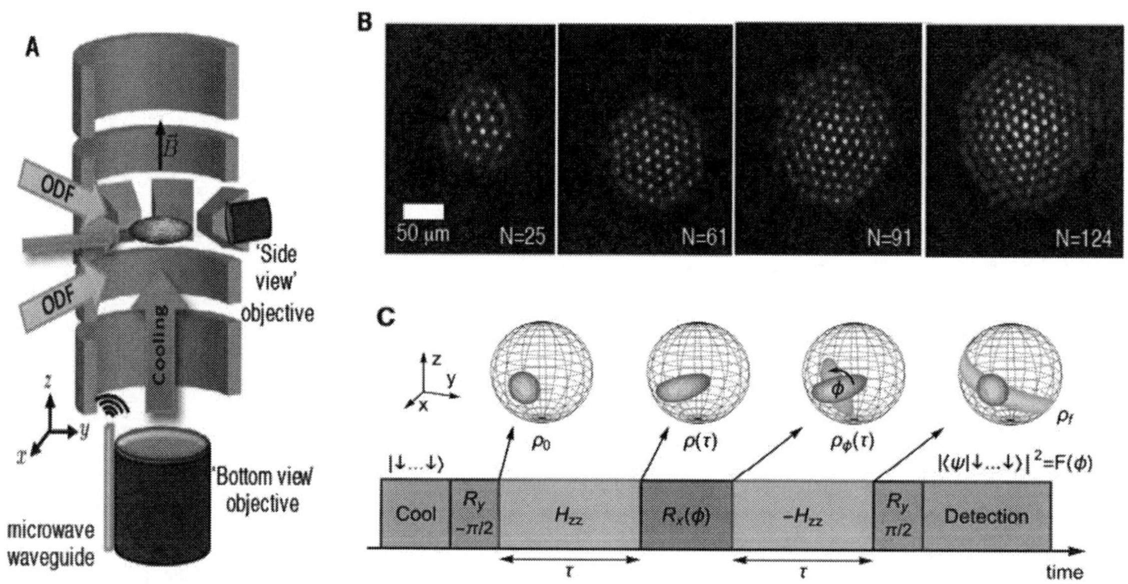


**Figure 1.** A) Penning trap quantum simulator, B) Coulomb crystal images, C) An experimental pulse sequence.

[1] J. G. Bohnet, B. C. Sawyer, Joseph W. Britton, Michael L. Wall, Ana Maria Rey, Michael Foss-Feig, John J. Bollinger, Science **352**, 1297 (2016).

# Advances in airbone and underground atom interferometers

Philippe Bouyer[1]

[1] LP2N, Laboratoire de Photonique Numérique et Nanosciences, Institut d'Optique Graduate School IOA, Rue François Mitterrand, 33400, Talence, France

The remarkable success of atom coherent manipulation techniques has motivated competitive resea rch and development in precision metrology. Matter-
wave inertial sensors accelerometers, gyrometers, gravimeters based on these techniques are all at the forefront of their respective measurement classes. Atom inertial sensors provide nowadays about the b est accelerometers and gravimeters [1] and allow, for instance, to make the most precise monitoring of gravity or to device precise tests of the weak equivalence principle (WEP). I present here some recen t advances in theses fields :

The outstanding developments of laser-cooling techniques and related technologies allowed the demonstration of an airborne matter-wave interferometers, which operated in the micro-gravity environment created during the parabolic flights of the Novespace Zero-g aircraft (see Figure 1). Using two atomic species (for instance $^{39}$K and $^{87}$Rb) allows to verify that two massive bodies will undergo the same gravitational acceleration regardless of their mass or composition, allowing a test of the Weak Equivalence Principle (WEP).

New concepts of matter-wave interferometry can be used to study sub Hertz variations of the strain tensor of space-time and gravitation. For instance, the MIGA instrument which is currently built in France, will allow the monitoring of the evolution of the gravitational field at unprecedented sensitivity, which will be exploited both for geophysical studies and for Gravitational Waves (GWs) detection.

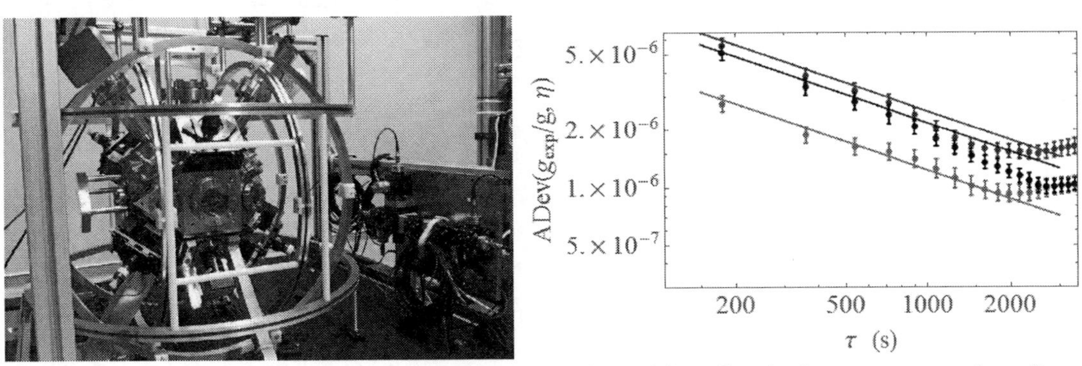


**Figure 1.** Left : the apparatus used in the Novestage 0-g airbus for dual matter-wave interferometry. Right : Allan deviation of the acceleration measurement with a $^{39}$K interferometer (blue), and a $^{87}$Rb interferometer (Red). In Black, the differential measurement used for the WEP test.

[1] B. Barrett, et al., **arXiv:1311.7033**, Proceedings of the International School of Physics "Enrico Fermi" on Atom Interferometry (2014) and references therein.

# Observation of two-species vortex lattices in a mixture of mass-imbalance Bose and Fermi superfluids

Xing-Can Yao, Hao-Ze Chen, Yu-Ping Wu, Xiang-Pei Liu, Xiao-Qiong Wang, Xiao Jiang, Youjin Deng, Yu-Ao Chen, Jian-Wei Pan

[1]CAS Center for Excellence and Synergetic Innovation Center in Quantum Information and Quantum Physics, University of Science and Technology of China, Shanghai, 201315, China

[2]CAS-Alibaba Quantum Computing Laboratory, Shanghai, 201315, China

The superfluid mixture of interacting Bose and Fermi species is a remarkable many-body quantum system. Dilute degenerate atomic gases, especially for two species of distinct masses, are excellent candidates for exploring fundamental features of superfluid mixture. However, producing a mass-imbalance Bose-Fermi superfluid mixture, providing an unambiguous visual proof of two-species superfluidity and probing inter-species interaction effects remain challenging. In this talk, I will present our recent experimental progress on the realization of a two-species superfluid [1]. We have employed a combination of state-of-the-art techniques, including advanced laser-cooling scheme and specially designed disk-like optical trap etc., and produced a two-species Bose-Fermi superfluid. According to our knowledge, the obtained superfluid mixture is also the largest one, containing more than $1.8\times10^5$ 41K atoms and $1.5\times10^6$ 6Li atoms. Moreover, vortex lattices—quantized topological excitations of superfluid—are successfully generated and directly visualized in both the single- and the two-species superfluids. Finally, a series of precise measurements on the formation and decay of the two-species vortices reveal pronounced effects of the inter-species interaction. We also stress that the vortex-vortex and the inter-species interactions may induce many unconventional behaviors in such a vortex matter of interacting species, remaining to be explored both theoretically and experimentally.

[1] X.-C. Yao *et al.*, arXiv:1606.01717 (2016).

# Quantum gas microscopy - a close-up of fermions, entanglement and quantum statistical physics

Markus Greiner[1]

[1]Harvard University and Harvard-MIT Center for Ultracold Atoms, Cambridge, MA, USA

With the advent of quantum gas microscopy of ultracold atoms in optical lattices, it is possible to detect and control all quantum degrees of freedom of each of the particles that compose a many-body system. I will first report on microscopic studies of the Fermi Hubbard model and the observation of strong antiferromagnetic correlations. I will then discuss how we harness quantum gas microscopy to directly measure entanglement entropy in a quantum many-body system. These measurements allow us to study the behavior of entanglement within the ground-state of the Bose-Hubbard model as we tune across the phase diagram, as well as the dynamics of entanglement when the system is excited. These latter measurements resulted in some of the first experiments on the foundations of quantum statistical physics, in which we directly observe quantum thermalization and the fundamental role of entanglement entropy.

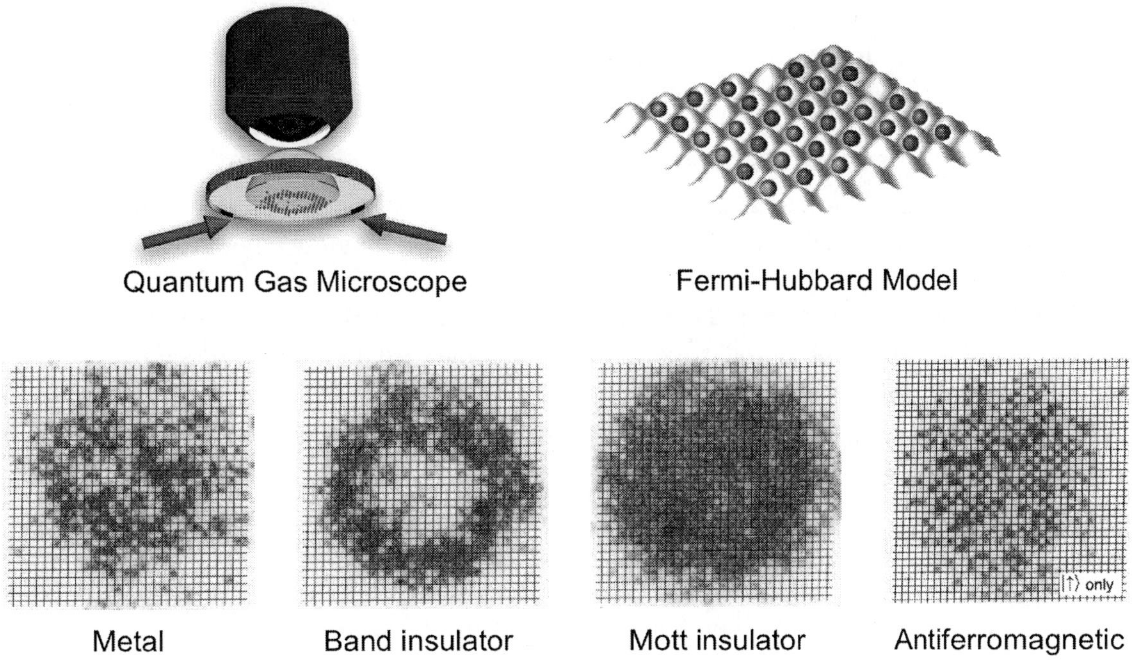


**Figure 1:** Quantum gas microscopy of a Fermi-Hubbard lattice system. We detect various states of matter and directly observe strong antiferromagnetic correlations.

# A Quantum-Gas Microscope for Fermionic 40-Potassium

Stefan Kuhr[1]

[1]University of Strathclyde, Department of Physics, Glasgow G4 0NG, United Kingdom

Single-atom-resolved detection in optical lattices using quantum-gas microscopes [1] has enabled a new generation of experiments in the field of quantum simulation. To achieve single-site-resolved detection of individual atoms on the lattice, it is desirable to maximize the fluorescence yield while at the same time maintaining a negligible particle loss rate and preventing the atoms from hopping between lattice sites. These conditions can be efficiently achieved by simultaneously laser cooling the atoms to sub-Doppler temperatures while detecting the fluorescence photons emitted during this process. However, cooling of fermionic alkaline atoms in optical lattices is challenging, as their low mass and small excited-state hyperfine splitting make it more difficult to obtain low temperatures using the standard technique of polarization-gradient cooling.

In this talk I will present how we achieved single-atom-resolved fluorescence imaging of 40K using electromagnetically-induced-transparency (EIT) cooling. This technique relies on the existence of a spectrally narrow, Fano-like line profile and dark resonances arising from quantum interference in a 3-like level scheme. In confining potentials with quantized vibrational levels, as is the case in our optical lattice, the narrow absorption line can selectively excite red-sideband transitions that cool the atomic motion by removing one vibrational quantum, while carrier and blue-sideband excitations are suppressed.

We present EIT cooling as an alternative to Raman sideband cooling techniques, which have been applied to achieve single-site imaging of fermionic potassium or lithium by other groups. In our setup, we detected on average 1000 fluorescence photons from a single atom within 1.5s, while keeping it close to the vibrational ground state of the optical lattice. Our fermionic quantum-gas microscope will provide new possibilities to probe strongly correlated fermionic systems at the single-atom level. It will allow the direct observation of spin-spin correlations, or, for example, the study of out-of-equilibrium dynamics and the entanglement build-up in many-particle fermionic quantum systems.

[1] S. Kuhr, Nat. Sci. Rev. **3**, 170 (2016); arXiv:1606.06990.
[2] E. Haller et al., Nature Physics **11**, 738 (2015).

# Deterministic quantum simulators with cold atoms

Selim Jochim
Heidelberg University, Germany

Experiments with ultracold gases have been extremely successful in studying many body physics, such as Bose Einstein condensates or fermionic superfluids. In general, they are deep in the regime of statistical physics, where adding or removing an individual particle does not matter. An essential challenge for current experiments - in particular with fermions - is to reach low enough entropies to observe low-temperature phases such as magnetically ordered states. In our work we deterministically prepare generic model systems containing a precise number of few ultracold fermionic atoms with tunable interaction in a well-defined quantum state. We have started the exploration of such few-body systems with a two-particle system that can be described with analytic theory. As we increase the system size atom by atom, we have been working in a one-dimensional framework allowing us to describe the system as a Heisenberg spin chain at strong repulsion. This allowed us to deterministically prepare a finite antiferromagnetically ordered state.

It is our vision to use our few-body systems as microscopic building blocks to assemble deterministic quantum systems that allow for the simulation of complex many-body models close to zero temperature.

a)

b)

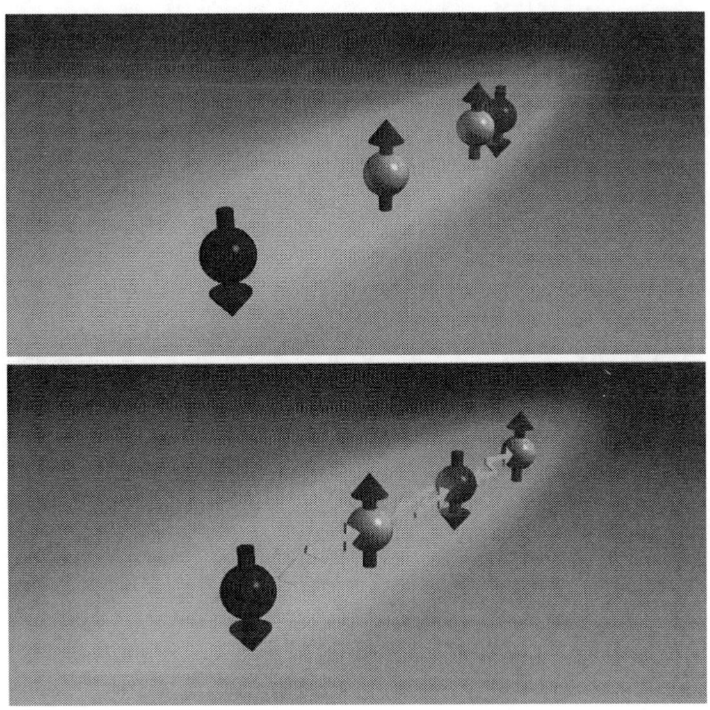

**Figure 1.** Illustration of the Heisenberg spin chain realized in our experiments in a 1D potential, a): for weak interactions, the particles' relative positions are uncorrelated, b) for very strong repulsion the atoms form a crystal that exhibits antiferromagnetic order.

[1] F. Serwane *et al.*, Science **332**, 6027 (2011).
[2] S. Murmann *et al.*, Phys. Rev. Lett. **115**, 215301 (2015).

# Dynamics of a Bose-Fermi Superfluid Mixture

Christophe Salomon, Marion Delehaye, Sébastien Laurent, Matthieu Pierce,
Shuwei Jin, Igor Ferrier-Barbut, Andrew Grier, Benno Rem, Frédéric Chevy

Laboratoire Kastler Brossel, Ecole Normale supérieure,
CNRS, UPMC, Collège de France, PSL, 24 rue Lhomond 75231, Paris, France

As pioneered by Bose, Einstein, Fermi, and Dirac about 90 years ago, quantum statistics play a central role in establishing properties of matter. Many-body theories describe atoms, molecules, gases, liquids, and solids from the very high temperature case of neutron stars to the low temperature domain of liquid helium and dilute quantum gases.

Thanks to the tunability of the atomic interactions offered by magnetic Feshbach resonances, ultracold Bose and Fermi gases can be brought to the quantum degenerate strongly interacting regime. Using lithium isotopes, we have recently produced and studied a novel mixture of Bose and Fermi gases where both species are superfluid, one of the remaining challenges in Helium 4 -Helium 3 mixtures [1]. We probe the collective dynamics of this system by exciting center-of-mass oscillations (dipole modes) that exhibit extremely low damping below a certain critical velocity. Using high precision spectroscopy of these low-lying modes we observe coherent energy exchange and measure the coupling between the two superfluids.

We have also measured the critical velocity for superfluid counterflow in this Bose-Fermi mixture as a function of interaction strength between fermions and temperature [2]. Near unitarity, in the phonon-dominated regime we find a critical velocity close to the speed of sound of the Fermi gas i.e much higher than the Bose gas speed of sound. We compare our findings with the recent prediction made for homogeneous Bose and Fermi gases and for weak Bose-Fermi coupling [3]. In this situation the critical velocity is predicted to be given by the sum of the sound velocities in the Bose gas and in the Fermi gas. Finally, raising the temperature of the mixture slightly above the superfluid transitions reveals an unexpected phase- locking of the Bose and Fermi center of mass oscillations. We interpret this phenomenon by mapping this system onto a Caldeira-Leggett model of two quantum harmonic oscillators coupled to a thermal excitation bath.

[1] I. Ferrier-Barbut, M. Delehaye, S. Laurent, A.T. Grier, M. Pierce, B.S. Rem, F. Chevy, C. Salomon, A Mixture of Bose and Fermi Superfluids, Science **345**, 1035 (2014)
[2] M. Delehaye, S. Laurent, I. Ferrier-Barbut, S. Jin, F. Chevy, and C. Salomon, Critical Velocity and Dissipation of an ultracold Bose-Fermi Counterflow, Phys. Rev. Lett., **115**, 265303, (2015)
[3] Y. Castin, I. Ferrier-Barbut, and C. Salomon, the Landau critical velocity for a particle in a Fermi superfluid, Comptes-Rendus Physique **16**, 241, (2015)

# Correlation-induced Bloch oscillations in a translationally invariant one-dimensional Bose liquid

Hanns-Christoph Nägerl[1]

[1]Institute for Experimental Physics, University of Innsbruck, Austria

Strongly interacting one-dimensional quantum many-body systems ("quantum wires") exhibit peculiar properties as a result of the interplay between confinement and interactions. Particle-particle correlations are greatly enhanced and conventional approaches to describe the system such as Landau's Fermi liquid theory of quasiparticles or Luttinger liquid theory break down. Probing and understanding the properties of quantum wires and in particular to study transport through such systems are of great relevance for potential future applications in view of the ongoing miniaturization of electronic devices.

We experimentally investigate the quantum motion of an impurity particle that is immersed in a strongly interacting one-dimensional bosonic quantum wire and that is subject to an external accelerating force. We observe characteristic oscillations in the evolution of the impurity momentum, which strongly modulate the relaxation toward finite drift velocities. The oscillatory motion arises from Bragg scattering of the impurity momentum at the edge of an emergent Brillouin zone. Our findings are interpreted as Bloch oscillations that, in contrast to conventional lattice systems, do not require a broken continuous translational symmetry but rather result from strong quantum correlations in the interacting quantum liquid. The experimental observations are supported by large-scale numerical simulations.

# Fermionic Mott Insulators under the Microscope

Martin W. Zwierlein[1]

[1]MIT-Harvard Center for Ultracold Atoms, Research Laboratory of Electronics, and Department of Physics, Massachusetts Institute of Technology, Cambridge, MA, USA

Ultracold atomic Fermi gases in optical lattices realize the Fermi-Hubbard model of interacting, itinerant fermions, believed to capture the essence of the physics behind cuprate high-temperature superconductivity. I will describe our recent experiments on two-dimensional Fermi gases of $^{40}$K atoms under a Fermi gas microscope [1], where metallic, Mott insulating and band insulating states of the Fermi-Hubbard model were observed with single-site, single-atom resolution (see Figure 1) [2]. The microscope allows the direct detection of spatial charge and spin correlations in this system [3], revealing antiferromagnetic spin correlations, the Pauli and correlation hole in the metallic regions, and strong doublon-hole bunching in the Mott insulating region near half-filling. The latter is expected in the presence of singlet bonds forming between adjacent lattice sites and should play an important role for transport in the Fermi-Hubbard model.

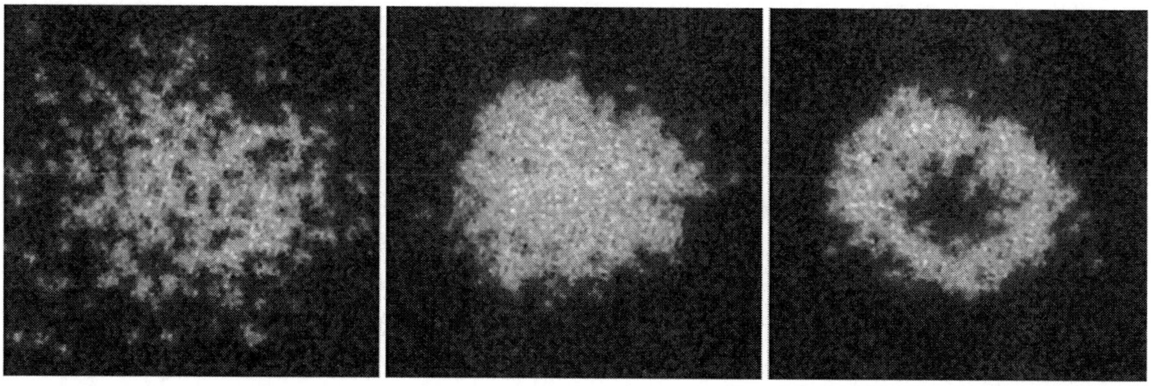

**Figure 1.** Observation of two-dimensional metals (left), Mott insulators (middle) and band insulators (right) realized with $^{40}$K atoms under the Fermi gas microscope. From [2].

[1] Lawrence W. Cheuk, Matthew A. Nichols, Melih Okan, Thomas Gersdorf, Vinay V. Ramasesh, Waseem S. Bakr, Thomas Lompe, Martin W. Zwierlein, Phys. Rev. Lett. **114**, 193001 (2015).
[2] Lawrence W. Cheuk, Matthew A. Nichols, Katherine R. Lawrence, Melih Okan, Hao Zhang, Martin W. Zwierlein, Phys. Rev. Lett. **116**, 235301 (2016).
[3] Lawrence W. Cheuk, Matthew A. Nichols, Katherine R. Lawrence, Melih Okan, Hao Zhang, Ehsan Khatami, Nandini Trivedi, Thereza Paiva, Marcos Rigol, Martin W. Zwierlein, arXiv:1606.04090 (2016).

# Low-lying excitations in strongly interacting Fermi gases

Sascha Hoinka[1], Paul Dyke[1], Cécile Carcy[1], Marcus Lingham[1], Jami Kinnunen[2], Georg Bruun[3], Peter Hannaford[1], Chris Vale[1]

[1]Centre for Quantum and Optical Sciences, Swinburne University of Technology, Melbourne 3122, Australia

[2]COMP Centre of Excellence and Department of Applied Physics, Aalto University, FI-00076 Aalto, Finland

[3]Department of Physics and Astronomy, University of Aarhus, DK-8000 Aarhus C, Denmark

Phase transitions that spontaneously break a continuous symmetry lead to the existence of collective excitations with vanishing energy in the long wavelength limit (Goldstone's theorem). Here we present an experimental and theoretical study of the low-lying excitations in a strongly interacting Fermi gas based on low-momentum Bragg spectroscopy. Below the superfluid transition temperature the Goldstone mode or Bogoliubov-Anderson (BA) phonon is the dominant low-energy excitation through the BCS to Bose-Einstein condensate crossover. At energies greater than twice the pairing gap, single-particle pair-breaking excitations become visible. Both the phonon and single-particle excitation branches display qualitatively different experimental signatures associated with sound propagation and incoherent scattering, respectively. Using Bragg lasers focused into the centre of a harmonically trapped cloud we are able to measure the response of an effectively homogeneous system. Our data show very good agreement with a theoretical model based on the quasiparticle random phase approximation and allow determination of the speed of sound and direct read-off of the pairing gap as a function of the interaction strength.

# Quantum Turbulence in BEC: overview and new perspectives

Vanderlei S. Bagnato
Instituto de Física de São Carlos - University of São Paulo - Brazil
P.O.Box- 369 - São Carlos -SP - Brazil

Quantum Turbulence in Superfluids is a well established field, which now received an important innovation through the use of the Bose-Condensate. The investigations in BEC create new and exciting possibilities that go beyond just similarities with the classical phenomena of turbulence. With trapped superfluids turbulence is observed with a smaller number of vortices, when compared with Superfluid-He, possibilities to control vortex lines, homogeneous and non-homogeneous systems, finite size effects and most important, a large range of densities to be explored. Investigations of 2D and 3D turbulence will be presented, with description of the main characteristics of the turbulent state in terms of hydrodynamic properties, stable configurations, momentum distributions. Special attention will be given for the view of the development of Turbulence in a BEC as a possible inverse Kibble - Zurek scenario. Using the point of view of matter waves in instead of fluid, we demonstrate that Quantum turbulence for a BEC is equivalent to a speckle light field for a Gaussian Coherent Beam of light. This correspond to the possibility to introduce a statistical atom-optics using the free expansion of a turbulent atomic cloud. The determination of second order correlation function allow to construct the equivalence between the two physical phenomena. New perspectives involve the spontaneous generation of turbulence from the decay of highly charged vortex and the mixture of superfluids.

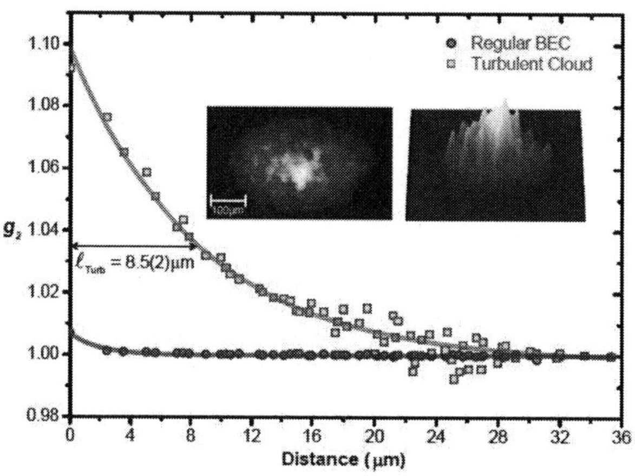


**Figure 1.** Turbulent cloud of trapped BEC and the second order correlation function for the normal and turbulent BEC. Equivalence with a Speckle field of light is determined.

This work is supported by FAPESP and CNPq ( Brazilian Agencies) and has the collaboration with the following scientists from different institutions : V. Yukalov, A. Fetter, C. Barenghi, M. Tsubota, R. Hulet, E. Santos, G. Telles, K. Magalhaes, D. Magalhaes, E. Henn, G. Roati and from the students ( some of them formal students) and Post-Docs: P. Tavares, A. Cidrin, A. Fritsch, M. Tsatsos, M. Caracanhas, A. Smaira, R. Shiozaki, P. Castilho, E. Pedroso, E. Mercado, J. Seman, F. Vivanco, A. Novikov

[1] A full list of article are presented in **http://cepof.ifsc.usp.br**

[2] Our recent publication Tsatsos et al - Physics Reports 622(2016)1-52, provides an overview of the field

# Orbital optical lattices

Andreas Hemmerich[1,2]

[1]Institut für Laser-Physik, Universität Hamburg, Luruper Chaussee 149, 222761 Hamburg, Germany

[2]Wilczek Quantum Center, Zhejiang University of Technology, Hangzhou 310023, China

Bosons condensed in the lowest Bloch band of an optical lattice possess a simple positive definite wave function. If, however, condensation is achieved in higher bands, the rich world of orbital degrees of freedom is accessed [1], which is known to be at the basis of the structural variety of electronic matter. Orbitals with different nodal geometry and orientation can join in order to form wave functions with highly complex local phase patterns. Energetic degeneracies between different orbitals at the same lattice site may lead to the consequence that even tiny energy scales as that of contact interactions can become determinant in deciding the shape of the minimal energy state within the band. This can lead to unconventional superfluid order, topological excitation bands with non-zero Chern invariants and topologically protected edge excitation modes.

In the past few years we have explored these new possibilities in a series of experiments. I will discuss, how bosons can be excited into higher bands of optical lattices and under which conditions long band life times allow one to condense these atoms in the band minima. As a specific example I will discuss how due to orbital degeneracies interaction-induced chiral superfluid order can arise in the second band of a square lattice [2] and how this scenario could be extended in order to engineer topological excitation bands [3].

[1] Xiaopeng Li and W. Vincent Liu, arXiv:1508.06285, to appear in Rep. Prog. Phys. (2016).

[2] T. Kock, et al., J. Phys. B: At. Mol. Opt. Phys. 49, 042001 (2016).

[3] Zhi-Fang Xu, Li You, Andreas Hemmerich, W. Vincent Liu, arXiv:1602.04555 (2016).

# What does the Golden Ratio have to do with friction? An answer atom by atom

Vladan Vuletić[1], Dorian Gangloff[1], Alexei Bylinskii[1], Ian Counts[1], K. Rajibul Islam[1], Joonseok Hur[1], Wonho Jhe[2]

[1]Department of Physics, MIT-Harvard Center for Ultracold Atoms and Research Laboratory of Electronics, Massachusetts Institute of Technology, Cambridge, Massachusetts 02139, USA

[2]Department of Physics and Astronomy, Seoul National University, Seoul 151-747, Korea.

Friction is the basic, ubiquitous mechanical interaction between two surfaces that results in resistance to motion and energy dissipation. In spite of its technological and economic significance, our ability to control friction remains modest, and our understanding of the microscopic processes incomplete. To test long-standing atomistic models of friction processes -at the nanoscale, we have implemented a synthetic nano-friction interface using laser cooled ions subject to the periodic potential of an optical standing wave [1]. We show that stick-slip friction can be tuned from maximal to nearly frictionless via arrangement of the ions relative to the periodic potential, and that friction at the nanoscale can substantially differ from the simple phenomenological laws observed at the macroscale [1,2]. We also establish a direct link [3] between Aubry's structural transition [4] for an infinite chain in an incommensurate periodic potential, and the vanishing of friction in nanocontacts (superlubricity) [5].

[1] A. Bylinskii, D. Gangloff, and V. Vuletić, Science **348**, 1115-1118 (2015).

[2] D. Gangloff, A. Bylinskii, I. Counts, W. Jhe, and V. Vuletić, Nature Physics **11**, 915-919 (2015).

[3] A. Bylinskii, D. Gangloff, I. Counts, and V. Vuletić, Nature Materials, DOI: 10.1038/NMAT4601 (2016).

[4] S. Aubry, Physica D **7**, 240-258 (1983).

[5] K. Shinjo and M. Hirano, Surf. Sci. **283**, 473-478 (1993).

# BEC with spin-orbit coupling

Jian-Wei Pan [1, 2], Shuai Chen [1, 2]

[1] Shanghai Branch, Hefei National Lab for Physical Science at Microscale and Department of Modern Physics, University of Science and Technology of China, Shanghai 201315, P.R. China

[2] CAS Center for Excellence and Synergetic Innovation Center of Quantum Information and Quantum Physics, University of Science and Technology of China, Hefei, Anhui 230026, China

We report the experiment of quantum simulations with synthetic spin-orbit coupled Bose gas. First, Spin-Orbit (SO) coupling in 1D with ultracold Bose gas of $^{87}$Rb was generated with Raman coupling technique. The fundamental properties of SO coupled Bose gas were systematically studied, including the dipole oscillation of the collective modes, the zero [1] and finite temperature [2] phase diagram of the ground states, and the observation of excitation spectrum with contains "phonon-maxon-roton" structure [3]. Very recently, we have proposed and experimentally realized, for the first time, 2D SO coupling and topological bands with $^{87}$Rb BEC through a 2D optical lattices together with a phase correlated Raman coupling lattice [4]. A controllable crossover between 2D and 1D SO couplings is studied, the SO effects and nontrivial band topology are also observed by measuring the atomic cloud distribution and spin texture in the first Brillouin zone. Our achievements open a new way in the field of quantum simulation to study the topological quantum matter.

[1] J. –Y. Zhang, S. –C. Ji, Z. Chen, L. Zhang, Z. –D. Du, B. Yan, G. –S. Pan, B. Zhao, Y. –J. Deng, H. Zhai, S. Chen, and J. –W. Pan, Phys. Rev. Lett. **109**, 115301 (2012).

[2] S. –C. Ji, J. –Y. Zhang, L. Zhang, Z. –D. Du, W. Zheng, Y. –J. Deng, H. Zhai, S. Chen, and J. –W. Pan, Nature Physics **10**, 315 (2014) doi: 10.1038/nphys2905.

[3] S. –C. Ji, L. Zhang, X. –T. Xu, Z. Wu, Y. –J. Deng, S. Chen and J. –W. Pan, Phys. Rev. Lett. **114**, 103601 (2015).

[4] Zhan Wu, Long Zhang, Wei Sun, Xiao-Tian Xu, Bao-Zong Wang, Si-Cong Ji, Youjin Deng, Shuai Chen, Xiong-Jun Liu, and Jian-Wei Pan, Preprint at http://arXiv.org/abs/1511.08170 (2015).

# Einstein-Podolsky-Rosen entanglement of narrowband photons from cold atoms

Jong-Chan Lee, Kwang-Kyoon Park, Tian-Ming Zhao, Yoon-Ho Kim
Department of Physics, POSTECH, Pohang 37673, Korea

The gedankenexperiment proposed by Einstein-Podolsky-Rosen (EPR) in 1935 involves a pair of particles that are entangled in their positions and momenta. The position-momentum-like continuous variable EPR entanglement was initially explored by using quadrature-phase amplitudes of two-mode squeezed states. Genuine EPR position-momentum entanglement of photon pairs, essential for quantum imaging and quantum metrology, was demonstrated later by spontaneous parametric down-conversion (SPDC) photons. Since SPDC photons have very large bandwidth (~ THz), they are not suitable for atom-photon interaction, which typically require ~ MHz bandwidth. While narrowband entangled photons can be generated via cavity-enhanced SPDC, the optical cavity necessary for bandwidth narrowing completely removes EPR position-momentum entanglement between the photons. In this work, we report, to the best of our knowledge, the first experimental demonstration of EPR position-momentum entanglement of narrowband photon pairs by using spontaneous four-wave mixing in a cold atom ensemble. We explicitly demonstrate position-momentum entanglement, with the help of quantum ghost imaging and interference effects, between the narrowband photons by showing that the photon pairs violate the separability criterion. We further demonstrate continuous-variable EPR steering by using the EPR position-momentum entangled narrowband photons [1].

[1] J.-C. Lee, K.-K. Park, T.-M. Zhao, and Y.-H. Kim, arXiv:1606.02036 (2016).

# Direct laser cooling and trapping of diatomic molecules

David DeMille[1]

[1]Yale University, Department of Physics, Yale University, New Haven, CT 06511 USA

The rich internal structures of diatomic molecules enable a wide range of experiments in regimes that are not accessible with atoms [1]. For example, the structure of molecules dramatically amplifies effects from fundamental physics phenomena such as parity- and time-reversal violating interactions due to new, exotic particles [2]. Production of cold and trapped molecular gases is of significant interest, since they could enable improved sensitivity in these types of precision measurements. Moreover, other novel and interesting phenomena can be investigated with ultracold molecules, such as study and control of quantum effects in chemical reactions, novel phases in strongly-interacting dipolar gases, etc.

Despite this broad interest, methods for cooling and trapping molecules have been far less advanced than those for atoms. In particular, direct laser cooling of molecules was long considered infeasible: the same complex internal structure that makes molecules useful also makes it difficult to engineer the sort of closed optical cycling transition needed for laser cooling. Over the past several years, however, our group and others have devised and implemented methods to overcome these difficulties [3-6]. Most of the standard tools of atomic laser cooling—including Doppler and sub-Doppler cooling, beam slowing, and (most recently) magneto-optical trapping—have now been demonstrated to work with molecules, in a manner very similar to the familiar cases of atoms. For example, we can now produce MOTs of strontium monofluoride (SrF) with up to $\sim 10^4$ molecules, temperature of $\sim 400$ μK, density of $\sim 10^6$ cm$^{-3}$, and lifetime of $\sim 0.5$ s (see Fig. 1). In this talk I will review progress in laser cooling and trapping of molecules, and give an outlook for future directions enabled by these rapidly-developing methods.

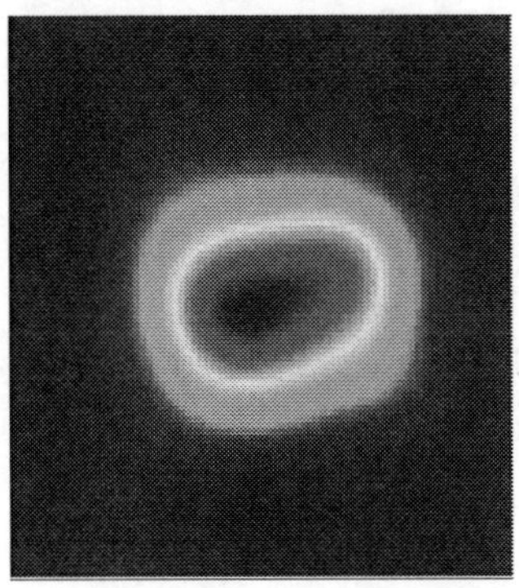

**Figure 1.** Image of fluorescence from a SrF magneto-optical trap.

[1] L.D. Carr, D. DeMille, R.V. Krems, and Jun Ye. New J. Phys. **11**, 055049 (2009).
[2] D. DeMille. Phys. Today **68**, 34 (2015).
[3] E.S. Shuman, J.F Barry, and D. DeMille. Nature **467**, 820 (2010).
[4] J.F. Barry, E.S. Shuman, and D. DeMille. Phys. Rev. Lett. **108**, 103002 (2012).
[5] J.F. Barry, D. McCarron, E.B. Norrgard, M.H. Steinecker, and D. DeMille. Nature **512**, 286 (2014).
[6] E.B. Norrgard, D.J. McCarron, M.H. Steinecker, M.R. Tarbutt, and D. DeMille. Phys. Rev. Lett. **116**, 063004 (2016).

# Laser cooling and trapping of molecules: experiments and models

Mike Tarbutt[1]

[1]Centre for Cold Matter, Blackett Laboratory, Imperial College London, UK

We slow and cool a beam of CaF molecules to low velocity using frequency-chirped or frequency-broadened counter-propagating laser light [1]. The velocity of the beam can be reduced to a few m/s, while compressing its velocity spread by a factor of ten and retaining its high brightness. The slow molecules are delivered to a magneto-optical trap (MOT) where we aim to cool them to μK temperatures.

Using multi-level rate equations and optical Bloch equations, we have studied how molecules are cooled and trapped in a 3D MOT [2]. The molecules tend to be pumped into dark Zeeman sub-levels, which weakens the cooling and trapping forces. The force is further weakened if the excited state has a small magnetic moment, which it often does for molecules. Both effects are avoided when the MOT beams contain two oppositely-polarized laser components, one red- and one blue-detuned [3]. We find that sub-Doppler processes are important in the molecule MOT. When the light is red-detuned, the temperature is raised by sub-Doppler mechanisms which oppose the Doppler cooling force at low velocities, even in the magnetic field of the MOT. The temperature can be lowered by lowering the light intensity, and by switching to a blue-detuned molasses or MOT.

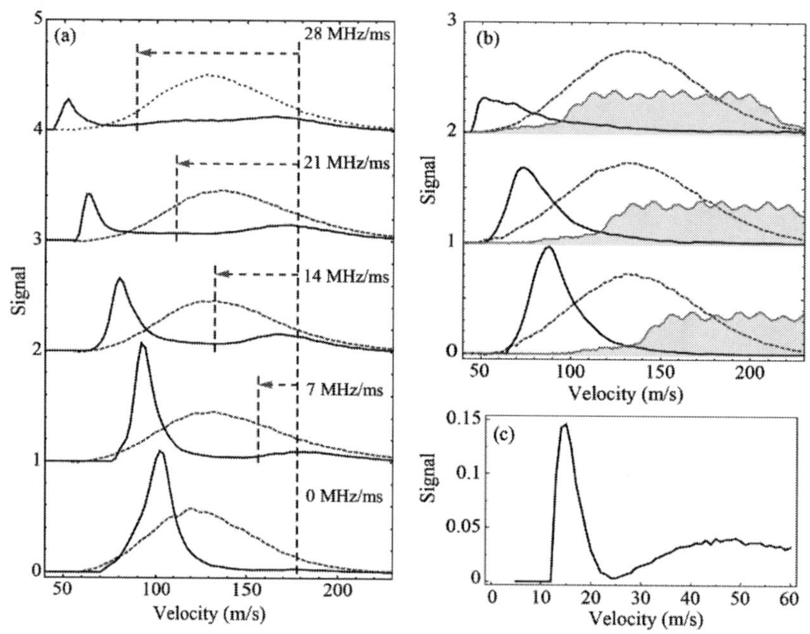


**Figure 1.** Laser slowing and cooling of a beam of CaF. Solid and dotted lines are velocity profiles measured with and without counter-propagating laser light. (a) Lasers are frequency chirped to sweep the resonant velocity from the initial to the final values shown by the dashed lines. (b) Lasers are frequency broadened to address the velocity range shown by the shaded regions. (c) Frequency-chirp slowing to low velocity, using a converging laser beam to reduce the transverse divergence.

[1] S. Truppe *et al.*, arXiv:1605.06055 (2016).
[2] M. R. Tarbutt, New J. Phys. **17**, 015007 (2015).
[3] M. R. Tarbutt and T. C. Steimle, Phys. Rev. A **92**, 053401 (2015).

# Coherent control of ultracold $^{87}Rb^{133}Cs$ molecules

Simon L. Cornish

Joint Quantum Centre (JQC) Durham-Newcastle, Department of Physics,
Durham University, South Road, Durham DH1 3LE, UK

The formation of ultracold heteronuclear molecules opens up many exciting areas of research spanning precision measurement, quantum computation, quantum simulation, ultracold chemistry and fundamental studies of quantum matter. Long-lived, trapped samples of molecules with full quantum control of the molecular internal state are crucial to many of these applications. Here we demonstrate coherent microwave control of the rotational and hyperfine state of ultracold, chemically stable $^{87}Rb^{133}Cs$ molecules. We create up to 4000 molecules in the rovibrational and hyperfine ground state at a temperature of ~1.2 μK and a density of $\sim 10^{11}$ $cm^{-3}$ using magnetoassociation on a Feshbach resonance [1] followed by optical transfer using stimulated Raman adiabatic passage [2,3]. We then use precision microwave spectroscopy of the rotational transition to probe the rich hyperfine structure of the molecule. Finally, we exploit coherent Rabi oscillations to transfer the total population of molecules between hyperfine levels and report measurements of the collisional lifetime for molecules in a single hyperfine level of both the ground state and the first rotationally excited state.

[1] M. P. Köppinger, D. J. McCarron, D. L. Jenkin, P. K. Molony, H.-W. Cho, C. R. Le Sueur, C. L. Blackley, J. M. Hutson and S. L. Cornish, Phys. Rev. A **89**, 033604 (2014).

[2] P. K. Molony, P. D. Gregory, Z. Ji, B. Lu, M. P. Köppinger, C. R. Le Sueur, C. L. Blackley, J. M. Hutson, and S. L. Cornish, Phys. Rev. Lett. **113**, 255301 (2014).

[3] P. D. Gregory, P. K. Molony, M. P. Köppinger, A. Kumar, Z. Ji, B. Lu, A. L. Marchant and S. L. Cornish, New J. Phys. **17**, 055006 (2015).

# Ultracold chemistry and asymptotic physics with diatomic strontium molecules

Robert Moszynski[1]

[1]Department of Chemistry, University of Warsaw, Pasteura 1, 02-093 Warsaw, Poland

During the last decade state-of-the-art ab initio methods of quantum chemistry have been applied with success to interpret precision experiments on two-body and many-body processes in atomic gases in the ultracold regime. In this talk I will present recent examples of successful applications of the ab initio methods to describe two-body processes in atomic optical lattices leading to the formation of unusual chemical bonds and to observations of exotic optical transitions in diatomic molecules [1]. I will also discuss the studies of ultracold chemistry enabled by photodissociation of diatomic strontium molecules, including the phenomena of resonant and nonresonant barrier tunneling, matter wave interference of reaction products, and forbidden reaction pathways [2].

[1] B.H. McGuyer *et al.,* Nature Phys. **11**, 32 (2015).
[2] M. McDonald *et al.*, Nature in press, (2016).

# Experimental realization of a two-dimensional synthetic spin-orbit coupling in ultracold Fermi gases

Jing Zhang

The State Key Laboratory of Quantum Optics and Quantum Optics Devices, Institute of Opto-Electronics, Shanxi University, Taiyuan 030006, P.R. China

A bottleneck in current studies is the lack of a two-dimensional (2D) synthetic SOC, which is crucial for realizing high-dimensional topological matters. Here, we report the experimental realization of 2D SOC in ultracold 40K Fermi gases using three lasers, each of which dresses one atomic hyperfine spin state. Through spin injection radio-frequency (rf) spectroscopy, we probe the spin-resolved energy dispersions of dressed atoms, and observe a highly controllable Dirac point created by the 2D SOC. Our work paves the way for exploring high-dimensional topological matters in ultracold atoms using Raman schemes.

[1] L. Huang, Z. Meng, P. Wang, P. Peng, S. Zhang, L. Chen, D. Li, Q. Zhou, J. Zhang, “Experimental realization of a two-dimensional synthetic spin-orbit coupling in ultracold Fermi gases”. Nature Phys. 12, 540 (2016); arXiv:1506.02861.

[2] Z. Meng, L. Huang, P. Peng, D. Li, L. Chen, Y. Xu, C. Zhang, P. Wang, J. Zhang, “Experimental observation of topological band gap opening in ultracold Fermi gases with two-dimensional spin-orbit coupling” arXiv: 1511.08492..

# Cat states and quantum metrology with Rydberg atoms

Jean-Michel Raimond[1], Adrien Signoles[1], Adrien Facon[1], Eva-Katharina Dietsche[1], Dorian Grosso[1], Serge Haroche[1], Michel Brune[1] and Sébastien Gleyzes[1]

[1]Laboratoire Kastler Brossel, Collège de France, CNRS, ENS-PSL Research University, UPMC-Sorbonne Universités, 11 place Marcelin Berthelot, 75005, Paris, France

The enormous Hilbert space of a Rydberg Stark manifold is a large scene for quantum games. Using radiofrequency fields with well-defined polarizations, we address a set of equidistant levels, equivalent to a large J=25 spin. The lower state in this ladder is directly accessible by laser excitation. The upper one is the circular state with maximum angular momentum. We observe coherent Rabi oscillations of this spin driven by a resonant radiofrequency [1], exhibiting an unprecedented control over this complex system. A continuous selective interrogation of the spin ('have you reached this precise level?') limits its evolution to an adjustable subspace. The resulting quantum Zeno dynamics prepares mesoscopic Schrödinger-cat states, superpositions of spins pointing in different directions [2]. Such non-classical cat states are the key to quantum-enabled measurements. We have used them for the measurement of a static electric field. The achieved precision exceeds the Standard Quantum Limit (SQL) and approaches the fundamental Heisenberg limit (HL) (Fig. 1). The single-shot sensitivity reaches 1.2 mV/cm for a 100 ns interaction time (30 μV/cm/√Hz at a 3 kHz repetition rate). This highly sensitive, non-invasive space- and time-resolved field measurement extends the realm of electrometric techniques and could have important applications. Detection of individual electrons in mesoscopic devices at a 100 μm distance with a MegaHertz bandwith is feasible. We are aiming to the preparation of other cat states, such as a superposition of circular states with opposite angular momenta, utterly sensitive to magnetic fields and impervious to electric ones.

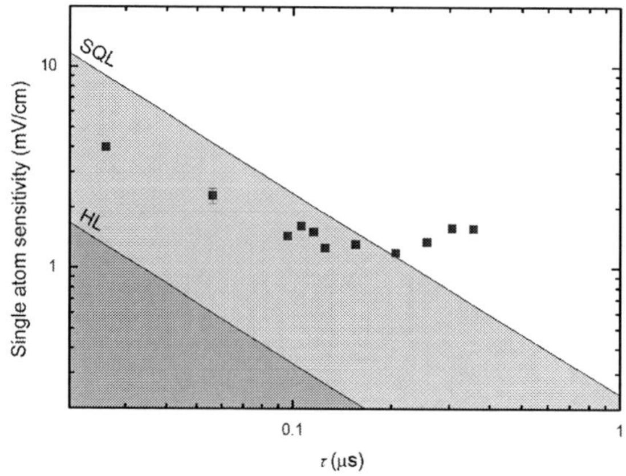


**Figure 1.** Blue points: single shot sensitivity as a function of the interrogation time τ (statistical error bars). The SQL and HL are depicted by black lines. The dark gray area is forbidden. The light-gray area corresponds to measurements beyond the SQL. Reprinted from [3].

[1] A. Signoles *et al.*, in preparation.

[2] A. Signoles *et al.*, Nature Physics **10**, 715 (2014) and ArXiv 1402.0111

[3] A. Facon *et al.*, Nature, in print (2016) and ArXiv 1602.02488

# Einstein-Podolsky-Rosen entanglement in spinor Bose-Einstein condensates

J. Peise[1], I. Kruse[1], K. Lange[1], B.Lücke[1], L. Pezzè[2,3] , J. Arlt[4], W. Ertmer[1], K. Hammerer[5], L. Santos[5], A. Smerzi[2,3], C. Klempt[1]

[1]Institut für Quantenoptik, Leibniz Universität Hannover, Welfengarten 1, D-30167 Hannover, Germany.

[2]QSTAR, Istituto Nazionale di Ottica, Largo Enrico Fermi 2, 50125 Firenze, Italy.

[3]LENS, Via Nello Carrara 1, 50019 Sesto Fiorentino, Italy.

[4]Institut for Fysik og Astronomi, Aarhus Universitet, Ny Munkegade 120, DK-8000 Århus C, Denmark.

[5]Institut für Theoretische Physik, Leibniz Universität Hannover, Appelstraße 2, D-30167 Hannover, Germany.

In 1935, Einstein, Podolsky and Rosen (EPR) questioned the completeness of quantum mechanics by devising a quantum state of two massive particles with maximally correlated positions and momenta. The EPR criterion [1] qualifies such continuous-variable entangled states, where a measurement of one subsystem seemingly allows for a prediction of the second subsystem with a precision beyond the Heisenberg uncertainty principle. We have created an EPR-correlated two-mode squeezed state in an atomic Bose-Einstein condensate [2]. The state shows an EPR entanglement parameter of 0.18(3), which is 2.4 s.d. below the threshold 1/4 of the EPR criterion. The data can be analyzed to obtain a full tomographic reconstruction of the underlying many-particle quantum state.

Finally, we show that the created state is useful for atom interferometry and demonstrate the operation of a prototypical atomic clock beyond the Standard Quantum Limit (SQL) [3]. The novel clock configuration surpasses the SQL by squeezing the vacuum in an empty input mode. The improvement of atom interferometers with squeezed vacuum offers exciting advantages which are also exploited in squeezed-vacuum optical interferometers, as they are developed for the detection of gravitational waves.

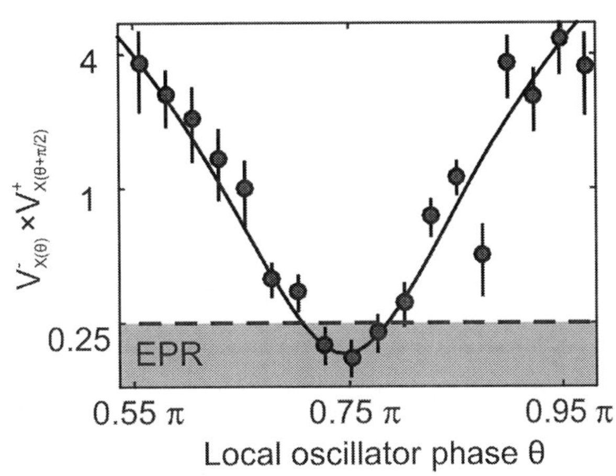


**Figure 1.** EPR parameter as a function of the local oscillator phase θ. Data points below the dashed line meet the EPR criterion [1].

[1] M. Reid, Phys. Rev. A **40**, 913–923 (1989).

[2] J. Peise, I. Kruse, K. Lange, B. Lücke, L. Pezzè, J. Arlt, W. Ertmer, K. Hammerer, L.Santos, A. Smerzi, C. Klempt, Nature Commun. **6**, 8984 (2015).

# Monday Posters

# Constraining Neutrino Electromagnetic Properties by Low Energy Atomic Response

Jiunn-Wei Chen,[1,2,3] Hsin-Chang Chi,[4] Cheng-Pang Liu,[4] Henry T. Wong,[5] and Chih-Pan Wu[1]

[1]Department of Physics and Center for Theoretical Sciences, National Taiwan University, Taipei 10617, Taiwan

[2]Leung Center for Cosmology and Particle Astrophysics, National Taiwan University, Taipei 10617, Taiwan

[3]Center for Theoretical Physics, Massachusetts Institute of Technology, Cambridge, MA 02139, USA

[4]Department of Physics, National Dong Hwa University, Shoufeng, Hualien 97401, Taiwan

[5]Institute of Physics, Academia Sinica, Taipei 11529, Taiwan

The processes of atomic ionization of detectors by scattering with neutrinos in a search of their electromagnetic properties and with light dark matter candidates (e.g. sterile neutrino) are interested for many experimental groups [1-3]. As current experimental searches for neutrino and dark matter have lowered the detector threshold down to the sub-keV regime [4, 5], accurate many-body calculations for atomic ionization are warranted for giving reliable results of experimental comparisons. This presentation briefly introduces an *ab initio* many-body method: the multi-configuration relativistic random phase approximation (MCRRPA) [6], and summarizes our recent efforts of applying this methods to neutrino-germanium scattering in particular [7-9].

[1] A. G. Beda *et al.* (GEMMA Collaboration), Phys. Part. Nucl. Lett. **10**, 139 (2013).

[2] H. T. Wong *et al.* (TEXONO Collaboration), Phys. Rev. D **75**, 012001 (2007).

[3] C. Arpesella *et al.* (The Borexino Collaboration), Phys. Rev. Lett. **101**, 091302 (2008).

[4] H. T. Wong, J. Phys. Conf. Ser. **309**, 012024 (2011).

[5] Q. Yue and H. T. Wong, Mod. Phys. Lett. A **28**, 1340007 (2013).

[6] K-N. Huang and W. R. Johnson, Phys. Rev. A **25**, 634 (1982).

[7] J. W. Chen *et al.*, Phys. Lett. B **731**, 159 (2014).

[8] J. W. Chen *et al.*, Phys. Rev. D **90**, 011301(R) (2014).

[9] J. W. Chen *et al.*, Phys. Rev. D **91**, 013005 (2015).

# A Velocimeter Based on Enhanced Light-Dragging Effect in a Moving Electromagnetically Induced Transparent Medium

Pei-Chen Kuan, Chang Huang, Wei-Sheng Chan, Sandoko Kosen, Shau-Yu Lan
Division of Physics and Applied Physics, School of Physical and Mathematical Sciences, Nanyang Technological University, Singapore 637371

Atoms based velocimeter typically relies on measuring the first order Doppler shift of individual atoms. To detect the center-of-mass motion of an atomic cloud, one usually needs to map out the velocity distribution of the cloud. Here, we demonstrate the light dragging effect in a moving electromagnetically induced transparent (EIT) medium and use it to sense the center-of-mass motion of an atomic cloud directly. The light dragging effect or the deviation from the speed of light $c$ in a moving medium was first observed by Fizeau in a flowing water experiment for the study of aether in the pre Einstein's special theory of relativity era [1, 2]. It was later explained by the Lorentz velocity addition to the first order [3]. The phase shift of light through a moving medium is proportional to the duration travelling in the medium. We enhance the dragging effect in a cold atomic medium under EIT condition and demonstrate a velocity sensor at a sensitivity two orders of magnitude smaller than the velocity width of the atomic medium used. This new type of sensor depends on the collective motion of the atomic cloud and only single shot measurement, and could lead to a new design of motional sensor beyond the limitation of Doppler broadening of atoms.

[1] H. Fizeau, C. R. Acad. Sci. 33, 349 (1851)

[2] A. A. Michelson, E. W. Morley, Am. J. Sci. 34, 333 (1887)

[3] H. A. Lorentz, Leiden - E. J. Brill (1895)

# Measurement and control of single nitrogen-vacancy centers in diamonds rotating at high speed

A. A. Wood[1], E. Lilette[1,2], Y. Fein, D. A. Simpson[1,3], J-. P Tetienne[1,3], A. D. Stacey[1,3], L.P. McGuinness[4], L. C. L. Hollenberg[1,3], R. E. Scholten[1] and A. M. Martin[1]

[1]School of Physics, The University of Melbourne, Australia

[2] Universitee Europeenne de Brettagne, France

[3]CQC2T, The University of Melbourne, Australia

[4]Intitute For Quantum Optics, Ulm University, Germany

Occupying the overlap between condensed matter science and AMO physics, the nitrogen vacancy (NV) centre in diamond allows the power and flexibility of atomic manipulation and measurement protocols to begin to unravel the solid state. It has found application as a nanoscale detector of magnetic fields [1], electric fields [2], crystal strain [3] and temperature [4]. The robust nature of the NV center also makes it suitable for application as a proposed sensor of motion and rotation in noisy, real-world environments. Measuring the Berry phase accumulated by the NV center during mechanical rotation of the diamond crystal [5] has been proposed as a means of detecting rotation, the NV center thus has the potential to realise a nanoscale gyroscope [6]. However, the technical requirements for mechanical rotation are challenging. The typical dephasing times $T_2^*{\sim}10\mu s$ and $T_2{\sim}300\mu s$ times for NVs hosted within high-purity diamonds require exceptionally stable rotational speeds of order kHz to observe appreciable Berry phase accumulation within the spin dephasing times while still allowing for the quantum phase of the NV to be read out. In this work, we discuss work towards measurement and control of single NV centers in a rotating diamond. We are able to create strobed confocal images of the NV and develop techniques to read the spin state during rotation with high fidelity. Our results bring novel applications of the NV center -proposed only theoretically- within experimental reach.

[1] L. Rondin, J.-P. Tetienne, T. Hingant, J.-F. Roch, P. Maletinsky and V. Jacques, Rep. Prog. Phys. 77, 056503 (2014).

[2] F. Dolde, H. Fedder, M. W. Doherty, T. Nbauer, F. Rempp, G. Balasubramanian, T. Wolf, F. Reinhard, L. C. L. Hollenberg, F. Jelezko, and J. Wrachtrup, Nat Phys 7, 459 (2011).

[3] 3M. W. Doherty, V. V. Struzhkin, D. A. Simpson, L. P. McGuinness, Y. Meng, A. Stacey, T. J. Karle, R. J. Hemley, N. B. Manson, L. C. Hollenberg, and S. Prawer, Phys. Rev. Lett. 112, 047601 (2014).

[4] V. M. Acosta, E. Bauch, M. P. Ledbetter, A. Waxman, L.-S. Bouchard, and D. Budker, Phys. Rev. Lett. 104, 070801 (2010).

[5] D. Maclaurin, M. W. Doherty, L. C. L. Hollenberg, and A. M. Martin, Phys. Rev. Lett. 108, 240403 (2012)

[6] M. P. Ledbetter, K. Jensen, R. Fischer, A. Jarmola, and D. Budker, Phys. Rev. A 86, 052116 (2012).

# Optically-pumped atomic magnetometry for magnetic nanoparticle detection and characterization

Simone Colombo[1], Victor Lebedev[1], Zoran D. Grujic[1], Vladimir Dolgovskiy[1], and Antoine Weis[1]

[1]Department of Physics, University of Fribourg, Ch. Du Musée 3, 1700 Fribourg, Switzerland

We describe an atomic magnetometer (AM) for magnetic nanoparticle (MNP) detection in view of biomedical imaging applications. Because of their superparamagnetic character, the nonlinear MNP response to an external magnetizing field $H_{drive}$ allows discrimination of MNP signals from the linear (diamagnetic) background response of biological tissues.

The AM operates in the Mx-mode based on the optical detection of the magnetic resonance of spin polarized Cs atoms confined in a paraffin-coated cell. A circularly-polarized D1 laser beam polarizes the Cs vapor by means of optical pumping. The Cs spin polarization precesses at the Larmor frequency $\omega_L = \gamma_F|\mathbf{B}_0|$ in an external DC magnetic field $\mathbf{B}_0$. A magnetic field oscillating at $\omega_{rf}$ resonantly forces the spin precession when $\omega_{rf} = \omega_L$. A phase-locked loop (PLL) that keeps the rf-frequency resonant with the Larmor frequency allows tracking changes of the $|\mathbf{B}_0|$ [2].

The MNPs are excited by a time varying external field $\boldsymbol{H}_{drive}(t)$, oriented such that the induced field $\delta\mathbf{B}_{NP}(t)$ produced by the MNPs at the AM position is parallel to $\mathbf{B}_0$. The PLL frequency is then proportional to $|\mathbf{B}_{tot}(t)| = |\mathbf{B}_0 + \delta\mathbf{B}_{NP}(t)|$.

The detected $\delta B_{NP}$ fields are in the range of 1 pT to 1 nT, depending on the excitation field amplitude and the sample's iron content. Here, we report on the characterization of MNP samples performed with our AM operated in a two-layer aluminum chamber. The sensitivity of 5 pT/Hz$^{1/2}$ allows a fast detection and analysis of MNP samples containing down to 1 μg of iron.

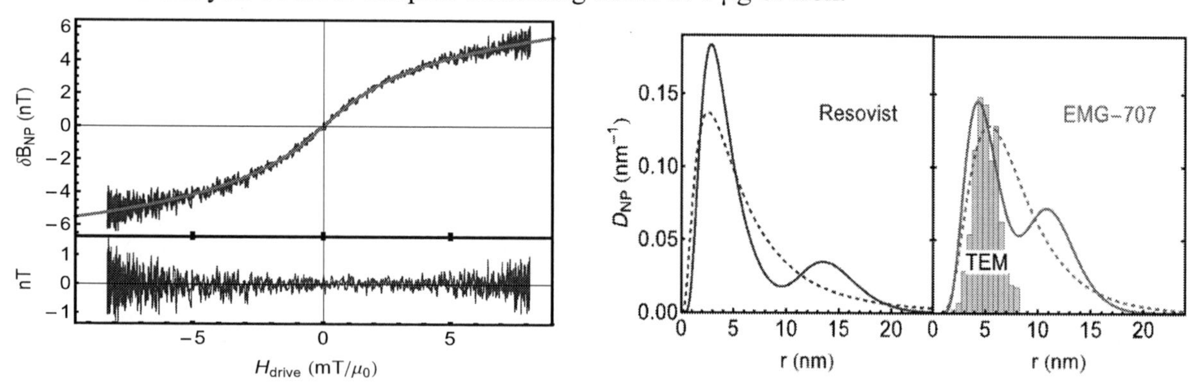


**Figure 1**: *Left graph: Recording of the field produced by a magnetized MNP sample as a function of the amplitude of the drive field $H_{drive}$, together with a fit by a single core MNP model. Right graph: MNP size dependence of the mass fraction distribution of two MNP samples inferred by fitting single core models to $\delta B_{NP}(H_{drive})$ recordings* [2].

[1] S. Groeger, G. Bison, J. L. Schenker, R. Wynands, and A. Weis, The European Physical Journal D-Atomic, Molecular, Optical and Plasma Physics, 38(2), 239-247 (2006).

[2] S. Colombo, V. Lebedev, Z. D. Grujic, V. Dolgovskiy, and A. Weis, International Journal On Magnetic Particle Imaging 2, 1 (2016). https://journal.iwmpi.org/index.php/iwmpi/article/view/28

# Extensive Study of the atomic Hong-Ou-Mandel experiment in momentum space with metastable Helium

Pierre Dussarrat[1], Maxime Perrier[1], Almazbeck Imanaliev[1], Raphaël Lopes[2], Marc Cheneau[1], Denis Boiron[1], Christopher Westbrook[1], Alain Aspect[1]

[1]Institut d'Optique *graduate school*, 2 Avenue Augustin Fresnel, 91127 Palaiseau, France
[2]Cavendish Laboratory, University of Cambridge, J. J. Thomson Avenue, Cambridge CB3 0HE, United Kingdom

The Hong-Ou-Mandel experiment is a milestone in quantum optics, first performed in 1987 [1] it is the fact that two indistinguishable photons impinging at both input ports of a 50:50 beam splitter always emerge together in one of the outputs ports, leading to a reduction of the coincidence counts between the output ports. We first reported its atomic analogue in 2015 [2] with metastable Helium atoms in momentum space and present here an extensive study of the phenomenon.

The achievement of this experiment open the way to perform even more fundamental experiments like the Bell's inequalities test with massive particles.

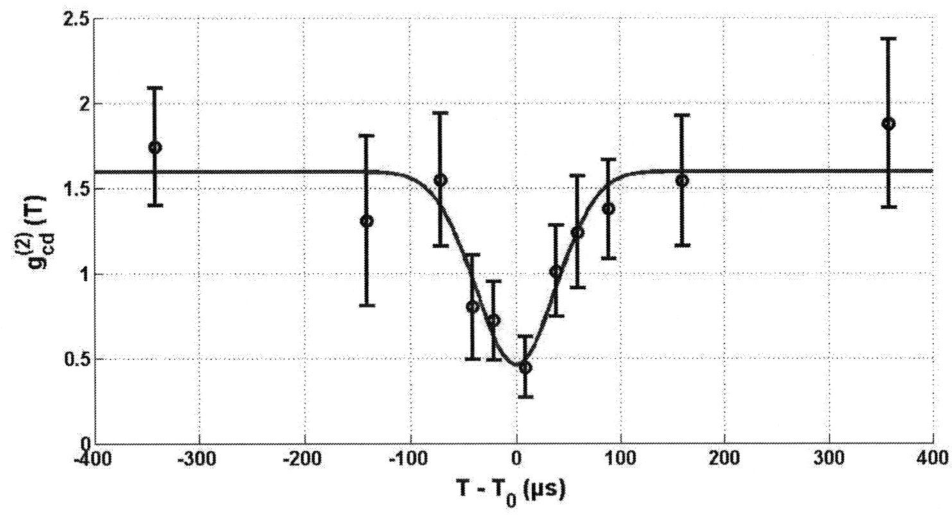


**Figure 1**. The atomic Hong-Ou-Mandel dip observed with the correlation function $g_{cd}^{(2)} = \frac{\langle N_c N_d \rangle}{\langle N_c \rangle \langle N_d \rangle}$ between the output ports (c and d) of the 50:50 atomic beam splitter. T is the application time of the optical lattice doing the beam splitter, applied at the right time ($T_0$) the two atoms impinging on both input ports are indistinguishable and leave always together leading to the reduction of the $g_{cd}^{(2)}$ function.

[1] C. K. Hong, Z. Y. Ou, and L. Mandel, PRL Vol. 59, 18 (1987).
[2] R. Lopes et al., Nature 520 66-68 (2015).

# Towards measuring parity violation in cold chiral molecules using vibrational spectroscopy

S.K. Tokunaga[1], B. Argence[1], M. Pierens[1], D.B.A. Tran[1], R. Santagata[1], A. Shelkovnikov[1], O. Lopez[1], C. Daussy[1], C. Chardonnet[1], A. Amy-Klein[1], D. Nicolodi[2], M. Abgrall[2], Y. Le Coq[2], R.J. Hendricks[3], J.D. Tandy[3], E.A. Hinds[3], M.R. Tarbutt[3], B. Darquié[1]

1. Laboratoire de Physique des Lasers, Université Paris 13, Sorbonne Paris Cité, CNRS, F-93430, Villetaneuse, France

2. LNE-SYRTE, Observatoire de Paris, PSL Research Univ, CNRS, Sorbonne Univ, UPMC Univ. Paris 06, Paris 75014, France

3. Centre for Cold Matter, Blackett Laboratory, Imperial College London, London SW7 2AZ, UK

Parity violation (PV) has never been observed in chiral molecules. Caused by the weak nuclear force, PV should lead to frequency differences in the rovibrational spectra of the two enantiomers of a chiral molecule. A successful PV measurement will shed some light on the origins of biomolecular homochirality. It can also constitute a test of the standard model in the low-energy regime and a probe of physics beyond it, and serve as a stringent benchmark in relativistic quantum chemistry calculations. However, the effect is small.

We present the results of preliminary investigations conducted on methyltrioxorhenium (MTO), an achiral test molecule whose chiral derivatives are expected to show a $\sim 10^{-14}$ level PV effect [1]. We report on the high-resolution spectroscopy of MTO [2,3,4], both in cells and in molecular beams. This work has enabled us to identify several key elements of the current experiment needing improvement prior to making a PV measurement.

The first is the lack of tuneability of our CO2 lasers. We present our on-going work towards their replacement with quantum cascade lasers (QCLs) [4,5], which offer broad and continuous tuning. We are now able to lock mid-IR radiation to a frequency comb stabilized to a near-IR reference. This frequency reference, generated at the French national metrology institute, is monitored against atomic frequency standards and transferred via an optical fibre. Stabilizing a QCL this way provides the ultimate frequency accuracy (potentially the $3\times10^{-16}$ of the Cs fountain clock) and stability ($\sim 10^{-15}$ after 1s of integration). We thus obtain a linewidth of 0.2-Hz [5].

Secondly, our current molecular beam source only yields a modest flux for species such as MTO. We therefore plan to switch to a buffer-gas-beam source. Towards this end, we show that cold gas phase MTO is produced by ablating a solid target in a cryogenic cell containing helium at ~5 K. By sending the QCL beam through the cell, we perform absorption spectroscopy of the Re=O stretching mode from which we estimate a temperature of ~6 K. The resolution is ~3 MHz, allowing the hyperfine structure in the excited vibrational state to be partially resolved and modeled.

[1] N. Saleh, S. Zrig, T. Roisnel, *et al.*, Phys. Chem. Chem. Phys. **15**, 10952 (2013).
[2] C. Stoeffler, B. Darquié, A. Shelkovnikov *et al.*, Phys. Chem. Chem. Phys. **13**, 854 (2011).
[3] S.K. Tokunaga, C. Stoeffler, F. Auguste *et al.*, Mol. Phys. **111**, 2363 (2013).
[4] P.L.T. Sow, S. Mejri, S.K. Tokunaga *et al.*, App. Phys. Lett. **104**, 264101 (2014).
[5] B. Argence, B. Chanteau, O. Lopez *et al.*, Nature Photon. **9**, 456 (2015).

# Analysis of effective electric fields of heavy molecules for search of the electron EDM

Ayaki Sunaga[1], Minori Abe[1], Bhanu Pratap Das[2], Masahiko Hada[1]

[1] Tokyo Metropolitan University, 1-1, Minami-Osawa, Hachioji-city, Tokyo 192-0397, Japan

[2] International Education and Research Center of Science and Department of Physics, Tokyo Institute of Technology, 2-12-1-H86 Ookayama, Meguro-ku, Tokyo152-8550, Japan

An important probe of physics beyond the Standard Model of particle interactions is the electric dipole moment of the electron (eEDM, $d_e$) [1]. It can be obtained by combining the observed energy shift ($\Delta E$) in a molecule with the calculated effective electric field ($E_{eff}$). Since $\Delta E$ is proportional to $E_{eff}$, molecules with large $E_{eff}$ are good candidates for the experiment of the search for the eEDM. Almost for a half century, it has been a kind of common sense that the molecules with strong electric polarization, such as the halides and the oxides, should have large $E_{eff}$ [2]. However, our work shows that molecules with relatively small polarizations like certain hydrides have larger $E_{eff}$ than their fluoride counterparts and analyzed the reason for this from molecular orbital theory.

We calculated $E_{eff}$ and the molecular electric permanent dipole moment (PDM) of YbH, YbF HgH and HgF using the four-component relativistic coupled-cluster single and doubles (RCCSD) method. Table I shows our calculated results. The hydrides have larger $E_{eff}$ than the fluorides, although their PDMs are smaller. It is in conflict with the expectation based on previous works [2].

Owing to the features of the electron EDM interaction Hamiltonian, only single occupied molecular orbital (SOMO) contribute to $E_{eff}$ at the Dirac-Fock level. Besides, the mixture of *s* and *p* orbitals in SOMO is needed for non-zero $E_{eff}$. Hence, the large contributions of *p* orbitals in SOMO would increase $E_{eff}$. We performed a Mulliken population analysis (MPA), which indicates that the number of electrons belonging to each atomic orbital in a molecule. MPA shows that the contributions of *p* orbitals in SOMO of the hydrides are larger than those of the fluorides. The reason is that the admixture of *s* and *p* in the valence orbital of the heavy atom can strongly interact with the valence orbital of the light atom. This strong interaction originates from the smaller energy gap ($\Delta\varepsilon$) (Table II) between the valence orbitals of each atom in the hydrides. The factor which increases $E_{eff}$ is not the electric polarization of the molecule, but $\Delta\varepsilon$.

Table I. Results at RCCSD(QZ) level

| | $E_{eff}$ (GV/cm) | PDM (D) |
|---|---|---|
| YbH | 31.3 | 2.93 |
| YbF | 23.2 | 3.59 |
| HgH | 118.5 | 0.15 |
| HgF | 114.4 | 2.97 |

Table II Energy gap of the valence orbitals of the atoms(a.u.)

| YbH | YbF | HgH | HgF |
|---|---|---|---|
| 0.30 | 0.53 | 0.17 | 0.41 |

From the above analysis, we propose new candidates for the experiment, which have large $E_{eff}$ and also an advantage for the orientation. Polarizing electric field $\varepsilon_{pol} = 2B/D$ is the quantity which is the required lab electric field to orient molecules along the magnetic field. $B$ is the rotational constant and $D$ is PDM. Molecules with small $\varepsilon_{pol}$ is good for the experiment and in this sense, HgH has a disadvantage though its $E_{eff}$ is quite large. Hence, instead of HgH, we propose RaX (X= Ag, I, Au, At) molecules which have small $\varepsilon_{pol}$ because of small $B$ and large $D$. The results of RaX are given in our presentation. Our results show that molecules with small electric polarization, including heavy-heavy molecules, can have large $E_{eff}$. It will require novel techniques to use our promising theoretical results in devising experiments to search for the electron EDM using these molecules.

[1] E. E. Salpeter, Phys. Rev. **112**, 1642 (1958).

[2] P. G. H. Sandars, Phys. Rev. Lett. 19, 1396 (1967); D. DeMille, Phys. Today 68(12), 34 (2015)

# Progress towards laser spectroscopy of antihydrogen in the ALPHA experiment

Steven A. Jones[1] and Stefan Eriksson[1] from the ALPHA collaboration[2]
[1]Department of Physics, College of Science, Swansea University, Swansea SA2 8PP, UK
s.j.eriksson@swansea.ac.uk
[2]CERN, CH-1211 Genene 23, Switzerland

Antihydrogen offers a unique way to test matter/antimatter symmetry. Antihydrogen can reproducibly be synthesised and trapped in the laboratory for extended periods of time [1, 2] offering an opportunity to study the properties of antimatter at high precision. New techniques to study antihydrogen have emerged; the ALPHA collaboration at CERN can now interrogate the bound state energy structure with resonant microwaves [3], determine the gravitational mass to inertial mass ratio [4] and measure charge neutrality [5,6]. The results are not yet sensitive enough to draw conclusions on matter/antimatter symmetry but recent progress shows that experiments with trapped antihydrogen are possible and the collaboration is firmly en-route towards precision measurements. Here, we present recent results and progress towards laser spectroscopy of trapped antihydrogen with the upgraded trapping apparatus which provides improved access for both laser beams and microwave radiation.

[1] G. B. Andresen et al. (ALPHA-Collaboration), Nature **468**, 673 (2010).
[2] G. B. Andresen et al. (ALPHA-Collaboration), Nature Physics **7**, 558 (2011).
[3] C. Amole et al. (ALPHA-Collaboration), Nature **483**, 439 (2012).
[4] A. E. Charman and C.Amole et al. (ALPHA-Collaboration), Nature Communications **4**, 1785 (2013).
[5] C. Amole et al. (ALPHA-Collaboration), Nature Communications **5**, 3955 (2014).
[6] M. Ahmadi et al. (ALPHA-Collaboration), Nature **529,** 373 (2016).

# High density nitrogen-vacancy centers magnetometry with a $MgF_2/TiO_2$ coated diamond

Lixia Xu[1], Heng Yuan[1], Chen Zhang[1], Ning Zhang[1], Jixing Zhang[1], Jiancheng Fang[1]

[1]School of Instrument Science and Opto-electronics engineering, Beihang University, Beijing, China

In recent years, the negatively charged nitrogen-vacancy (NV) centers in diamond have attracted attention owing to their promise in quantum information and metrology. However, the low fluorescence collection efficiency in bulk diamond limits the sensitivity of NV based sensors and various methods have been explored [1,2]. In this work, a $MgF_2/TiO_2$ double-layer antireflection(AR) coating is designed to enhanced the fluorescence collection efficiency. The membrane can be deposited on the diamond substrate using electron beam and thermal evaporation techniques. An optical cavity is optimized to collect the fluorescence transmitted from the sides of the diamond. Based on these techniques, a small-sized magnetometer is designed using diamonds with high density NV centers. Results show that the reflectivity for the wavelength ranging from 600nm to 800nm is below 1% for the coated diamond surface assuming near-normal incidence light. Combined with the optical cavity we can achieve nearly 100% fluorescence collection efficiency.

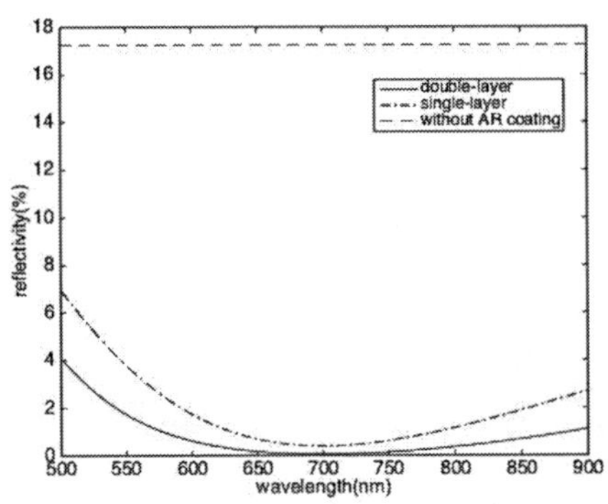


(a)

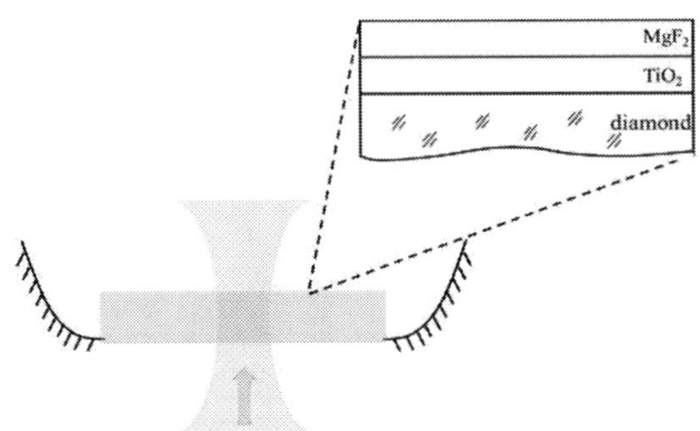


(b)

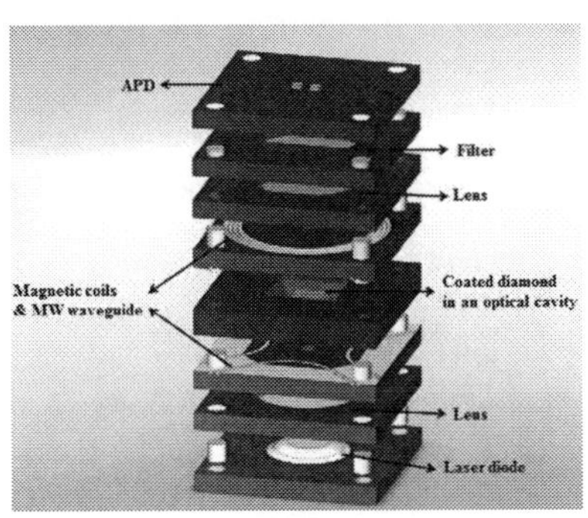


(c)

**Fig 1.** (a) Simulated reflection spectra for different AR coatings (b) Schematic diagram of the fluorescence collection structure. (c) The small-sized diamond device

[1] T. K. Yeung, D. Le Sage, L. M. Pham, et al., Applied Physics Letters **100**,251111(2012);
[2] T. Wolf, P. Neumann, K. Nakamura, et al., Physical Review X **5**,041001(2015).

# Ultra-sensitive all-optical magnetometer for analyzing magnetization produced by hyperpolarized noble gas

Sheng Zou[1, 2*], Hong Zhang[1, 2], Wei Quan[2], Ming Ding[2], Zhaohui Hu[2]

[1] School of Instrument Science and Engineering, Southeast University, Nanjing 210096, China
[2]Science and Technology on Inertial Laboratory, Beihang University, Beijing 100191, China

According to working principle of all-optical atomic magnetometer, relevant experimental equipment has been set up for theoretical verification and sensitive tests. Weak magnetic analysis is of great significance in many fields, including the mechanism analysis of weak magnetic biological molecular, the mechanism analysis and performance evaluation of weak magnetic material. The all-optical atomic spin magnetometer is a new type of magnetic field detection device based on the quantum effects, and its extreme sensitivity is expected to reach the order of aT magnitude [1, 2]. The sensitivity of this magnetometer can reach to the highest level compared with other magnetometers. Aiming at the principle of the magnetometer, the relevant model based on Bloch equation has been built up, which confirms the relationship between the input signal and the output signal, as shown in following figure.

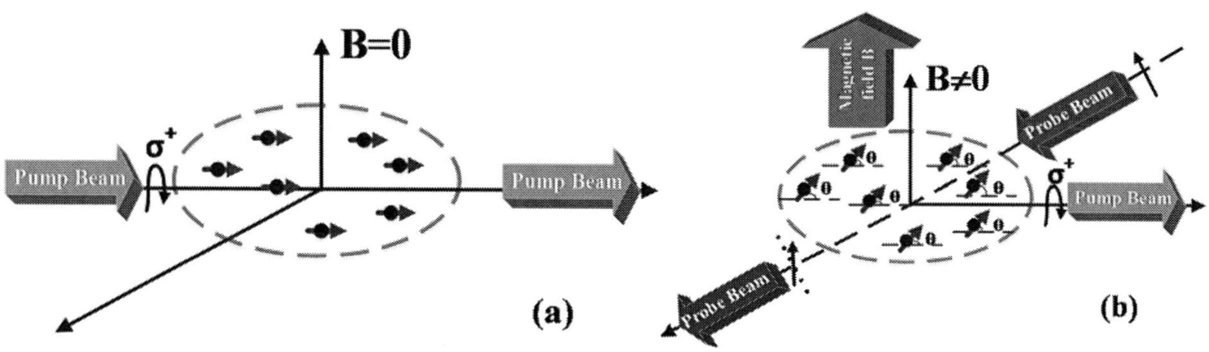


**Figure 1.** Basic principle of ultra-sensitive all-optical magnetometer.

According to principle of all-optical atomic magnetometer, relevant experimental equipment has been set up for measuring the magnetization produced by spin polarized Xe-129, as shown in following figure.

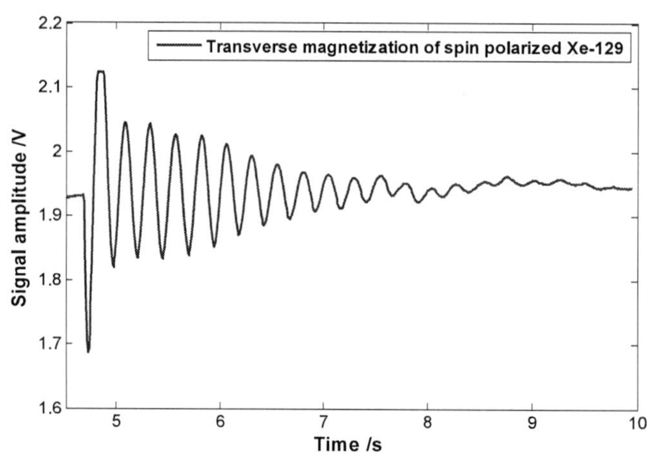


**Figure 2.** Transverse magnetization produced by spin polarized xenon-129 nuclei.

[1] I. Savukov, T. Karaulanov, and M. G. Boshier, Appl. Phys. Lett. **104**, 023504 (2014).
[2] B. Patton, E. Zhivun, D. C. Hovde, and D. Budker, Phys. Revi. Lett. **113**, 013001 (2014).

# Simultaneous direct spectral purity transfer at three optical clock transitions using an all-PM Er-doped fiber frequency comb

Wolfgang Hänsel[1], Michele Giunta[1,2], Marc Fischer[1], Ronald Holzwarth[1,2]

[1]Menlo Systems GmbH, Am Klopferspitz 19a, 82152 Martinsried, Germany
[2]Max-Planck-Institut für Quantenoptik, Hans-Kopfermann-Str. 1, 85748 Garching, Germany

Frequency combs are taking a central role in the comparison of optical frequency standards and in the operation of optical clocks. As relative accuracy and stability of optical frequency standards are tackling the border of $10^{-18}$ [1, 2], the demands for the comb as central clockwork are increasing. A very low short-term noise is required to study systematics within reasonable time. Even more, spectral purity transfer [3] can be used to compare optical transition frequencies starting from a single cavity-stabilized laser, thus rejecting laser noise common-mode by synchronous interrogation [4]. While intrinsic noise of the frequency comb can in principle be mixed out using refined RF techniques, robustness and flexibility are strongly enhanced by direct transfer.

Here we report about the stability of the comb lines as inferred from the comparison of two ultra-low noise frequency combs stabilized to a common stable reference laser at 1542 nm. Optical beat signals between the two combs are simultaneously generated at three different wavelengths (1396.9 nm, 1156.8 nm, 1062.6 nm) as used for the interrogation of the clock transitions in Sr, Yb and Hg, respectively. The wavelengths are provided by different fiber sections each containing an Er-doped fiber amplifier followed by a highly-nonlinear fiber.

Figure 1 shows the observed fractional instabilities measured with pi-mode counters with 1 MHz bandwidth. The Allan deviation starts below $10^{-16}$ at 1s, thus rivaling best Ti:Sa lasers.

a)
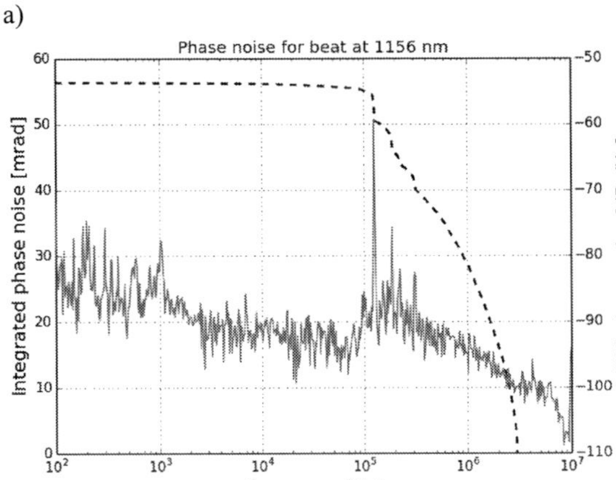


b)
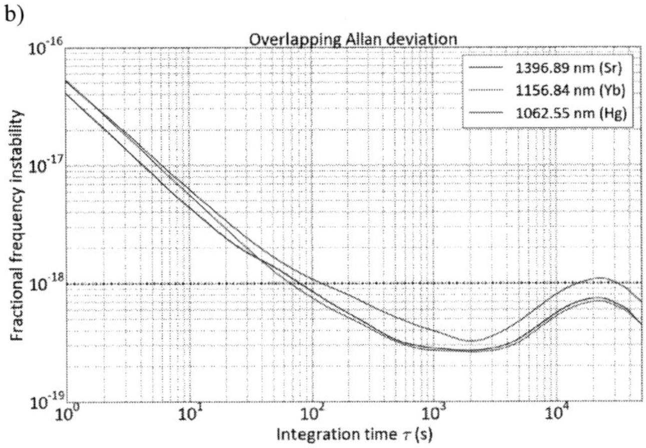


**Figure 1.** a) Phase noise of optical beat at 1156 nm, b) Overlapping Allan deviation of the fractional optical frequency instability for the three observed wavelengths. The peak around 20'000 s stems from day-night temperature variations affecting the connecting fibers between the frequency combs.

[1] I. Ushijima et al., Nature Photonics **9**, 185 (2016).
[2] T.L. Nicholson, et. al., Nature Communications **6**, 6896 (2015).
[3] D. Nicoldi et al., Nature Photonics **8**, 219 (2014).
[4] N. Nemitz et al., Nature Photonics **10**, 258 (2016).

# Progress towards a cold, slow beam of YbF molecules for measuring the electron's electric dipole moment

Jongseok Lim[1], James Almond[1], James Bumby[1], Noah Fitch[1], Ben Sauer[1], Mike Tarbutt[1], Ed Hinds[1]

[1]Centre for Cold Matter, Blackett Laboratory, Imperial College London, London, UK

YbF molecules have been used to measure the electron's electric dipole moment (eEDM) [1]. The precision of this eEDM experiment can be improved by using more molecules and increasing the time they spend in the experiment. This can be achieved by slowing the molecules down or increasing the length of the experiment, and then using transverse laser cooling to reduce the divergence of the molecular beam.

We produce a cryogenically-cooled buffer gas beam of YbF with speeds between 100 and 300 m/s shown in Fig. 1 (a). We have experimented with single-stage and two-stage buffer gas cells and we compare their performance. The YbF beam then passes through a 20 cm long region of transverse laser light in a multi-pass configuration [2]. The laser light contains the three wavelengths shown in Fig. 1 (b), each with the rf sidebands needed to address the hyperfine components of the transitions. This gives a nearly-closed laser cooling transition. From multi-level rate equation simulations, we expect to reduce the velocity range from $\pm 3.6\ \mathrm{m/s}$ to $\pm 0.08\ \mathrm{m/s}$. We will present our progress in producing a slow, laser-cooled beam of YbF molecules, and discuss how this beam will be used to improve the measurement of the eEDM.

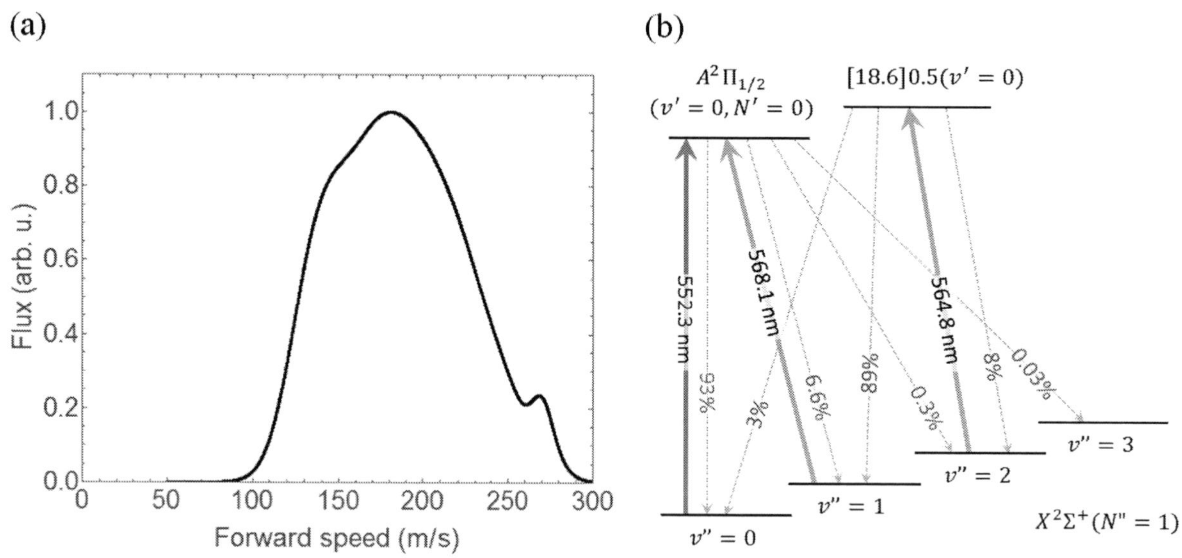


**Figure 1.** (a) A normalized forward speed distribution of the cryogenic YbF beam source. (b) The molecular transitions we use for laser cooling and the required wavelengths of the lasers. *v"* and *v'* are the vibrational quantum numbers of the ground and excited states respectively.

[1] J. J. Hudson *et al.*, *Nature* **473**, 493 (2011).

[2] E. Shuman, J. Barry and D. DeMille, *Nature* **467,** 820 (2010).

# Lifetime of surface plasmon produced in femtosecond laser-structuring

Haiying Song*, Shibing Liu
Strong-field and Ultrafast Photonics Lab, Institute of Laser Engineering
Beijing University of Technology, Beijing 100124, China
*corresponding e-mail: hysong@bjut.edu.cn

The periodic nanostructures (ripples or gratings) induced by femtosecond-laser (fs) exposing solid surface is of great attention in surface engineering field and nevertheless the formation process is very complex. Here, we divide the interaction process of laser-surface into three stages in temporal domain. Firstly, at the leading edge of incident pulse the absorbed energy of metal electrons from the incident laser cannot reach the ionization threshold but the electrons are excited to t-heir energetic bound states in the skin-layer of target surface, in this stage, no plasmons or free electrons are generated on the target surface. Subsequently, in the peak region of incident pulse the energy absorbed by electrons exceed the work function (ionization potential for dielectric media) following the laser-target interaction, the electron ionization is rapidly triggered and sustained, during which a slab plasmon is formed and consequently the surface two-plasmon resonance (STPR) [1] is astatically excited due to the electron density perturbation in this stage. Finally, at the lagging edge of incident pulse the parametric resonant surface two-plasmon is phase-locked and steadily maintained due to the supply of damped pulse energy as a traditional pendulum, during which the Coulomb ablation induced by resonant two-plasmon surface wave occurs on the target surface. Thus, the lifetime of the surface plasmon produced by femtosecond laser pulse, which provides the required time for Coulomb ablation, is crucial to induce periodic surface nanostructures.

As we know that the laser-produced plasma (charged particles) states would emit abundant electromagnetic radiations continuously (continuous spectrum, CS) due to their random motions, while the electrons begin to recombine with their parent ions the characteristic line spectra (CLS) radiate because of the state transition between energy levels in atoms. The duration from CS emitting to first CLS appearing is generally acknowledged as the surface plasmon lifetime. Based on this feature, in this paper, we present a time-resolved spectra method to measure the lifetime by employing an ICCD spectroscopic spectrum analysis system. Here, we use a moderate intensity power fs-laser pulse to excite metallic copper targets as a demonstrating example to investigate the evolution of copper plasmon and the physical processes. The result shows that the lifetime of surface plasmon (electron) is far more than the acceleration time of the target-ion.

[1] H. Y. Song, S. B. Liu, H. Y. Liu, et al., Opt. Express 24, 12151 (2016)

# Precision spectroscopy of the 2S-4P transition in atomic hydrogen

Lothar Maisenbacher[1], Axel Beyer[1], Arthur Matveev[1], Ksenia Khabarova[2], Randolf Pohl[1], Dylan C. Yost[1], Alexey Grinin[1], Nikolai N. Kolachevsky[2], Theodor W. Hänsch[1,3], Thomas Udem[1]

[1] Max-Plank-Institut für Quantenoptik, Garching, Germany
[2] P. N. Lebedev Physical Institute, Moscow, Russia
[3] Ludwig-Maximilians-Universität, München, Germany

Precision measurements of atomic hydrogen have long been successfully used to extract fundamental constants and to test bound-state QED. However, both these applications are limited by measurements of hydrogen lines other than the very precisely known 1S-2S transition [1]. Moreover, the proton r.m.s. charge radius $r_p$ extracted from electronic hydrogen measurements currently disagrees by $4\,\sigma$ with the much more precise value extracted from muonic hydrogen spectroscopy [2].

We have measured the 2S-4P transition in atomic hydrogen using for the first time a cryogenic beam of hydrogen atoms optically excited to the initial 2S state [3]. This strongly suppresses the first order Doppler shift of the one-photon 2S-4P transition, which is further suppressed by actively stabilized counter-propagating laser beams and time-of-flight resolved detection. Quantum interference arising from spontaneous emission [4] can lead to significant line distortions in our system. We use an experimentally verified, simple line shape model to take these distortions into account. With this, we can extract a new value for $r_p$ and the Rydberg constant $R_\infty$ with comparable accuracy as the combined previous H world data [5].

[1] C. G. Parthey *et al.*, Physical Review Letters **107** 203001 (2011).
[2] R. Pohl *et al.*, Nature **466** 213 (2010).
[3] A. Beyer *et al.*, Physica Scripta **165** 014030 (2015).
[4] M. Horbatsch and E. A. Hessels, Physical Review A **82** 052519 (2010).
[5] A. Beyer, L. Maisenbacher, A. Matveev *et al.*, *in preparation* (2016).

# A new search for the permanent electric-dipole moment of $^{129}$Xe

N. Sachdeva[1], E. Babcock[2], M. Burghoff[3], T. Chupp[1], S. Degenkolb[1], I. Fan[3], P. Fierlinger[4], F. Gong[1], E. Kraegeloh[4], W. Kilian[3], S. Knappe-Grüneberg[3], F. Kuchler[4], T. Lins[4], M. Marino[4], J. Meinel[4], B. Niessen[4], Z. Salhi[2], A. Schnabel[3], F. Seifert[3], J. Singh[5], S. Stuiber[4], L. Trahms[3], J. Voigt[3]

[1]University of Michigan, USA

[2]Juelich Center for Neutron Science, Germany

[3]Physikalisch-Technische Bundesanstalt, Germany

[4]Excellence Cluster Universe and Technische Universität München, Germany

[5]Michigan State University and NSCL, USA

Sources of flavor diagonal CP-violation beyond what is currently provided in the Standard Model are necessary to explain the observed matter-antimatter imbalance in the universe. CP-violating sources in beyond-the-standard-model theories give rise to permanent electric-dipole moments (EDMs) in particles. Precise EDM measurements of the neutron, electron, paramagnetic and diamagnetic atoms constrain CP-violating parameters. At the FRM-II (Munich Research Reactor) a new and improved search for the $^{129}$Xe EDM is underway. The previous limit for the $^{129}$Xe EDM is $6 \times 10^{-27}\ e \cdot \mathrm{cm}$ (95% CL). The HeXeEDM experiment utilizes an ultralow magnetic field in a high-performance magnetically shielded room and $^{3}$He comagnetometer to improve the limit by up to three orders of magnitude. $^{3}$He and $^{129}$Xe are hyperpolarized in the same measurement volume using spin-exchange optical pumping and the Larmor precession of the two species is detected with a SQUID magnetometer array. Results from the May-June 2016 experimental run will be presented. $^{129}$Xe and $^{3}$He spin precession with high voltage applied was observed with $1\ \mu$Hz frequency resolution. Results suggest a current sensitivity on the order of $10^{-27} e \cdot \mathrm{cm}$ in one day.

# In-beam measurement of the hydrogen hyperfine structure and prospects for antihydrogen spectroscopy

Eberhard Widmann[1], Martin Diermaier[1], Christian B. Jepsen[2], Bernadette Kolbinger[1], Chloé Malbrunot[1,3], Oswald Massiczek[1], Clemens Sauerzopf[1], Martin C. Simon[1], Johann Zmeksal[1]

[1]Stefan Meyer Institute, Vienna, Austria

[2]Department of Physics, Princeton University, USA

[3]CERN, Geneva, Switzerland

The ASACUSA collaboration at the Antiproton Decelerator of CERN aims at a precise measurement of the ground-state hyperfine splitting (GS-HFS) of antihydrogen [1]. Comparing the GS-HFS of hydrogen and antihydrogen offers one of the most sensitive tests of the fundamental CPT symmetry, that states that all particles and antiparticles have exactly equal or exactly opposite properties. Since CPT is strictly conserved in the Standard Model of particle physics, a violation, if found, would point directly to theories behind this framework.

In order to characterize the setup devised to measure the GS-HFS and to evaluate its potential, a beam of cold, polarized, monoatomic hydrogen was built and used together with the microwave cavity and sextupole magnet designed for the antihydrogen experiment. A quadrupole mass spectrometer to detect the hydrogen atoms completed the Rabi-type atomic beam setup [2]. The $(F,M)$=(1,0) to (0,0) transition was measured to a precision of several ppb, more than a factor 10 better than in the previous measurement using a hydrogen beam. This result shows that the apparatus developed is capable of making a precise measurement of the GS-HFS of antihydrogen provided a beam of similar characteristics (velocity, polarization, quantum state) becomes available.

In a recent publication on the non-minimal Standard Model Extension (SME), describing possible violations of Lorentz and CPT invariance, Kostelecky and Vargas [3] conclude that the in-beam hyperfine measurements of hydrogen alone can be used to constrain certain coefficients of their model, which are not accessible to hydrogen maser experiments. The sensitivities reachable with our apparatus will be discussed.

[1] E. Widmann et al., Hyperfine Interactions **215**, 1 (2013).
[2] M. Diermaier et al., Hyperfine Interactions **233**, 35 (2015).
[3] V.A. Kostelecky and A.J. Vargas, Physical Review D **92**, 056002 (2015).

# Cold highly charged ions for novel optical clocks and the search for α variation

Lisa Schmöger[1,2], Oscar O. Versolato[1,2], Maria Schwarz[1,2], Matthias Kohnen[2], Alexander Windberger[1], Baptist Piest[1], Stefanie Feuchtenbeiner[1], Jofre Pedregosa[3], Tobias Leopold[2], Peter Micke[1,2], Anders K. Hansen[4], Thomas M. Baumann[1], Michael Drewsen[4], Thomas Pfeifer[1], Joachim Ullrich[2], Piet O. Schmidt[2,5], José R. Crespo López-Urrutia[1]

[1] Max-Planck-Institut für Kernphysik, Saupfercheckweg 1, 69117 Heidelberg, Germany

[2] Physikalisch-Technische Bundesanstalt, Bundesallee 100, 38116 Braunschweig, Germany

[3] Aix-Marseille Université, Ave Escadrille Normandie-Niémen, 13397 Marseille cedex 20, France

[4] Aarhus University, Bygning 1520, 8000 Aarhus C, Denmark

[5]Institut für Quantenoptik, Leibniz Universität, Welfengarten 1, 30167 Hannover, Germany

With the ability to simultaneously control both the excitation and the motional degrees of freedom of individual quantum objects, atomic physics has reached a degree of accuracy without peer among the experimental sciences. Immediate practical consequences are the development of atomic and optical clocks. These are meanwhile capable of measuring subtle effects of relativity and even of setting upper limits to possible variations of fundamental constants. In principle, the binding energy of atomic electrons in their ground state is sensitive to all levels of Standard Model physics. Their wavefunction adapts to all small contributions arising from all known interactions. This universal pattern of sensitivity is expected to appear again if forces beyond the Standard Model were to exist.

In highly charged ions (HCIs), the wavefunction of the optically active electron has an enhanced sensitivity to electron-nucleus interactions and QED terms in general, and an extremely suppressed sensitivity to external field perturbations. Further, E1 forbidden optical transitions near level crossings in HCIs are extremely sensitive to possible drifts in the fine-structure constant. These favorable properties have been widely recognized as an advantage for precision tests of fundamental physics and the development of HCI based clocks in many recent theoretical works. However, all known sources of HCIs produce them at high temperatures - typically in the MK regime - which severely limits the achievable spectral resolution of photonic studies.

We have developed an experiment for retrapping, cooling and high-precision laser spectroscopy of HCIs [1]. It is based on continuously laser-cooled $Be^+$ Coulomb crystals in a linear cryogenic Paul trap [2] for stopping the motion of externally produced HCIs and sympathetically cooling them below 250 mK. This cooling induces the formation of stable mixed crystals – down to a single HCI cooled by a single co-trapped $Be^+$ ion [3]. The strongly suppressed thermal motion of the embedded HCIs offers novel possibilities for investigation of questions regarding the time variation of fundamental constants, parity non-conservation effects, and quantum electrodynamics. Our current work aims at high-precision spectroscopy of the $^2P_{1/2}$ -$^2P_{3/2}$ M1 transition at 441 nm in cold $Ar^{13+}$ ions. Adding HCIs to the quantum toolbox is one important goal within the scope of next-generation experiments, which are currently being set up. One aims at applying quantum logic schemes to HCIs and developing an HCI optical clock, the other one at direct VUV frequency comb spectroscopy of transitions in HCIs.

[1] L. Schmöger *et al.*, Rev. Sci. Instrum. **86**, 103111 (2015).

[2] M. Schwarz *et al.*, Rev. Sci. Instrum. **83**, 083115 (2012).

[3] L. Schmöger *et al.*, Science **347**, 1233 (2015).

# A method to measure the nuclear anapole moment using ultracold Fr

T. Aoki[1], Y. Torii[1], B. K. Sahoo[2], B. P. Das[3], K. Harada[4], T. Hayamizu[4], K. Sakamoto[4], H. Kawamura[4,5], T. Inoue[4,5], A. Uchiyama[4], S. Ito[4], R. Yoshioka[4], K. S. Tanaka[4], M. Itoh[4], A. Hatakeyama[6], Y. Sakemi[4]

[1]*Institute of Physics, The University of Tokyo, Tokyo 153-8902, Japan*
[2]*Theoretical Physics Division, Physical Research Laboratory, Ahmedabad-380009, India*
[3]*International Education and Research Center of Science and Department of Physics, Tokyo Institute of Technology, Tokyo 152-8550, Japan*
[4]*Cyclotron and Radioisotope Center, Tohoku University, Sendai, Miyagi 980-8578, Japan*
[5]*Frontier Research Institute for Interdisciplinary Sciences, Tohoku University, Sendai, Miyagi 980-8578, Japan*
[6]*Department of Applied Physics, Tokyo University of Agriculture and Technology, Tokyo 184-8588, Japan*

The two sources of parity nonconservation (PNC) in atomic systems are the neutral current weak (NCW) interactions due to the exchange of the Z boson between the nucleus and the electrons and the nuclear anapole moment. The NCW interactions can give rise to nuclear spin independent (NSI) as well as nuclear spin dependent (NSD) PNC, while the interaction of the nuclear anapole moment with the electrons results in PNC of only the NSD kind. NSI PNC is a probe of new physics beyond the Standard Model and NSD PNC from the nuclear anapole moment provides important information about PNC in nuclei. In heavy atomic systems, the dominant contribution to NSD PNC comes from the nuclear anapole moment. The only system for which a non-zero value of NSD-PNC has been reported so far is atomic Cs on which the experiment was performed using the Stark mixing technique [1]. However, there are discrepancies between the values of the nucleon-nucleon PNC coupling constants obtained from Cs and nuclear experiments. Therefore, other atomic experiments to detect the nuclear anapole moment are desirable to resolve this controversy.

We report precise calculations of NSD PNC transition amplitude between $7S_{1/2}$ and $6D_{5/2}$ states of Fr atom, and detection scheme of this effect as light shifts induced by E2 and PNC-E1 transitions using ultracold Fr atom [2]. The advantage of utilizing this transition is absence of NSI PNC amplitude, which is typically much larger than NSD PNC amplitude and affects the detection of the small NSD PNC effect. We consider the light shifts for all the magnetic sublevels, which are determined by angular factors for E2 and PNC E1 transitions [3]. These light shifts will be useful to experimentally study the nuclear anapole moment of $^{210}$Fr in CYRIC, Tohoku University in the near future.

[1] C. S. Wood, S. C. Bennett, D. Cho, B. P. Masterson, J. L. Roberts, C. E. Tanner, and C. E. Wieman, Science **275**, 1759 (1997).

[2] B. K. Sahoo, T. Aoki, B. P. Das, and Y. Sakemi, Phys. Rev. A **93**, 032520 (2016).

[3] T. Aoki *et al.*, Asian J. Phys. submitted.

# Light intensity stabilization based on the second harmonic of the photoelastic modulator detection in the atomic magnetometer

Lihong Duan[1], Rujie Li[1], Liwei Jiang[1], Wenfeng Fan[1], Ming Ding[1], Wei Quan[1]
[1]School of Instrument Science and Opto-electronics Engineering, Beihang University, Beijing 100191, China

In the spin exchange relaxation free (SERF) atomic magnetometer, the fluctuations of the probe light intensity caused by laser source and components in the optical path seriously affect the performance of the high sensitive atomic magnetometer [1]. We proposed a method to stabilize the light intensity based on the second harmonic of the PEM detection, where the power feedback loop and probe signal of the system share the same optical path. According to the PEM detection theory, the second harmonic is proportional to the light intensity. The second harmonic obtained by the lock-in amplifier was regarded as feedback signal of the control loop. Comparing the second harmonic term with the set value in real-time, the light intensity could be adjusted to the desired value and the fluctuation of the light intensity in the whole optical path would be accurately controlled. As the feedback signal is extracted from the downstream experimental, the fluctuations of the laser source and the whole optical path are eliminated effectively. The stability of the atomic magnetometer is greatly improved especially at long-term time. Furthermore, it does not need to add additional control devices and is apt to application fields which require compact size. It is also useful for other applications based on the PEM detection. The light intensity stability under three different conditions in the same monitoring point were compared: (a) free running mode that the control loop of the light power is turned off, (b) traditional stabilization method and (c) the scheme based on the second harmonic of the PEM detection (PEM SH method). The performance of three different stabilization mode is shown in Figure1.

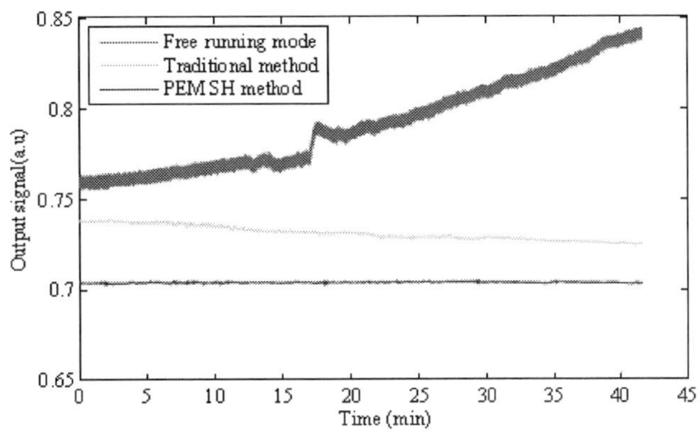


**Figure 1.** The fluctuations of the output signal of the atomic magnetometer applying different stabilization methods: free running mode, traditional method, and PEM SH method: the method based on the second harmonic component of the PEM detection. They reflect the stability of the atomic magnetometer detction system.

[1] J. Fang, S. Wan, J. Qin, C. Zhang, W. Quan, H. Yuan, and H. Dong, Rev. Sci. Instrum. **84**, 083108 (2013).
[2] J. M. Brown, Ph.D. dissertation (Princeton University, 2011).

# Photon and neutrino emission from quantum ions in circular motion

N. Sasao[1], H.Hara[1], T.Hiraki[1], Y.Honda[4], Y.Ichikawa[2], O.Kamigaito[2], Y.Kanai[3], T.Nagatomo[2], T.Nakagawa[2], T.Masuda[1], Y.Miyamoto[1], K.Sakaue[5], S. Uetake[1], K.Yokoya[4], M.Yoshida[4], A.Yoshimi[1], K. Yoshimura[1], M. Yoshimura[1]

[1]1Research Institute for Interdisciplinary Science, Okayama University, Okayama , 700-8530, Japan

[2] RIKEN Nishina Center for Accelerator-Based Science, Wako, Saitama 351-0198 Japan

[3]Physics Research Unit, RIKEN, Wako, Saitama 351-0198, Japan

[4]High Energy Accelerator Research Organization (KEK), Tsukuba, Ibaraki, 305-0801, Japan

[5] Waseda Institute for Advanced Study, Waseda University, Shinjuku-ku, Tokyo 169-8050, Japan

We present a new idea of creating intense beams of neutrino pair and photon, and describe a proposed pilot experiment planned at RIKEN, Japan. The idea is based on a new concept of quantum heavy ion beam in circular motion. Quantum ion is a pure state of mixed ground and excited state formed by laser irradiation. These quantum heavy ions in circular motion emit photons and neutrino pairs, having novel and interesting features quite distinct from the usual synchrotron radiation: emitted particle energies, either neutrino pair or photon, can reach the GeV region if the ion's Lorentz boost $\gamma$-factor is large. The sharp energy cutoff is given by twice of the $\gamma$-factor times the level spacing of ion states. Applications of the beams extend to many fields in fundamental science. For example, the intense gamma ray beam in MeV region may be used for generation of high intensity neutron flux, which may drastically change future of neutron science. The neutrino pair beam may also have important implications on neutrino physics, in particular CP violating measurement. We describe a pilot experiment (Fig.1) with the RIKEN ECR ion source in detail

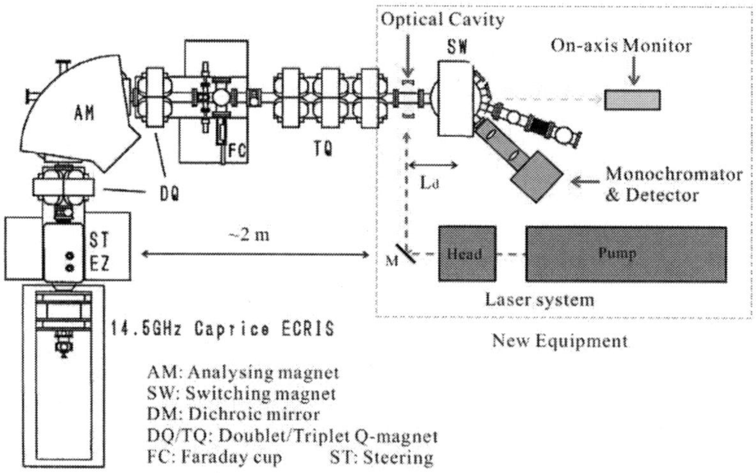


**Fig.1** Setup of the pilot experiment at RIKEN.

[1] M. Yoshimura and N. Sasao,"Neutrino pair and gamma beams from circulating excited ions", Phys . Rev. **D92**, 073015 (2015). See also arXiv:1505.07572v2 [hep-ph] 4 Jun 2015

# High density NV center ensembles coupling with uniform microwave magnetic field

Ning Zhang[1*], Chen Zhang[1], Lixia Xu[1], Jixing Zhang[1], Heng Yuan[1], Jiancheng Fang[1]
[1]Beihang University, Beijing, China
[*]Corresponding author: ningzhang@buaa.edu.cn

The negatively-charged nitrogen-vacancy (NV) center in diamond has become a developing trend in nanoscale solid state sensors[1]. Besides the number of the active centers in a single shot, the magnetic sensitivity is also dependent on the microwave magnetic filed homogeneity among the ensemble spins volume. Better homogeneity benefits linewidth in the ODMR which enhances the DC magnetic sensitivity. We coupled an ensemble of $10^{12}$ NV centers strongly with a designed resonator operating with a 150 MHz bandwidth[2]. With the configurated resonator and optimized decoupling sequences, this work would achieve a uniform microwave magnetic field with homogeneity better than 5% over an effective $1\times1\times0.3$ $mm^3$ sensor volume at room-temperature.

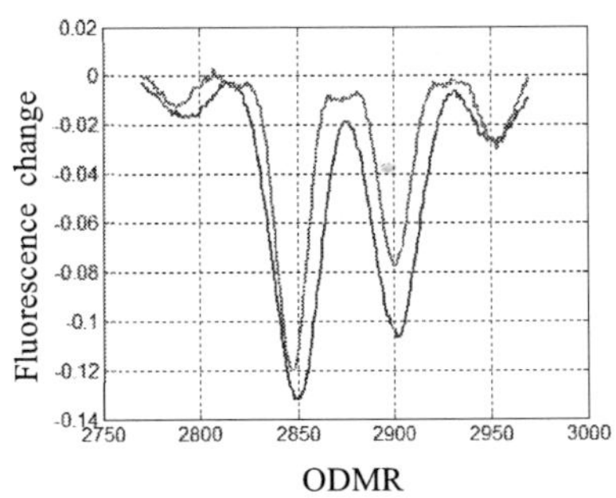


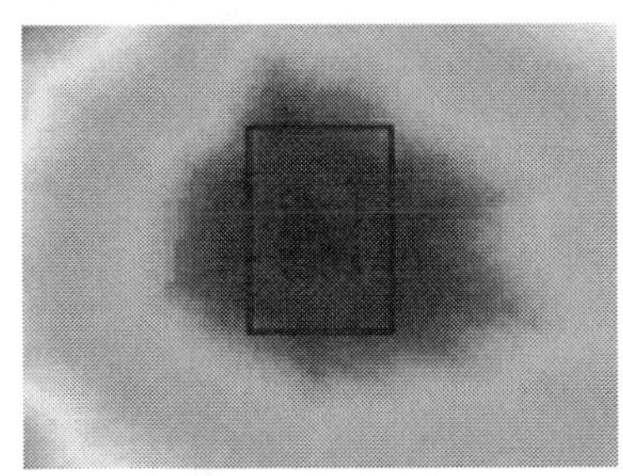

**Figure 1.** (a) ODMR experiments of different microwave resonators with various linewidth and contrast under external static field 30G. (b) Simulated uniform microwave magnetic filed.

[1] L. M. Pham, D. Le Sage, P. L. Stanwix, et al. New J. Phys., **13**(4), 045021 (2011).
[2] N. Zhang, C. Zhang, L. Xu, et al. Appl. Magn. Reson., **47**(6), 589-599 (2016).

# Micro-integrated semiconductor laser module platform for precision quantum optical experiments on Earth and in space

Max Schiemangk[1,2], Bassem Arar[1], Ahmad Bawamia[1], Heike Christopher[1,2], Anja Kohfeldt[1,3], Mandy Krüger[1], Christian Kürbis[1], Wojciech Lewoczko-Adamczyk[1], Erdenetsetseg Luvsandamdin[1], Achim Peters[1,2], Andreas Wicht[1,2], Günther Tränkle[1]

[1]Ferdinand-Braun-Institut, Leibniz-Institut für Höchstfrequenztechnik, Gustav-Kirchhoff-Str. 4, 12489 Berlin, Germany

[2]Institut für Physik, Humboldt-Universität zu Berlin, Newtonstr. 15, 12489 Berlin, Germany

[3]Institut für Luft- und Raumfahrt, Technische Universität Berlin, Marchstr. 12, 10587 Berlin, Germany

Narrow linewidth lasers with high output power and appropriate modulation capabilities are essential for high precision measurement setups such as atom interferometers and optical atomic clocks. For some of these experiments, a mobile setup on Earth or even a deployment in space is highly desirable or necessary to meet the stringent requirements on measurement accuracy and precision. This requires compact, robust, energy-efficient, and space-qualified laser modules, which also meet the stringent electro-optical requirements of the experiments. To suit these needs, micro-integrated diode laser modules have been developed at FBH. We will give a brief review on the modules, which were successfully used in quantum gas experiments in the Bremen drop tower [1] and on sounding rocket missions [2, 3].

Based on this technological heritage we developed a new technology platform for micro-integrated semiconductor modules, which provides the means to utilize arbitrary combinations of two semiconductor chips. This includes active components (e.g. lasers or amplifiers) as well as passive components (e.g. modulators). In addition micro-optics, DC and RF electronics, and fiber-coupling into a single-mode, polarization maintaining fiber are integrated on a structured aluminum nitride (AlN) substrate with a footprint of 80 x 30 mm². The AlN substrate is encapsulated into a hermetically sealed housing with custom-made feedthroughs for all DC, RF, and optical signals. The packaged lasers fit into a volume of 125 x 75 x 22.5 mm³. Some of the first modules built on this platform will be used on a sounding rocket mission that plans to test Einstein's Equivalence Principle with cold rubidium and potassium atoms.

Finally, we will give an outlook on possibilities to transfer our new integration platform for the use in further applications. This includes a laser system for an optical atomic clock operating in the UV range and key elements for precision measurements like an ultra-narrow linewidth laser and a spectroscopy setup.

[1] M. Schiemangk, K. Lampmann, A. Dinkelaker, A. Kohfeldt, M. Krutzik, C. Kürbis, A. Sahm, S. Spießberger, A. Wicht, G. Erbert, G. Tränkle, and A. Peters, Appl. Opt. **54**, 5332-5338 (2015).

[2] T. Wilken, M. Lezius, T. W. Hänsch, A. Kohfeldt, A. Wicht, V. Schkolnik, M. Krutzik, H. Duncker, O. Hellmig, P. Windpassinger, K. Sengstock, A. Peters, and R. Holzwarth, in CLEO: 2013, OSA Technical Digest, Af2H.5 (2013).

[3] E. Luvsandamdin, C. Kürbis, M. Schiemangk, A. Sahm, A. Wicht, A. Peters, G. Erbert, and Günther Tränkle, Opt. Express **22**, 7790-7798 (2014).

# PENTATRAP - An advanced high-precision Penning-trap mass spectrometer

A. Rischka[1], H. Bekker[1], J.R. Crespo López-Urrutia[1], M. Door[2], S. Eliseev[1], P. Filianin[1], Yu. N. Novikov[3], R. X. Schüssler[1], C. Schweiger[2], S. Sturm[1], S. Ulmer[4], K. Blaum[1]

[1]Max-Planck-Institut für Kernphysik, Saupfercheckweg 1, 69117 Heidelberg, Germany
[2]Universität Heidelberg, Fakultät für Physik und Astronomie, Im Neuenheimer Feld 226, 69120 Heidelberg, Germany
[3]Petersburg Nuclear Physics Institute, 188300 Gatchina, Russia
[4]RIKEN, Ulmer Initiative Research Unit, Wako, Saitama 351-0198, Japan

Aiming at a relative mass uncertainty of better than $10^{-11}$ for stable nuclides allows the PENTATRAP mass spectrometer to make contributions to neutrino physics, fundamental test of special relativity and tests of quantum electrodynamics (QED) in high-fields [1]. With its currently unique connection to a SUPER-EBIT and two table top EBITs, PENTATRAP has access to a wide variety of nuclides in different charge states, e.g. hydrogen-like lead for stringent tests of QED, the rare long-lived $^{163}$Ho isotope for neutrino physics and short-lived radionuclides like $^{36}$Cl for a test of special relativity. The key features of the mass spectrometer are the five Penning traps each equipped with an axial detection system and three traps equipped with a cyclotron cooling resonator. The traps are placed in a highly stable superconducting 7 T magnet. The 4K cryogenic vacuum setup will provide a vacuum of better than $10^{-16}$ mbar for long storage times which ensure efficient measurement cycles. This contribution will give an overview on the PENTATRAP mass spectrometer, on its present status as well as on the applications of high-precision mass data.

[1] J. Repp et al., Appl. Phys. B **107**, 983-996 (2012).

# Rotation measurement coupling in a K-Rb-$^{21}$Ne comagnetometer with large light shift

Liwei Jiang[1], Rujie Li[1], Lihong Duan[1], Wenfeng Fan[1], Ming Ding[1], Wei Quan[1]
[1]School of Instrumentation Science and Opto-electronics Engineering,
Beihang University, Beijing 100191, China

Alkali-metal-noble-gas comagnetometers operating in the spin-exchange relaxation free (SERF) regime can be exploited for rotation sensing [1] [2]. In this paper, we discuss the implementation of a K-Rb-$^{21}$Ne comagnetometer, where K atoms are pumped along z axis by an external cavity diode laser centered at D1 line. Rb atoms are polarized by collision with K atoms. The probe beam is along x axis and the rotation along y axis is measured. Due to the large light shift of Rb atoms in the hybrid optical pumping regime, the rotation along x axis is coupled to the output which is expressed by Equation (1). To validate this theory, another probe beam along y axis which is generated by the same probe laser is added to the set-up, which is with the same optic components to guarantee the equal scale factor. By applying different rotations along x axis, the coupling coefficient is derived from the two probe beams, which is 18.4 %. The light shift is 1.6 nT and the total alkali relaxation rate is 288.4 1/s. The theory coupling coefficient is 15.5 %, which is consistent with the experiment result.

$$P_x^e \approx \frac{-P_z^e \gamma^e R_{tot}^e}{\left(R_{tot}^e\right)^2 + \left(\gamma^e\right)^2 \left(L_z\right)^2} \left( \frac{\Omega_y}{\gamma^n} + \frac{\gamma^e L_z}{R_{tot}^e} \frac{\Omega_x}{\gamma^n} \right) \quad (1)$$

[1] T. Kornack, R. Ghosh, and M. Romalis, Phys. Rev. Lett. **95**, 230801 (2005).
[2] J. Fang, Y. Chen, S. Zou, X. Liu, Z. Hu, W. Quan, H. Yuan, and M. Ding, J. Phys. B: At. Mol. Opt. Phys. **49**, 065006 (2016).

# Polarizability and tune-out wavelength for the helium 1s2s $^3$S state

G.W.F. Drake, Jacob Manalo
Department of Physics, University of Windsor, Windsor, Ontario N9B 3P4 Canada

There has been considerable recent interest in measurements of the tune-out wavelength in atomic systems as a high-precision test of quantum electrodynamics (QED), other than the transition frequencies such as the Lamb shift. The tune-out wavelength is the wavelength at which the frequency-dependent polarizability vanishes. It has recently been measured for the helium 1s2s $^3S_1$ to be 413.0938(20) nm (5 ppm) [1] and calculated by a relativistic configuration-interaction method [2]. In this paper we present an alternative method of calculation based on the nonrelativistic Schroedinger equation and with relativistic corrections included by perturbation theory. The advantage is that fully correlated Hylleraas coordinates can be used to obtain higher accuracy for the contributions from electron correlation. Our result for the static scalar polarizability of the $M = 0$ state, including relativistic corrections of order $\alpha^2$ and finite nuclear mass corrections, is 315.716 092 95(1) $a_0^3$, as compared with 315.7165(4) $a_0^3$ from Ref. [2], where $\alpha$ is the fine structure constant, $a_0$ is the Bohr radius, and the uncertainty is the convergence uncertainty in our final figure quoted. Further relativistic recoil corrections and comparisons with experiment for the tune-out wavelength will be discussed at the conference.

[1] B. M. Henson, R. I. Khakimov, R. G. Dall, K. G. H. Baldwin, L.-Y. Tang, and A. G. Truscott, Phys. Rev. Lett. **115**, 043004 (2015).

[2] Y.-H. Zhang, L.-Y. Tang, X.-Z. Zhang, and T.-Y. Shi, Phys. Rev. A **93**, 052516 (2016).

# Detecting dark matter waves with a network of AMO precision measurement tools

Andrei Derevianko

University of Nevada, Reno, Nevada 89557, USA

Virialized Ultra-Light Fields (VULFs) while being viable cold dark matter candidates can also solve the standard model hierarchy problem. Direct searches for VULFs due to their non-particle nature require low-energy precision measurement tools. Here we consider scalar VULF candidates. While the previous proposals have focused on detecting coherent oscillations of the measured signals at the VULF Compton frequencies at the device location, here we point out that VULFs also have a distinct spatial signature, forming dark matter *waves*. Thereby the discovery reach can be improved by using distributed networks of precision measurement tools. We find the expected dark-matter wave signal by deriving spatio-temporal two-point dark matter fields correlation function. Based on the developed formalism for coherence properties of dark-matter fields, we propose several experiments for dark matter wave detection. In the most basic version, the modifications to already running experiments are minor and only require GPS-assisted time-stamping of data [1]. We also derive the expected dark matter line profile for individual detectors [2].

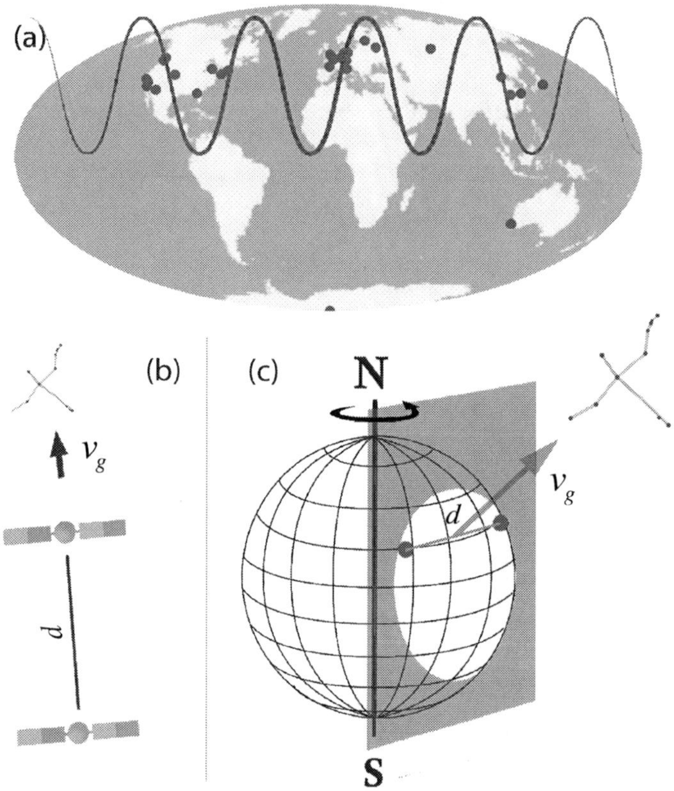


**Figure 1.** (a) Dark-matter wave observatory based on a global network of existing low-energy precision measurement laboratories (dots) around the globe.

(b) Satellite mission for probing VULF dark matter correlation function; both the distance between the satellites and the angle between galactic velocity $\vec{v}_g$ and separation $\vec{d}$ vectors can be varied.

(c) Terrestrial experiment with fixed nodes utilizes the daily variation of the angle between galactic velocity and two-node separation vector.

[1] D. Budker and A. Derevianko, Phys. Today **68**, 10 (2015).

[2] Details can be found in A. Derevianko arXiv:1605.09717 (2016).

# A precision measurement of the electron EDM using trapped ions

William Cairncross, Daniel N. Gresh, Kia Boon Ng, Yan Zhou, Eric A. Cornell, Jun Ye
JILA, NIST & University of Colorado, and Department of Physics, University of Colorado, USA

Precision measurements of the permanent electric dipole moment of the electron (eEDM) constrain extensions to the Standard Model. The most sensitive measurements of the eEDM to date have used neutral molecular beams, and thus are susceptible to similar classes of systematic errors. We perform an eEDM measurement using $HfF^+$ ions confined in an RF trap, attaining long coherence times and similar sensitivity to an eEDM, and observing sensitivity to different sources of systematic error. We will present our experimental setup, known sources of systematic error and our efforts to suppress them, and the results of our current eEDM measurement.

# A novel atomic precession detection method based on the optical fiber interferometer

Xuejing Liu[1], Yanhui Hu[1], Ming Ding[1*], Wei Jin[1,2], Jiancheng Fang[1]
[1]School of Instrumentation Science and Opto-electronics Engineering, Beihang University, Beijing 10091, China
[2]Department of Electrical Engineering, The Hong Kong Polytechnic University, Hong Kong
E-mail:mingding@buaa.edu.cn

In this paper, we propose a method to detect atomic precession in spin exchange relaxation free (SERF) atomic magnetometer based on a fiber interferometer. The atomic precession in SERF magnetometer can be detected by interference between orthogonal polarization lights in a single polarization maintaining fiber (PMF). Compared with classical methods[1,2], the system is of all fiber components as shown in Fig. 1. The light from a laser source becomes linearly polarization after passing through the circulator and fiber polarizer. The highly linear birefringence pigtail of the polarizer is connected to another polarization maintaining fiber whose birefringence axes are oriented at 45° with respect to those of the polarizer. The linearly polarization converts into two linearly polarizations propagating in the orthogonal axes of the PMF. An alkali cell is followed the PMF and a reflector is positioned at the end which is to improve the signal amplitude. The light intensity output of the system is expressed by Equation (1). The simulation results of all fiber atomic precession detection system interference spectra shown in Fig. 2 suggests the feasibility of a portable, low-power and high performance atomic precession detection device. Such a device can greatly expands the range of applications for atomic magnetometers, such as the detection in extreme environment[3].

$$E_{out} = \frac{1}{4} I_0 (1 + \cos 4F)(1 + \cos 2\delta) \quad (1)$$

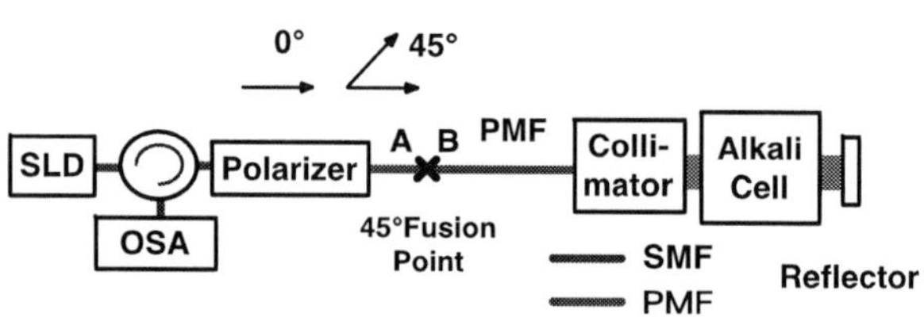


**Fig. 1 The all fiber atomic precession detection system**

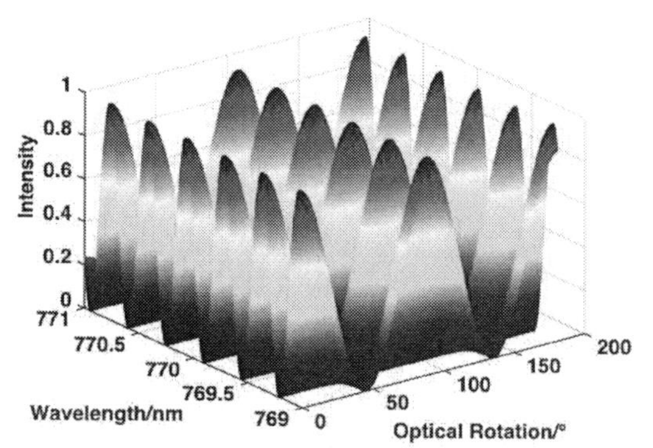


**Fig. 2 The interference spectra of the all fiber atomic precession detection system**

## Reference

[1] R. Wyllie, M. Kauer, G. S. Smetana, et al, Physics in Medicine and Biology, **57**, 2619 (2012).
[2] H. B. Dang, A. C. Maloof, M. V. Romalis, Applied Physics Letters, **97**, 151110(2010).
[3] D. Sheng, S. Li, N. Dural, M. V. Romalis, Physical review letters, **110**, 160802(2013).

# High-precision evaluation of Wigner's *d* matrix by exact diagonalization

Xiaomin Feng[1], Ping Wang[2,3], Wen Yang[2], Guangri Jin[1]
[1]Department of Physics, Beijing Jiaotong University, Beijing 100044, China
[2]Beijing Computational Science Research Center, Beijing 100084, China
[3]Department of Modern Physics, Hefei National Laboratory for Physics Sciences at Microscale, University of Science and Technology of China, Hefei, Anhui 230026, China

The Wigner's *d*-matrix, closely related to spherical harmonics and Legendre polynomials, is of interest in various fields of quantum mechanics, chemical physics, and mathematical physics. Due to the presence of large numbers, direct calculations of the matrix using Wigner's formula suffer from a loss of precision. Here, we put forward a very simple method to avoid this problem by expanding the *d* matrix into a complex Fourier series and calculate the Fourier coefficients by exactly diagonalizing the angular momentum operator $J_y$ in the eigenbasis of $J_z$. This method allows us to compute the *d* matrix and its various derivatives for spins up to a few thousand. The precision of the *d* matrix from our method is about $10^{-14}$ for spins up to 100.

[1] X. M. Feng P. Wang, W. Yang and G. R. Jin, Phys. Rev. E **92**, 043307 (2015).
[2] N. Tajima, Phys. Rev. C **91**, 014320 (2015).

# Mercury alkalis as candidates for electron electric dipole moment searches

V. S. Prasannaa[1], A. Sunaga[2], A. C. Vutha[3], M. Abe[2], B. P. Das[4]
[1]Indian Institute of Astrophysics, Koramangala, Bangalore-560034, India
[2]Tokyo Metropolitan University, 1-1 Minami-Osawa, Hachioji-city, Tokyo 192-0397, Japan
[3]Department of Physics, University of Toronto, 60 St. George Street, Toronto ON M5S 1A7, Canada
[4]International Education and Research Centre of Science and Department of Physics, Tokyo Institute of Technology, 2-12-1-H86, Ookayama, Meguro-ku, Tokyo 152-8550, Japan

Searches for the electron electric dipole moment (eEDM) are important for constraining Beyond-Standard-Model theories that attempt to explain the matter antimatter asymmetry in the universe [1,2].The sensitivity of these experiments can be improved by using large ensembles of molecules with long coherence times [3]. We identify mercury alkali molecules as interesting candidates for future eEDM experiments. We present calculations of the molecular effective electric field, $E_{eff}$, that acts to amplify the effect of an eEDM in these molecules. In addition to having large effective electric fields, mercury alkali molecules could also be produced in large numbers, directly from trapped ultracold atomic samples, enabling measurements with longer coherence times. The identification of these molecules opens up a viable path towards high sensitivity eEDM experiments with ultracold molecules.

[1] K. Fuyoto, J. Hisano, and E. Senaha, Phys Lett B, **755**, 491 (2016).

[2] The ACME Collaboration, J. Baron et al, Science, **343**, 269 (2014).

[3] V. S. Prasannaa, A. C. Vutha, M. Abe, B. P. Das, Phys Rev Lett, **114**, 183001 (2015).

# Spin orientation dependent universality of free fall test using atom interferometry

Xiao-Bing Deng, Xiao-Chun Duan, Min-Kang Zhou, Feng Xiong, Cheng-Gang Shao, Zhong-Kun Hu

MOE Key Laboratory of Fundamental Physical Quantities Measurement, School of Physics, Huazhong University of Science and Technology, Wuhan 430074, China

We present a test of the universality of free fall (UFF) using spinful atoms, which is an extension of our former work on the UFF test using rotating gyroscopes and polarized molecules. A Mach-Zehnder-type atom interferometer is exploited to measure and compare the gravity acceleration of the $^{87}$Rb atoms in $m_F$=+1 versus those in $m_F$=−1, of which the corresponding spin orientations are opposite. The interferometer using atoms in $m_F$=±1 is highly sensitive to the magnetic field inhomogeneity, and a double differential measurement method is developed to alleviate the corresponding influence. The resultant Eötvös ratio is $\eta_S$=(0.2±1.2)×10$^{-7}$, which also gives an upper limit of 5.4×10$^{-6}$ m$^{-2}$ for the gradient of possible spacetime torsion field. This is the first direct spin-orientation related UFF test on quantum basis, which also gives an upper constraint of the torsion gradient for the first time.

[1] X. C. Duan, X. B. Deng, M. K. Zhou, K. Zhang, W. J. Xu, F. Xiong, Y. Y. Xu, C. G. Shao, J. Luo, and Z. K. Hu, Phys. Rev. Lett., in press (2016).

[2] M. K. Zhou, Z. K. Hu, X. C. Duan, B. L. Sun, J. B. Zhao, and Jun Luo, Phys. Rev. A **82**, 061602R, (2010).

[3] Z. K. Hu, B. L. Sun, X. C. Duan, M. K. Zhou, L. L. Chen, S. Zhan, Q. Z. Zhang, and J. Luo, Phys. Rev. A **88**, 043610 (2013).

# Highly efficient optomechanical excitation of nonlinear mechanical oscillation in high-Q micromechanical resonator

Da In Song[1], Dae Seok Han[1], Myeong Soo Kang[1]

[1]Department of Physics, Korea Advanced Institute of Science and Technology
291 Daehak-ro, Yuseong-gu, Daejeon 34141, Republic of Korea

Nonlinear responses of micromechanical resonators have been recently explored for various kinds of application, e.g., synchronization of multiple oscillators [1], high-sensitivity mass detection [2], energy harvesting [3], information storage and operation [4], and frequency stabilization [5]. While in most of the previous works mechanical oscillations were electrically driven, optical excitation of such the nonlinear vibration would give more functionality. Here, we report our experimental demonstration of the optical excitation of nonlinear mechanical oscillations in a high-Q micromechanical resonator. We use a 30-nm-thick $0.5\times0.5$-mm$^2$-wide silicon nitride square membrane [6] placed in vacuum ($<2\times10^{-5}$ torr) at room temperature. The frequency and quality factor of the fundamental mechanical mode (FMM) are measured as 619.6 kHz and $4.4\times10^5$, respectively. The FMM is driven by an intensity modulated laser beam, and the resulting mechanical oscillation amplitude is measured interferometrically as a function of the modulation frequency. Although the peak-to-peak modulation of optical force is estimated to be smaller than ~7 pN, the mechanical oscillation amplitudes larger than 60 nm can be easily produced (Figure 1(a)). Furthermore, we clearly observe a nonlinear Duffing-type mechanical oscillation (Figure 1(b)). We also observe piezoelectrically driven nonlinear coupling between the FMM and the higher order (3,3) mode, which we are trying to implement using an optical driving force. Our results open up possibilities of new types of optically controllable nonlinear optomechanical devices.

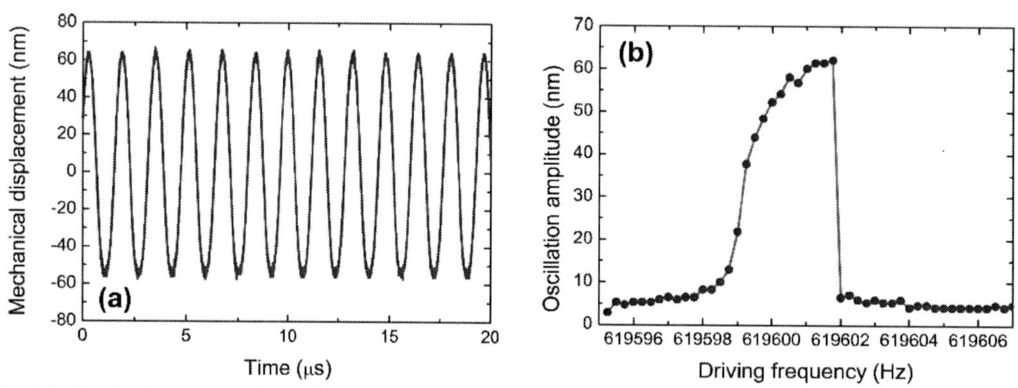


**Figure 1.** (a) Typical mechanical displacement of the membrane. (b) Mechanical oscillation amplitude as a function of the modulation frequency of the driving laser light. In this measurement, the driving frequency is gradually increased, while the peak-to-peak power modulation is fixed at 33 mW.

[1] M. H. Matheny, M. Grau, L. G. Villanueva, R. B. Karabalin, M. C. Cross, and M. L. Roukes, *Phys. Rev. Lett.* **112**, 014101 (2014).

[2] J. Atalaya, J. M. Kinaret, and A. Isacsson, *EPL* **91**, 48001 (2010).

[3] F. Cottone, H. Vocca, and L. Gammaitoni, *Phys. Rev. Lett.* **102**, 080601 (2009).

[4] M. Bagheri, M. Poot, M. Li, W. P. H. Pernice, and H. X. Tang, *Nat. Nanotechnol.* **6**, 726 (2011).

[5] D. Antonio, D. H. Zanette, and D. López, *Nat. Commun.* **3**, 806 (2012).

[6] J. D. Thompson, B. M. Zwickl, A. M. Jayich, F. Marquardt, S. M. Girvin, and J. G. E. Harris, *Nature* **452**, 72 (2008).

# First results from the GPS.DM observatory: dark matter search using atomic clocks and the GPS constellation

B. M. Roberts[1], G. Blewitt[1], C. Dailey[1], N. Lundholm[1], M. Pospelov[2], A. Rollings[1], J . Sherman[3], W. Williams[1], A. Derevianko[1] (GPS.DM Collaboration)

[1]University of Nevada, Reno, USA

[2]University of Victoria, BC & Perimeter Institute, Canada

[3]NIST, Boulder, USA

Despite the overwhelming cosmological evidence for the existence of dark matter, and the considerable effort of the scientific community over decades, there is no evidence for dark matter in terrestrial experiments. The GPS.DM observatory uses the existing GPS constellation as a 50,000 km-aperture sensor array, analyzing the satellite and terrestrial atomic clock data for exotic physics signatures. In particular, the collaboration searches for evidence of transient variations of fundamental constants correlated with the Earth's galactic motion through the dark matter halo. This type of search is particularly sensitive to exotic forms of dark matter, such as topological defects [1] and virialized ultra-light field (VULF) dark matter waves [2]. We will present an update on the search.

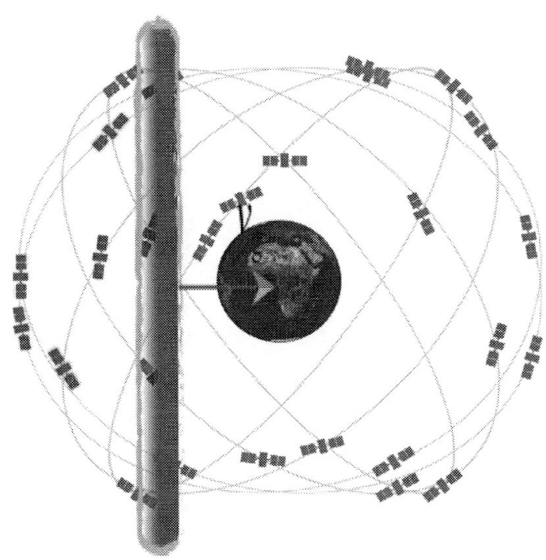

**Figure 1.** Dark-matter signal propagation through the GPS constellation.

[1] A. Derevianko and M. Pospelov, Nat. Phys. **10**, 933 (2014).

[2] A. Derevianko, arXiv:1605.09717 (2016).

# Cold atoms trapped in dipole trap for quantum nondemolition measurement

Xiaoxiao Ma[1], Fei Zi[1], Kaikai Huang[1], Xuanhui Lu[1]

[1] Department of physics, Zhejiang University, Hangzhou, China

Quantum nondemolition (QND) measurement is the way to generate spin squeezed states of atomic ensembles [1, 2]. It has played an important role in precision measurements such as atom clock etc [3]. In our experiments, we achieved the atoms which were trapped in the magnetic optical trap (MOT). And through the sub-Doppler cooling stage, the atoms were cooled in low temperature about ~30μk. Then we loaded these atoms into the far-off resonant optical dipole trap (FORT) in order to confine the atoms in high density. In this work, the Mach-zehnder interferometer (MZI) is the main experimental tool for QND measurement. In order to cancel the classical phase noise of the interferometer, we need to lock the interferometer to the white-light position. Here, we show the features of the atoms in FORT and the white-light alignment of the MZI preparing an applicable condition for the imminent QND measurement.

[1] D. J. Wineland, J. J. Bollinger, W. M. Itano, F. L. Moore, and D. J. Heinzen, Phys. Rev. A **46**, R6797(1992).

[2] M. Kitagawa and M. Ueda, Phys. Rev. A **47**, 5138(1993).

[3] J. Appel, P. J. Windpassinger, D. Oblak, U. B. Hoff, N. Kjærgaard, and E. S. Polzik, PNAS, vol. **106** no. 27 10961(2009).

# Spin-exchange relaxation free regime discriminant of atomic magnetometer

Hong Zhang[1, 2], Sheng Zou[1, 2], Wei Quan[2], Zhaohui Hu[2], Ming Ding[2*]

[1] School of Instrument Science and Engineering, Southeast University, Nanjing 210096, China
[2]Science and Technology on Inertial Laboratory, Beihang University, Beijing 100191, China

Magnetometers have a wide range of applications, for example, biomagnetism, geomagnetism, neuroscience, materials science and basic physics [1-2].The spin exchange optical pumping technique and the photoelectric detection technique promote the developments of the ultra-high sensitive magnetic field measurement devices, such as the atomic spin magnetometer based on the theory of the spin-exchange relaxation free (SERF) [3]. The most obvious feature of such an atomic spin magnetometer is that the alkali-metal atoms work in SERF regime. In this regime, not only atomic spin relaxation time increases with several orders, but also the atomic spin coherence is improved drastically.

In this work, a discriminant method of atomic spin magnetometer SERF regime was demonstrated. For the K atomic spin magnetometer, the realization condition of its SERF regime was given referring to the simulation results.

Usually, the total relaxation time can be constituted by the longitudinal relaxation time $T_1$ and the transverse relaxation time $T_2$. When the atomic spins are in SERF regime, one of the most important characteristics is that spin exchange relaxation, which can induce the total relaxation to decrease, is suppressed adequately. At this time, $T_2$ is approximate equal to $T_1$. This is the key criterion to judge whether the atomic spin magnetometer is operated in SERF regime. The simulation result of the K atomic spin magnetometer realizing SERF regime is shown in Fig. 1.

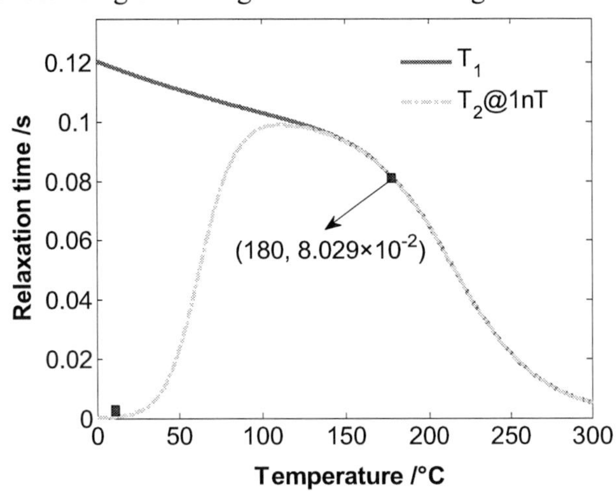


**Figure 1.** Simulation result of the K atomic spin magnetometer realizing SERF regime.

Figure 1 shows that when the external magnetic field with 1 nT exists and the heating temperature is about 180 °C, the longitudinal relaxation time $T_1$ is approximately equal to the transverse relaxation time $T_2$, and the spin-exchange relaxation is suppressed. At this moment, the K atomic spin magnetometer achieves the SERF regime.

[1] B. R. Lancor, Thesis, MADISON: Wisconsin University (2011).
[2] Kiwoong Kim, Samo Begus, H. Xia, Seung-Kyun Lee, Vojko Jazbinsek, Zvonko Trontelj, and M. V. Romalis, NeuroImage, **62**, 133201 (2013).
[3] J. C. Fang, T. Wang, W. Quan, H. Yuan, H. Zhang, Y. Li, and Sheng Zou, Rev. Sci. Instrum. **85**, 063108-1 (2014).

# A reflected probe method for detecting the polarization rotating angle of the SERF magnetic measurement

Hongwei Cai[1], Ming Ding[1], Weiren Wu[1, 2], Yang Li[1], Xuejing Liu[1], Hong Zhang[1]

[1]School of Instrumentation Science and Opto-electronics Engineering, Beihang University, Beijing 10091, China

[2]Lunar Exploration Engineering Center, Beijing 100037, China

The atomic spin exchange relaxation free (SERF) magnetometer uses a pump-probe structure to measure ultra-weak magnetic field [1, 2]. The sensitivity is calculated using a formula expressed by a signal to noise ratio. In order to improve the probe performance, a reflected measure method that use s one mirror at the backside of the alkali vapor cell is adopted. The setup is shown in figure 1 in whi ch the probe beam is reflected right backward to the cell and a beam splitter is utilized to separate the incident and reflected beam. By reflecting the probe beam, the probe beam propagates twice the optical length through the vapor cell, which makes its rotating angle of the polarization twice as well, which results in the effective signal would also be almost twice. The experiment is carried out using SERF magnetic measurement platform. The experiment results showed that with the same probe beam power, the reflected probe method has a more effective signal portion, i.e. a higher signal to noise ratio. The sensitivity of the experiment reached ~25 $fT/Hz^{1/2}$@30Hz as shown in figure 2.

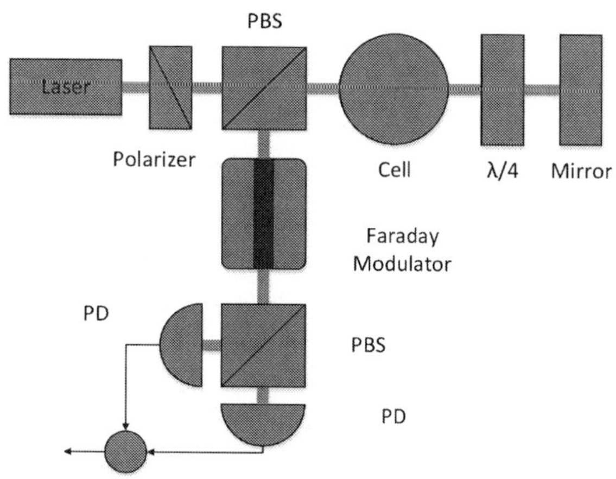


Figure 1 reflected probe setup

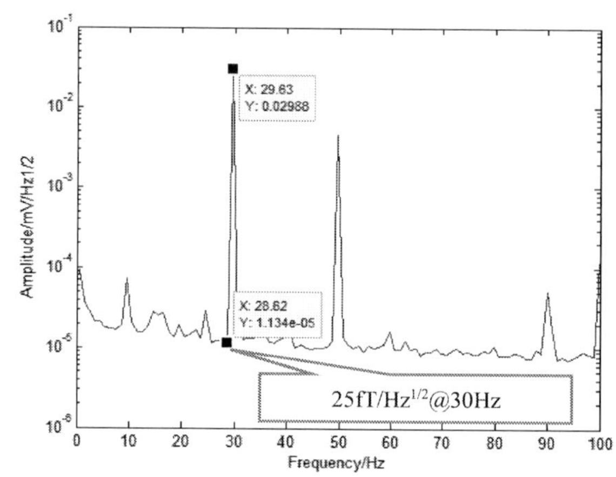


Figure 2 sensitivity result

[1] K.Kamada, Y.Ito, S.Ichihara, N.Mizutani, T.Kobayashi, OPTICS EXPRESS 6976 Vol. 23, No. 5 (2015).

[2] K.Kim, S.Begus, H.Xia, S.K.Lee, V.Jazbinsek, Z.Trontelj, M.V.Romalis, NeuroImage 89, 143–151(2014).

# Displacement measurement using an optoelectronic oscillator with an intra-loop Michelson interferometer

Sooyoung Park, Jehyun Lee, D.Cho

Department of Physics, Korea University, Seoul 136-713, Korea

We report on precise measurement of small displacement using an optoelectronic oscillator (OEO) with an intra-loop Michelson interferometer. In comparison with homodyne or heterodyne detection, where displacement shows up as either a power change or a phase shift, respectively, in the OEO detection displacement produces a shift in oscillation frequency. Because frequency is usually the easiest quantity to measure, OEO detection method provides unique advantages.

Our apparatus is similar to that of heterodyne detection: a special modulator produces a sideband which is polarized orthogonal to the carrier. A polarizing beam splitter sends the two beams to orthogonal arms of a Michelson interferometer. Unlike the heterodyne detection, however, because the interferometer is a part of an OEO loop, the phase shift induced by mirror displacement results in a shift in oscillation frequency. In order to avoid nonlinear response of the OEO to a displacement, the OEO is phase locked to a stable oscillator and the correction signal is measured to extract displacement.

We operate our apparatus in both heterodyne and OEO detection scheme to compare performances. Operating frequency is 80 MHz with a free spectral range of 1.8 MHz. The device is placed on an active vibration isolation stage, and housed in a vacuum chamber. Allan deviations in both schemes are similar and they are largely limited by environmental perturbation in spite of our isolation efforts. With 10 ms measurement time, Allan deviation from both measurements is smaller than 10 pm.

# Quantum technology developments at the University of Birmingham

L. Zhu, W. He, A. Hinton, J. Hordell, A. Lamb, M. Menchetti, C. Rammeloo, L. Smith, D. Swierad, Q. Ubaid, S. Viswam, G. Voulazeris, J. Hughes, A. Kaushik, J. Malcolm, A. Niggebaum, P. Petrov, M. Perea-Ortiz, S. Plant, A. Stabrawa, Y.-H. Lien, V. Boyer, M. Holynski, Y. Singh, and K. Bongs

Midlands Ultracold Atom Research Centre, UK National Quantum Technology Hub in Sensors and Metrology, School of Physics and Astronomy, University of Birmingham, UK

The UK National Quantum Technology Hub in Sensors and Metrology is working to bring cold atom based sensing into real world applications, ranging across brain imaging, detecting features under the ground and space-based Earth observation. This involves the combination of fundamental research, industry expertise and end-user needs. In this abstract, three examples of quantum technology activities at the University of Birmingham will be discussed, including the development of clocks, quantum optical squeezing and portable gravimetry.

The unparalleled precision of strontium optical lattice clocks has been demonstrated in several groups across the world, but despite their excellent performance these are often confined to the laboratory. Enabling their transition into portable devices would open applications in time referencing, relativistic geodesy, data encryption, navigation, and even the definition of the second. In this poster we report our progress and results of the robust transportable Space Optical Clock (SOC2) [1]. So far, the robust preparation of cold $^{88}$Sr atoms and clock transition with 3.4 Hz linewidth are achieved with SOC2. The most recent instability measurement results in $3.4 \times 10^{-17}$ in 7000 seconds time.

The hub's quantum imaging activities revolve around the creation of spatially squeezed light for increased optical resolution in certain optical super-resolution schemes. We have produced light fields which are locally squeezed at each point of their transverse profiles and we are working towards the realization of portable sources of spatially squeezed light.

The hub gravity sensing activity focuses on creating portable demonstrators for applications such as underground mapping, and creating test-beds aiming at ultimate precision to both facilitate knowledge and technology development within industry and new scientific techniques. In this poster we focus on progress in the development of our test-bed systems.

We acknowledge the funding by EPSRC Grant EP/M013294/1, the Space Optical Clock (SOC-2) Grant EU-FP7-SPACE-2010-1 No. 263500 and the FUTURE ATOMIC CLOCK TECHNOLOGY (FACT) Grant FP7-PEOPLE-2013-ITN No. 607493.

[1] K. Bongs, Y. Singh, L. Smith, W. He, O. Kock, D. Świerad, J. Hughes *et al.*, C R Phys **16**, 553-564 (2015).

# Measuring electric fields beyond the SQL with Rydberg Schrödinger cat states

Eva-Katharina Dietsche[1], Adrien Facon[1], Dorian Grosso[1], Frederic Assemat[1], Serge Haroche[1], Jean-Michel Raimond[1], Michel Brune[1], Sebastien Gleyzes[1]

[1]Laboratoire Kastler Brossel, College de France, ENS-PSL, UPMC-Sorbonne Universite, CNRS, 11 Place Marcelin Berthelot 75005 Paris, France

In the classical world there is no fundamental limit to the precision of a measurement. In quantum mechanics, however, the precision of a measurement is ultimately limited by quantum fluctuations. For instance, the direction of a large angular momentum $J$ prepared in a coherent spin state cannot be determined with a precision better than $1/\sqrt{J}$, the standard quantum limit (SQL) for this system. A measurement uncertainty below the SQL can only be attained by the use of quantum-enabled metrology techniques. It is then possible to reach the ultimate limit, the Heisenberg limit, which scales as $1/J$.

Here, we present a quantum-enhanced measurement of the electric field using mesoscopic Schrödinger-cat-like superpositions of Rydberg states. The atom behaves like a large angular momentum $J$ whose precession frequency depends on the electric field. Instead of performing a standard Ramsey experiment, we prepare the atom in a quantum superposition of two different coherent spin states and observe the evolution of their relative phase. As the atom is in a Schrödinger cat superposition of two states with very different polarizabilities, this quantum phase is highly sensitive to variations in the amplitude of the electric field. With this single-atom-electrometer we succeed in measuring field variations in the order of 1mV/cm in 100ns, beating the SQL [1]. The extreme sensitivity of this measurement paves the way to many practical applications.

[1] A sensitive electrometer based on a Rydberg atom in a Schrödinger-cat state, A. Facon, E.K. Dietsche, D. Grosso, S. Haroche, J.M. Raimond, M. Brune, S. Gleyzes, doi:10.1038/nature18327

# Precision spectroscopy with single $^{40}Ca^{+}$ ions in a Paul trap

Hua Guan, Yao Huang, Wu Bian, Ke-lin Gao

State Key Laboratory of Magnetic Resonance and Atomic and Molecular Physics, Wuhan Institute of Physics and Mathematics,Chinese Academy of Sciences, Wuhan 430071, China

Precision measurements of the 4s $^2S_{1/2}$-3d $^2D_{5/2}$ clock transition based on $^{40}Ca^{+}$ ion at 729 nm are reported. Single $^{40}Ca^{+}$ ion is trapped and laser-cooled in a ring Paul trap, and the storage time is more than one month. The linewidth of a 729 nm laser is reduced to about 1 Hz by locking to a super cavity for longer than one month uninterruptedly. In order to carry out the comparison of two $^{40}Ca^{+}$ optical frequency standards, another similar $^{40}Ca^{+}$ optical frequency standard is constructed. Two optical frequency standards exhibit stabilities of $1\times10^{-14}\tau^{-1/2}$ with 3 days of averaging. The overall systematic uncertainty of the clock transition is evaluated to be better than $5.5\times10^{-17}$ for both frequency standards. The absolute frequency of the clock transition is measured at the Hz level by using an optical frequency comb referenced to a hydrogen maser which is calibrated to the SI second through the global positioning system (GPS). The frequency value is 411 042 129776 401.7(1.1) Hz with the correction of the systematic shifts [1]. The magic wavelengths for the 4s $^2S_{1/2}$-3d $^2D_{5/2}$ clock transitions: $\lambda_{|m\,j|=1/2}$ = 395.7992(7) nm and $\lambda_{|m\,j|=3/2}$ = 395.7990(7) nm are reported, and it's the first time that two magic wavelengths for the $^{40}Ca^{+}$ clock-transition have been reported [2].

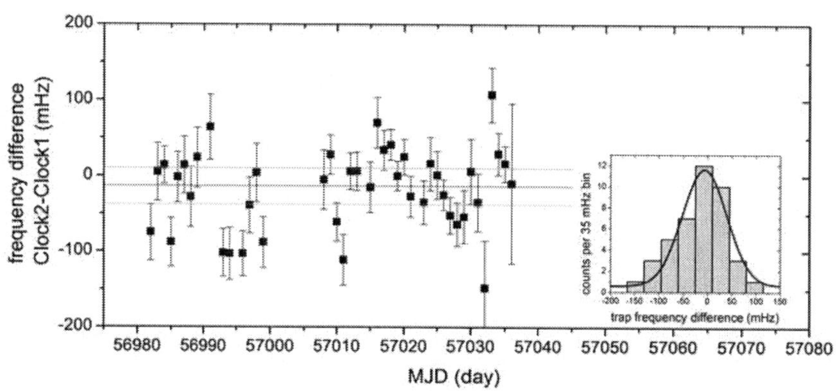


**Figure 1.** Frequency comparison of the two clocks. The mean frequency is indicated by the red line and the 3σ uncertainties (σ represents the standard deviation of the mean calculated for the whole data set) by green lines. Inset: A histogram of the individual frequency measurements shown in the figure with a Gaussian fit toshow the normally distributed data.

[1] Y. Huang, H. Guan, P. Liu, W. Bian, L. Ma, K. Liang, T. Li, K. Gao, Phys. Rev. Lett. **116**, 013001 (2016).

[2] P. Liu, Y. Huang, W. Bian, H. Shao, H. Guan, Y. Tang, C. Li, J. Mitroy, K. Gao, Phys. Rev. Lett. **114**, 223001 (2015).

# Photoionization parameters of Double-Excitation Resonances in Mg

Cheng-Liang Lu[1], Hsin-Chang Chi[2], and Keh-Ning Huang[3]

[1]Kaohsiung Municipal Kaohsiung Senior High School,
Kaohsiung, Taiwan 807, Republic of China
[2]Department of Physics,
National Dong Hwa University,
Hualien, Taiwan 974, R.O.C.
[3]Institute of Atomic and Molecular Physics,
Sichuan University,
Chengdu 610065, China

Photoionization parameters in magnesium are calculated in the multiconfiguration relativistic random-phase approximation to resolve the interplay between relativistic and correlation effects. The angular distribution and spin polarization of photo-electrons from valence subshell 3s show strong dependence on core-shielding effects. Precise energies and widths of all five Rydberg series of doubly-excited states $(3pns)\,^{1,3}P_1^O$, $(3pnd)\,^{1,3}P_1^O$, and $(3pnd)\,^{3}D_1^O$ between the $Mg^+[3s]$ and $Mg^+[3p]$ thresholds are given. New experiments have shown better resolution to clarify the high energy portion of Rydberg series amid peaks corresponding to Rydberg-series excitations can be distinguished theoretically in comparisons with our predictions. In addition, we have calculated the quantum defect parameters associated with distinct Rydberg series converging to fine-structure ionization thresholds $Mg^+[(3p)\,^2P_{1/2}^O]$ and $Mg^+[(3p)\,^2P_{3/2}^O]$, respectively. Systematic analyses of the behavioral characteristics in the series quantum defect parameters are performed to demonstrate the interplay of relativistic effects and double-excitation correlations.

# Scalar and Tensor Light Shift in Rb and Cs.

J. C. Camparo[1] G. A. Costanzo[2,3], B. Francois[2], S. Micalizio[2], C. E. Calosso[2], A. Godone[2], and F. Levi[2]

[1] The Aerospace Corporation 2310 E. El Segundo Blvd., El Segundo, CA 90245, USA
[2] Istituto Nazionale di Ricerca Metrologica Strada delle Cacce 91, 10135 Torino, Italy
[3] Politecnico di Torino, Corso duca degli Abruzzi 24, 10125, Italy

The ac Stark shift or light shift is ubiquitous in atomic physics: it is the basis for a host of basic physics applications including laser cooling and lattice clocks at the magic wavelength, and it is one of the leading sources of instability and inaccuracy in both terrestrial laboratory time standards and space-based atomic clocks. In order to better test theory regarding this important fundamental process, we have begun a series of experiments focusing (at present) on the near-resonant light shift in Cs and Rb atomic clock systems. Specifically, according to theory [1] three (and only three) distinct contributions to the light shift make their appearance: the scalar, the vector, and the tensor components. The scalar contribution to the light shift is always present; the vector contribution only makes its appearance when the perturbing field's photons carry net angular momentum along the atoms' quantization axis, and the tensor component only manifests itself when the excited state hyperfine splitting is resolved. In our two sets of experiments this latter condition is achieved: 1) a light shift test of the $D_1$ Rb transition in a vapor phase system, and 2) a light shift test of the $D_2$ Cs transition in a cold-atom fountain.

The first set of measurements was performed using a pulsed vapor cell clock [2], and perturbing the $^{87}$Rb atoms of the clock with a laser tuned near the $1^{st}$-resonance $D_1$ transition (*i.e.*, $5^2P_{1/2} - 5^2S_{1/2}$ at 795 nm). Our second set of measurements was performed using a cold atom fountain primary frequency standard, and perturbing the atoms with a very narrow linewidth, singlemode laser tuned near the $1^{st}$-resonance $D_2$ transition (*i.e.*, $6^2P_{3/2} - 6^2S_{1/2}$ at 852 nm). The cold atom fountain provides an almost ideal experimental test of the light shift, since the atoms are in a motionless and collisionless environment; Doppler broadening is virtually absent, and the hyperfine splitting of the excited states is much larger than the transition's natural linewidth. In this presentation we discuss both sets of experiments and our results: 1) verification of the theoretical light shift predictions with an unprecedented level of accuracy and 2) evidence for the unique contribution of the tensor component of the light shift in two quite different systems.

This work has been funded by the EMRP project IND55 MClocks. EMRP is jointly funded by the EMRP participating countries within EURAMET and the European Union. The activities of J. Camparo were supported by The Aerospace Corporation's Sustained Experimentation and Research for Program Applications program, and funded by U.S. Air Force Space and Missile Systems Center under Contract No. FA8802-09-C-0001.

[1] W. Happer and B. S. Mathur, “Effective operator formalism in optical pumping,” Phys. Rev. 163(1), 12-25 (1967); B. S. Mathur, H. Tang, and W. Happer, “Light shifts in the alkali atoms,” Phys. Rev. 17191), 11-19 (1968).

[2] F. Levi et al. Precision test of the ac Stark shift in a rubidium atomic vapor, PRA 93, 023433 (2016).

# Towards Precision Measurement of the $2^1S_0 \rightarrow 3^1D_2$ Two-Photon Transition in Atomic Helium

Yi Jan Huang[1], Te-Hwei Suen [2], Li-Bang Wang[2], and Jow-Tsong Shy[1,2]

[1]Institue of Photonics Technologies, National Tsing Hua University, Hsinchu 30013, Taiwan

[2]Department of Physics, National Tsing Hua University, Hsinchu 30013, Taiwan

We report our progress towards a precision measurement of the $2^1S_0 \rightarrow 3^1D_2$ two-photon transition in atomic helium at 1009 nm. A master oscillator power amplifier (MOPA) seeded by an external cavity diode laser (ECDL) is constructed to generate more than 500 mW of laser power with $TEM_{00}$ at 1009 nm. The frequency of the ECDL is determined and controlled by a precision wavelength meter (HighFinesse WSU10). To observe the two-photon transition, a helium cell is placed inside an optical cavity and the helium atoms at $2^1S$ metastable level are prepared by a pulsed RF discharge. The two-photon spectrum is observed by monitoring the 668 nm $3^1D_2$ to $2^1P_1$ fluorescence after RF discharge is turned off [1]. Before, we have measured the $2^1S_0 \rightarrow 2^1P_1$ [2] and $2^1P \rightarrow 3^1D_2$ [3] transition frequencies. Due to the short lifetime of the $2^1P$ state (0.57 ns), the accuracy is limited to ~ 200 kHz .The first observation of the $2^1S_0 \rightarrow 3^1D_2$ two-photon transition is reported and our experimental progress towards absolute frequency measurement of this transition will be discussed.

[1] C. Dorrer, F. Nez, B. de Beauvoir, L. Julien and F. Biraben, Phys. Rev. Lett. 78, 3658 (1997).

[2] P-L Luo, J-L Peng, J-T Shy, and L-B Wang, Phys. Rev. Lett. 13, 54321 (2014).

[3] P-L Luo, Y-C Guan, J-L Peng, J-T Shy, and L-B Wang, Phys. Rev. A. 88, 054501 (2013).

# Observation of vibrational overtones by single molecule resonant photodissociation

Ncamiso B. Khanyile[1], Gang Shu[2], Kenneth R. Brown[2]
[1]Stellenbosch University, South Africa
[2]Georgia Institute of Technology, United States

Molecular ions sympathetic cooled by a chain of laser-cooled atomic ions are ideal for performing high-precision molecular spectroscopy with applications in astrochemistry and fundamental physics. The same system can be coupled with a broadband laser to discover new molecular transitions. Here we present our result of using three-ion chains of $Ca^+$ and $CaH^+$ to observe vibrational transitions via resonance enhanced multiphoton dissociation detected by $Ca^+$ fluorescence [1]. Based on theoretical calculations, the observed peaks are assigned to the transition from the ground vibrational state, $v = 0$ to $v = 9$ and $v = 10$. Our method allows to track single molecular events, and it can be extended to handle any molecule by monitoring normal mode frequency shifts to detect the dissociation. This survey spectroscopy serves as a bridge to the precision spectroscopy required for molecular ion control. We also present ongoing work in quantum feedback control with unsharp measurements at the University of Stellenbosch.

[1] N. B. Khanyile, G. Shu, and K. R. Brown, Nat. Comm. **6**, 7825 (2015)

# Towards Highly Sensitive Cavity-Enhanced Polarimeter with NICE-OHMS Spectroscopy

Tzu-Ling Chen and Yi-Wei Liu*

Department of Physics, National Tsing Hua University, Hsinchu 30013, Taiwan

*Electronic address: ywliu@phys.nthu.edu.tw

Chirality and parity violation reveal the left and right asymmetry of the world with a common phenomenon, optical rotation. While chirality plays a fundamental part in biological molecules, atomic parity non-conservation is to probe new physics beyond Standard Model. However, detecting and quantifying optical rotation remains challenging particularly in weak absorption region, which promotes the developments of highly sensitive cavity-enhanced polarimeters in recent years [1].

In this work, a novel cavity-enhanced polarimeter is proposed to detect the small optical rotation. With a non-orthogonal molecular beam system, the reciprocal optical rotation can be converted to non-reciprocal to be enhanced in a linear cavity. Furthermore, the molecular beam system allows the determination of the optical rotation of individual rotational states [2]. When inserting such an enhancement cavity between crossed polarizers, this configuration can be a highly sensitive polarimeter on the basis of the highly sensitive absorption techniques "NICE-OHMS (noise-immune cavity-enhanced optical heterodyne molecular spectroscopy)" . Take the advantage of noise immunity by combining frequency modulation with enhancement cavity, NICE-OHMS results in an ultrasensitive spectrometer with the sensitivity close to the shot noise limit [3]. In current work, a quantum-dot laser based NICE-OHMS apparatus at 1.28 $\mu$m using a high finesse cavity (F=18500) has been constructed [4]. As a preliminary result shown in Fig. 1, a minimum noise-equivalent absorption coefficient, $\alpha_{min}$, $3\times10^{-10}$ $cm^{-1}Hz^{-1}$, had been reached with $N_2O$ weak molecular overtone transitions in gas phase.

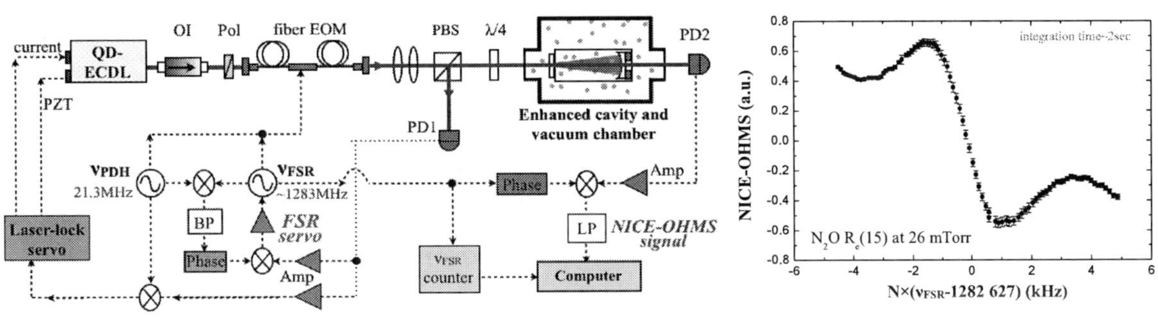


Figure 1: *Left: Experimental setup for NICE-OHMS. Right: NICE-OHMS dispersion signal with a time constant of 2 sec, where the frequency scale is derived from counted modulation frequency.*

[1] D. Sofikitis, L. Bougas, G.E. Katsoprinakis, A.K. Spiliotis, B. Loppinet, and T.P. Rakitzis, Nature **514**, 76 (2014).

[2] A.J. MacDermott and R.A. Hegstrom, Chem. Phys. **305**, 47 (2004)

[3] J. Ye, L.S. Ma, J.L. Hall, J. Opt. Soc. Am. B **15**, 6 (1998)

[4] T.L. Chen and Y.W. Liu, Opt. Lett. **40**, 4352 (2015).

# Spectroscopy of the 2S-6P transition in atomic hydrogen

Ksenia Khabarova[1,2*], Lothar Maisenbacher[3], Axel Beyer[3], Arthur Matveev[3], Alexey Grinin[3], Thomas Udem[3], Theodor W. Hänsch[3,4], N.N. Kolachevsky[1,2]

[1]P.N Lebedev Physical Institute, 119991, Moscow, Russia

[2]Russian Quantum Center, 143025, Skolkovo, Russia

[3]Max Plank Institute of Quantum Optics, 85748, Garching, Germany

[4]Ludwig-Maximilians University, 80539 Munich, Germany

Recent progress of one-photon spectroscopy in atomic hydrogen [1] opens possibility to measure transition frequencies of 2S-nP transitions with uncertainties in the low kHz range. Using spectroscopy of the 2S-4P transition at 486 nm on a cryogenic hydrogen beam we showed that leading systematic effects (Doppler effect and quantum interference) can be characterized at the 1-2 kHz level. Combining the measured frequency of the 2S-4P transition with the 1S-2S frequency [2] one gets independently values for the Rydberg constant and the proton charge radius. The corresponding uncertainty competes with the uncertainty of the combined world result obtained from hydrogen spectroscopy, which critically contributes to the CODATA analysis [3]. New independent measurements of 2S-nP transitions in atomic hydrogen will be an important input for resolving the "proton radius puzzle" coming from the discrepancy of muonic and electronic hydrogen data [4]. The 2S-6P transition at 410 nm possesses a three times narrower natural line width compared to the 2S-4P transition, which can improve statistical uncertainty and suppress systematic effects of an absolute frequency measurement. At the same time sensitivity of the transition to electric and magnetic fields is still small enough such that corresponding shifts can be controlled at sub kHz level. We have upgraded the apparatus described in [1] to accommodate the new transition frequency and further improve statistics, which should allow a precision measurement of the 2S-6P transition in the near future. Here, we report on preliminary measurements of the 2S-6P transition (Fig 1).

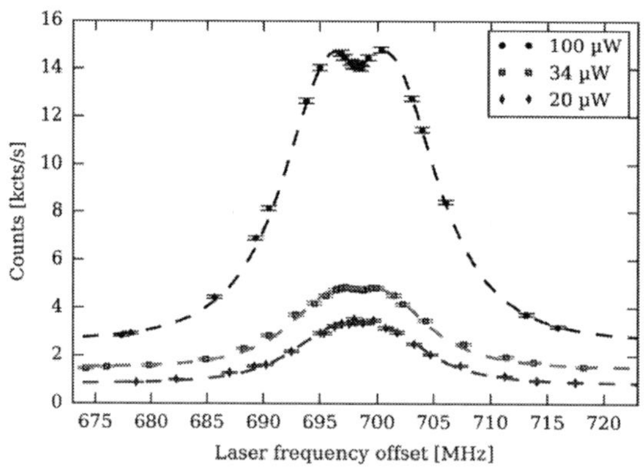


**Figure 1**: Preliminary measurements of the 2S-6P transition. Shown are single resonance scans for three different excitation powers (per beam). The dashed lines are the corresponding simulations, taking into account the atomic beam divergence and excitation dynamics.

[1] A. Beyer et al., Annalen der Physik **525**, 671 (2013).

[2] C. G. Parthey, et al., Phys. Rev. Lett., **107**(20):203001 (2011).

[3] P. J. Mohr, et al., Rev. Modern Phys., **84**(4):1527–1605 (2012).

[4] Pohl et al., Nature **466**, 7303 (2010).

# Analytical solutions of susceptibility for Doppler-broadened two-level atoms

Seung-Chul Yang, Hyun-Jong Kang, Heung-Ryoul Noh
Department of Physics, Chonnam National University, Gwangju 61186, Korea

We present analytical solutions of susceptibility in pump-probe laser spectroscopy for an atomic medium composed of two-level atoms in the Doppler-broadened limit. We derive an accurate analytical formula for susceptibility up to the first order at the Rabi frequency of the probe beam for the pump beam of arbitrary intensity. We find the analytical form of the susceptibility in a succinct single term. The analytical solutions approach the familiar form obtained solely from the incoherence contribution as the intensity of the pump beam becomes weaker. Using the analytical solution of the susceptibility, analytical forms of the visibility (normalized peak intensity) and linewidth of the saturated absorption spectroscopy (SAS) spectrum are obtained.

[1] H. R. Noh, J. Opt. Soc. Am. B 33, 308 (2016).

# Analytical calculation of line shape in polarization spectroscopy for the $J_g=0 \rightarrow J_e=1$ transition

Heung-Ryoul Noh

Department of Physics, Chonnam National University, Gwangju 61186, Korea

We present an analytical solutions of the line shape in polarization spectroscopy (PS) for the $J_g=0 \rightarrow J_e=1$ transition in the Doppler-broadened limit. The analytical solutions are obtained up to the first order of the Rabi frequency of the probe beam while the Rabi frequency of the pump beam is arbitrary. The line shape is influenced by the coherent effect resulting from nonlinear interactions between laser fields via atoms. We also calculate the line shape by neglecting the nonlinear wave mixing which is usually called the results of rate equation calculation. We find that the difference between the results of the accurate and rate equation calculations decreases as the pump beam Rabi frequency decreases or transverse decay rate increases. We also find the analytical solutions of the slope at the resonance and amplitude of the line shape in PS.

# Precision spectroscopy of $^{87}$Rb Rydberg states using an optical frequency synthesizer based on an optical frequency comb

Naoto Watanabe, Hikaru Tamura, Nobunaga Naka, Mitsuru Musha, Ken'ichi Nakagawa
Institute for Laser Science, Univ. of Electro-Communications, Tokyo, Japan

Highly stable frequency-stabilized lasers in the wide wavelength region are required for recent atom physics experiments, including laser cooling, atom interferometry, and quantum information. Optical frequency combs can provide highly accurate optical frequency references from near infrared to visible region. Especially optical frequency combs based on Er-doped mode-locked fiber lasers are reliable and relatively compact, thus these combs have been widely used for various atom physics experiments. In our laboratory, we have studied the high-lying Rydberg states of $^{87}$Rb atoms using two frequency-stabilized lasers at 780 nm and 480 nm. For this application, it is needed to realize frequency stable and tunable laser sources at 480 nm. We have developed an optical frequency synthesizer based on a high repetition frequency (325 MHz) Er-fiber optical frequency comb [1]. In this paper, we report the measurement of absolute transition frequencies of $^{87}$Rb to nD (n=60-70) Rydberg states. We observed electromagnetically induced transparency (EIT) spectra on a Rb vapor cell with probe (780 nm) and coupling (480 nm) lasers. The probe laser at 780 nm was stabilized on the $^{87}$Rb $5S_{1/2}$ F=2 – $5P_{3/2}$ F=3 transition, whereas the coupling laser was offset-locked to a tooth of the comb. The frequency of the coupling laser was precisely scanned over more than 100 MHz with approximately 1 MHz frequency step by changing the repetition rate of the comb. We measured two Rydberg spectra of $61D_{3/2}$, and $61D_{5/2}$, which were fitted by a Lorentz function to determine the center frequency (Fig. 1). The present measured transition frequencies are in good agreement with the theoretically predicted frequencies within 2 MHz [2]. Further measurement of the transition frequencies for other Rydberg states will allow for more precise determination of the quantum defects.

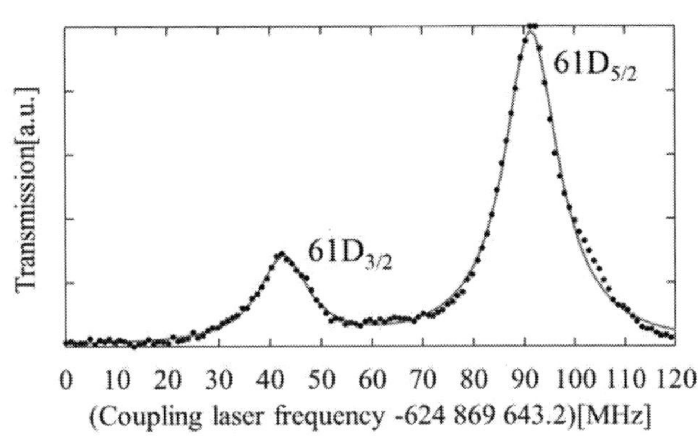


Figure 1: The EIT spectra of $61D_{3/2}$ and $61D_{5/2}$. Black circles represent the absorption signal with 50 ms averaging time. A solid line is a fitted Lorentz curve.

[1] H. Qi, J. Zhang, G. Zhou, A. Wang, and Z. Zhang, Chin. Opt. Lett. **11**, 061402 (2013).

[2] M. Mack, F. Karlewski, H. Hattermann, S. Höckh, F. Jessen, D. Cano, and J. Fortágh, Phy. Rev. A **83,** 052515 (2015).

# Spectroscopy of the Hydrogen 1s-3s Transition with chirped Laser Pulses

D.C. Yost, A. Matveev, A. Grinin, E. Peters, L. Maisenbacher, A. Beyer, R. Pohl, N. Kolachevsky, K. Khabarova, T.W. Hänsch, and Th. Udem

Max-Planck-Institute of Quantum Optics, Garching Germany

We identify a systematic present in two-photon direct frequency comb spectroscopy (DFCS) which is a result of chirped laser pulses and is a manifestation of the first-order Doppler effect [1]. We carefully analyze this systematic and propose methods for its mitigation within the context of our measurement of the hydrogen 1s-3s transition. We also report on our determination of the absolute frequency of this transition, which is comparable to a previous measurement using continuous-wave spectroscopy [2], but was obtained with a different experimental method.

[1] D.C. Yost *et al.* Phys. Rev. A 93, 042509 (2016)
[2] O. Arnoult *et al.*, Eur. Phys. J. D 60, 243 (2010)

# Photodissociation of ultracold, quantum state-selected $^{88}Sr_2$ molecules

Mickey McDonald[1], Bart H. McGuyer[1], Florian Apfelbeck[1], Chih-Hsi Lee[1], Iwona Majewska[2], Robert Moszynski[2], and Tanya Zelevinsky[1]

[1]Department of Physics, Columbia University, 538 West 120th Street, New York, NY 10027-5255, USA

[2]Quantum Chemistry Laboratory, Department of Chemistry, University of Warsaw, Pasteura 1, 02-093 Warsaw, Poland

When a molecule is subjected to an energetic photon it can break apart into fragments through a process called *photodissociation*. These fragments expand outward in a sphere possessing a complex angular distribution determined by the “shapes” of the rotational wavefunctions of the allowed output channels, as well as their interferences. For over 70 years this basic reaction has been employed by chemists to study the character of dissociative transitions in a semi-classical way. However, efforts to compare theory to experiment at the most fundamental quantum level have been hindered by the imperfect state control, brief interaction times, and relatively high temperatures inherent to molecular beams.

We present recent experiments [1] involving the photodissociation of ultracold and fully quantum state-resolved $^{88}Sr_2$ molecules. Molecules are prepared in a 1D optical lattice and photofragments are detected via optical absorption imaging (Fig. 1). Accurate state preparation and narrow-linewidth laser spectroscopy permits the observation of many novel effects requiring fully quantum mechanical explanations, including tunneling through electronic and rotational barriers and matter-wave interference of the reaction products. This interference results in photofragment angular distributions with a strong breaking of cylindrical symmetry, and whose exact shapes reveal information about not just the relative strengths of dissociative output channels, but their relative phases as well.

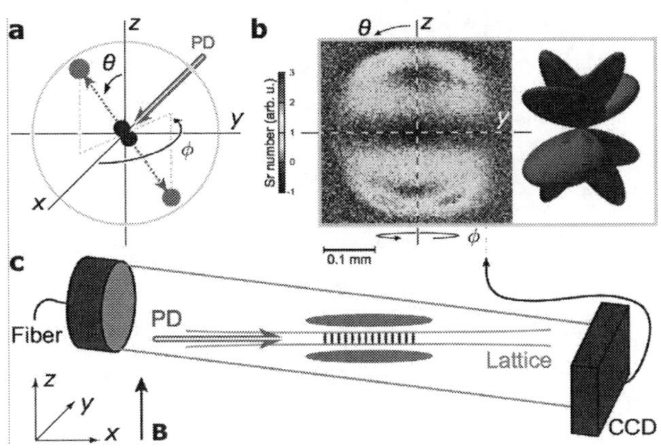


**Figure 1.** (a) Molecules subjected to energetic laser light can break apart into photofragments, with an angular distribution parameterized by angles $\theta$ and $\phi$. (b) The full 3D photofragment angular distribution is projected onto a 2D plane via absorption imaging onto a CCD (left), with a predicted distribution which is highly cylindrically asymmetric (right) (c) Molecules are trapped and cooled in a 1D optical lattice.

[1] M. McDonald et al., “Photodissociation of ultracold diatomic strontium molecules with quantum state control,” Nature (in press).

# Phase noise effects of optical fields in four-wave mixing process

Taek Jeong, Han Seb Moon
*Department of physcis, Pusan National University, Busan, 609-735, Korea*

We performed experiment for measurement of phase noise of a generated conjugate field generated through four-wave mixing(FWM) process in double lambda scheme by using co-propagating three coherent lights, which are a probe, coupling, pump field, in rubidium vapor cell. Fig. 1 shows frequency component of beat-signal between the generated conjugate field and the probe field when the pump field is locked with the others or not. We could confirm that phase noise of the generated conjugate field follow the pump field. The results indicate that two-photon process is dominant key factor for FWM process

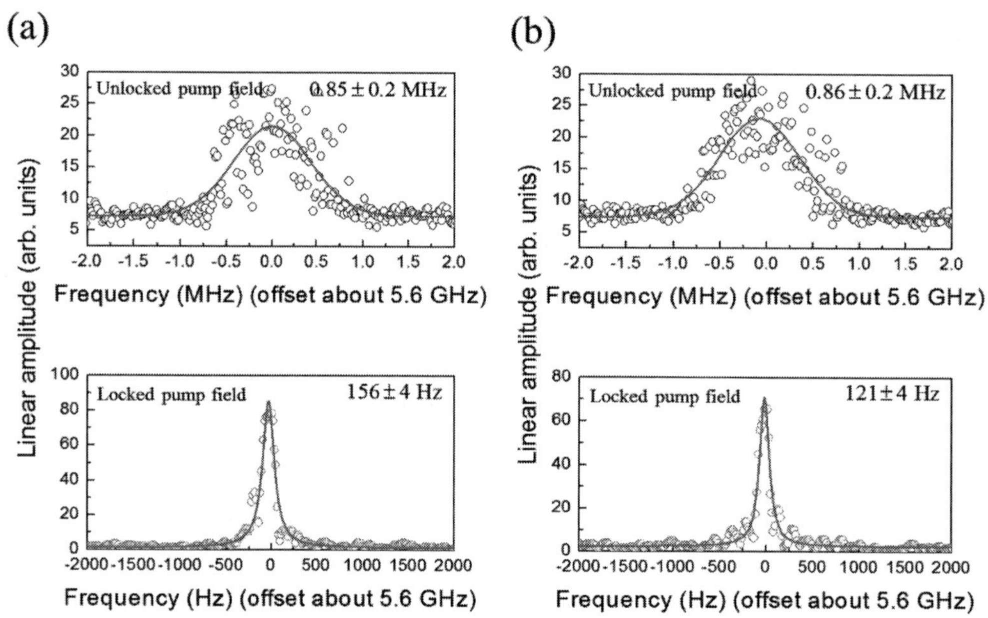


**Figure 1.** (Color online) (a) Beat signal in spectrum analyzer between the generated conjugated field and the probe field in EIT-scheme when pump field is locked (bottom) or not (upper). (b) in two-photon absorption(TPA)-scheme.

# Doppler-free spectroscopy of the $5P_{3/2} \rightarrow 6P_{3/2}$ electric quadrupole transition in atomic rubidium

F. Ponciano-Ojeda[1], S. Hernández-Gómez[1], O. López-Hernández[1], C. Mojica-Casique[1], R. Colín-Rodríguez[1], F. Ramírez-Martínez[1], J. Flores-Mijangos[1], D. Sahagún[2], R. Jáuregui[2], and J. Jiménez-Mier[1]

[1]Instituto de Ciencias Nucleares, UNAM. Circuito Exterior, Ciudad Universitaria, 04510, Cd. Mx., México.

[2]Instituto de Física, UNAM, Circuito de la Investigación Científica, Ciudad Universitaria, 04510, Cd. Mx., México.

Electric dipole forbidden transitions have been traditionally used in astrophysical and plasma studies [1], they now play a fundamental role in metrology [2] and have been employed to test parity non-conserving interactions in atoms [3]. However, they are yet to become a broadly used tool due to their characteristic low excitation probability, which is typically of the order of $10^6$ times smaller than that of their electric dipole counterparts. For the alkali, most of the studies have been done either in cold or ultra-cold atoms [4] or by employing very strong magnetic fields and nanometer size cells [5]. Only recently, studies of electric dipole forbidden transitions have been performed in room temperature atoms with more accessible Doppler-free techniques [6, 7]. Here we report the observation, for the first time, of the $5p_{3/2} \rightarrow 6p_{3/2}$ electric quadrupole (E2) transition in atomic rubidium [6]. Experiments were performed with a room-temperature rubidium cell and with continuous-wave external cavity diode lasers. Optical-optical double-resonance spectroscopy with counter-propagating beams allows the detection of the non-dipole transition. The $5p_{3/2}$ state is prepared by excitation with a laser locked to the maximum F cyclic transition of the D2 line; while the $5p_{3/2} \rightarrow 6p_{3/2}$ forbidden transition is excited by a 911 nm laser, which in turn is monitored by detection of the 420 nm fluorescence that results from decay of the $6p_{3/2}$ state. The spectra reveal three narrow lines ($\approx$13 MHz FWHM) reflecting the E2 transition selection rules and with the characteristic $F - 1$, $F$, and $F + 1$ splitting of the $6p_{3/2}$ hyperfine structure in both rubidium isotopes. The results are in very good agreement with a direct calculation that takes into account the $5s_{1/2} \rightarrow 5p_{3/2}$ preparation dynamics, the $5p_{3/2} \rightarrow 6p_{3/2}$ non-dipole transition probability, and the $6p_{3/2} \rightarrow 5s_{1/2}$ decay. The comparison shows that the electric-dipole-forbidden transition is a very sensitive probe of the preparation dynamics.

[1] E. Biémont and C. J. Zeippen, Phys. Scr., T65, 192 (1996)

[2] R. van Rooij, *et al.*, Science 333, 196 (2011)

[3] B. Sayer, R. Wang, J. C. Jeannet, and M. Sassi, J. Phys. B 4, L20 (1971)

[4]M. Bhattacharya, C. Haimberger, and N. P. Bigelow, Phys. Rev. Lett. **91**, 213004 (2003); R. Pires, *et al.*, Phys. Rev. A **80**, 062502 (2009); D. Tong, S. M. Farooqi, E. G. M. van Kempen, *et al.*, Phys. Rev. A **79**, 052509 (2009)

[5] A. Sargsyan, A. Tonoyan, G. Hakhumyan, *et al.*, Laser Phys. Lett. 11, 055701 (2014)

[6] Ponciano-Ojeda, F. *et al.*, Phys. Rev. A, 92(4), 042511 (2015)

[7] E. A. Chan, S. A. Aljunid, N. I. Zheludev, *et al.*, Opt. Lett. 41, 2005 (2016).

# Control of electronic magnetic state populations via light polarization in the $5P_{3/2} \rightarrow 6P_{3/2}$ electric quadrupole transition in atomic rubidium

C. Mojica-Casique[1], F. Ponciano-Ojeda[1], S. Hernández-Gómez[1], O. López-Hernández[1], R. Colín-Rodríguez[1], J. Flores-Mijangos[1], F. Ramírez-Martínez[1], D. Sahagún[2], R. Jáuregui[2], and J. Jiménez-Mier[1]

[1]Instituto de Ciencias Nucleares, UNAM. Circuito Exterior, Ciudad Universitaria, 04510, Cd. Mx., México.

[2]Instituto de Física, UNAM, Circuito de la Investigación Científica, Ciudad Universitaria, 04510, Cd. Mx., México.

In this work we demonstrate the control in the preparation of electronic magnetic states by exploiting the selection rules of an electric quadrupole transition. We use Doppler-free optical double-resonance spectroscopy in room-temperature rubidium atoms to study the excitation sequence of a $5s_{1/2} \rightarrow 5p_{3/2}$ electric dipole preparation step followed by the $5p_{3/2} \rightarrow 6p_{3/2}$ electric quadrupole excitation. Here we present a detailed experimental and theoretical study of the 420 nm fluorescence decay ($6p_{3/2} \rightarrow 5s_{1/2}$) dependence on the polarization of the excitation beams. Firstly, when a circularly polarized preparation beam is used, it produces a strongly oriented $5p_{3/2}$ intermediate state. In this case, a linearly polarized quadrupole excitation beam transfers the oriented state to the $6p_{3/2}$ hyperfine states. Next, for linearly polarized preparation and quadrupole excitation beams, the spectra of the $6p_{3/2}$ hyperfine lines follow a cosine-squared dependence on the angle between the polarization directions of the two radiation components. As a consequence, we demonstrate that the choice of polarization configuration allows direct use of the electric quadrupole transition selection rules to control the populations of the $6p_{3/2}$ hyperfine magnetic sublevels in the absence of external fields. This is achieved by independently enhancing or suppressing either $\Delta M_F = \pm 1$ or $\pm 2$ electric quadrupole transitions.

# Investigation of the $A^1\Sigma^+$ states of the NaH and NaD molecules

Thou-Jen Whang[1], Chia-Ching Chu[1], Yu-Nien Chuang[1] and Chin-Chun Tsai[2]
[1]Department of Chemistry, National Cheng Kung University, Tainan, Taiwan
[2]Department of Physics, National Cheng Kung University, Tainan, Taiwan

The laser induced fluorescence excitation spectra of $A^1\Sigma^+$-state of NaH and NaD were obtained by using tunable single beam dye laser in the range of 362 to 445 nm. We have observed the 839 rovibrational levels for v' = 2-15 and J' up to 26 of NaH and 1404 rovibrational levels for v' = 5-22 and J' up to 30 of NaD. New Dunham coefficients for the first excited $A^1\Sigma^+$ states of NaH and NaD were deduced by combining these data and previously published experimental results. A broad range of rovibrational quantum number ( $0 \leq v' \leq 25$ and $0 \leq J' \leq 38$ of NaH; $4 \leq v' \leq 21$ and $0 \leq J' \leq 30$ of NaD ) are observed and hence the spectral positions can be reproduced with very good accuracy.

# Laser absorption spectroscopy of iodine: 915 nm to 985 nm in a single scan

Stephan Falke, Christian Nölleke, Christoph Raab and Rudolf Neuhaus
TOPTICA Photonics AG, Lochhamer Schlag 19, 82166 Gräfelfing, Germany

Iodine gas cells serve as frequency reference for lasers thanks to the good understanding of their rich molecular spectrum. This is particularly true for the wavelength range 500 nm to 900 nm, where the theoretical model of the B-X system and experimental findings agree on a level of a few MHz [1-2]. Hence, iodine cells provide frequency references with a line spacing of typically a few GHz only.

In order to extend this knowledge towards longer wavelengths, a deeper understanding of the A-X system of iodine is required. Recent developments in laser technology enable high-resolution spectroscopy with unprecedented ease. In this work, we record absorption spectra with a 75 cm long iodine cell, in which vapor pressure and temperature of the molecular gas are controlled independently. We utilize the vast mode-hope-free tuning range of the DLC CTL laser system [3] developed within our company to investigate the distribution of observable molecular lines as well as their signal-to-noise ratio in the range 915 nm to 985 nm. We derive the frequency of the observed absorption lines using a calibrated wavelength meter, which is capable of a precision of better than 10 MHz across the full range.

We find excellent agreement with the experimental study by Gerstenkorn *et al.* [4], which used Fourier transform spectroscopy. Due to the higher experimental resolution and better signal-to-noise ratio, we are able to identify lines that could not be resolved before. This, along with the reduced uncertainty of the frequency of the reference lines, may motivate studies on the A-X system aiming on improving or detailing the analysis of the molecular states [5].

The recorded data already represents a denser and higher precision atlas of iodine reference lines recorded in laser absorption spectroscopy, which is typically used to stabilize the frequency of lasers.

[1] Knöckel *et al.*, Eur. Phys. J. D **28**, 199 (2004).

[2] Salumbides *et al.*, Eur. Phys. J. D **47**, 171 (2008).

[3] Neuhaus, Paasch-Colberg, "Continuously tunable diode lasers enable mode-hop-free wavelength tuning", Exploring the Micro, Nano and Quantum World. Optik & Photonik **10**, 25 (2015). DOI 10.1002/opph.201500014

[4] Gerstenkorn et al., "Atlas du Spectre d'Absorption de la Molecule d'Iode", Laboratoire Aimé Cotton, CNRS II, Orsay (1993).

[5] Appadoo *et al.*, J. Chem. Phys. **104**, 903 (1996).

# Vibrational quantum defect and rotational constants resonances analyzed in a 2-channel model to measure the coupling between molecular series

Haikel Jelassi[1], Laurence Pruvost[2]

[1]Laboratoire de Recherche Energie et Matière pour les Développement des Sciences Nucléaires & Direction de la Recherche sur l'Energie et la Matière, National Center for Nuclear Science and Technology, Sidi Thabet Technopark 2020 Ariana Tunisia

[2] Laboratoire Aimé Cotton, CNRS, Université de Paris Sud, ENS-Cachan, Université Paris Saclay, 91405, Orsay, France.

The quantum defect theory is a well-known approach used for Rydberg atom, to characterize a Rydberg series and to reveal any coupling between Rydberg series. The signature appears in the change of the quantum defect progression of the levels.

In order to identify coupling between molecular potentials, available to give routes to form molecules from cold atoms via weakly-bound molecules, we have investigated the spectrum analysis by introducing the vibrational quantum defect (VQD). This quantity issues from the analogy between the Rydberg law and the LeRoy-Bernstein one. VQD-graphs obtained with data of high-resolution photoassociation spectroscopy exhibit resonances that have been associated to coupling between molecular potentials [1, 2, 3] and analyzed by a 2 channel model, providing the coupling parameters.

Another signature of the coupling exists in the progression of the rotational constants $B_v$ ($B_v$-spectrum) with resonances in the vicinity of the perturbing levels. We propose an improved-$B_v$-Formula associated to a 2-channel model to explain both the Lorentzian-shape of the resonances and the background and to fit the $B_v$-spectrum of $6s_{1/2}$-$6p_{1/2}$ $0_g^-$ given in [4]. Fitting parameters give the coupling parameters in agreement with the VQD method [5].

[1] H. Jelassi, B. Viaris de Lesegno, and L. Pruvost,, Phys. Rev. A **73**, 032501 (2006).

[2] H. Jelassi, B. Viaris de Lesegno, and L. Pruvost, M. Pichler, W. C. Stwalley, Phys. Rev. A. **78**, 022503 (2008).

[3] L. Pruvost and H. Jelassi, J. Phys. B: At. Mol. Opt. Phys. **43**, 125301 (2010).

[4] H. Lignier, A. Fioertti, R. Horchani, C. Drag, N. Bouloufa, M. Allefrini, O. Dulieu, L. Pruvost, P. Pillet, and D. Comparat, Phys. Chem. Chem. Phys. **13** (2011)18910.

[5] H. Jelassi and L. Pruvost, submitted to EPL.

# Effect of rotational-state-dependent molecular alignment on the optical dipole force

Lee Yeong Kim[1], Ju Hyeon Lee[2], Hye Ah Kim[2], Sang Kyu Kwak[3], Bretislav Friedrich[4], Bum Suk Zhao (조범석)[1,2]

[1]Department of Physics, National Institute of Science and Technology, Ulsan 44919, Korea

[2]Department of Chemistry, Ulsan National Institute of Science and Technology, Ulsan 44919, Korea

[3]School of Energy and Chemical Engineering, Ulsan National Institute of Science and Technology, Ulsan 44919, Korea

[4]Fritz-Haber-Institut der Max-Planck-Gesellschaft, 14195 Berlin, Germany

The properties of molecule-optical elements such as lenses or prisms based on the interaction of molecules with optical fields depend in a crucial way on the molecular quantum state and its alignment created by the optical field. However, in previous experimental studies, the effects of state-dependent alignment have never been included in estimates of the optical dipole force acting on the molecules while previous theoretical investigations took the state-dependent molecular alignment into account only implicitly. Herein, we consider the effects of molecular alignment explicitly and, to this end, introduce an effective polarizability which takes proper account of molecular alignment and is directly related to the alignment-dependent optical dipole force. We illustrate the significance of including molecular alignment in the optical dipole force by a trajectory study that compares previously used approximations with the present approach. The trajectory simulations were carried out for an ensemble of linear molecules subject to either propagating or standing-wave optical fields for a range of temperatures and laser intensities. The results demonstrate that the alignment-dependent effective polarizability can serve to provide correct estimates of the optical dipole force, on which a state-selection method applicable to nonpolar molecules could be based. We note that an analogous analysis of the forces acting on polar molecules subject to an inhomogeneous static electric field reveals a similarly strong dependence on molecular orientation.

# Zeeman spectroscopy and polarization dependence of 3S-4D two-photon transition in Na atoms

Kotaro Shijo[1], Noriki Koga[1], Hiroki Kodama[1], Yuhi Tani[1], Masaharu Mitsunaga[1]

[1] *Graduate School of Science and Technology, Kumamoto University, Kumamoto 860-8555, Japan*

We have theoretically and experimentally investigated the fine- and hyperfine-Zeeman structures of the $3S_{1/2}$-$4D_{3/2,5/2}$ two-photon transition [1] in Na vapor. The ground $3S_{1/2}$ state is split into F=1 and F=2 levels by the hyperfine coupling and the excited 4D states are split into $4D_{3/2}$ and $4D_{5/2}$ states by the spin-orbit coupling. Therefore, at small magnetic fields, 3S levels are described by $|F, m_F\rangle$ having 8 sublevels whereas 4D levels can be denoted as $|J, m_J\rangle$ having 10 levels, and we will then have 80 possible $|F, m_F\rangle$-to-$|J, m_J\rangle$ two-photon transitions. In this case, it is difficult to obtain the selection rules because the ground and the excited states have different angular momenta unlike 3S-5S two-photon transition [2]. We have thoroughly investigated all these Zeeman spectra theoretically including the signal peak position, the relative peak intensity, and the polarization dependence of each Zeeman component. Experimentally, Zeeman spectra critically depended on the polarization configuration of the excitation laser. All these polarization-dependent Zeeman spectra are compared with the theoretical predictions, and excellent agreement between the theory and the experiment was obtained.

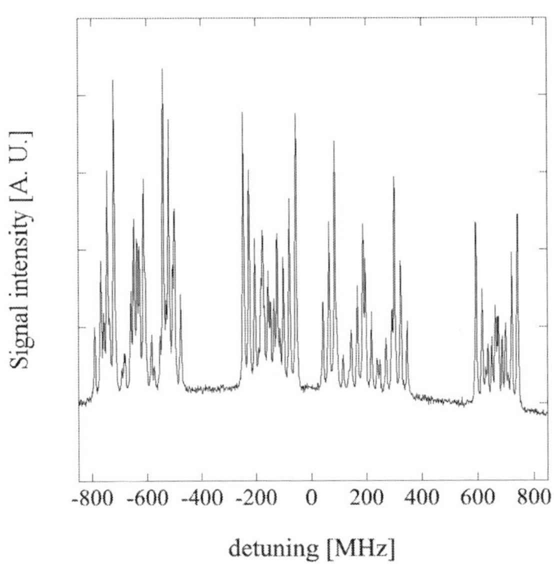


**Figure 1.** Whole spectrum of the Na two-photon 3S-4D transition. B=74 G.

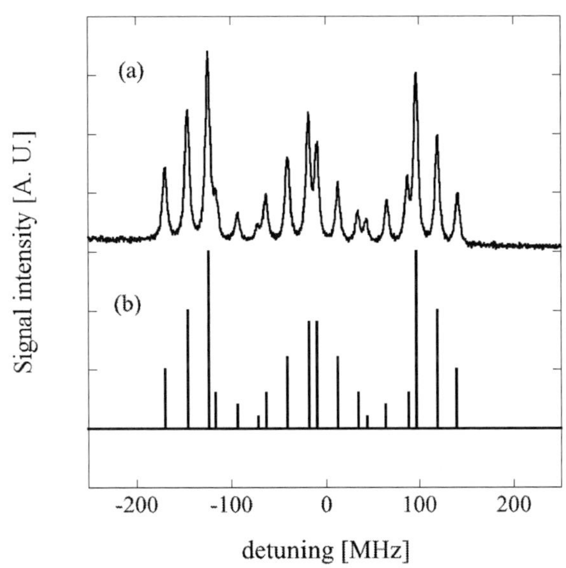


**Figure 2.** (a) Zeeman spectra of the Na $3S_{1/2}$, F=1-to-$4D_{5/2}$ transition. B = 74 G. (b) Theoretically predicted signal peak positions and relative peak intensities.

[1] W. Kaiser and C. G. B. Garrett, Phys. Rev. Lett. **7**, 229-231 (1961).

[2] N. Bloembergen, M. D. Levenson, and M. M. Salour, Phys. Rev. Lett. **32**, 867-869 (1974).

# Compare the neutron skin size data obtained from simulation and experiment of Pb-208 and Ca-48

D. Ghimire[1], B. Paudel[2], Dr. Nilam Shrestha[1,2]
[1]Tribhuvan University, Nepal
[2]Tribhuvan University, Nepal

Nuclear parameters are defined on the basis of weak charge radius and point neutron radius. Almost uncertainties in the size of neutron skin (radii difference between proton and neutron). Distribution vary according to the size and number of neutron in the nuclei. Comparing the value of weak charge radius and the point neutron radius of Pb-208 and Ca-48 calculated from simulation and experiment by using the statistical tool. We apply Monte Carlo software for the simulation and data used from the distorted wave electron scattering calculations of the Experiment. We used R and SAS for the analytical calculation of the obtained values. We find a correlation coefficient between two set of data is +0.89. Our results can easily claims the size of neutron skin can vary by number 0.138 and 0.019 due to experimental setup and statistical model respectively. Now our experiment allows us to predict the exact position of neutron and protons in the nucleus.

[1] G.Hagen, A. Ekstrom et.al Nature physics 12, 186-190 (2016)

[2] Z.Lin, C.J.Horowitz APS vol. 92,ISS.1-July 2015

[3] J.Chem. phys.107,8083 (1997)

# Towards Precision Laser Spectroscopy of Lithium 2S-3P Transition

Yao-Chin Huang and Li-Bang Wang
Department of Physics, National Tsing Hua University, Hsinchu 30013, Taiwan

The lithium 2S-3P transition has relatively narrow linewidth of only 764 kHz and is suitable for testing QED atomic calculation using laser spectroscopy. We present our progress towards laser spectroscopy of lithium 2S-3P transition in a collimated atomic beam. Laser radiation at 323 nm is produced by frequency doubling of 646 nm laser system, whose frequency is stabilized to iodine saturation spectroscopy. Laser-induced fluorescence of the lithium transition is observed as shown in Fig. 1. Our main experiment scheme is similar to the previous setup [1][2]. A microwave counter determines the beat frequency between two lasers. The absolute frequency of the lithium transition can then be determined by measuring the iodine laser frequency using an optical frequency comb.

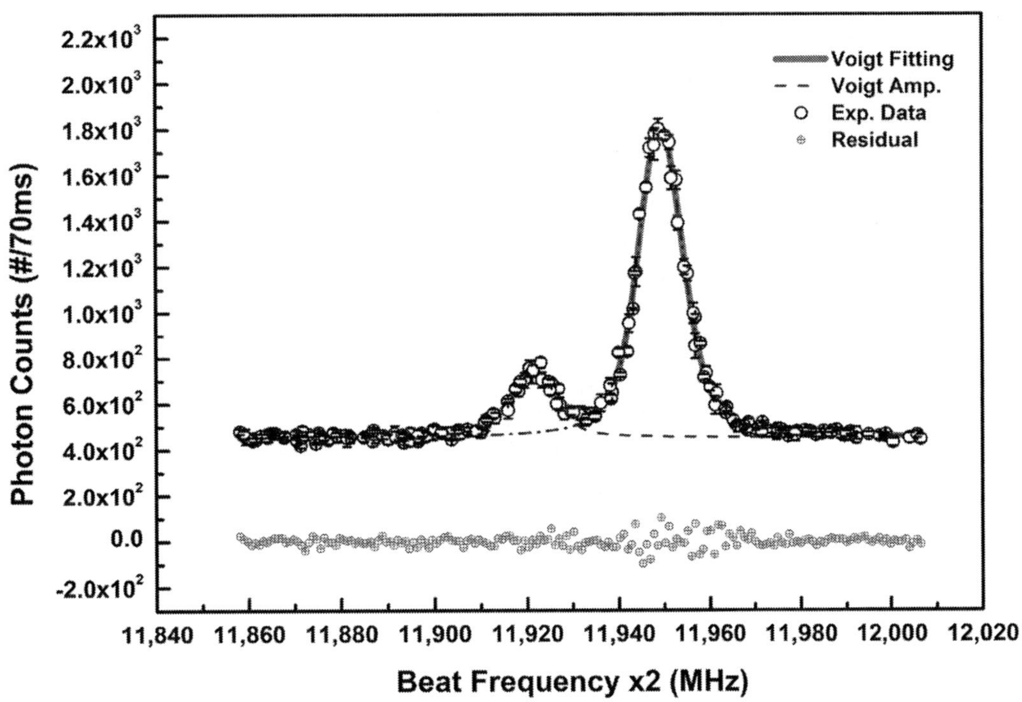


**Figure 1.** The fluorescence of the $2S_{1/2} - 3P_{1/2}$ state.

[1] Y.-C. Huang, W.-J Luo, Y.-T. Kuo and L.-B. Wang, Journal of Physics B **46**, 075004 (2013).
[2] Y.-C. Huang, H.-C. Chen, S.-E. Chen, J.-T. Shy and L.-B. Wang, Applied Optics **52**, 1448 (2013).

# Two-electron atomic systems: Calculations and properties of the angular components of the Fock expansion

Evgeny Z. Liverts[1]

[1]Racah Institute of Physics, The Hebrew University, Jerusalem 91904, Israel

The angular coefficients $\psi_{k,p}(\alpha,\theta)$ of the Fock expansion characterizing the $^1S$-state wave function of the two-electron atomic system are studied [1]. To simplify the calculation problem, the Fock recurrence relations are separated into the independent individual equations associated with definite power $j$ of the nucleus charge $Z$. Effective technique for deriving the solutions (components) $\psi_{k,p}^{(j)}(\alpha,\theta)$ of the individual equations with the simplest nonseparable and separable right-hand sides is proposed. All $j$ components of $\psi_{4,1}$ and the majority of components and subcomponents of $\psi_{3,0}$ are calculated and presented for the first time. The explicit analytic representations for the edge components $\psi_{k,0}^{(0)}$ and $\psi_{k,0}^{(k)}$ are derived for $k \leq 8$. The methods developed enable such a calculation for arbitrary $k$ [2]. The exact values of the presently determined components of the Fock coefficients at the two-particle coalescences are obtained and presented [3]. The Green functions approach was successfully applied to simplify the most complicated calculations. The boundary conditions for the components $\psi_{k,p}^{(j)}(\alpha,\theta)$ in the hyperspherical angular coordinates $\alpha$ and $\theta$ are derived from the Kato cusp conditions. The additional boundary conditions are not arising from the Kato's ones are obtained as well. The validity of the obtained boundary conditions was verified on examples of all the presently determined components. The Wolfram *Mathematica* was used intensively.

[1] E. Z. Liverts and Nir Barnea, Phys. Rev. A **92**, 042512 (2015).

[2] E. Z. Liverts, arXiv: physics.atom-ph/1604.01254

[3] E. Z. Liverts, arXiv: physics.atom-ph/1605.00222

# Experimental Study on Spectra and Lifetimes of Rare-earth Atoms

Chang-jian Dai, Shi-wei Ye, Hua Jing
School of Science, Tianjin University of Technology, Tianjin 300384, China

Many bound Rydberg states of both Eu and Sm atoms are detected with either photoionzation or electric field ionization(EFI) method[1]. The Eu $4f^7$ 6snp Rydberg states are populated with three-step laser excitation, whose radiative lifetimes are measured with a pulsed electric field by controlling the time delay relative to the laser pulses. The lifetimes for the Rydberg states with n=20-70 are ranged from 1μs to 30μs. Besides, the spectra of many Eu $4f^7$ $6p_{1/2}$ ns and $4f^7$ $6p_{1/2}$ nd autoionizing states are investigated with the so called isolated-core-excitation (ICE) technique[2]. Some disturbance in the spectra, either by the mixing or by interference effect, are observed, where the former results in some sharp features on the top of spectra[3], while the latter leads to the Fano line shape. Two examples of spectra are shown in the Fig.1.

Spectra of Sm atom in both bound and autoionization regions are found much more comp lex than the counterparts of Eu atom, due to the complex nature of its atomic structure. In other words, unlike the Eu atom, whose ground state has only a single J value, the Sm atom has a ground state with 7 fine structures. There are four fine-structures for the $6s^+$ ionization limit for the Sm atom, which is twice of that for the Eu ion. It is much more difficult for one to apply the EFI method to the Sm atom, as it is so sensitive to the time delay between the pulses of electric field and lasers. Furthermore, two polarization configurations of three lasers are selected for one to observe the impact on the spectra of Eu atom. It is found that mixing of M with several values makes the identification of spectral features with a unique J value more difficult than that of the alkaline-earth atoms[4]. Alternatively, as it has a ground state with J≠0, the difference between the spectra of $4f^7$ $6p_{1/2}$ ns autoionizing states obtained with different polarization configurations is less obvious due to the above mentioned effect of admixture.

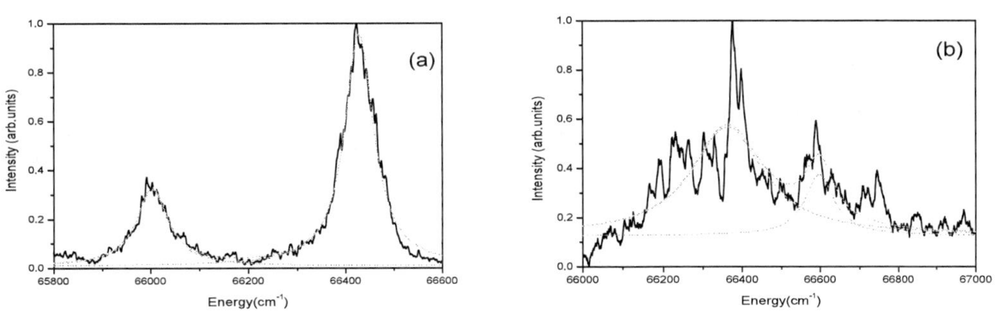


Fig. 1. Observed spectra of Eu $4f^7$ $6p_{1/2}$ nd autoionizing state with the best fit (dashed line). (a) n=8 (b) n=9

References

[1] C. J. Dai, S. Zhang, X. W. Shu, Journal of Quant. Spectro. and Radiative Transfer **53,** 179(1995).

[2] C. J. Dai, G. W. Schinn, T. F. Gallagher, Physical Reviews **A42,** 223(1990).

[3] X. Wang, L. Shen, C. J. Dai, Journal of Physics, **B45,** 165001(2012).

[4] C. J. Dai and J. Lu, Journal of Physics **B29,** 2473(1996).

# Cavity-enhanced Raman Microscopy of Individual Carbon Nanotubes

Thomas Hümmer[1,2], Jonathan Noe[2,3], Matthias S. Hofmann[2,3], Theodor W. Hänsch[1,2], Alexander Högele[2,3], David Hunger[1,2]

[1] Fakultät für Physik, Ludwig-Maximilians-Universität, Schellingstraße 4, 80799 München, Germany

[2]Max-Planck-Institut für Quantenoptik, Hans-Kopfermann-Str. 1, 85748 Garching, Germany80799 München, Germany

[3]Center for NanoScience (CeNS), Ludwig-Maximilians-Universität, Schellingstraße 4, 80799 München, Germany

We use a tunable high-finesse optical microcavity[1] to demonstrate Purcell enhancement of Raman scattering in combination with high-resolution scanning-cavity imaging[2]. We detect cavity-enhanced Raman spectra[3] of individual carbon nanotubes and co-localize measurements with cavity-enhanced absorption microscopy. We observe a 320-times enhanced Raman scattering spectral density and an effective Purcell factor of 6.2, together with a collection efficiency of 60%. We expand the technique to hyperspectral imaging, where we can deduce information such as the diameter and the metallic or semiconducting character of the nanotubes. Potential for significantly higher enhancement, inherent spectral filtering and absence of intrinsic background render the technique a promising tool for molecular imaging. Furthermore, cavity-enhanced Raman transitions involving localized excitons could potentially be used for gaining quantum control over nanomechanical motion and open a route for molecular cavity optomechanics.

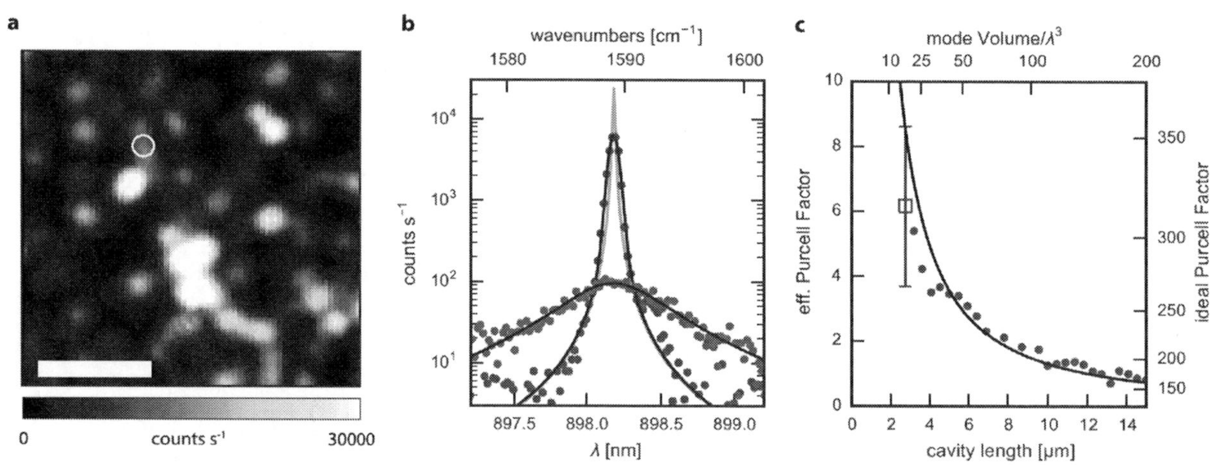


**Figure 1.** a, Cavity enhanced Raman scan showing the strength of the G+ Raman feature of carbon nanotubes (scale bar 10μm). b, Comparison of confocally collected (blue data points with Lorentzian fit) and cavity-enhanced detected spectra (red data points with Voigt fit) of the G+Raman feature of a carbon nanotube. c, Scaling of the integrated cavity-enhanced Raman intensity (red datapoints) as a function of the effective length of the cavity. The solid line is the theoretically predicted ideal and effective Purcell factor.

[1] Hunger et al., New J. Phys. **12**, 065038 (2010).
[2] Mader et al., Nat. Commun. **6**, 7249 (2015).
[3] Hümmer et al., Nat. Commun. (in press), arXiv:1508.06810 (2015).

# MCDHF calculations of He-like ions for rare gases

Dhia Elhak Salhi[1,2], Haikel Jelassi[1]

[1]Laboratoire de Recherche Energie et Matière pour les Développement des Sciences Nucléaires & Direction de la Recherche sur l'Energie et la Matière
National Centre for Nuclear Science and Technology Sidi Thabet Technopark 2020 Ariana, Tunisia.

[2] Faculty of Sciences of Tunis University Tunis El Manar Tunis, Tunisia.

Atomic data such as energy levels, oscillator strength (f-values) and radiative rates (A-values) are required for the modeling of plasmas, in astrophysical media and nuclear fusion Tokamak. Among species, light atoms are for great importance in plasma. Its transition data can be used in the determination of element abundances and plasma diagnostics [1,2]. Moreover, the ITER need for more accurate atomic data for a wide range of ions has become even greater with the ongoing ITER project. However, He-like ions are the most important atomic species in high-temperature astrophysical and laboratory sources.
In the present work, we report preliminary calculations of energy levels among the levels of $1s^2$ and 1snl (n=2-6) configurations, weighted oscillator strength and electric-dipole transitions E1 for He-like neon (Ne IX). We treat also the case for He-like argon (Ar XVII) by calculating the transitions probabilities for all types of transitions probabilities (E1, E2, M1 and M2). The GRASP2K (general-purpose relativistic atomic structure package) was adopted for calculating energy levels, weighted oscillator strength and radiative rates among the lowest 71 levels. Comparisons have been made with similar data obtained with FAC (flexible atomic code) to assess the accuracy of the results. We have also considered relativistic effects by incorporating quantum electrodynamics (QED) and Breit corrections. The present results are compared with other available experimental and theoretical data and good agreement has been found which confirms the reliability of our results.

[1] D. L. Lennon, P. L. Dufton, A. Hibbert and A. E. Kingston, Astrophys. J. **294**, 200 (1985).

[2] F. P. Keenan, P. L. Dufton, W. A. Feibelman, K. L. Bell, A. Hibbert and R. P. Stafford, Astrophys. J. **423**, 882 (1994).

# Transition from doubly excited states to singly excited states in LiH

Hyun-Jin Kim[1], Chun-Woo Lee[2]

[1]Department of Energy Systems Research, Ajou University, Suwon, 16499, Korea

[2]Department of Chemistry, Ajou University, Suwon 16499, Korea

The electronic structure of LiH is characterized by the presence of low-lying doubly excited states only at small *R*, which is a typical behavior in four-electron molecular systems unobserved in 1-3 electron molecular systems. Transition from the doubly-excited configuration dominant states to the singly-excited configuration dominant states is found to take place in a narrow range of *R*. Rydberg series at small *R* and in the spectroscopically observable region [1] are different because of this different nature of two regions. The transition of the doubly excited $2p^2\ ^1D_\sigma$ and $2p^2\ ^1D_\delta$ acting intruders at small *R* into singly excited Rydberg states in the spectral region is explored with diabatization by Smith's method [2] and by examining the dominant configurations (see Figure 1).

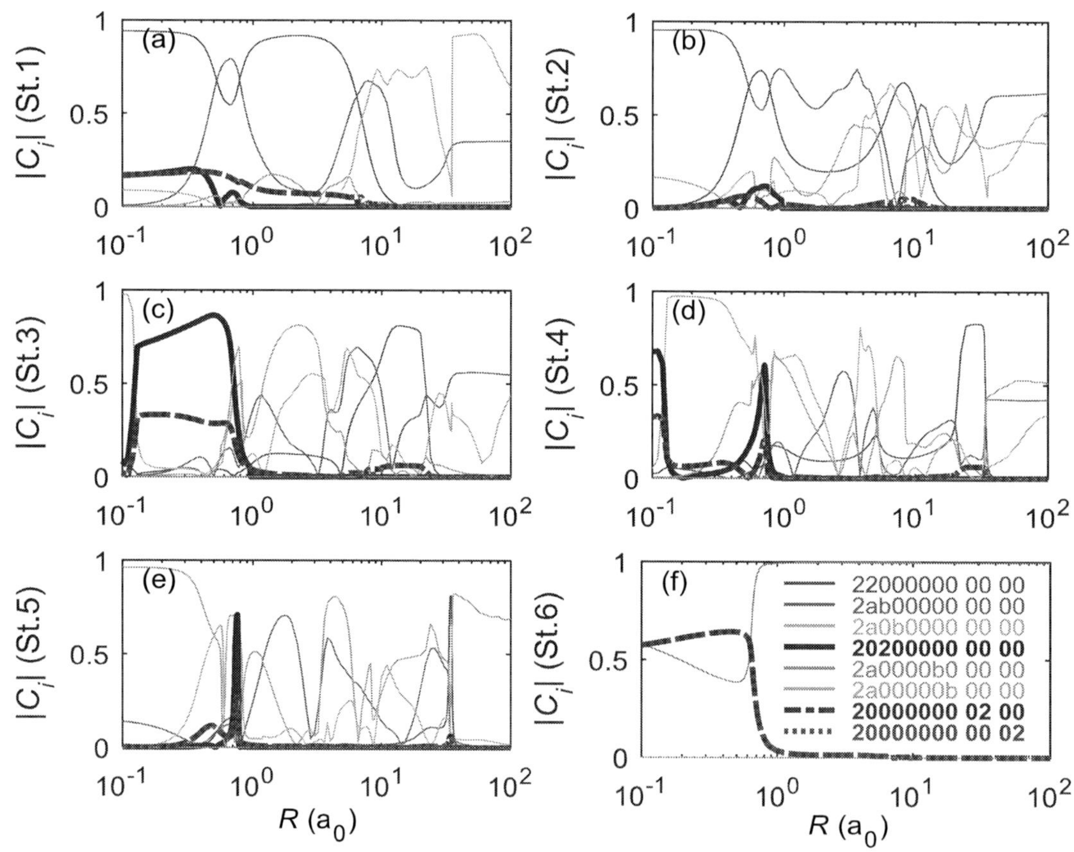


**Figure 1**. The absolute value graphs of expansion coefficients of the six lowest states into the configurations labelled in the legend in (f). Doubly excited configurations are drawn in thick lines.

[1] Y. Gim and C.-W. Lee, J. Chem. Phys. **141**, 144313 (2014).

[2] F. T. Smith, Phys. Rev. **179**, 111 (1969).

# Study of nonadiabatic couplings in $HeH^+$ and HeH

Hyun-Jin Kim[1], Chun-Woo Lee[2]
[1]Department of Energy Systems Research, Ajou University, Suwon, 16499, Korea
[2]Department of Chemistry, Ajou University, Suwon 16499, Korea

We made calculations of the non-adiabatic coupling matrix elements (NACMEs) over a wide range of $R$ for $HeH^+$. The NACMEs for HeH are already available [1]. NACMEs are known to depend on the choice of coordinate systems, known as the electron-translation-factor problem in atomic collisions [2]. Their calculation needs special care besides difficulty in obtaining good quality values, in particular at small $R$. We performed the double-checking on the calculated NACMEs with the Butler's formula [3]. With good quality NACMEs (Figure 1), diabatization by Smith's method is performed on these molecules. Interesting reinterpretation to the Fermi model [4] in terms of avoided crossing interactions is obtained. Correlations between the UA and SA limits are studied.

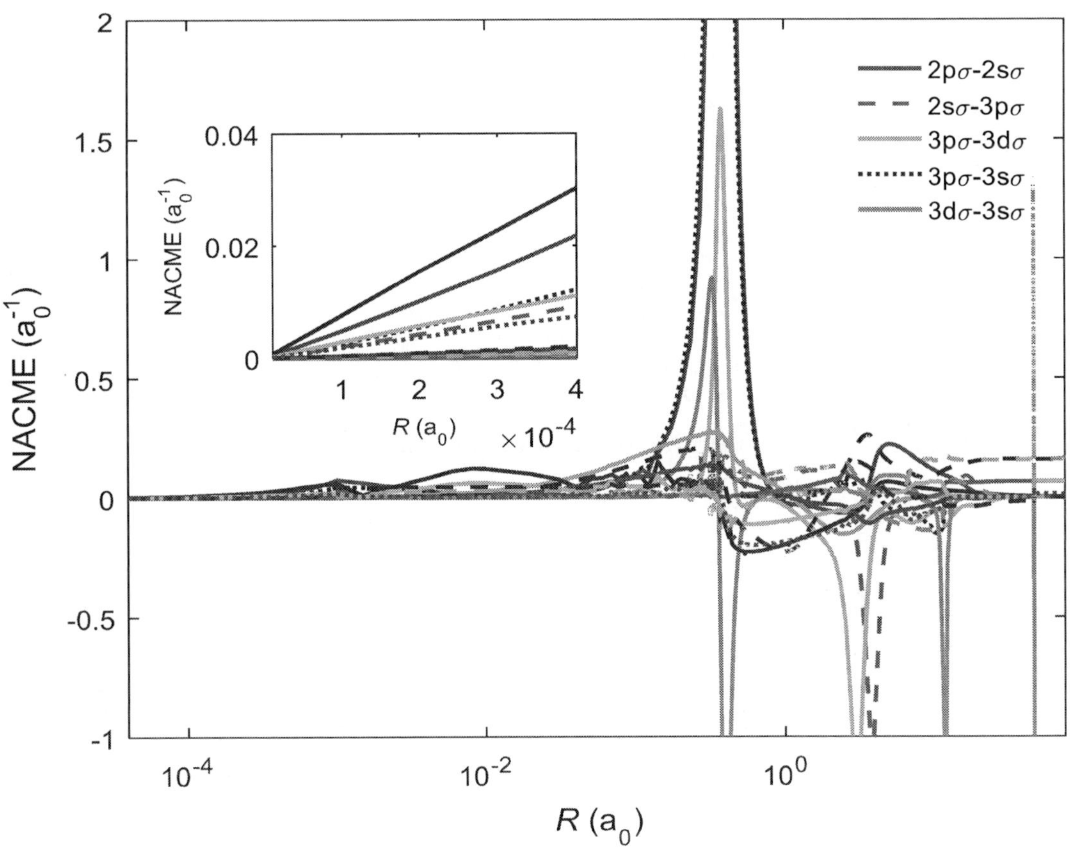


**Figure 1.** Non-adiabatic coupling matrix element curves of $HeH^+$.

[1] H.-J. Kim, C.-W. Lee, Bull. Korean Chem. Soc. **37**, 469 (2016).
[2] D. R. Bates and R. McCarroll, Proc. R. Soc. London, Ser. A **245**, 175 (1958).
[3] S. E. Butler, Phys. Rev. A **23**, 1 (1981).
[4] E. Fermi, Nuovo Cimento **11**, 157 (1934).

# Fate of ionic states and corridors to the doubly excited $(2p\sigma)^2$ F $^1\Sigma_g^+$ state provided by the avoided-crossings in $H_2$

Hyun-Jin Kim[1], Chun-Woo Lee[2]

[1] Department of Energy Systems Research, Ajou University, Suwon, 16499, Korea

[2] Department of Chemistry, Ajou University, Suwon 16499, Korea

$H^-(1s^2) + H^+$ has two ionic states corresponding to the *gerade* and *ungerade* combinations. Evolution of two ionic states from the united atom (UA) to the separated atoms (SA) limits is very exciting story and studied by examining the dominant configurations and non-adiabatic coupling matrix element curves. The *ungerade* counterpart $\sigma_u 1s$ $^1\Sigma_u^+$ of the $2p\sigma$ state does not exist in the SA limit due to the Pauli-exclusion principle. It is correlated to the *ungerade* ionic state, instead. On the other hand, the *gerade* ionic state has no counterpart in the singly excited states in the UA limit and correlated to the doubly excited $(2p\sigma)^2$ F $^1\Sigma_g^+$ state [1,2] causing extensive avoided crossing interactions with all the singly excited states lying on the way to the F state (see Figure 1). Corridors are formed by these extensive avoided crossings.

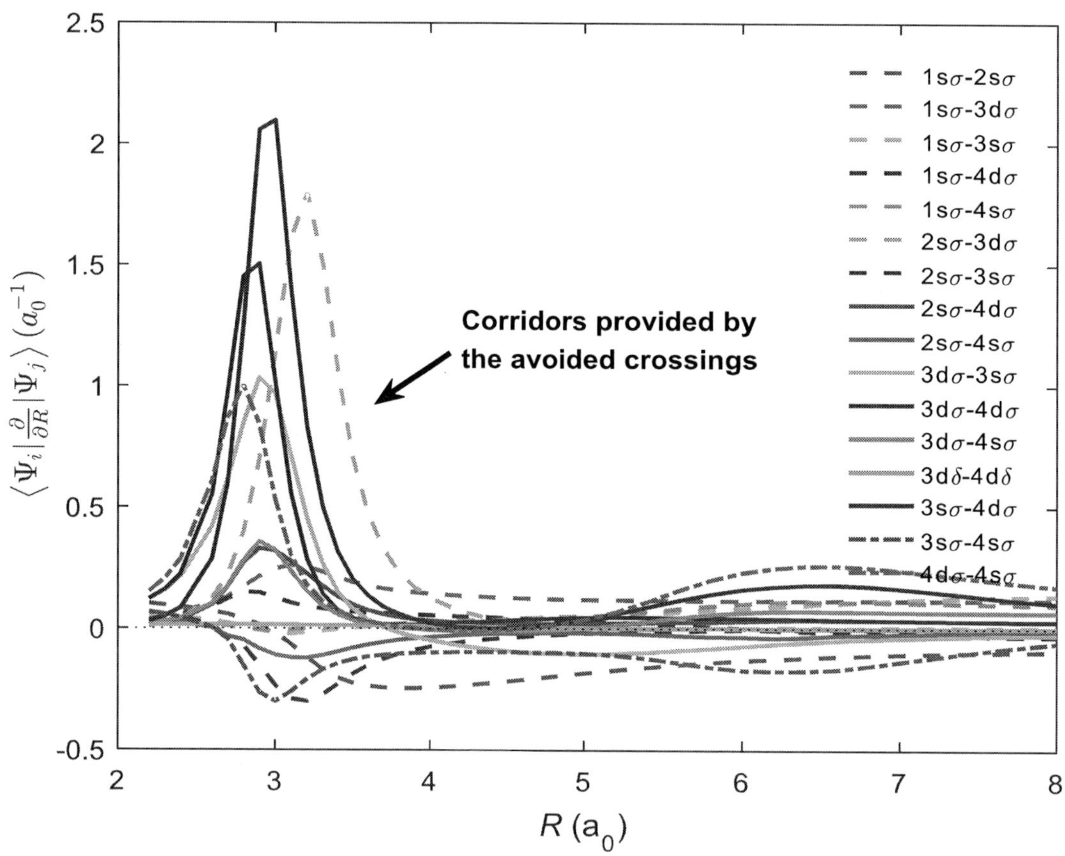


**Figure 1.** Radial non-adiabatic coupling matrix elements of $H_2$.

[1] E. R. Davidson, J. Chem. Phys. **35**, 1189 (1961).

[2] T. E. Sharp, At. Data Nucl. Data Tables **2**, 119 (1970).

# Time-resolved spectroscopy for $5s'^4D_{7/2}$ state transition from electron-ion recombination in femtosecond laser-produced copper plasma

Haiying Song, Peng Gu, Hui Li, Shibing Liu
Strong-field and Ultrafast Photonics Lab, Institute of Laser Engineering
Beijing University of Technology, Beijing 100124, China
*corresponding e-mail: hysong@bjut.edu.cn

The selectable time-resolved probe methods have been widely used for the investigation of various ultrafast phenomena, such as the time-resolved shadowgraphs coming of the material ejection[1], the electron transition process and population in the excited states by probing the time-resolved atoms/molecule emission spectrum[2], the ultrafast dynamics of electron transfer from the femtosecond stimulated Raman spectroscopy[3], and etc.

In this work, we have studied copper by orthogonal femtosecond (fs) laser pulse (70 fs, 800 nm, 6.37 $*10^{14}$ W/cm$^2$) at atmospheric pressure by the time-resolved ultrafast spectrum method, the time resolution in the present experiment was about 2 ns. The transient dynamics for $5s'^4D_{7/2}$ state transition to $4p'^4D^0_{7/2}$ (465.11 nm) and $4p'^4F^0_{9/2}$ (529.25 nm) states in Cu atoms, resulted from the electron-ion recombination process in the fs laser-produced Cu-plasma. The results show that the lifetime of two spectral of excited state ($5s'^4D_{7/2}$) linearly fit with the increased laser intensity and the peak intensity of two spectral of excited state ($5s'^4D_{7/2}$) quadratic nonlinearly fit with the increased laser intensity. The ratio of the peak intensity of two spectral tend to 3.5, very closed to the theoretical value 3.49.The transition process of CuI 465.11 nm and CuII 529.25 nm having a correlation and exists in sequence, the excited state ($5s'^4D_{7/2}$) began to have populations after 7.9 ns relaxed. At the same time the transition channel I of $5s'^4D_{7/2} \rightarrow 4p'^4D^0_{7/2}$ started. With further relaxed of excited state, another transition channel II of $5s'^4D_{7/2} \rightarrow 4p'^4D^0_{7/2}$ began after 10 ns. When the time 260 ns<t<290 ns, the intensity of transition channel I began decreased, but intensity of the transition channel II still increased. When the time t> 290 ns, the intensity of transition channel I start decreased. With the reduced of population of excited stated, after approximately 2.4 μs the transition channel II first stop occurring and the transition channel I also stop at 100 ns later.

It can be concluded that the lifetime of $5s'^4D_{7/2}$ population state and its maximum population probability are all associated with the applied laser intensity and the population springs from the electrons in the continuous states by the process of electron-ion recombination, based on which a significant fact is confirmed that the rate of random transition from continuous state to bound state is faster than the rate of selective transition between the bound states.

[1] N. Zhang, X. Zhu, J. Yang, PRL **99**, 167602 (2007).
[2] H.C. Joshi, A. Kumar, R.K. Singh, V. Prahlad, Spectrochimica Acta Part B: Atomic Spectroscopy **65**, 415 (2010).
[3] D. P. Hoffman, O. P. Lee, J. E. Millstone, J. Phys. Chem. C **117**, 6990 (2013).

# Selective Photodissociation of Formaldehyde using Fiber Laser System

Yonghee Kim, Min-Kook Nah, Yong-Ho Cha, Lim Lee, Hyunmin Park, Do-Young Jeong
Quantum Optics Division, Korea Atomic Energy Research Institute,
989-111 Daedeok-daero, Yuseong-gu, Daejeon, 305-353, Korea

It is well-known that the formaldehyde molecule($CH_2O$) photodissociates by the ultraviolet light. Especially, at 350 nm near-ultraviolet range, formaldehyde is decomposed into hydrogen and carbon monoxide with almost unity quantum efficiency [1]. The ro-vibrational spectrum of formaldehyde which contains different isotopes($^{12}C$, $^{13}C$, $^{14}C$, H, D, T, $^{16}O$, $^{17}O$, and $^{18}O$) are different from each other, we can photodissociate formaldehyde of certain combination of isotopes selectively. In 1970-80, laser isotope separation of carbon, hydrogen and oxygen based on formaldehyde photodissociation was reported [2,3], but the laser system was too complicated and expensive for real applications.

Nowadays, high power, narrow linewidth, wavelength tunable UV laser can be possible because of dramatic development of laser technologies, laser isotope separation will be feasible compare to conventional distillation or centrifugal method. We demonstrate the carbon-isotope selective photodissociation of formaldehyde molecules using fiber laser system. The fiber laser system consists of MOPA configuration multi-stage Yb-doped fiber amplifiers and nonlinear frequency conversion stages. We can obtain about 10-20 W of 352 nm UV lasers. Using this laser system, we have realized the selective photodissociation of $^{13}C$ and $^{12}C$-formaldehyde with separation factor about 200-1000 starting from natural formaldehyde. We expect that the results will be applied for $^{14}C$ waste problem in nuclear industry and high purity $^{12}C$ for source of NV-center diamond production.

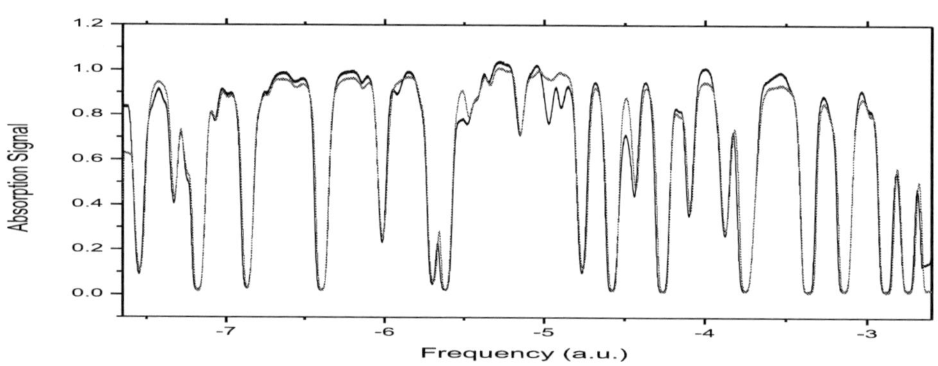


**Figure 1.** Spectrum of formaldehyde at near UV range before(black) and after(red) photo-dissociation.

[1] P. L. Houston and C. B. Moore, J. Chem. Phys. **65**. 757 (1976).
[2] J. Marling, J. Chem. Phys. **66**, 4200 (1977).
[3] L. Mannik and S. K. Brown, Appl. Phys. B **37**, 79 (1985).

# ppb-level detection of $^{14}CO_2$ isotopologue using an OA-ICOS

Kwang-Hoon Ko, Yonghee Kim, Lim Lee, Sungman Lee, Hyunmin Park, Do-Young Jeong
Quantum Optics Division, Korea Atomic Energy Research Institute, 1045 Daedeok-daero, Yuseong-gu, Daejeon, 305-353 KOREA

$^{14}CO_2$ is one of gaseous effluents which are released from nuclear power plants to the environment [1]. Carbon is one of the main constituents of the living organics, and therefore the monitoring of the $^{14}C$ emission has been strengthened so far. Carbon dioxide in nature consists of 18 stable isotopologues and one radio-isotopologue $^{14}CO_2$ whose abundance is approximately $10^{-12}$. In nuclear industry, the permitted amount of $^{14}CO_2$ for the emission is limited to $10^4$ Bq/m$^3$, and the corresponding isotope ratio of $^{14}CO_2$ is approximately $10^{-9}$.

Typically, in nuclear industry, $^{14}CO_2$ is measured by the radiation detection method. However, the periodical exchanges of chemicals for capturing carbon dioxide and the long process time are required for the detection. These make it difficult to realize the real-time and automated detection. Recently, the sensitive detection method of gaseous molecules has been developed successfully by using a spectroscopic method in mid-infrared region. The method employed the cavity ring down spectroscopy for the $^{14}CO_2$ detection is achieved the detection less than 50 parts per trillion, but this technique requires complicated electronic devices and sensitive optical adjustment [2-3].

In this paper, we demonstrated the $^{14}CO_2$ detection by using an OA-ICOS (Integrated Cavity Output Spectroscopy), which requires less sensitive adjustment and simple electronic/optical configuration. By using the schematic diagram shown in Fig. 1, the spectroscopy is performed at fundamental ro-vibration lines in mid-infrared region. We present the results on the mid-infrared beam generation by using the wave-mixing technique and ~ppb level detection of $CO_2$ isotopologue. We will further discuss the future experiments and applications.

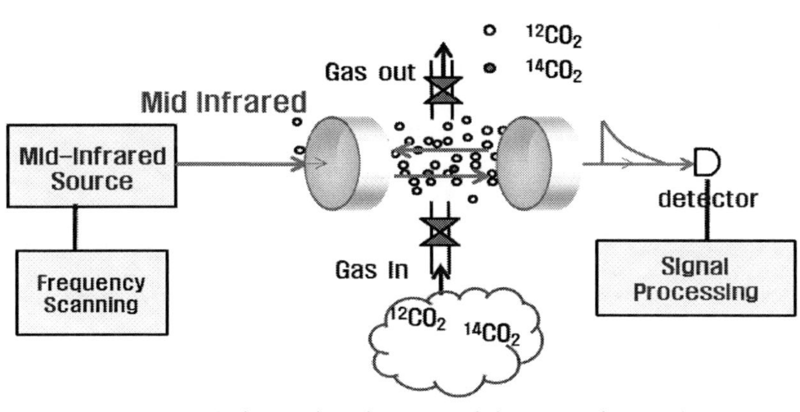


**Figure 1.** Schematic Diagram of the Experimental Setup.

[1] W. Sohn, D.-W. Kang, W.-S. Kim, J. Nucl. Sci. Technol. **40**, 604 (2003).

[2] G. Genoud *et al.*, Opt. Lett. **40**, 1342 (2015).

[3] I. Galli *et al.*, Phys. Rev. Lett. **107**, 270802 (2011).

# Doppler-free spectroscopy of metastable Sr atoms using a hollow cathode lamp

Yusuke Hayakawa, Takatoshi Aoki, Yoshio Torii
*Institute of Physics, The University of Tokyo, Tokyo 153-8902, Japan*

We report on the demonstration of Doppler-free spectroscopy of metastable Sr atoms using a hollow cathode lamp for the $5s5p\ ^3P_2 - 5s5d\ ^3D_2$ (497 nm), $5s5p\ ^3P_2 - 5s5d\ ^3D_3$ (496 nm), $5s5p\ ^3P_1 - 5s6s\ ^3S_1$ (688 nm), and $5s5p\ ^3P_0 - 5s6s\ ^3S_1$ (679 nm) transitions, which are relevant to the repumping processes in laser cooling of Sr atoms. We employed a special Sr hollow cathode lamp filled with a mixture of neon (0.5 Torr) and xenon (0.5 Torr) as a buffer gas to reduce the total buffer-gas pressure and suppress the effect of velocity changing collisions [1]. Depending on the nature of the transitions, frequency modulation or modulation transfer [2] technique was used for Doppler-free spectroscopy. We have proved that all the $5s5p\ ^3P_j (j = 0,1,2)$ metastable states can have sufficient populations inside the hollow cathode to generate error signals for laser frequency stabilization for the transitions starting from these metastable states. The method explored in this work would significantly facilitate laser cooling of Sr and is readily applicable to spectroscopy of other group-II metastable atoms, such as Ca, Er, and Cd.

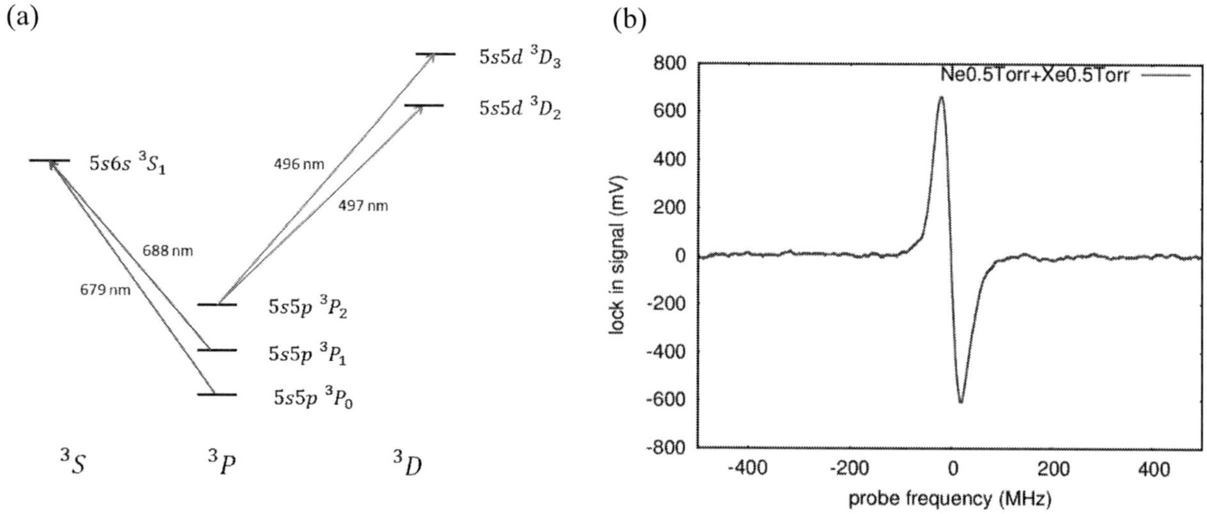


Fig. 1 (a) Relevant energy levels of Sr. Transitions indicated by arrows are investigated in this work. (b) Background-free error signal for the $^3P_2 - {}^3D_3$ transition obtained by using a modulation transfer technique in a hollow cathode lamp.

[1] R. L. Cavasso-Filho, A. Mirage, A. Scalabrin, D. Pereira, and F. C. Cruz, J. Opt. Soc. Am. B **18**, 1922-1927 (2001).

[2] J. H. Shirley, Opt. Lett. **7**, 537-539 (1982).

# Development of a novel deep-UV laser system for laser cooling of Hg atom

R-C. Zhao[1,2,3], X-H. Fu[1,2,3], J-M. Hu[1,2,3], L Zhang[1,2,3], J-F. Sun[1,2], Z. Xu[1,2,3,4], Y. Feng[1,2,3,4]

[1]Key Laboratory of Quantum Optics, Chinese Academy of Sciences, Shanghai 201800, China

[2]Shanghai Institute of Optics and Fine Mechanics, Chinese Academy of Sciences, Shanghai 201800, China

[3]University of Chinese Academy of Sciences, Beijing 100049, China

[4]Center of Cold atom physics, Chinese Academy of Sciences, Shanghai 201800, China

Mercury atom is one of good candidates for optical lattice clock, due to its insensitive to blackbody radiation [1]. It is also important for the testing on the variation of natural physical constants for its large atomic number. Laser cooling of mercury atoms requires high power deep-UV tunable laser source at 253.7 nm, which is difficult to obtain. So far, the 253.7 nm lasers are reported by frequency quadrupling 1014.8 nm lasers based on Yb-doped crystal [2-3], semiconductor materials [1], or Yb-doped fiber with liquid-nitrogen cooling [4-5]. Here we give a presentation of our novel deep-UV laser system, including the 1014.8 nm external cavity diode laser (ECDL), fiber laser amplifier working at room temperature and frequency quadrupling system.

The 1014.8 nm single-mode laser is a homemade ECDL. Its output power is 50 mW, hope-free tuning range is 4 GHz, linewidth is 120 kHz, and amplified spontaneous emission (ASE) suppression is 54 dB. The ASE suppression can be further decreased to over 70 dB by a bandpass filter before incident to the fiber amplifier. Finally, about 20 mW power is injected into fiber laser amplifier.

The fiber laser amplifier uses Yb-doped polarization-maintaining double-clad single-mode fibers as gain media. Up to 8 W and 19.3 W laser power are achieved with single-stage and two-stage amplifiers, respectively [6-7]. By two commercial frequency doubling cavities, output power of 1 W at 507.4 nm and 75 mW at 253.7 nm are achieved with laser input of 4 W at 1014.8 nm [6]. The ASE suppression of fiber amplifier could reach 45 dB even though working at room temperature. In addition, the following two frequency doubling cavities will lead a further cut of ASE suppression.

The frequency quadrupling system is composed of two high efficient frequency doubling cavities, which are both bow-tie cavities, with a LBO crystal and a BBO crystal respectively. Up to 3W second harmonic (507.4 nm) is achieved with 4.9 W fundamental laser (1014.8 nm) by the first cavity, which has cavity finesse of 150 and mode-matching efficiency of 98%. Up to 500 mW forth harmonic (253.7 nm) is achieved with 2 W input power of second harmonic (507.4 nm) through the second cavity.

[1] H. Hachisu, K. Miyagishi, S. G. Porsev, *et al.*, Phys. Rev. Lett. **100**, 053001 (2008).

[2] M.Scheid, F. Markert, J. Walz, *et al.*, Opt. Lett. **32**, 955-957 (2007).

[3] M. Petersen, R. Chicireanu, S. T. Jacobs, *et al.*, phys. Rev. Lett.**101**, 183004 (2008).

[4] A. Seifert, M. Sinther, T. Walther, *et al.*, Appl. Opt. **45**, 7908-7911 (2006).

[5] R. Steinborn, A. Koglbauer, P. Bachor, *et al.*, Opt. Express. **21**, 22693-22698 (2013).

[6] J-M. Hu, L. Zhang, H-L. Liu, *et al.*, Opt. Express **21**, 30958-63 (2013).

[7] J-M. Hu, L. Zhang, H-L. Liu, *et al.*, Appl. Opt. **53**, 4972-4977 (2014).

# Stepwise transition of atomic distribution in a moving optical lattice

Kyeong Ock Chong[1], Jung-Ryul Kim[1], Jinuk Kim[1], Myounggyu Hwang[1] and Kyungwon An[1]
[1]Department of Physics and Astronomy, Seoul National University, Seoul, Korea

We investigated atomic localization as well as tunneling effect in a moving optical lattice of a phase-stabilized magneto-optical trap (MOT) by observing the spectrum of resonant fluorescence. A 3-dimentional optical lattice is generated in a phase-stabilized MOT [1]. By changing the detuning of a trap laser, the lattice is made to move with a velocity proportional to the detuning. The motion of the atoms is determined by sub-Doppler cooling as well as dipole forces in the moving lattice. We observed that atoms undergo a transition from localized states in the lattice to delocalized states as the speed of the lattice increases. In addition, the population of delocalized states changes stepwise due to the existence of vibrational states around the local minima of the lattice potential. In order to explain the experimental results, a rate-equation model of atom populations has been introduced. The model describes quantum mechanical transitions among the bound states of lattice potential and the thermal states of MOT. Our experimental data is well described the rate-equation model.

[1] W. Kim *et al.*, Nano Letters **11**, 719 (2011).

# Cooperative motion and cooling with superradiant and subradiant atoms

Er-Siang Tang, Guin-Dar Lin
CQSE and Department of Physics, National Taiwan University, Taiwan

The technique of Doppler laser cooling has become ubiquitous in all sub-fields of atomic and molecular physics and quantum optics. In this work we explicitly calculate the optical forces on atoms that are cooperatively correlated. By varying the inter-atom spacing, the dipole-dipole interaction leads to different degrees of mixtures of superradiant and subradiant states, and introduces more subtleties to both the center-of-mass and relative coordinates. In the optical molasses setup, we find that, along the axis of two atoms, there present mechanically stable and unstable regions alternatively as the separation increases. We also investigate the classical trajectories and map out the stability diagram.

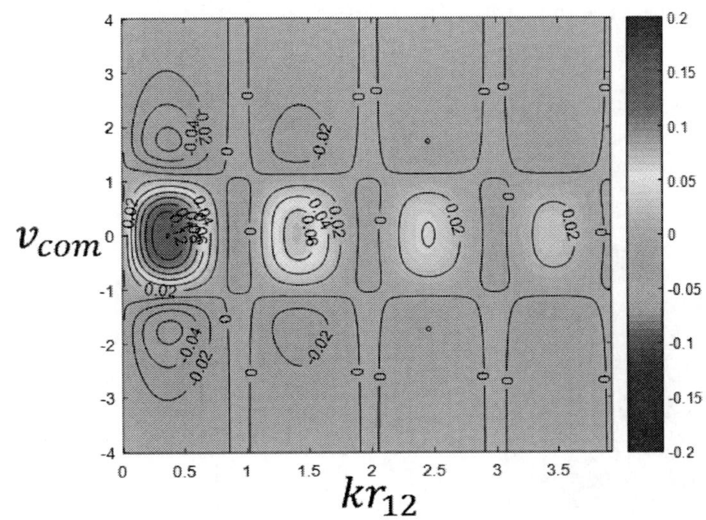


**Figure 1.** Force diagram (for the relative coordinate) for various center-of-mass velocities and distances. The color shows the relative force in units of $\gamma^{-1}\hbar k$ between two atoms with separation $r_{12}$.

# Tailored Optical Potentials for Atomtronics

Filip Auksztol[1], Tin Nghia Nguyen[2], Paul Condylis[1], Koon Siang Gan[2], Rainer Dumke[1,2]
[1] Centre for Quantum Technologies, National University of Singapore, Singapore
[2] Nanyang Technological University, Singapore

Fully tailored optical potentials with finely controlled parameters are essential to the realization of modern atomtronic devices. We present our work using Spatial Light Modulators (SLM) and Digital Micromirror Devices (DMD) to produce such potentials in both static and dynamic domains. In particular, we demonstrate the experimental realization of a ring lattice system with a weak link using the SLM device in which a live feedback loop corrects for the optical system imperfections and ensures smoothness of the potential (see the comparison in Figure 1). We also show the technological barriers present when moving into the dynamically evolving trapping potentials and our ideas to overcome them. The lattices obtained will be used to trap the Bose Einstein Condensate (BEC) of Rb-87 atoms in later stages of our experiment and the theoretical groundwork for that project has been lined out in our recent paper [1].

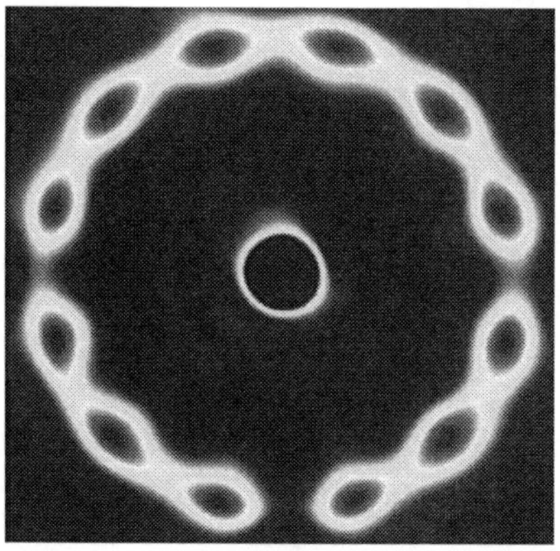

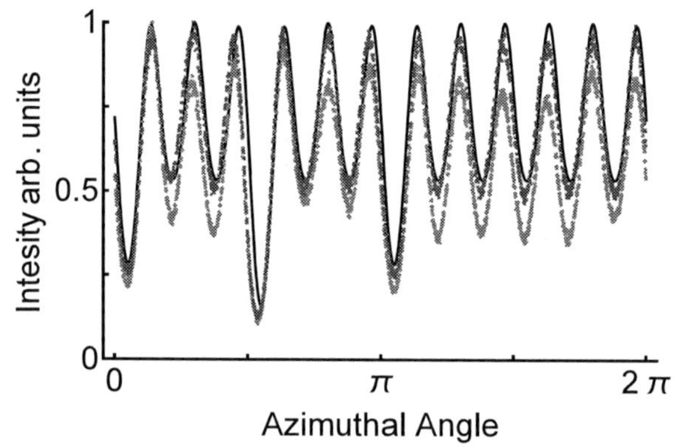


**Figure 1.** Solid line in the graph indicates the target profile, red points show the SLM output after the first iteration and the blue ones after the 5th iteration.

[1] D. Aghamalyan, N.T. Nguyen, F. Auksztol, K. S. Gan, M. Martinez Valado, P. C. Condylis, L.-C. Kwek, R. Dumke, and L. Amico , arxiv: 1512.08376 [quant-ph] (2016)

# Reformation of holographic single-atom arrays for 3D quantum system architectures

Woojun Lee, Hyosub Kim, Jaewook Ahn
Department of Physics, KAIST, Daejeon 305-701, Korea

Formulating and architecting single-atom arrays are an essential cornerstone for neutral atom-based quantum computing and quantum simulation. One of the most universal ways to generate single-atom arrays with a high degree of freedom in positioning is holographic trapping, in which far-detuned trap beams programmed by spatial light modulators provide the trap array configuration [1, 2]. We show that the positioning freedom with this kind of scheme can be even extended to three dimensions [3]. With the first images of the moving atoms in 3D trap arrays, we demonstrate transportation of individual atoms in the full 3D degree of freedom. It is hoped that this method of reforming single-atom arrays in arbitrary way becomes useful for high dimensional architectures for quantum many-body physics.

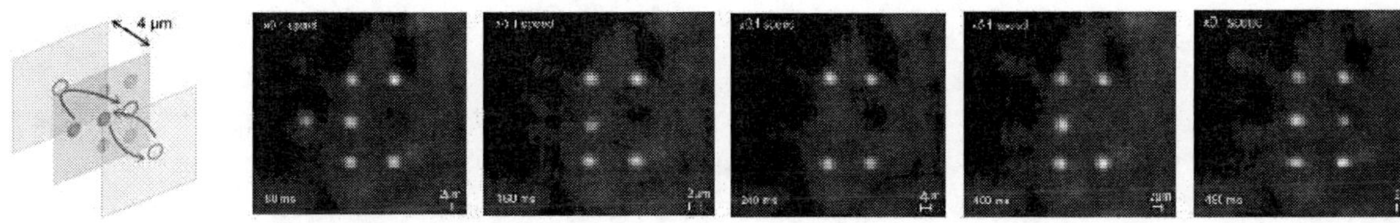


**Figure 1.** Demonstration of three-dimensional reconfiguration of a holographic single atom array. The atom in the left side contours the neighboring atom out of plane while the neighboring atom travels and comes back in the opposite direction.

[1] F. Nogrette, H. Labuhn, S. Ravets, D. Barredo, L. Béguin, A. Vernier, T. Lahaye, and A. Browaeys, Physical Review X **4** (2), 021034 (2014).
[2] Hyosub Kim, Woojun Lee, Han-gyeol Lee, Hanlae Jo, Yunheung Song, Jaewook Ahn, http://arxiv.org/abs/1601.03833.
[3] Woojun Lee, Hyosub Kim, and Jaewook Ahn, Optics Express **24** (9), 9816 (2016).

# Sub-Micron Period Magnetic Lattices for Ultracold Atoms

Y. Wang[1], T. Tran[1], I. Herrera[2], P. Surendran[1], S. Jose[1], S. Whitlock[3], R. McLean[1], A. Sidorov[1], P. Hannaford[1]

[1]Centre for Quantum and Optical Science, Swinburne University of Technology, Melbourne, Australia 3122

[2]Dodd-Walls Centre for Photonic and Quantum Technologies, Department of Physics, University of Auckland, Private Bag 92019, Auckland, New Zealand

[3]Physikalisches Institut, Universität Heidelberg, Im Neuenheimer Feld 226, 69120 Heidelberg, Germany

Magnetic lattices produced by permanent magnetic microstructures patterned on a magnetic film provide a promising complementary tool to optical lattices that allow a high degree of design flexibility, low technical noise and state-selective trapping of atoms. We report here progress towards using ultracold atoms trapped in magnetic lattices as a platform for simulating quantum many-body phenomena.

We have realized Bose-Einstein condensates of $^{87}$Rb |F=1, $m_F$=-1> atoms trapped in multiple sites of a 1D 10 μm-period magnetic lattice [1]. Small site-to-site variations of the trap bottom, atom temperature, condensation fraction and chemical potential indicate that the magnetic lattice is remarkably uniform [2]. Successful loading of ultracold atoms into 2D 10 μm-period square and triangular magnetic lattices has recently been reported by the Amsterdam group [3].

To achieve significant tunnelling and hence interaction between atoms on neighbouring sites, lattices with sub-micron period are required. We have recently fabricated high-quality 0.7 μm-period magnetic lattice structures including 2D square and triangular geometries by patterning a Co/Pd multi-atomic layer magnetic film using electron-beam lithography and reactive ion-etching [4]. Such magnetic lattices would be useful for the simulation of Hubbard physics and quantum spin models [5]. A discussion of the initial steps for loading ultracold atoms into 0.7 μm-period magnetic lattices will be presented.

[1] S. Jose *et al.*, Phys. Rev. A **89**, 051602(R) (2014).

[2] P. Surendran *et al.*, Phys. Rev. A **91**, 023605 (2015).

[3] V.Y.F Leung *et al.*, Rev. Sci. Instrum. **85**, 053102 (2014).

[4] I. Herrera *et al.*, J. Phys. D **48**, 115002 (2015).

[5] M. Lewenstein *et al.*, Adv. Phys. **56**, 243 (2007).

# Towards the production of a quantum gas near a tapered optical nanofibre

C.A. Alampounti, R.A. Jenkins and S. Eriksson
Department of Physics, College of Science, Swansea University, Singleton Park, SA2 8PP, UK

We are working towards a light-quantum degenerate gas interface using tapered optical nanofibres (TNFs). Since a TNF does not impose severe restrictions on laser cooling techniques, we can probe macroscopic quantum systems by placing the atom cloud close to a TNF. The setup would enable observation of phenomena via fluorescence into the TNF, or atomic interaction via resonant nanofibre photons. The scheme can also be extended to detect atoms in different internal states. We magnetically trap $^{87}$Rb atoms routinely, and we have recently incorporated a hybrid "dimple" trap technique for a Bose-Einstein condensate production whilst maintaining compatibility with mounting and working with TNFs. The experimental setup also includes provision for independent detection of atoms via a high resolution single photon sensitive optical system based upon a four component fluorescence collection lens. Here we present an overview of our work to date.

# Ion-atom collsions are enhanced by the generation of plasma during photoionization of laser cooled atoms

Shuang-Fei Lv [1], Jin-Yun Liu [1], Feng-Dong Jia [1*], Ya-Ping Ruan [2], Long-Wei Liu [1], Zhen Sun [1], Wei Huang [3], Ping Xue [4], Xiang-Yuan Xu [4,5], Jian-Guo Wang [6], Zhi-Ping Zhong [1]

[1] *School of Physics, University of Chinese Academy of Sciences, PO Box 4588, Beijing 100049, China*

[2] *College of Engineering and Applied Scieces, Nanjing University, 210093*

[3] *Laboratory of Quantum Engineering and Quantum Materials, SPTE, South China Normal University, Guangzhou 510006, China*

[4] *State Key Laboratory for Low-Dimensional Quantum Physics, Department of Physics, Tsinghua University, Beijing 100084,China*

[5] *Department of Physics, Capital Normal University, Beijing 100037,China*

[6] *Data Center for High Energy Density Physics, Institute of Applied Physics and Computational Mathematics, PO Box 8009, Beijing 100088, China*

The temperature and the number of trapped $^{87}$Rb atoms after two-step photoionization in a magneto-optical trap (MOT) as a function of photoionization light intensity have been measured. As the photoionization light intensity increases, the number of steady-state MOT atoms deviates from the predicted linear dependence on the photoionization laser intensity and shows a significant reduction. Moreover, the temperature of trapped atoms after two-step photoionization shows plateau character, thus, the relationship between the temperature and number of trapped atoms deviates from a power law. A combination of experimental measurements and a rate-equation model characterize the deviations of the linear dependence of the steady-state MOT atoms numbers on the photoionization laser intensity and the plateau of trapped atoms temperature with the increasing of photoionization laser intensity, and confirm that the creation of an ultracold neutral plasma formed by photoionizing laser-cooled atoms accounts for the above experimental results: the creation of ultracold neutral plasma results in an accumulation of ions and would enhance the atom-ion collision. The atom-ion collisions lead to atom heating and escaping from the MOT.

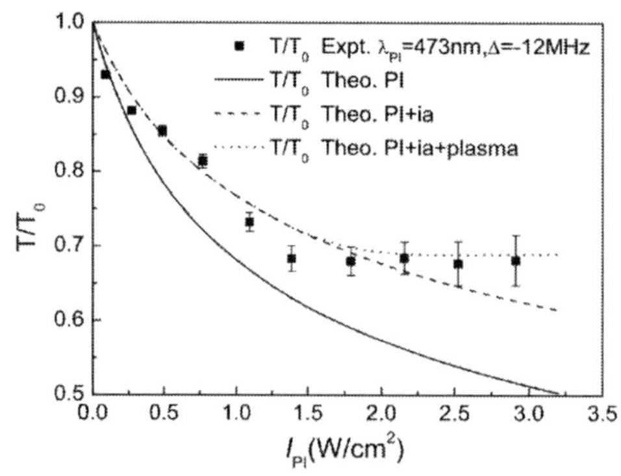


**Figure 1.** The comparison of experimental and theoretical results of $T/T_0$ as a function of different average intensity $I_{PI}$.

[1] RUAN YaPing , JIA FengDong , LIU LongWei ,SUN Zhen , HUANG Wei , XUE Ping , XU XiangYuan ,DAI XingCan and ZHONG Zhi-Ping, CHIN. PHYS. LETT. **31**, 07341 (2014).

# Anomalous Heating in Surface Ion Traps

Maya Lewin-Berlin[1], Crystal Noel[1], Ishan Talukdar[1], William Tokumaru[1], Nikos Daniilidis[1], Hartmut Haeffner[1]
[1]University of California, Berkeley, United States

Trapped ions provide a suitable platform for quantum information applications due to their long coherence times and controllable quantum states. The planar Paul trap uses a flat surface with microfabricated electrodes to suspend ions in a harmonic potential. The benefit of this architecture is its scalability; it can be extended to create a fast, compact, multi-qubit system. However, an ion trapped close to a surface has increased sensitivity to electric field noise. This leads to so called 'anomalous heating' of the ions, thus limiting our ability to perform computations. We report here studies seeking to identify sources of the noise using analysis techniques such as FTIR and Auger spectroscopy. We present recent results on the effect of exposing the surface to near UV light on anomalous heating. We also discuss initial studies of improved surface treatment methods.

# Fast and optimal transport of atoms in non-harmonic traps

Qi Zhang[1], Xi Chen[1]
[1]Department of Physics, Shanghai University, People's Republic of China

We investigate the fast transport of an atom or a packet of atoms by different kinds of non-harmonic traps including power-law traps [1]. The study is based on the reverse engineering method. First of all, exact results are obtained and applied to design robust transport protocols. The optimization of the transport trajectory is performed with classical trajectories, and remains valid for the transport of a wave packet. Secondly, we combine the inverse engineering and optimal control theory to design the fast and optimal transport with respect to anharmonic perturbation, by minimizing the average anharmonic energy [2]. The fast and optimal transport of cold atoms may have potential applications in atom optics and quantum information processing.

[1] Qi Zhang, Xi Chen, and D. Guéry-Odelin, Phys. Rev. A **92**, 043410 (2015).
[2] Qi Zhang, J. G. Muga, D. Guéry-Odelin, and Xi Chen, J. Phys. B. **49**, 125503 (2016).

# Coherent Rydberg Excitation of a Single Atom by Lasers Stabilized on a Tunable High Finesse Transfer Cavity

Yong Zeng[1,2,3], Kun-Peng Wang[1,2,3], Yang-Yang Liu[1,2,3], Xiao-Dong He[1,2], Min Liu[1,2] , Peng Xu[1,2], Jin Wang[1,2], Ming-Sheng Zhan[1,2]

[1]State Key Laboratory of Magnetic Resonance and Atomic and Molecular Physics, Wuhan Institute of Physics and Mathematics, Chinese Academy of Sciences, Wuhan 430071, China

[2]Center for Cold Atom Physics, Chinese Academy of Sciences, Wuhan 430071, China

[3]School of Physics, University of Chinese Academy of Sciences, Beijing 100049, China

With two-photon Raman transition, we coherently excite single Rb atom from the hyperfine ground state to highly excited Rydberg state. The excitation lasers are frequency stabilized by using a new method with a tunable cavity whose finesse is about 100000. The cavity length is stabilized by an Iodine-frequency-stabilization-laser. After the cavity being locked, the long term drift of the cavity is less than the FWHM (full width at half maximum) of the Iodine-frequency-stabilization-laser transmission peak. The FWHM is about 17.4 kHz. Then, the Rydberg excitation lasers of 780nm and 480nm are locked to the stabilized cavity. In this way, the long term frequency drifts of the excitation lasers are less than 20 kHz for 24 hours, meanwhile the line widths are suppressed to be less than 15 kHz. Using this technique, we have realized Rabi oscillation between the state $|5s_{1/2},F=+2,m_F=0\rangle$ and the state $|43d_{5/2},m_J=+5/2\rangle$ with oscillating frequency about 1.6 MHz, and the maximal Rydberg exciting efficiency can go up to over 90%. The short exciting time and high exciting efficiency make it a crucial step of single Rydberg atom used for quantum computing.

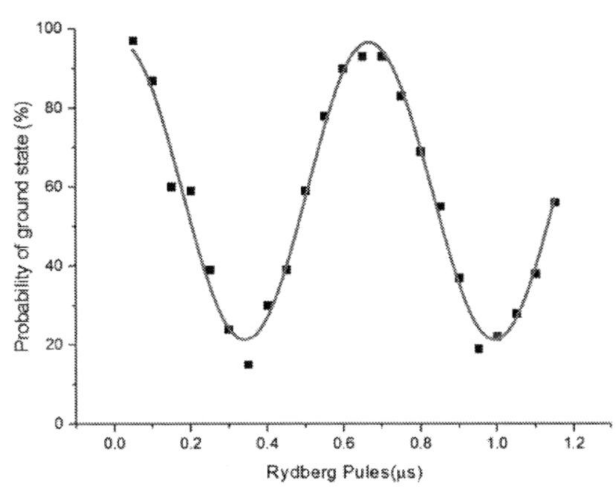


**Figure 1.** Rabi oscillation between $|5s_{1/2},F=+2,m_F=0\rangle$and $|43d_{5/2},m_J=+5/2\rangle$.

# Microwave sideband spectrum of single atoms in a strongly confined optical dipole trap

Kunpeng Wang[1,2,3], Xiaodong He[1,2], Peng Xu[1,2], Ruijun Guo[1,2,3], Cheng Sheng[1,2], Min Liu[1,2], Jin Wang[1,2], Mingsheng Zhan[1,2]

[1]State Key Laboratory of Magnetic Resonance and Atomic and Molecular Physics, Wuhan Institute of Physics and Mathematics, Chinese Academy of Sciences, Wuhan 430071, China.

[2] Center for Cold Atom Physics, Chinese Academy of Sciences, Wuhan 430071, China.

[3] School of Physics, University of Chinese Academy of Sciences, Beijing 100049, China.

Coherent manipulation of motional states of atoms confined in an optical dipole trap, is one of the key issues of quantum simulation and quantum information processing [1-2]. Two-photon Raman transition is a common technique to address motional states of neutral atoms. In state dependent optical lattices, single atoms can be manipulated with microwave transitions sensitive to atomic motion and have been cooled to the vibrational ground state [3-5]. Here we demonstrate motion-sensitive microwave transitions of single atoms in an optical dipole trap. In our experiment we utilize the polarization gradient effect (PGE) under strong confinement with a high numeric aperture (NA=0.6) microscope [1-2, 6]. Microwave sideband spectra are measured for different atomic states up to second order sideband. The result is helpful for performing microwave sideband cooling and manipulating motional states after ground state cooling without the companion of photon recoils.

[1] A. M. Kaufman, B. J. Lester, and C. A. Regal, Phys. Rev. X **2**, 041014 (2012).

[2] J. D. Thompson, T. G. Tiecke, A. S. Zibrov, V. Vuletic, and M. D. Lukin, Phys. Rev. Lett. **110**, 133001 (2013).

[3] L. Förster, M. Karski, J. Choi, A. Steffen, W. Alt, D. Meschede, and A. Widera, Phys. Rev. Lett. **103**, 233001 (2009).

[4] X. Li, T. A. Corcovilos, Y. Wang, and D. S. Weiss, Phys. Rev. Lett. **108**, 103001 (2012).

[5] N. Belmechri, L. Förster, W. Alt, A. Widera, D. Meschede and A. Alberti, J. Phys. B: At. Mol. Opt. Phys. **46**, 104006 (2013).

[6] B. Richards and E. Wolf, Proc. R. Soc. A **253**, 358 (1959).

# Laser cooling $^{171}$Yb atoms with the $^1S_0$ -$^3P_1$ transition below the Doppler cooling limitation

Qi Gao, Min Zhou, Xiaohang Zhang, Chengyin Han, Yilin Xu, Peng Xu, Shangyan Li, Shuang Zhang, and Xinye Xu

State Key Laboratory of Precision Spectroscopy and Department of Physics, East China Normal University, Shanghai 200062, China

The ytterbium optical clock [1-3] has been regarded as one of the promising candidates for the next generation of optical frequency standards, and precooling atoms to low temperature is an essential requirement for making optical clocks. In our experiment, after the blue magneto-optical trap (MOT) is turned off, the green MOT is turned on immediately, at the same time the B-field gradient is turned down from 50 Gs/cm to 15 Gs/cm. The green MOT uses the 556-nm $^1S_0$ - $^3P_1$ intercombination transition with a natural linewidth of 182 kHz, and the 556-nm light with several kilohertz linewidth is generated by the use of the Pound-Drever-Hall laser frequency stabilization system as discussed in our previous work [4], followed by frequency doubling the 1111.6-nm using a commercial SHG frequency doubler. We have studied the dependences of the temperature, the number and density of atoms in the green MOT with the magnetic field gradient, the intensity and detuning of the trapping laser respectively. Finally, we have achieved the temperature of $^{171}$Yb atoms below the Doppler cooling limitation. Furthermore the cooled atoms have been loaded into the optical lattice for developing ytterbium optical lattice clocks.

[1] X. H. Zhang, M. Zhou, N. Chen, Q. Gao, C. Y. Han, Y. Yao, P. Xu, S. Y. Li, Y. L. Xu, Y. Y. Jiang, Z. Y. Bi, L. S. Ma, and X. Y. Xu, Laser Phys. Lett. **12**, 025501 (2015).
[2] C. Y. Park, D. H. Yu, W. K. Lee, S. E. Park, E. B. Kim, S. K. Lee, J. W. Cho, T. H. Yoon, J. Mun, S. J. Park, T. Y. Kwon, and S. B. Lee, Metrologia **50**, 119 (2013).
[3] H. Inaba, K. Hosaka, M. Yasuda, Y. Nakajima, K. Iwakuni, D. Akamatsu, S. Okubo, T. Kohno, A. Onae, and F. L Hong, Opt. Express **21**, 184042 (2013).
[4] M. Zhou, L. Y. H, and X. Y. Xu, Chin. Opt. Lett. **11**, 121402 (2013).

# Laser cooling to 20μK with an inverted crossover resonance in $^{171}$Yb

Liam A. Salter, Ljerka Nenadović & John J. McFerran

School of Physics, University of Western Australia, 6009, Crawley, Australia

Explorations in degenerate Fermi gases [1,2], strongly correlated states [3,4], quantum computation, and atomic clocks [5,6] all make use of neutral ytterbium. A common means of producing atomic cloud temperatures at a few tens of micro-K is to employ first- and second-stage laser cooling with 399nm and 556nm light, where the latter exploits the $6s^2\ ^1S_0\ (F{=}1/2)$ - $6s6p\ ^3P_1\ (F{=}3/2)$ inter-combination line. The natural linewidth of which is 184kHz and has a corresponding Doppler cooling limit of 4.4μK. To generate adequately stable 556nm light we employ saturated absorption spectroscopy (SAS) upon a thermal Yb beam that leads to a Zeeman slower and magneto-optical trap. Serendipitously, we have identified an inverted Lamb dip in saturated absorption spectroscopy of the $^1S_0\ (F{=}1/2)$ - $^3P_1\ (F{=}3/2)$ and $^1S_0\ (F{=}1/2)$ - $^3P_1\ (F{=}1/2)$ transitions in $^{171}$Yb. The signal is most apparent with these two lines where the nuclear spin is $I{=}1/2$. For both hyperfine lines the application of a dc magnetic field transforms a two-level scheme into a four-level scheme, with degeneracy only marginally lifted in the ground state. The atoms are excited through $\pi$ transitions, while the de-excitation process completes a simple optical pumping scheme. We have identified the enhanced absorption signal as an inverted crossover resonance, where the principal lines involved in the crossover are transitions to symmetrically separated Zeeman substates. FWHM linewidths of 660kHz are observed. This may be the first instance of an inverted crossover resonance where $\Delta F{=}0$ (at least in divalent atoms). The resonance is exploited to generate a dispersive curve for 556 nm laser frequency stabilization, producing a frequency instability of ~10kHz/$\sqrt{\tau}$ (but also dependent on the atomic flow rate). This light is used to cool $^{171}$Yb ($F'{=}3/2$) atoms in a dual-colour magneto-optical trap, demonstrating a mean temperature of 18μK with cycle to cycle variations of 2μK [7]. The low temperatures have been achieved without the use of an optical reference cavity for high-frequency noise suppression (the master laser at 1112nm is a fibre laser).

We have used the saturated absorption scheme to perform absolute optical frequency measurements of the $6s^2\ ^1S_0\ (F{=}1/2)$ - $6s6p\ ^3P_1\ (F{=}3/2)$ transition in $^{171}$Yb [8]. Frequency modulation is applied to the 556nm beams that are orthogonal to the atomic beam, and the green fluorescence is detected with a photomultiplier cell whose signal is sent to a lock-in amplifier for third harmonic detection. Helmholtz coils in close vicinity of the SAS zone are used to null the magnetic field in the vertical direction. Sub-harmonic light at 1112nm has its frequency measured by use of a frequency comb that is referenced to a hydrogen maser. We have validated our measurements by performing clock transition ($^1S_0$ - $^3P_0$) spectroscopy on $^{171}$Yb atoms held in a magneto-optical trap (MOT), and finding agreement with previously reported values to within 10kHz.

Frequency measurements of both the inter-combination line and the clock line were made almost daily over the course of a month as a test of the reproducibility. The statistical uncertainty of the $^1S_0$ - $^3P_1$ [$\sigma/\sqrt{(N-1)}$] frequency measurements is below 20kHz. The mean frequency of the $^{171}$Yb ($F'{=}3/2$) line is 539 390 406 830 (310) (16) kHz [8]. The SAS scheme is also used to carry out the highest resolution measurements to date of the isotopic shifts and hyperfine separations of the $6s^2\ ^1S_0$ - $6s6p\ ^3P_1$ transition in ytterbium [7]. We have resolved the $^{171}$Yb ($F'{=}3/2$) and $^{173}$Yb ($F'{=}3/2$) lines at zero $B$-field for the first time, to our knowledge.

[1] T. Fukuhara, Y. Takasu, M. Kumakura, Y. Takahashi, Phys. Rev. Lett., **98**, 030401, (2007).
[2] M. Schreiber, *et al.*, Science, **349**, 842 (2015).
[3] J. Dalibard, F. Gerbier, G. Juzeliunas, and P. Ohberg, Rev. Mod. Phys. **83**, 1523 (2011).
[4] S. Taie, R, Yamazaki, S. Sugawa, and Y. Takahashi, Nat. Phys., **8**, 825 (2012).
[5] N. Nemitz *et al.*, Nat. Photon., **10**, 258 (2016)
[6] A.D. Ludlow, M.M. Boyd, J. Ye, E. Peik and P.O Schmidt, Rev Mod. Phys., **87**, 637 (2015).
[7] J.J. McFerran, J. Opt. Soc. Am. B, **33**, 1278 (2016).
[8] L. Nenadovic and J.J. McFerran, J. Phys. B, **49**, 065004 (2016).

# Magnetically insensitive ground state cooling of single atom

Pimonpan Sompet[1], Yin Hsien Fung[1], Eyal Schwartz[1], Mikkel F. Andersen[1]
[1]Dodd-Walls Centre for Photonic and Quantum Technologies, QSO Centre for Quantum Science,
Department of Physics, University of Otago, Dunedin, New Zealand

Ultimate control of pure quantum states provides an excellent platform for various experiments in quantum science and engineering. In this work, we carry out quantum manipulation of individual Rubidium 85 atoms trapped in a tightly focus optical tweezer in order to cool them into their vibrational ground state via Raman sideband cooling. By using blue-detuned light assisted collision method [1], a single atom is trapped with 82% loading efficiency and cooled using a combination of Polarized Gradient Cooling (PGC) and Sisyphus cooling in order to localize the atomic wave function before initiating further Raman sideband cooling. Our cooling scheme involves a combination of optical pumping of the atom and Raman sideband transitions, which couples two magnetic field sublevels (|F=3,m=0> and |F=2,m=0>) that are indifferent to magnetic noise (Fig. 1a). In our latest scheme, we achieved 99% efficiency of optically pumping an atom to the designated quantum state which allowed us to achieve 2D cooling of the radial plane with a ground state population of 85% (Fig. 1b), on the way to create a full near nil entropy state of single atoms and single molecules. Our procedure therefore prepares a single atom in particular internal and 2D vibrational state with a fidelity of 69%. This would lead to longer coherence times compared to magnetic sensitive cases [2, 3].

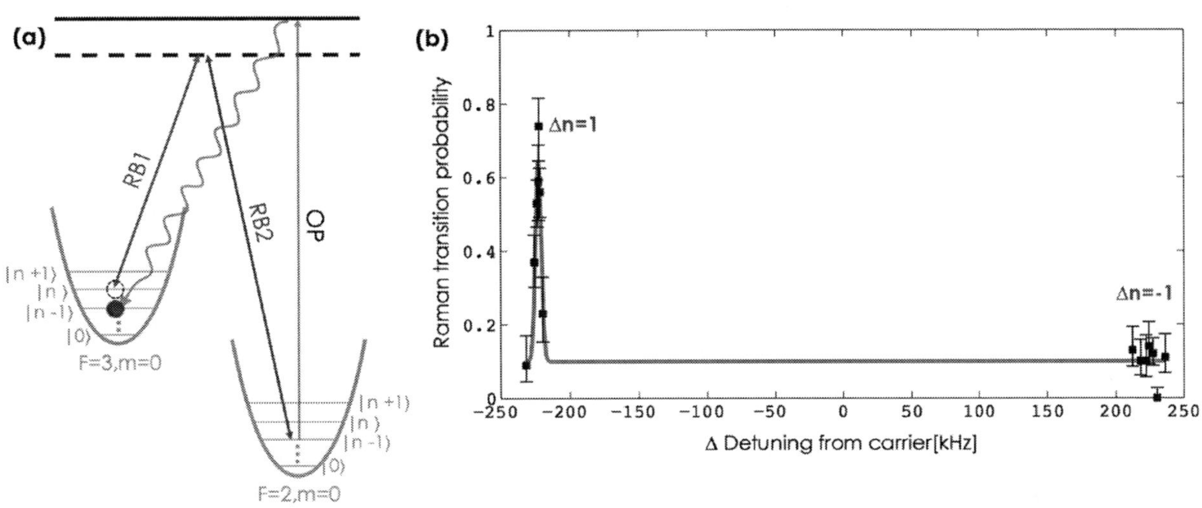


**Figure 1.** (a) Energy level diagram showing transitions relevant to Raman sideband cooling. (b) Sidebands showing final occupations in the radial direction after Raman sideband cooling.

[1] A. V. Carpentier, Y. H. Fung, P. Sompet, A. J. Hilliard, T. G. Walker, and M. F. Andersen, Laser Phys. Lett. **10**(12), 125501 (2013).
[2] J. D.Thompson, T. G. Tiecke, A. S. Zibrov, V. Vuleti ́c , and M. D. Lukin, Phys. Rev. Lett. **110**, 133001 (2013).
[3] A. M. Kaufman, B. J. Lester, and C. A. Regal, Phys. Rev. X **2**, 041014 (2012).

# Doppler-Free multi-photon spectroscopy in MOT

Pyeong Woo Kim, Taek Jung, Han Seb Moon
Department of physics, Pusan National University, Busan, Korea

We report the optical-optical double resonance (OODR) spectroscopy in the $5S_{1/2}$–$5P_{3/2}$–$4D_{5/2}$ transition of the cold $^{87}$Rb atoms. Since we want to investigate multi-photon spectroscopy in atomic ensemble system for future work, we build Magneto Optical Trap (MOT) using conventional method [1]. First of all, we made Master Oscillator Power Amplifier (MOPA) to increase laser power because MOT needs high optical power to trap atoms. And then we locked both MOPA system and re-pump laser for stability. After that we also used AOM to shift frequency of both laser for trapping $^{87}$Rb atoms. In this situation, $^{87}$Rb atoms in the octagon glass cell feel σ+ or σ- transition, thus the atoms which move in the opposite direction of laser propagation feel optical force. In this MOT system, we investigated the Doppler-free OODR absorption spectra of the $5P_{3/2}$ – $4D_{5/2}$ transition according to the powers of the cooling laser. In our experiment, the cooling laser for MOT is used for the first excitation from the $5S_{1/2}$ state to the $5P_{3/2}$ state. We compared the OODR absorption spectra with the Double Resonance Optical Pumping (DROP) spectrum [2], as shown in the bottom graph of Fig. 1. The spectral feature of the OODR spectra was changed according to the powers of the cooling laser. In this work, we discussed about the spectral feature of the OODR in MOT.

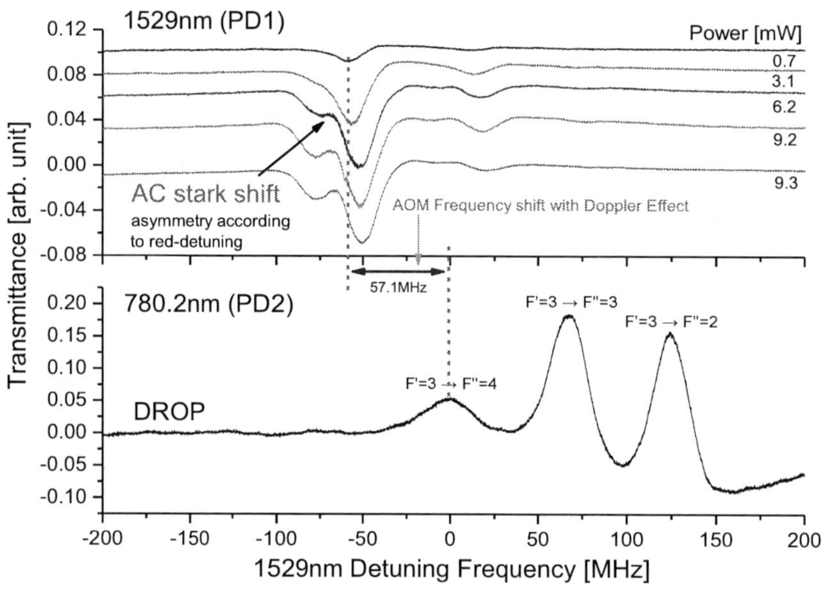


**Figure 1.** Doppler-free optical-optical double resonance (OODR) in $5S_{1/2}$–$5P_{3/2}$–$4D_{5/2}$ transition of the cold $^{87}$Rb atoms

[1] Hidetoshi Katori, Tetsuya Ido, Yoshitomo Isoya, and Makoto Kuwata-Gonokami, "Magneto-Optical Trapping and Cooling of Strontium Atoms down to the Photon Recoil Temperature", Phys. Rev. Lett. **82**, 1116 (1999).

[2] H. S. Moon, L. Lee, and J. B. Kim, "Double resonance optical pumping in Rb atoms", J. Opt. Soc. Am. B **24**, 2157-2164 (2007).

# All-optical production and transport of a large $^6$Li quantum gas in a crossed optical dipole trap

Christian Gross[1], Jaren Gan[1], Kai Dieckmann[1,2]
[1]Centre for Quantum Technologies, Singapore
[2]Department of Physics, National University of Singapore

We report on an efficient production scheme for a large quantum degenerate sample of fermionic lithium. The approach is based on our previous work on narrow-line, $2S_{1/2} \rightarrow 3P_{3/2}$, laser cooling resulting in a high phase-space density of up to $3 \times 10^{-4}$. This allows utilizing a large volume crossed optical dipole trap with a total power of 45W, leading to high loading efficiency and $8 \times 10^6$ trapped atoms. The same optical trapping configuration is used for rapid adiabatic transport (Figure 1) over a distance of 25cm in 0.9s, and subsequent evaporative cooling. With optimized evaporation we achieve a degenerate Fermi gas with $1.7 \times 10^6$ atoms at a temperature of 60nK, corresponding to $T/T_F=0.16(2)$. Furthermore, the performance is demonstrated by evaporation near a broad Feshbach resonance creating a molecular Bose-Einstein condensate of $3 \times 10^5$ lithium dimers.

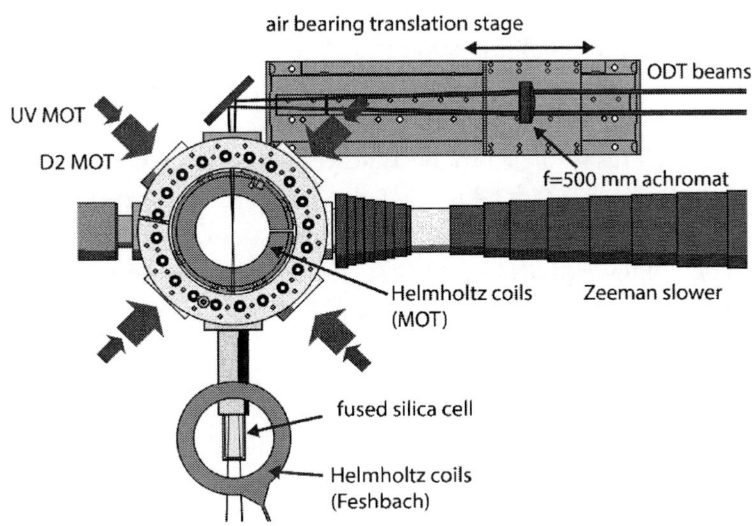


**Figure 1.** Transport system with optical dipole trapping beams crossing under a narrow angle.

# Magneto-optical trapping of metastable europium atoms

Yuki Miyazawa, Ryotaro Inoue, Keiji Nishida, Takuya Ogata, Mikio Kozuma
Department of Physics, Tokyo Institute of Technology, 2-12-1 O-okayama, Meguro-ku, Tokyo, Japan

Recently several groups have succeeded in laser cooling chromium (Cr), erbium (Er) and dysprosium (Dy) atoms with large dipole moments, which are 6, 7, and 10 Bohr magneton ($\mu_B$) respectively. Their dipole moments cause dipole-dipole interaction (DDI) that is long-range and anisotropic. The behavior of these magnetic dipolar quantum gases is drastically different from that for atomic species with only s-wave scattering interaction. So far, many fascinating phenomena have been observed, such as d-wave symmetric explosion of Bose-Einstein condensate [1], deformation of a fermi surface [2], and Rosensweig instability [3]. Moreover, as the DDI can couple electronic spins and orbital angular momenta of atoms, both spin textures and quantum vortices are predicted to appear in ground states [4].

In order to investigate such quantum phases, it is desirable to be able to control the s-wave scattering length through Feshbach resonance, because characteristics of such quantum phases are determined by competition between the s-wave scattering interaction and the DDI. Feshbach resonance is usually induced by applying external magnetic field to the atoms, which kills the spin degrees of the freedom. Another possibility to control the s-wave scattering length is inducing Feshbach resonance by applying microwave to the atoms [5], which enables us to control the scattering length with maintaining the spin degrees of the freedom. The microwave Feshbach resonance is applicable only to the atoms having hyperfine structures in the ground states. Unfortunately, bosonic isotopes of Cr, Er, and Dy atoms do not have the hyperfine structures in their ground states. We thus planned to cool down europium (Eu) atoms that have the large dipole moment ($7\mu_B$) and also the hyperfine structure.

In order to implement the Zeeman slowing and the magneto-optical trapping for the europium atoms, we need a strong dipolar optical transition. While the europium atoms have an optical transition ( $a^8S_{7/2} \rightarrow y^8P_{9/2}$ ) with the natural linewidth of 27MHz, the excited state ( $y^8P_{9/2}$ ) decays spontaneously to the metastable states other than the ground state with relatively high probability. The probability of the optical leaks was estimated to be $\sim 10^{-3}$ through our fluorescence spectroscopy. To avoid this optical leak problem, we transferred the atoms to the long-lived metastable state ($a^{10}D_{13/2}$) and implemented both the Zeeman slowing and the magneto-optical trapping using $a^{10}D_{13/2} \rightarrow z^{10}F_{15/2}$ cyclic optical transition (wavelength: 583nm, the natural linewidth: 8.2MHz). Figure 1 shows a typical image of the fluorescence from the magneto-optical trapping for the cold europium atoms.

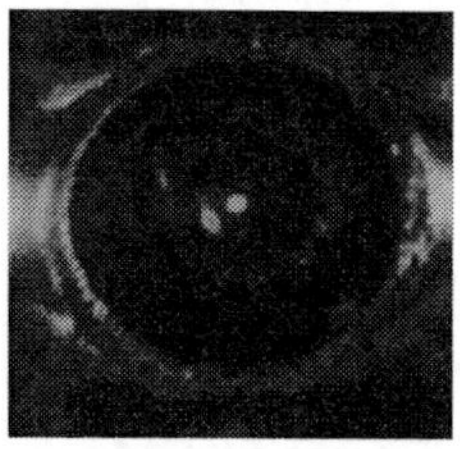

**Figure 1.** Magneto-optical trapping of metastable Eu atoms

[1] T. Lahaye et al., Phys. Rev. Lett. **101**, 080401(2008). [2] K. Aikawa et al., Science **345**, 1484-1487 (2014). [3] H. Kadau et al., Nature **530**, 194-197 (2016). [4] Y. Kawaguchi et al., Phys. Rev. Lett. **97**, 130404 (2006). [5] D. J. Papoular et al., Phys. Rev. A **81**, 041603 (2010).

# UHV glass chamber with mirrored windows to form a Fabry-Perot cavity

Meungho Seo, Sooyoung Park, Seokchan Yoon, D. Cho
Department of Physics, Korea University, Seoul, Korea

A Fabry-Perot cavity provides an optical lattice with a well-defined and stable mode compared with a lattice formed by a retro-reflecting mirror. It also produces deep potential wells from the resonant power buildup. It is challenging, however, to put the cavity inside a vacuum chamber without compromising ultrahigh vacuum required for experiments like Bose-Einstein condensation or quantum information processing. Placing the cavity mirrors outside a chamber complicates the situation with windows between the mirrors.

We report our construction of an octagonal glass chamber, whose diagonally placed windows have engraved mirrors at the center to form a Fabry-Perot cavity. The mirror has 3.25 cm radius of curvature and reflectivity of 98% at 980 nm. They are attached to the octagonal chamber by using UHV-compatible glue (EPO-TEK H77). We achieved pressure as low as $5 \cdot 10^{11}$ Torr. We trap a few million rubidium atoms to the lattice formed by the cavity. The lifetime is longer than 5 s and we will report further progress.

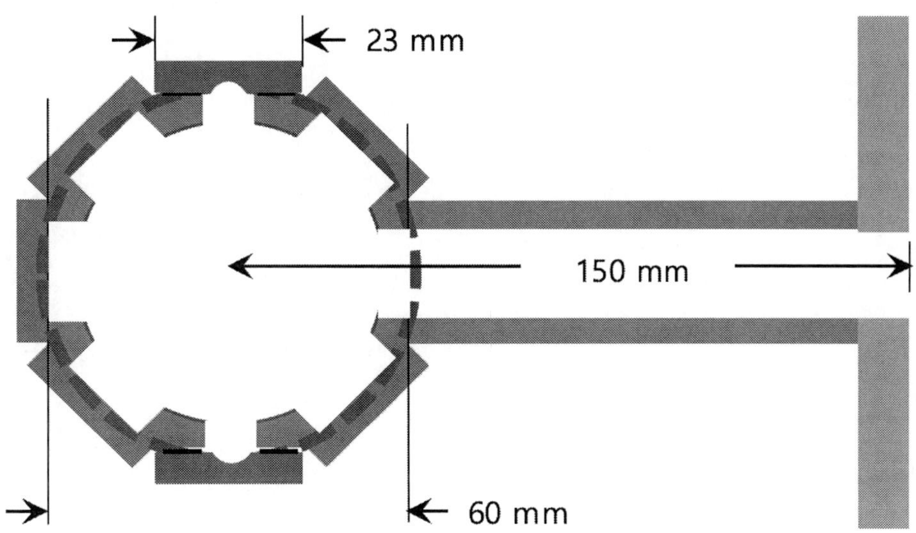


**Figure 1.** Dimensions of the UHV glass chamber with mirrored windows.

# Simulation of cold atomic beam from pyramidal magneto-optical trap funnel

Dongjin Lee[1], Yoonsung Jeon[1], Hoon Yu[2], Sugtae Kang[1], Donghoon Min[1], Sangmin Jung[1], Jungbog Kim[1]

[1]Physics Education, Korea National University of Education, Chung-buk, Republic of Korea

[2]Hanwha Corporation Defense R&D center, Dae-jeon, Republic of Korea

Monochromatic and slow atomic beam is applicable for atom optics such as atomic clock and atom interferometry. Compared to typical six beam Magneto-Optical Trap (MOT), pyramidal MOT is a simple and robust tool to achieve atomic beam [1]. We simulate flux, divergence, longitudinal velocity distribution of atomic beam from $^{87}$Rb vapor-loaded pyramidal mirror (18mm x 18mm) with truncated at its apex. We also investigate characteristics according to the size of pyramidal mirror.

[1] J.J. Arlt, O. Marago, S. Webster, S. Hopkins, C.J. Foot, Opt Commun. **157**, 303–309 (1998)

# Phase-engineered light patterns for ultracold atom experiments

David Bowman, Graham D. Bruce, Tiffany Harte, Philip Ireland, Jonathan Keeling, Donatella Cassettari

SUPA School of Physics and Astronomy, University of St Andrews, North Haugh, St Andrews, Fife KY16 9SS, UK

In recent years, there has been increasing interest in spatially-tailored optical patterns for use in manipulating ultracold neutral atoms. We demonstrate our methods to produce smooth, high-accuracy and multi-wavelength light patterns through the use of a Spatial Light Modulator (SLM), as well as our progress towards light patterns with arbitrary amplitude and phase.

The calculation of phase-modulating holograms is an inverse problem, and is commonly performed by variants of the Gerchberg-Saxton Iterative Fourier Transform Algorithm. As an alternative to this approach, we have developed a method for the generation of holograms based on the direct minimisation of a cost function by a conjugate gradient local search algorithm [1]. This algorithm offers a high degree of control, as demonstrated by controlling optical vortex formation and can be extended to manipulate the phase as well as the amplitude of the light pattern.

Multi-wavelength light patterns are experimentally realised using a single SLM illuminated by up to three overlapped laser beams at 670nm, 780nm and 1064nm, where the light profile of each wavelength is independently controlled [2]. In order to compensate for experimental aberrations, we have developed a simple and robust feedback-enhanced algorithm [3] to improve the accuracy of the light patterns, which reduces the RMS error to the percent level. Furthermore we demonstrate the applicability of these holographic traps for two experiments: ring traps for rotation sensing [4] and the creation of a cold atom device that can be used to investigate the topological Kondo effect [5].

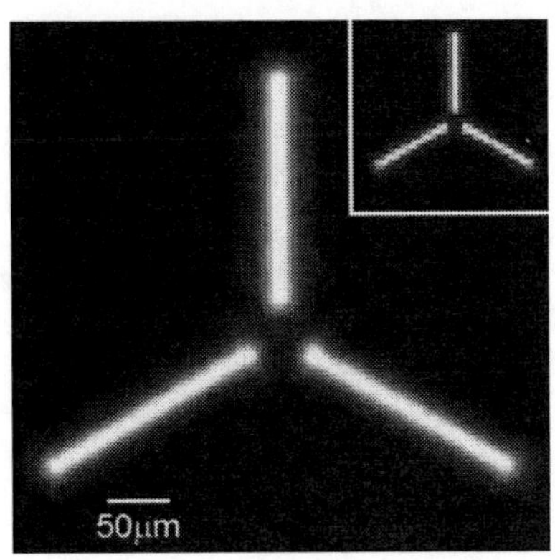


**Figure 1.** Holographic realization of a Y-junction

[1] T. Harte *et al.*, Opt. Express **22** 26548-26558 (2014)
[2] D. Bowman, *et al.*, Opt. Express **23** 8365-8372 (2015)
[3] G. D. Bruce *et al.*, J. Phys. B: At. Mol. Opt. Phys. **48** 115303 (2015)
[4] G. D. Bruce *et al.*, Phys. Scr. **T143**, 014008 (2011)
[5] F. Buccheri *et al.*, accepted in New J. Phys. (2016), arXiv:1511.06574

# Laser cooling of $^{87}$Rb atoms using $5S_{1/2}$-$6P_{3/2}$ transition

Tetsuo Kishimoto[1, 2], Yuta Yoshino[1]

[1] Department of Engineering Science and [2] Institute for Advanced Science,
University of Electro-Communications,
1-5-1 Chofugaoka, Chofu, Tokyo 182-8585, Japan

Recently, narrow-line cooling using nS-(n+1)P transitions for alkali atoms have been reported [1][2][3]. The advantage of using these transitions is that the natural linewidth of such transitions is narrower than the nS-nP transitions, leading to lower Doppler temperature. In addition to these transitions having shorter wavelength, these are open transitions, so effectively the absorption cross-section becomes smaller and the MOT can be compressed further compared to ordinary D2-transition MOTs. Due to these effects, phase space density can be increased and improved before the evaporation stages in quantum degenerate gas experiments.

In this work, we will discuss on laser cooling using the $5S_{1/2}$-$6P_{3/2}$ transition of $^{87}$Rb atoms. We first cool and trap these atoms with the normal $5S_{1/2}$-$5P_{3/2}$ transition (780nm). Then we switch the cooling laser to the $5S_{1/2}$-$6P_{3/2}$ transition (420nm). Three dimensional cooling and trapping was observed within 10ms timescale and reached nearly 40μK for the horizontal axis (perpendicular to gravity) [4].

In the presentation further details of the experiment and possible future plans using this transition will be discussed [5].

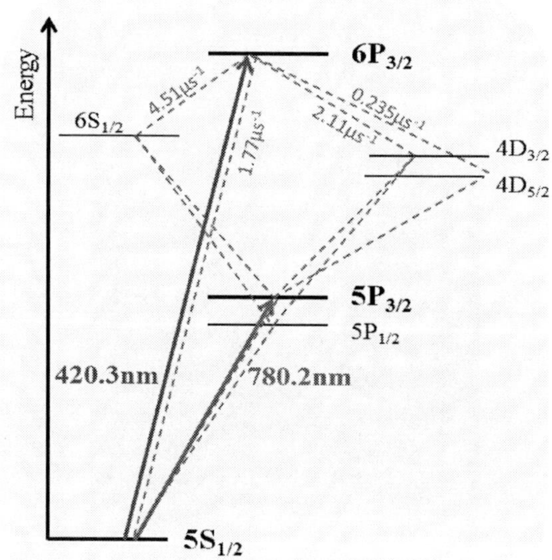


**Figure 1.** Energy level diagram for Rubidium 87.

[1] D. C. McKay *et al.*, Phys. Rev. A **84**, 063420 (2011).

[2] P. M. Duarte *et al.*, Phys. Rev. A **84**, 061406(R) (2011).

[3] J. Sebastian *et al.*, Phys. Rev. A **90**, 033417 (2014).

[4] Y. Yoshino and T. Kishimoto, in preparation.

[5] Y. Yoshino and T. Kishimoto, in preparation.

# Towards laser cooling of BaF

Wenhao Bu[1], Dizhou Xie[1], Bo Yan[1]
[1]Department of Physics, Zhejiang University, Hangzhou, Zhejiang, China

Laser cooling of polar molecules has been made great progress in recent years, magneto-optic al trap of polar molecules has been demonstrated with SrF[1] and YO[2]. Our project focus o n the BaF molecules, which has a very good Franck-Condon factor. We will give some detail s about the energy levels of this molecules, the design of the cooling machine, the target, an d the laser stabilization.

[1] J. F. Barry, *et al.*, Nature, **512**, 286 (2014).
[2] M. T. Hummon, *et al.*, Phys. Rev. Lett., **110**. 143001 (2013).

# Dynamic holographic single-atom arrays for deterministic single-atom loading

Hyosub Kim, Woojun Lee, Hangyeol Lee, Hanlae Jo, Yunheung Song, Kyungtae Kim, and Jaewook Ahn

KAIST, 291 Daehak-ro, Yuseong-gu, Daejeon, South Korea

As a scalable neutral-atom platform for quantum computation and quantum simulation, we consider micro-optical trap arrays generated by a liquid-crystal spatial light modulator (LCSLM). Using the flicker-free moving trap solution [1] newly devised for seamless rearrangement of N atoms, we demonstrate real-time feedback-control for single-atom array formation: cold rubidium atoms were simultaneously transported in the image plane with 2N moving degrees of freedom and the high success rate of 99% for 10 um translation [1, 2]. This holographic moving-trap is well suited to design arbitrary trap arrays, and thus we expect this method will be useful for deterministic loading of N single-atoms (currently 40% for N=4 atoms), especially on arbitrary lattice locations, and also for real-time shuttling of atomic qubits useful for selective pairing of Rydberg CNOT gating.

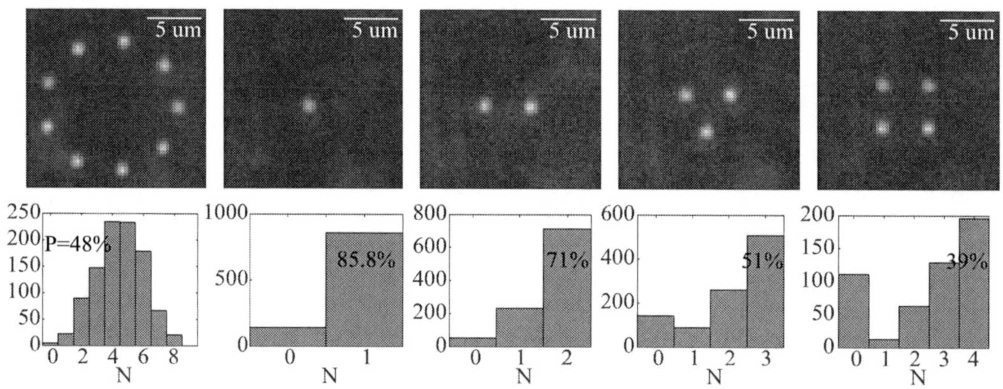


**Figure 1.** (Left to right) Initially loaded single atom array is fed-back to pre-determined position and number, respectively. The below histogram shows success rate of the single-feed-back loop.

[1] Hyosub Kim, *et al.*, arxiv 1601.03833

[2] Woojun Lee, Hyosub Kim, and Jaewook Ahn, Opt. Express **24** (9), 9816 (2016).

# Light shift of cesium atom in the magneto-optical trap

Chin-Chun Tsai, Zong-Syun He, Li-Ren Liu, Jyun-Yan Ye, Meng-Huang Sie, Yu-Han Chen, Jing-Yuan Su

Department of Physics, National ChengKung University, 1 University Road, Tainan, Taiwan 70101

The frequency of electromagnetically induced transparency (EIT) has shifted by varying the probe and coupling laser Rabi frequency as shown in Fig. 1. This frequency shift could be caused by the interactions of cold collisions at long range behaviors. Cold collision has a long coherence properties due to low collision rate. Therefore, using EIT in a magneto-optical trap (MOT) to probe the quantum phenomena is the best choose. In this study, the Cs atoms are confined in a MOT trap with a diameter of 5mm and temperature around 200 μK. Probing the cascade-type EIT of $|6^2S_{1/2}$ F=3> -> $|6^2P_{3/2}$F=3>-> $|8^2S_{1/2}$ F=4> transition simultaneously in the cold atom (MOT) and room temperature Cs cell, one can investigate the energy shift due to the different pump or probe laser intensities. The line width of EIT signals narrowed by the thermal averaging is also presented.

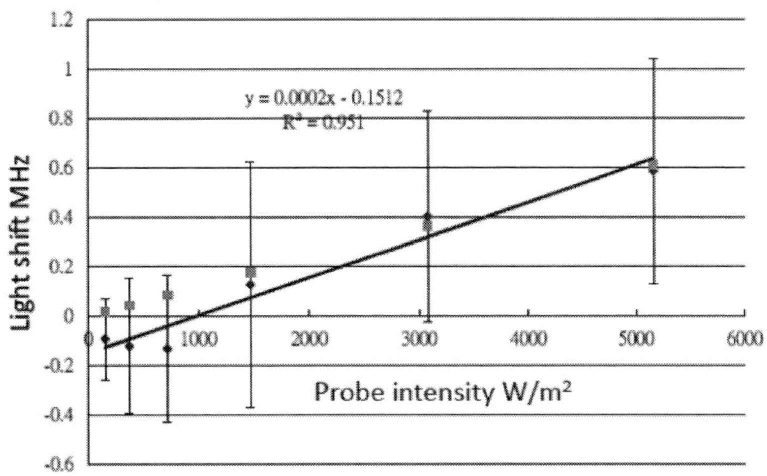


**Figure 1.** The figure displays the observed line shift (the solid symbols with their error bars for two runs) varying with the intensity of the coupling field.

# Scaling of thermal hysteretic behavior in a parametrically modulated cold atomic system

Wanhee Lee[1], Ji-Hyoun Kim[1], Heung-Ryoul Noh[2] , Wonho Jhe[1]

[1]Department of Physics and Astronomy, Seoul National University, Seoul, Korea

[2]Department of Physics, Chonnam National University, Gwangju, Korea

We observed hysteretic behavior of spontaneous symmetry breaking (SSB) transition in the modulated atomic system by sweeping the total number of atoms linearly, and determined the universality class of the hysteresis by obtaining the scaling exponent of hysteresis loops. In the mean-field model the closed hysteresis loop area ($A$) scales with a rate of temperature $R$ as

$$A = A_0 + bR^n \tag{1}$$

where $n$ is the scaling exponent of the hysteresis and $A_0$ and $b$ are constants. It is known that $n$ approaches two thirds which is universal for both the mean-field and field theoretical models concerned [1-3]

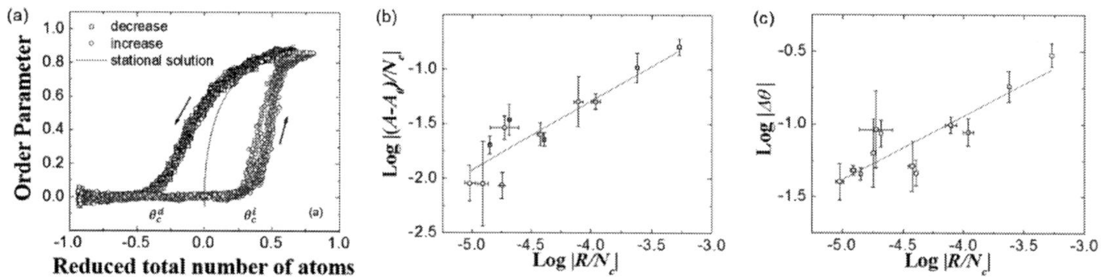


**Figure 1.** (a) The measured hysteresis loop by number sweeping, the red solid curves are the stationary solution of the SSB transtion. And log-log plot of (b) the hysteresis area $A$ and (c) the hysteresis width $\Delta\theta$ versus number sweeping rate $R$. The solid red curves in (b) and (c) are logarithmic fits of Eqs. (1) and (2), respectively.

In addition, we define the hysteresis width as $\Delta\theta = |\theta_c^i - \theta_c^d|$. The parameters, $\theta_c^i$ and $\theta_c^d$ are critical values of the reduced number for the symmetric state to the SSB state transition (upon increasing) and SSB state to Symmetric state transition (upon decreasing), respectively. It is well known that the hysteresis width is consistent with the following scaling law

$$\Delta\theta \approx R^{\beta}\,, \tag{2}$$

where $\beta$ is a scaling exponent [4]. The obtained values $n, \beta$ are quite close to the value given in the mean field theory and the theory of the kinetic Ising model [1-3, 5].

[1] F. Zhang and J. X. Zhang, Phys. Rev. E **51**, 2898 (1995).
[2] F. Zhong and J. X. Zhang, Phys. Rev. Lett. **75**, 2027 (1995).
[3] J. X. Zhang, F. Zhong, and G. G. Siu, Solid State Commun. **97**, 847 (1996).
[4] S. Yildz, O. Pekcan, A. N. Berker, and H. Ozbek, Phys. Rev. E. **69**, 031705 (2004).
[5] K. A. Takeuchi, Phys. Rev. E **77**, 030103(R) (2008).

# Light-induced atom losses and heating in magneto-optical trap of Dysprosium atoms

Chayma Bouazza, Thomas Chalopin, Davide Dreon,
Leonid A. Sidorenkov, Jean Dalibard, Sylvain Nascimbène
Laboratoire Kastler Brossel, Collège de France, CNRS, ENS-PSL Research University,
UPMC-Sorbonne Universités, 11 place Marcelin Berthelot, 75005 Paris, France

We report on comprehensive study of magneto-optically trapped atoms of $^{164}Dy$. Similarly to results obtained in Sr [1], Er [2], and Dy [3,4] MOTs, we observe a spin-polarized sample in a far-detuned, so-called narrow-line MOT regime ($\Delta$>10$\Gamma$). Decreasing detuning of the cooling light and approaching the resonance (10$\Gamma$ >$\Delta$>5$\Gamma$), however, shows a clear onset of spin depolarization. In this regime, we find a deviation of the MOT equilibrium properties, which in the limiting case of $\Delta \approx \Gamma/2$ should reveal those of typical alkaline-atoms MOT.

We mainly focus our study on atomic loss and heating mechanisms in spin-polarized and intermediate regimes. We have up to $3.5\times10^8$ atoms at the temperature around 15μK in the narrow-line MOT. In the intermediate regime, the observed temperature shows a clear trend to deviate from the equilibrium 'one-body' model of narrow-line MOT, and the atom loss increases dramatically. Empirical fits to our data show a typical increase of two-body loss coefficient β from few $10^{-10}$ to few $10^{-9}$ $cm^3/s$ while going from narrow-line to intermediate regime.

In order to better understand a significant impact of two-body effects, we follow the approach developed for the MOTs of alkaline atoms [5] and introduce a simplified model which involves a possible 'two-body' mechanism accounting for additional heating and atom loss. We consider formation of molecule in an attractive-branch excited state induced by nearly-resonant cooling light. Corresponding increase of kinetic energy during the time spent in excited molecular state, being released after spontaneous decay, leads to either heating of the sample or escape of the atoms. This model is in good qualitative agreement with our data on temperature and two-body loss coefficient in both above-mentioned regimes. A quantitative comparison would, however, require a better knowledge on excited-state molecular potentials and might be the subject of future work.

[1] T. H. Loftus, T. Ido, M. M. Boyd, A. D. Ludlow, and J. Ye, Phys. Rev. A **70**, 063413 (2004).
[2] A. Frisch, K. Aikawa, M. Mark, A. Rietzler, J. Schindler, E. Zupanič, R. Grimm, and F. Ferlaino, Phys. Rev. A **85**, 051401(R) (2012).
[3] S. H. Youn, M. Lu, U. Ray, and B. L. Lev, Phys. Rev. A **82**, 043425 (2010).
[4] T. Maier, H. Kadau, M. Schmitt, A. Griesmaier, and T. Pfau, Opt. Lett **39**, 3138 (2014).
[5] J. Weiner, V. S. Bagnato, S. Zilio, and P. S. Julliene, Rev. Mod. Phys **71.1**, 1 (1999).

# Realization of highly uniform microtrap arrays for quantum simulation of many-body systems using Rydberg atoms

Hikaru Tamura, Yusuke Yamaguchi, Ken'ichi Nakagawa

Institute for Laser Science, University of Electro-Communications,
1-5-1 Chofugaoka, Chofu, Tokyo, Japan

Single neutral atoms trapped in optical microtraps are an attractive platform for implementing multi-qubit quantum information processing and quantum simulation with Rydberg atoms [1, 2]. It is a challenging for such a bottom-up approach to scale up to a large scaling quantum system. The arrays of microtraps with high degree of freedom can be now generated using techniques including acousto-optic deflection [3] and phase-modulation [4] of trapping light. The required conditions for their implementation are realizing arrays with a single atom per site and also detecting individual atoms in each trap with high fidelity. A common technique for the detection of trapped atoms is sub-Doppler cooling by using a homogeneous red-detuned light. However, poor uniformity of trap depths in arrays increases the detection error due to the variance of cooling efficiency and the photon scattering rate from each atom.

In our experiment, we trap individual $^{87}$Rb atoms in two-dimensional arrays of microtraps separated by few micrometers. The arrays are generated by holographic methods of using a programmable spatial light modulator. Figure 1 shows averaged fluorescence images of single atoms trapped in various arrays [5]. In order to optimize the trap depth uniformity, we experimentally demonstrate iterative feedback methods with the measurements not only of light intensities but also of in-trap fluorescence of individual atoms [5]. By applying the latter method, we can reduce the trap depth variance from 20.8 % to 1.7 % for 4×4 arrays. The detection error of individual atoms is also reduced to less than 0.01 % for all various arrays with up to 62 traps. A feedback system such as the one presented here with holographic trap characteristics obtained by using trapped atoms is useful for creation of finely optimized microtrap arrays.

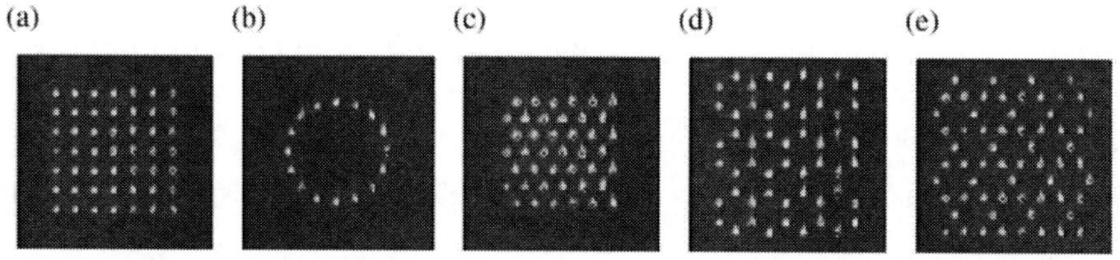


**Figure 1.** Averaged fluorescence images of single atoms trapped in (a) square, (b) ring, (c) triangle, (d) honeycomb, and (e) kagome arrays. The inter-atomic spacing is 5 μm.

[1] M. Saffman, T. Walker, and K. Mølmer, Rev. Mod. Phys. **82**, 2313 (2010).

[2] H. Labuhn, D. Barredo, S. Ravets, S. Léséleuc, T. Macrì, T. Lahaye, and A. Browaeys, Nature **534**, 667-670 (2016).

[3] B. J. Lester, N. Luick, A. M. Kaufman, C. M. Reynolds, and C. A. Regal, Phys. Rev. Lett. **115**, 073003 (2015).

[4] F. Nogrette, H. Labuhn, S. Ravets, D. Barredo, L. Béguin, A. Vernier, T. Lahaye, and A. Browaeys, Phys. Rev. X **4**, 021034 (2014).

[5] H. Tamura, T. Unakami, J. He, Y. Miyamoto, and K. Nakagawa, Opt. Express **24** 8132-8141 (2016).

# Optical specifications of pyramidal micro-mirror fabricated by bulk etching of silicon with Au and $MgF_2$ coating for MOT

Juntae Jin[1], Junseong Eom[1], Seung S. Lee[1]

[1]Department of Mechanical Engineering, KAIST, Daejeon, 305-701, Korea

Pyramidal MOT system size was hard to reduce by limitation of mechanical fabrication. So, we fabricate pyramidal micro-mirror using bulk etching of silicon. KOH solution with 44 %wt was used as etchant. We measured reflectance of mirror through indirect and direct way. First, we measured surface roughness Ra value using surface profiler. The average value of Ra was 12.5 nm for bulk etched plane, 70 nm for seed layer, and 18 nm for plane after remove seed layer. These roughness values are all smaller than tenth of wavelength of laser beam, 780 nm. So, we verified possibility of pyramidal micro-mirror for MOT. Secondly, we directly measured reflectance of optically coated mirror. Coating material and their thickness was decided for targeted wavelength, 780 nm. So, we chose Au with 195 nm thickness and $MgF_2$ with 195 nm thickness. That materials was coated on pyramidal mirror plane by E-beam evaporator. Protected Au mirror made by THORLABS was used as reference, and only Au 195 nm coated on pyramidal mirror to examine effect of $MgF_2$ coating. Optical measurement setup consisted of $\lambda/2$ plate, PBS, lens, and manufactured zig for inclined mirror. Relative reflectance of fabricated mirror to reference mirror was over 95 %. And relative reflectance of fabricated mirror to only Au coated mirror was over 96 %. As a result, developed pyramidal micro-mirror, bulk etched silicon plane coated with Au and $MgF_2$, has enough usability for MOT.

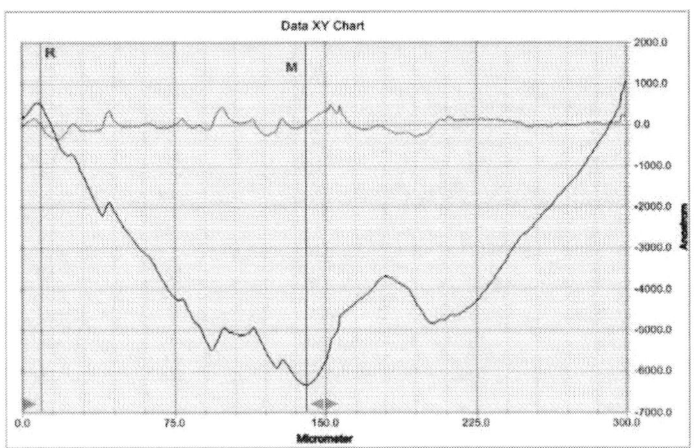


**Figure 1.** Result of surface roughness measurement using surface profiler

**Table 1.** Result of reflectance measurement

| | Reflectance of S-polar | Reflectance with P-polar |
|---|---|---|
| Reference mirror | 91.71 % | 93.76 % |
| Au + MgF2 coated pyramidal mirror | 88.04 % | 89.59 % |
| Au coated pyramidal mirror | 91.25 % | 93.38 % |

**[1]** M. Trupke *et al*, Appl. Phys. **88**, 071116 (2006)

**[2]** S. Pollock *et al*, New. J. Phys. **13**, 043029 (2011)

**[3]** J. M. Bennet, and E. J. Ashley, Appl. Optics. **2**, pp.221-224 (1965)

# A pulsed matter-wave interferometer for probing the mass limits of quantum mechanics

Jonas Rodewald[1], Nadine Dörre[1], Philipp Haslinger[1], Philipp Geyer[1], Andrea Grimaldi[1], Armin Shayeghi[1], Markus Arndt[1]

[1]University of Vienna, Vienna Center for Quantum Science and Technology (VCQ), Vienna, Austria

We demonstrate a matter-wave interferometer in the time domain (OTIMA) as a tool for testing the validity of quantum theory for increasingly heavy and large particles [1,2]. The interferometer operates in the near-field regime and utilizes three pulsed standing laser wave gratings. These imprint, on the one hand, a periodic phase pattern on to the traversing matter waves and, on the other hand, the photon depletion probability is modulated periodically with the distance from the reflecting mirror. Depending on the particle's ionization or fragmentation cross section and optical polarizability such gratings thus act as absorptive masks and phase gratings with an exceptionally small grating period of less than 80nm [3,4]. The pulsed scheme of the experiment facilitates interference measurements in the time domain which offers high count rate, visibility and measuring precision [5]. Since the action of optical gratings is non-dispersive the experiment is well suited for interference experiments on an increasingly large mass scale in the quest for novel decoherence effects and objective collapse mechanisms such as continuous spontaneous localization [2]. Experiments with various organic clusters and monomers have demonstrated the functionality of the interferometer and serve as a motivation for investigating the wave-particle character of nanoparticles with masses up to $10^5$ amu and beyond.

[1] P. Haslinger *et al.* Nat. Phys. **9**, 144 (2013).

[2] S. Nimmrichter *et al.* Phys. Rev. A **83**, 43621 (2011).

[3] N. Dörre *et al.* Phys. Rev. Lett. **113**, 233001 (2014).

[4] N. Dörre *et al.* J. Opt. Soc. Am. B **32**, 114 (2014).

[5] S. Nimmrichter *et al.* New J. Phys. **13**, 75002 (2011).

# Optical bichromatic force slowing of MgF molecules

Xiuxiu Yang[1], Yong Xia[1,2], Jianping Yin[1]

[1] State Key Laboratory of Precision Spectroscopy and Department of Physics, East China Normal University, Shanghai 200062, China

[2] NYU-ECNU Institute of Physics at NYU Shanghai, Shanghai, China

In the past few years, direct laser cooling and trapping of diatomic molecules has become possible. Besides certain polar molecules (SrF, YO, CaF) that have been demonstrated, there are much more candidates appropriate for laser cooling such as the ongoing YbF, BH, CaH, BaH, MgF, and even polyatomic SrOH. In order to produce a diverse set of dense, ultracold diatomic molecular species, it would be interesting and worthwhile to explore how to deliver much larger numbers of slow molecules for loading into a MOT; nowadays the bichromatic force (BF) slowing is a promising path. In this paper, we present a study of the BF exerted on MgF using three theoretical models based on direct numerical solution for the time-dependent density matrix in the rotating-wave approximation. Among three models, we propose a simplified numerical model to study the dynamic process of MgF slowing by neglecting the effect of the high vibrational levels and a skewed magnetic field is applied to destabilize dark state, which yields results very similar to the exact treatment in most cases. We discuss the feasibility on the optical BF to decelerate the MgF molecular beam, get some innovative research results in the following aspects: (1) We deduce the relation between the per-bichromatic-component irradiance $I$ and the quadrature sum of the Rabi frequency amplitudes of the addressed transitions which should be equal to $\sqrt{3/2}\,\delta$. (2) We compare our proposed simplified model (a) with a full 16-level with an additional high vibrational level simulation of MgF $A^2\Pi \rightarrow X^2\Sigma^+$ transition systems involving a vibrational repumping laser (b), and with a numerical model based on the lambda-type subsystems (c). A cryogenic MgF molecular beam can be decelerated nearly to 4 m/s, which can be captured by molecular MOT. (3) We calculate the BF for the $B^2\Sigma^+ \rightarrow X^2\Sigma^+$ system in MgF. With the limitation of the saturation irradiance $I_s$ and the value of $\Gamma/k$, we find that $A \rightarrow X$ transition in MgF is more suitable for the BF cooling and slowing. It guides us that an atomic or molecular system with larger $\Gamma/k$ and smaller $\pi hc\Gamma/3\lambda^3$ is more appropriate to be the candidate of the BF slowing.

[1] M. T. Hummon, M. Yeo, B. K. Stuhl, A. L. Collopy, Y. Xia, and J. Ye, Phys. Rev. Lett. 110, 143001 (2013).

[2] J. F. Barry, D. J. McCarron, E. B. Norrgard, M. H. Steinecker, and D. DeMille, Nature 512, 286 (2014).

[3] D. P. Dai, Y. Xia, Y. F. Fang, L. Xu, Y. N. Yin, X. J. Li, X. X. Yang, and J. P. Yin, J. Phys. B 48, 085302 (2015).

[4] L. Xu, Y.N. Yin, B. Wei, Y. Xia, and J. P. Yin, Phys. Rev. A 93, 013408 (2016).

[5] L. Aldridge, S. E. Galica, and E. E. Eyler, Phys. Rev. A 93, 013419 (2016).

# Spin-Orbit-Coupled Bose-Einstein Condensates of Polar Molecules

Y. Deng[1], L. You[1,2], and S. Yi[3]

[1]State Key Laboratory of Low Dimensional Quantum Physics,
Department of Physics, Tsinghua University, Beijing 100084, China

[2] Collaborative Innovation Center of Quantum Matter, Beijing 100084, China and

[3]Key Laboratory of Theoretical Physics, Institute of Theoretical Physics,
Chinese Academy of Sciences, P.O. Box 2735, Beijing 100190, China

The recent great advancement in experiment realization of ultracold polar molecules in rovibrational ground state open remarkable new frontiers for many areas of science in a highly controllable way. Particularly, the polar molecules possess the large permanent electric dipole moment which gives rise to an interesting nonlinear spin-orbit (SO) coupling of anisotropic dipole-dipole interaction (DDI) with exchanges rotational spin and orbital angular momentum. Meanwhile, the successful realizations of the SO coupling for ultracold atomic gases offer an ideal platform for investigating strongly correlated many-body physics.

We propose an experimental scheme to generate SO coupling for a pseudospin-1 condensates of the bialkali polar molecules between two rotational levels with N=0 and 1. We show that the SO coupling is realized through the hyperfine resolved two-photon Raman processes by employing two elaborately selected excited states which are coupled by strong singlet-triplet mixing of inherent SO interaction. Moreover, we investigate the mean-field phase diagram of this SO-coupled dipolar spinor model. It shows that the interplay of the linear and quadratic Zeeman shift gives rise to a rich quantum phases, including the phase of a doubly quantized vortex-standing wave, a doubly quantized vortex-plane wave, and two singly quantized vortex-plane waves. In addition, the obvious competition effects and a tricritical point of phase transitions are observed among these phases. Of particular interest, our proposal contains tunable linear Raman induced SO coupling and nonlinear anisotropic rotation-orbit coupling of DDI as well, which may facilitate to provide a unique platform for studying a variety of exotic topological quantum phases in dipolar spinor condensates of polar molecules.

[1] Y. Deng and S. Yi, Phys. Rev. A 92, 033624 (2015).

[2] Y. Deng, L. You, and S. Yi, (in preparation).

# Coherent control of ground state $^{23}Na^{40}K$ molecules

Zoe Yan[1], Jee Woo Park[1], Huanqian Loh[1,2], Sebastian Will[1], Martin Zwierlein[1]

[1]MIT-Harvard Center for Ultracold Atoms, Research Laboratory of Electronics and Department of Physics, Massachusetts Institute of Technology, Cambridge, Massachusetts, 02139, USA

[2]Center for Quantum Technologies, National University of Singapore, 3 Science Drive 2, Singapore, 117543

Ultracold molecules with tunable dipolar interactions are promising candidates for studies of quantum information science and interaction-dominated many-body physics. We demonstrate the creation of a spin-polarized ensemble of fermionic $^{23}Na^{40}K$ molecules in their rovibronic and hyperfine ground state. We perform microwave spectroscopy of the ground and first rotationally excited states, allowing our control over all internal and external degrees of freedom. The control afforded by microwave manipulation opens the door to engineering dipolar interactions via microwave dressing. Within the ground rotational state, we create superpositions of nuclear spin states and observe second-scale coherence times. A long coherence time is critical to quantum information processing and precision measurement schemes for probing novel physics.

# Investigation of collisional losses in ultracold Yb($^3P_2$)-Li mixtures

Florian Schaefer[1], Hideki Konishi[1], Shinya Ueda[1], Andreas Kell[1,2], Yoshiro Takahashi[1]
[1]Department of Physics, Graduate School of Science, Kyoto University, Kyoto 606-8502, Japan
[2]Physikalisches Institut, University of Bonn, Wegelerstrasse 8, 53115 Bonn, Germany

We present the results of our comprehensive study of inelastic collisional losses in quantum degenerate mixtures of ytterbium (Yb) in the metastable $^3P_2$ state and lithium (Li) in its electronic ground state. In a deep 3D optical lattice a regime is accessed where Yb is in its Mott insulating state and well localized. Facilitated by the large mass difference between Yb and Li, $m_{Yb}/m_{Li} \approx 29$, Li is only weakly affected by the lattice and remains in its metallic state. This realizes the paradigmatic situation of an impurity system [1] with Yb taking the role of the impurities. Recent experiments on impurity systems have added greatly to our understanding of many-body condensed-matter physics in general and given insights to the physics of polaron quasi-particles [2]. Further studies such as in the realm of Anderson localization [3], Anderson's orthogonality catastrophe [4] and on the Kondo effect [5] are desirable.

For further experimental investigations good control on the inter-species interactions is necessary. By the use of inter-species Feshbach resonances (FRs) such control should be attainable. Of particular interest are FRs between metastable Yb($^3P_2$) and ground state Li, where the existence of several broad FRs is predicted for bosonic $^{174}$Yb [6] and fermionic $^{171}$Yb [7] isotopes. Presently, only experimental results on $^{174}$Yb are available [8, 9]. Here, we review our current results on inelastic collisional properties between $^6$Li and $^{174}$Yb, $^{173}$Yb or $^{171}$Yb. The magnetic field region up to 9 G and a variety of Yb($^3P_2$) Zeeman levels are considered. Due to our high-resolution spectroscopic approach two, three, and four-body inelastic collisions can be probed individually.

While our results did not yet reveal any FRs, they provide data for a refinement of existing model calculations and lay the foundation for a large scale investigation at stronger bias fields using the full variety of isotope and spin richness Yb has to offer.

[1] P. Massignan, M. Zaccanti, and G. M. Bruun, Rep. Prog. Phys. **77**, 034401 (2014).
[2] A. Schirotzek, C. H. Wu, A. Sommer, and M. W. Zwierlein, Phys. Rev. Lett. **23**, 230402 (2009).
[3] S. Ospelkaus *et al.*, Phys. Rev. Lett. **96**, 180403 (2006).
[4] M. Knap *et al.*, Phys. Rev. X **2**, 041020 (2012).
[5] Y. Nishida, Phys. Rev. A **93**, 011606(R) (2016).
[6] M. L. González-Martínez and J. M. Hutson, Phys. Rev. A **88**, 020701 (2013).
[7] T. Chen, C. Zhang, X. Li, and J. Qian, New J. Phys. **17**, 103036 (2015).
[8] W. Dowd *et al.*, New J. Phys. **17**, 055007 (2015).
[9] H. Konishi, F. Schaefer, S. Ueda, and Y. Takahashi, in preparation.

# An ultracold gas of internal-state controlled polyatomic molecules

Alexander Prehn[1], Martin Ibrügger[1], Rosa Glöckner[1], Martin Zeppenfeld[1], Gerhard Rempe[1]
[1]Max-Planck-Institut für Quantenoptik, Hans-Kopfermann-Str. 1, 85748 Garching, Germany

Applications of ultracold ($T$<1mK) polar molecules including ultracold chemistry, quantum simulation, high-precision spectroscopy, and hybrid quantum devices for quantum information processing exploit the rich internal level structure and the electric dipole moment of the molecules. These applications demand a high degree of control over both external and internal degrees of freedom triggering the development of novel cooling and control techniques for polar molecules. Focusing on chemically diverse, polyatomic molecules, our group was recently able to achieve two important goals: a large ensemble of formaldehyde ($H_2CO$) molecules was prepared at both sub-millikelvin motional temperature and with high rotational-state purity.

Specifically, we present direct cooling of formaldehyde to the microkelvin regime [1]. Our method of optoelectrical Sisyphus cooling, which was demonstrated previously with methyl fluoride ($CH_3F$) [2], proceeds in an electric trap [3]. By cooling the external motion by a factor of thousand and increasing the phase-space density by a factor of $10^4$ we generate an ensemble of 300,000 molecules at a temperature of about 400 μK. Additionally, the molecules are prepared in a single rotational state with more than 80% purity employing optical pumping via a vibrational transition [4]. These results bring into reach studies of ultracold collisions or spectroscopy with a molecular fountain.

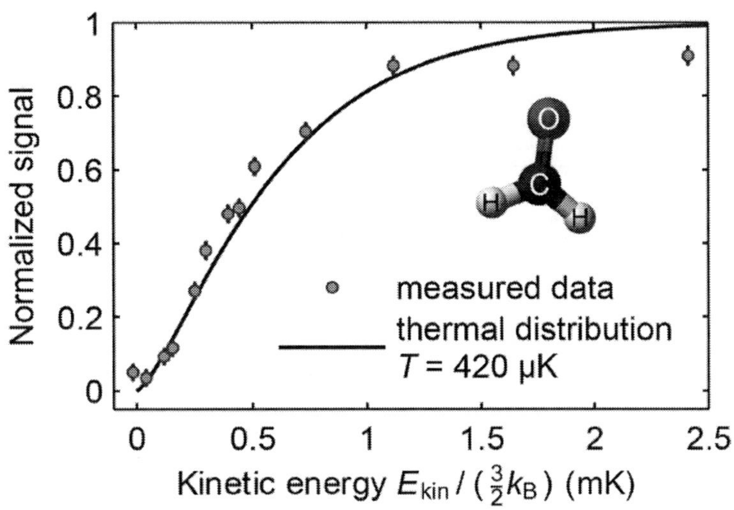


**Figure 1.** Measured integral of the kinetic energy distribution of the trapped gas of formaldehyde. The solid curve shows the expected distribution for a thermal ensemble at 420μK in a perfect box potential.

[1] A. Prehn *et al.*, Phys. Rev. Lett. **116**, 063005 (2016).

[2] M. Zeppenfeld *et al.*, Nature **491**, 570 (2012).

[3] B.G.U. Englert *et al.*, Phys. Rev. Lett. **107**, 263003 (2011).

[4] R. Glöckner *et al.*, Phys. Rev. Lett. **115**, 233001 (2015).

# Sound Propagation of a Bose-Fermi Mixture Superfluid at Finite Temperatures

Yosuke Ono, Ryohei Sakamoto, Rei Hatsuda, Emiko Arahata, Hiroyuki Mori
Faculty of Science and Engineering, Tokyo Metropolitan University, Japan

Sound propagation in ultracold atomic gases is currently attracting considerable attention, as it demonstrates basic properties of superfluid dynamics. Recent experiments of sound propagation in superfluid Bose and Fermi gases[1] are expected to be explained with Landau's two-fluid hydrodynamic equations[2], which predict the existence of the first- and second-sound modes when collisions are strong enough to maintain local thermodynamic equilibrium. The first sound is an in-phase mode in which a superfluid and normal fluid components move together, while the second sound is an out-of-phase mode in which a superfluid and normal fluid components move against each other. This general feature of the first- and second-sounds can be observed for both dilute Bose gases and superfluid $^4$He. However, in an ultracold gas, characteristics of the sound modes are quite different from those of superfluid $^4$He.

Recently, a mixture of Bose and Fermi superfluids was realized by ENS group[3]. The experiment has renewed interest in Bose-Fermi mixture superfluid. We studied the sound modes of a Bose-Fermi mixture superfluid at finite temperatures in the collisional hydrodynamic regime. We extended Landau's hydrodynamic theory to treat the Bose-Fermi mixture superfluid and showed that three sound modes existed. We calculated the hydrodynamic sound velocities numerically using the Nozières and Schmitt-Rink's theory in unitary limit. We found hybridization between the three sound modes[4] in contrast with the two sound modes exhibited by $^3$He and $^4$He superfluids[5].

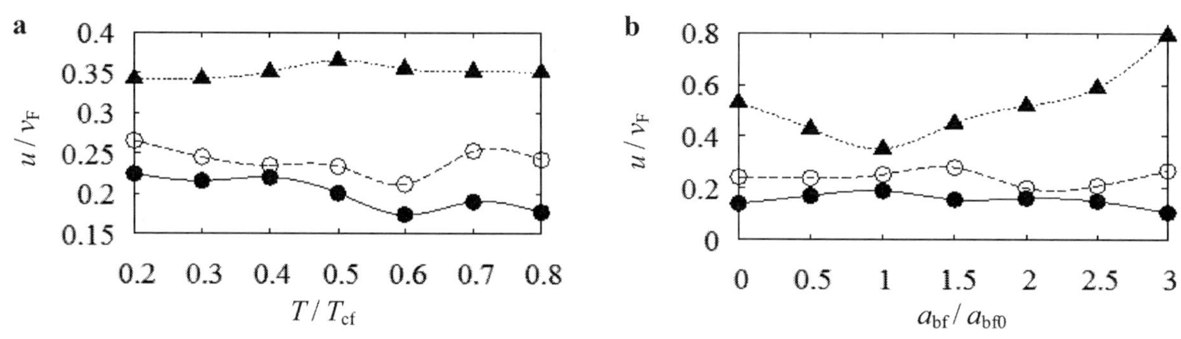


Fig. 1. The sound velocities in the superfluid of Bose-Fermi mixture. **a**: Temperature dependence of the sound velocities. **b**: Bose-Fermi interaction dependence of the sound velocities.

[1] R. Meppelink, *et al.*, Phys. Rev. A **80**, 043605 (2009); L. A. Sidorenkov, *et al.*, Nature **498**, 78 (2013).

[2] L. D. Landau, J. Phys. (USSR) **5**, 71 (1941).

[3] I. Ferrier-Barbut, *et al.*, Science **345**, 1035 (2014).

[4] Y. Ono, R. Sakamoto, H. Mori, and E. Arahata, J. Phys. Soc. Jpn. **85**, 065004 (2016).

[5] G. Volovik, *et al.*, Sov. Phys. JETP **69**, 675 (1975).

# Ground state phase diagram of Bose-Fermi mixture with synthetic spin-orbit coupling

Ryohei Sakamoto, Yosuke Ono, Rei Hatsuda, Emiko Arahata, and Hiroyuki Mori

Faculty of Science and Engineering, Tokyo Metropolitan University

In recent years, a spin-orbit (SO) coupled Bose-Einstein condensate (BEC) was experimentally realized and then an SO coupled degenerate fermions became available[1, 2]. To explore rich properties of SO coupled BECs and degenerate Fermi gases, the SO coupled cold atoms have been studied theoretically and experimentally[3]. Another effort was made for simultaneous cooling of bosonic and fermionic atoms[4]. The mixture systems where atoms of different quantum statistics coexist could have a variety of aspects and have been studied from different viewpoints. With this background, SO coupled Bose-Fermi mixtures are expected to open up another new field and exhibit various interesting phenomena. However, little has been reported so far on the properties of the SO coupled Bose-Fermi mixtures.

We studied the ground state of the SO coupled Bose-Fermi mixtures focusing on effects of the fermions. We introduced fermions into SO coupled BECs and analyzed the ground state of the bosons by using variational method and tight-binding approximation. As a result, we obtained ground state phase diagrams (Fig. 1) with two major characteristic results[5]. One is that the fermions, even if the number is much smaller than that of the bosons, could have a large influence in the ground state, and the other is that the phase diagrams significantly change with the density of the fermions or the strength of Bose-Fermi interactions.

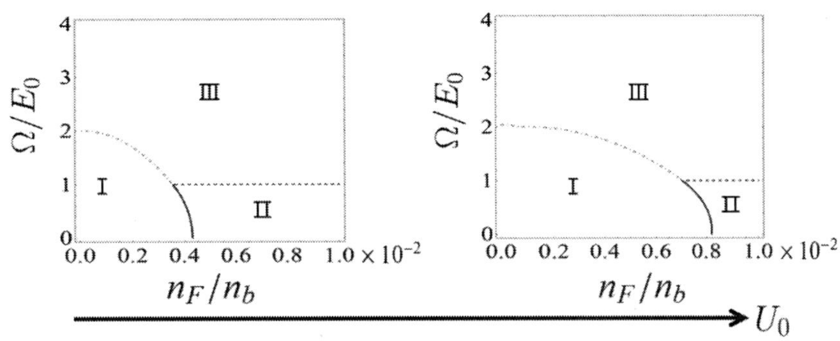


Fig. 1: Phase diagrams of the SO coupled Bose–Fermi mixture at zero temperature. The vertical axis shows the strength of magnetic field and the horizontal one the density of the fermions. The right diagram shows a case of stronger Bose-Fermi interactions.

[1] Y.-J. Lin, K. Jiménez-García, and I. B. Spielman, Nature 471, 83 (2011).

[2] P. Wang, Z.-Q. Yu, Z. Fu, J. Miao et al., Phys. Rev. Lett. 109, 095301 (2012).

[3] Y. Li, L. P. Pitaevskii, and S. Stringari, Phys. Rev. Lett. 108, 225301 (2012).

[4] Z. Hadzibabic, C. A. Stan, K. Dieckmann et al., Phys. Rev. Lett. 88, 160401 (2002).

[5] R. Sakamoto, Y. Ono, E. Arahata, and H. Mori, J. Phys. Soc. Jpn. 85, 064401 (2016).

# Persistent Superfluid Flow Arising from the He-McKellar-Wilkens Effect in Molecular Dipolar Condensates

A. A. Wood[1], B. H. J. McKellar[2], A. M. Martin[1]

[1]School of Physics, The University of Melbourne, Australia

[2]ARC Centre of Excellence for Particle Physics at the Terascale, School of Physics, The University of Melbourne, Australia

We show [1] that the He-McKellar-Wilkens effect can induce a persistent flow in a Bose-Einstein condensate of polar molecules confined in a toroidal trap, with the dipolar interaction mediated via an electric dipole moment. For Bose-Einstein condensates of atoms with a magnetic dipole moment, we show that although it is theoretically possible to induce persistent flow via the Aharonov-Casher effect, the strength of the electric field required is prohibitive. We also outline an experimental geometry tailored specifically for observing the He-McKellar-Wilkens effect in toroidally trapped condensates, as shown below.

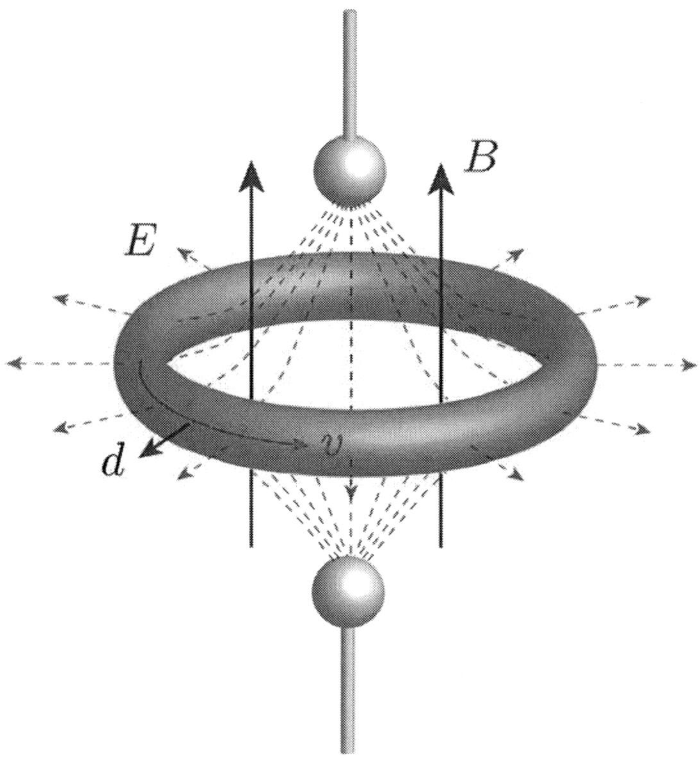


*Figure 1. Proposed experimental geometry to observe the generation of persistent currents in a toroidally trapped Bose-Einstein condensate of polar molecules.*

[1] A. A. Wood, B. H. J. McKellar and A. M Martin, arXiv:1604.04996v1 (to appear in *Phys. Rev. Lett.)*

# Phase separation and dynamics of 2-component Bose-Einstein condensates

K. L. Lee[1], N. B. Jørgensen[2], I.-K. Liu[1,3], L. Wacker[2], J. J. Arlt[2] and N. P. Proukakis[1]

[1]Joint Quantum Centre (JQC) Durham-Newcastle, School of Mathematics and Statistics, Newcastle University, UK

[2]Institut for Fysik og Astronomi, Aarhus Universitet, Denmark

[3]Department of Physics, National Changhua University of Education, Taiwan

The miscibility of two interacting quantum systems is an important testing ground for the understanding of complex quantum systems. Two-component Bose-Einstein condensates enable the investigation of this scenario in a particularly well controlled setting. In a homogeneous system, the transition between mixed and separated phases is fully characterised by a miscibility parameter, based on the ratio of intra- to inter-species interaction strengths. We show, however, that the experimental agreement [1] can be attributed to the repulsion developed during expansion dynamics rather than the in-situ phase separation. Hence, we show [2] that this parameter is no longer the optimal one for trapped gases, for which the location of the phase boundary depends critically on the atom numbers. We demonstrate how monitoring of damping rates and frequencies of dipole oscillations enables the experimental mapping of the phase diagram by numerical implementation of a fully self-consistent finite-temperature kinetic theory [3] for binary condensates. The change in damping rate is explained in terms of surface oscillations in the immiscible regime, and counterflow instability in the miscible regime, with collisions becoming only important in the long time evolution.

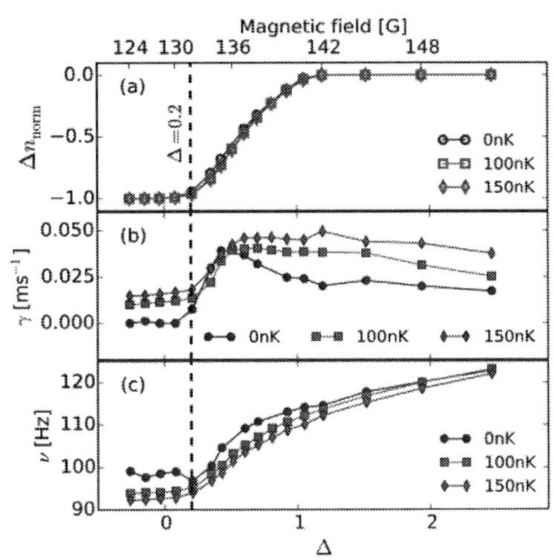


**Figure 1.** (a) Difference in normalised trap-center density $\Delta n_{norm}$, (b) damping rate $\gamma$ and (c) oscillation frequency $\nu$ of COM of $^{39}K$ atoms as a function of miscibility $\Delta$ (bottom axis) or Feshbach magnetic field (top axis) at different temperature.

[1] S. B. Papp, J. M. Pino and C. E. Wieman, PRL **101**, 040402 (2008); L. Wacker, N. B. Jørgensen, D. Birkmose, R. Horchani, W. Ertmer, C. Klempt, N. Winter, J. Sherson, and J. J. Arlt, PRA **92**, 053602 (2015); F. Wang, X. Li, D. Xiong and D. Wang, J. Phys. B **49**, 015302 (2016).

[2] K. L. Lee, N. B. Jørgensen, I.-K. Liu, L. Wackers, J. J. Arlt, N. P. Proukakis, arXiv: 1604.08063 (2016).

[3] M. J. Edmonds, K. L. Lee and N. P. Proukakis, PRA **91** 011602R (2015); *ibid.* **92** 063607 (2015).

# Towards ultracold $^{23}Na^{40}K$ ground-state polar molecules

Jun Rui[1], Huan Yang[1], Lan Liu[1], De-Chao Zhang[1], Ya-Xiong Liu[1], Bo Zhao[1], and Jian-Wei Pan[1]

[1]Shanghai Branch, National Laboratory for Physical Science at Microscale and Department of Modern Physics, University of Science and Technology of China

Ultracold heteronuclear molecules have seen growing interests in the past decades [1]. Fermionic ground state molecules, with large controllable electric dipole moment, would allow access to novel quantum phases in the dipolar Fermi gases [2]. The absolute ground-state molecule of $^{23}Na^{40}K$ is the only chemically stable fermionic molecule that has been experimentally realized by far [3,4]. Its long lifetime is favorable for further evaporative cooling towards quantum degenerate dipolar Fermi gases.

We present here our ongoing efforts towards the ultracold ground-state $^{23}Na^{40}K$ polar molecules. Experimentally, we load the atoms from dual species dark SPOTs which can reduce the light-assisted collisional losses, and then sympathetically evaporate them in an Ioffe-Prichard type magnetic trap. Before reaching quantum degeneracy, the mixture is transferred into a crossed optical trap for creating Feshbach molecules via radio-frequency association. The created Feshbach molecules has a number of $1.5\times10^4$~$2.0\times10^4$, with a conversion efficiency of 10%~15%. The Feshbach molecules realized in our work provides the basis for further creating the ground-state molecules.

[1] L. D. Carr, D. DeMille, R. V. Krems, and J. Ye, New J. Phys. **11**, 055049 (2009).
[2] M. A. Baranov, M. Dalmonte, G. Pupillo, and P. Zoller, Chem. Rev. **112**, 5012 (2012).
[3] C. H. Wu, J. W. Park, P. Ahmadi, S. Will, and M. Zwierlein, Phys. Rev. Lett. **109**, 085 301 (2012).
[4] J. W. Park, S. A. Will, and M. Zwierlein, Phys. Rev. Lett. **114**, 205302 (2015).

# The theoretical study of two-stage crossed beam cooling(TSCBC) in microgravity

Tian Luan, Chen Li, Tianwei Zhou, Kaixiang Yang, Wei Xiong, Xuzong Chen

School of Electronics Engineering and Computer Science,Peking University, Beijing 100871, People's Republic of China

Applying the direct Monte Carlo method developed for ultra-cold quantum gases research, we study the cooling process of Bose-Fermi mixture atoms in a crossed optical dipole trap[1] [2]. The obstacles to producing Fermi degenerate gas via direct sympathetic cooling with Bose gas are also analyzed, by which we find that the side-effect of the gravity is one of the main obstacles. Based on the dynamic nature of the Bose-Fermi mixture atoms, we suggest a two-stage cooling process with two pairs of crossed beams in microgravity environment. According to our simulations, the temperature of $^{6}$Li atoms can be cooled to tens of pico-Kelvin with several thousand atoms via sympathetic cooling with $^{133}$Cs, which propose a novel way to get ultra-cold Bose-Fermi mixture with quantum degeneracy near pico-Kelvin[2].

We also compare the two-stage crossed beam cooling with another cooling method delta-kick cooling (DKC) which could provide a path to pico-Kelvin temperatures of 3D atom gas. We compare the characteristics of TSCBC and DKC for producing a pico-Kelvin system in microgravity. Using a direct simulation Monte Carlo (DSMC) method, we simulate the cooling process of $^{87}$Rb using the two different cooling techniques. Under the same initial conditions, $^{87}$Rb can reach 7 pK in 15 s using TSCBC and 75 pK in 5.1 s with DKC. According to our analysis, TSCBC can reach lower temperatures compared with DKC, but requires more time and a more stable laser. There were no distinct differences in the radius and PSD of the atom clouds between the two methods. With a sympathetic cooling process, the Fermi gas can reach lower temperatures with TSCBC, while DKC takes less time. Moreover, we also analyzed the reasons why a microgravity environment is a better condition for both techniques[3].

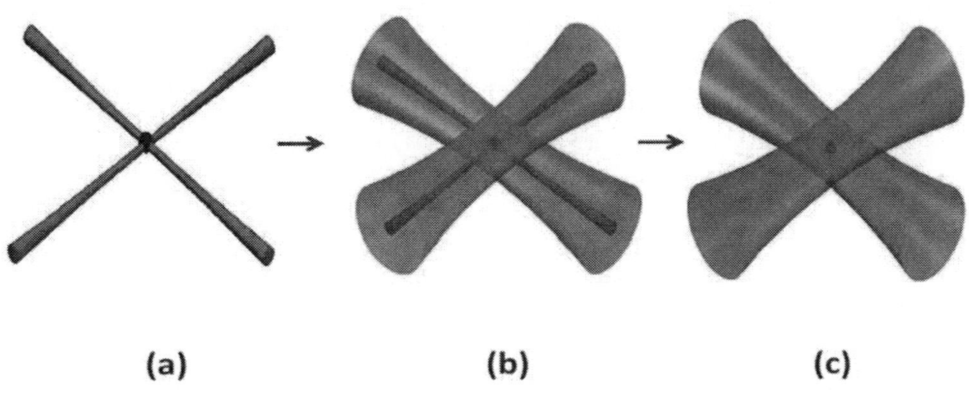


The proposed setup of the multi-beam optical dipole trap, which is designed for the two-stage sympathetic cooling.

[1] T. Luan, T. Jia, X Z Chen, Z. Ma, Chin. Phys. Lett. **31**,043401(2014).

[2] T. Luan, H. Yao, L. Wang, L. Chen et al. Optics Express, **23**, 011378 (2015).

[3] H. Yao, T. Luan, L. Chen, Y. Zhan, et al. Optics Comm. **359**, 123-128 (2016).

# Phase diagram of a strongly interacting Bose-Fermi mixture

Digvijay Kharga, Daisuke Inotani, Ryo Hanai, Yoji Ohashi
Faculty of Science and Technology, Keio University, 3-14-1 Hiyoshi, Yokohama, 223-8522, JAPAN

We theoretically determine the phase boundary between the normal state and the BEC phase i n a three-dimensional Bose-Fermi mixture with a pairing interaction between Bose and Fermi atoms. For this purpose, we extend the previous work in the ordinary $T$-matrix approximation (TMA) in terms of Bose-Fermi pairing fluctuations [1], to include the Hugenholtz-Pines's relati on $\Sigma_B(q=0, i\omega_\nu=0) = \mu_B$ for all the Bose Green's function involved in TMA self-energy diagrams (wh ere $\Sigma_B(q=0, i\omega_\nu=0)$ is the boson self-energy, and $\mu_B$ is the Bose chemical potential). We show that this improvement enhances Bose-Fermi pairing fluctuations, leading to the suppression of the BEC phase, compared to the ordinary TMA case, as shown in Figure 1.

Using this improved TMA, we also examine how the formation of composite fermions affe cts Bose and Fermi single-particle excitation spectra, from the weak-coupling regime to the strong -coupling limit. We show that the existence of composite fermions is seen in both the Bose and Fermi branches in the strong-coupling regime.

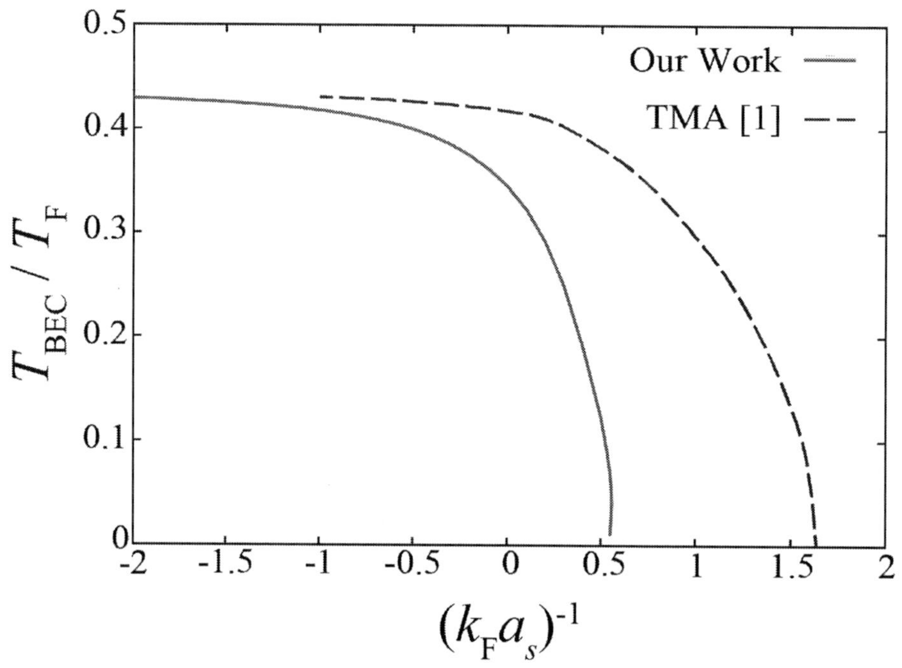


**Figure 1.** Phase diagram of a Bose-Fermi mixture. $T_{BEC}$ is the BEC phase transition temperat ure, which is normalized by the Fermi temperature $T_F$. The Bose-Fermi pairing interaction is measured in terms of the Bose-Fermi scattering as, normalized by the Fermi momentum $k_F$.

[1] E. Fratini and P. Pieri, Phys. Rev. A **81**, 051605 (2010)

# Hyperfine analysis of newly observed photoassociation spectra of the $2\ ^1\Pi(\Omega=1)$, $3\ ^3\Sigma^+(\Omega=1)$, and $2\ ^3\Pi(\Omega=1)$ states of ultracold $^{85}Rb^{133}Cs$ molecule

[1]Jin-Tae Kim, [2]Toshihiko Shimasaki, [2]David DeMille
[1]Department of Photonic Engineering, Chosun University, Gwangju 61452, Korea
[2]Department of Physics, Yale University, New Haven, CT 06511-8499, USA

We present hyperfine analysis of newly observed short-range photoassociation (PA) spectra of the $2\ ^1\Pi(\Omega=1)$, $3\ ^3\Sigma^+(\Omega=1)$, and $2\ ^3\Pi(\Omega=1)$ states of ultracold $^{85}Rb^{133}Cs$ molecule. Those rich hyperfine structures in the low N have been explained using various coupling schemes [1] of Hund's case $b_{\beta S}$ mainly considering Fermi contact terms for Cs and Rb atoms and electronic spin-spin interactions contributed from dipole-dipole and second-order spin-orbit interactions. Through triplet-singlet spin-orbit electronic mixings, the singlet state often has larger HFS splittings compared to the triplet state and shows HFS structure like a triplet state unexpectedly although there is no electronic spin. Transition probabilities between initial free atomic scattering states starting with $^{85}Rb$ and $^{133}Cs$ atoms trapped in their $|F_{Rb}=2\rangle$ and $|F_{Cs}=3\rangle$ hyperfine states in dark SPOT MOTs and final bound HFS lines with Hund's case $b_{\beta S}$ will be discussed using angular momentum recoupling methods. RbCs molecules in the rovibronic ground state have been produced through these PA lines via one-photon decay showing an efficient direct path to the rovibronic ground state.

[1] Li *et al.*, J. Chem. Phys. **116**, 10704 (2002).

# Photoassociation spectrum and photoionization spectrum of ultracold RbCs molecules

Zhonghua Ji, Jinpeng Yuan, Zhonghao Li, Yanting Zhao, Liantuan Xiao, Suotang Jia

Institute of Laser Spectroscopy, State Key Laboratory of Quantum Optics and Quantum Optics Devices, Shanxi University, Taiyuan, 030006, China

Ultracold heteronuclear molecules with anisotropic long-range dipole-dipole interaction have attracted intensive attentions in recent years for the versatile and promising applications in controlled chemistry, precision measurement, quantum computation and quantum simulation. Our group focus on the RbCs molecules due to the stability property of mlecular absolute ground state, large electric dipole moment, and possibility of heteronuclear molecular Bose-Einstein Condense. The experimental procedure [1] can be simply described as below. The excited RbCs molecules are produced by photoassociation (PA) laser in a dual-species dark spontaneous force optical trap and then spontaneously decay to ground state molecules, which can be ionized by photoionization (PI) laser and then detected by a high sensitive time-of-flight mass spectrum. Until now, the investigated PA spectrum in our group contain $(2)0^+$ state correlated to $Rb(5S_{1/2})+Cs(6P_{1/2})$ [2], $(5)0^+$ state correlated to $Rb(5P_{3/2})+Cs(6P_{1/2})$ [3], and $(4)0^-$ state correlated to $Rb(5P_{1/2})+Cs(6S_{1/2})$ [4]. The PI spectrum of metastable ground state [5] have been be investigated while the PI spectrum of absolute ground state molecules are under investigation. These abundant PA spectrum and PI spectrum have guided us to obtain absolute ground state RbCs molecules in the lowest vibrational level [6]. In the future, the possible researches in our group contains optical trapped molecules, atom-molecules collision, coherent transfer of ground state molecules, precise measurement of rotational and molecular hyperfine structures, microwave manipulation of molecular quantum states and so on.

[1] Z. H. Ji, H. S. Zhang, J. Z. Wu, J. P. Yuan, Y. G. Yang, Y. T. Zhao, J. Ma, L. R. Wang, L. T. Xiao, S. T. Jia, Phys. Rev. A **85**, 013401 (2012).

[2] J. P. Yuan, Z. H. Ji, Z. H. Li, Y. T. Zhao, L. T. Xiao, S. T. Jia, J. Chem. Phys. **143**, 044311 (2015).

[3] J. P. Yuan, Y. T. Zhao, Z. H. Ji, Z. H. Li, J. T. Kim, L. T. Xiao, S. T. Jia, J. Chem. Phys. **143**, 224312 (2015).

[4] Y. T. Zhao, J. P. Yuan, Z. H. Ji, C. L. Li, Z. H. Li, L. T. Xiao, S. T. Jia, accepted by J. Quant. Spectraosc. Ra. (2016).

[5] Z. H. Ji, J. P. Yuan, Y. G. Yang, Y. T. Zhao, L. T. Xiao, S. T. Jia, J. Quant. Spectraosc. Ra. **166**, 36 (2015).

[6] Y. T. Zhao, J. P. Yuan, Z. H. Li, Z. H. Ji, L. T. Xiao, S. T. Jia, Chin. Phys. Lett. **32**, 113301 (2015).

# Laser cooling of a polyatomic radical SrOH

Louis Baum[1,2], Ivan Kozyryev[1,2], Kyle Matsuda[1,2], Alex Sedlack[1,2], John M. Doyle[1,2]
[1]Harvard-MIT Center for Ultracold Atoms
[2]Department of Physics, Harvard University

Polyatomic molecules beckon with novel physics and chemistry [1-3]. Strontium monohydroxide (SrOH) is a promising candidate for laser cooling due to its linear geometry, diagonal Franck-Condon factors, short radiative lifetimes, and unresolved hyperfine splittings. SrOH is also a candidate backbone molecule for the laser cooling of much more complex molecules, SrO-R, where R is a radical such as CH3 or CH2CH3. We begin our approach to ultracold SrOH with production in a buffer-gas beam source [4] followed by one-dimensional laser cooling. Previously, we have reported deflection by radiation pressure force with optical cycling on the $\tilde{X}^2\Sigma^+ \leftrightarrow \tilde{A}^2\Pi_{1/2}$ transition with a single stretching vibrational repumper [5]. Now we report one dimensional laser cooling of molecules from a buffer-gas beam source. A transverse temperature of ~5 mK is achieved with ~330 scattered photons limited by vibrational branching into the bending mode. With the addition of a laser to repump the bending vibrational mode we anticipate scattering ~1000 photons which allows cooling to ultracold temperatures. We will describe our results on laser cooling and outline future efforts towards trapping of SrOH.

[1] M. Kozlov, Phys. Rev. A **87**, 032104 (2013).

[2] M. G. Kozlov and S. A. Levshakov, Ann. Phys. (Berlin) **525**, 452 (2013).

[3] H. Sabbah, L. Biennier, I. R. Sims, Y. Georgievskii, S. J. Klippenstein, and I. W. Smith, Science **317**, 102 (2007).

[4] I. Kozyryev, L. Baum, K. Matsuda, P. Olson, B. Hemmerling and J. M. Doyle, New J. Phys. **17** (2015) 045003.

[5] I. Kozyryev, L. Baum, K. Matsuda, B. Hemmerling and J. M. Doyle, J Phys. B **49**.13 (2016) 134002.

# Trapping and Cooling of Polar Molecules: YO and OH

Tim Langen[1], David L. Reens[1], Hao Wu[1], Alejandra L. Collopy[1], Yewei Wu[1], Shiqian Ding[1], Jun Ye[1],

[1]JILA, NIST and University of Colorado, Boulder, CO 80309, USA

We present two experiments progressing towards directly trapping and cooling polar molecular gases to the millikelvin regime and below. First, using microwave mixing of rotational states and two vibrational repump lasers, we implement a cycling transition in the yttrium (II) monoxide (YO) molecule that is closed to the $10^6$ level [1]. With this cycling transition, a beam of YO from a two-stage cryogenic buffer gas cell is decelerated to less than 5 m/s, slow enough to be loaded into a three-dimensional magneto-optical trap. We plan to implement a second stage of cooling to reach below 100 microkelvin [2]. Second, we present our progress in the magnetic trapping and cooling of a gas of the dipolar hydroxyl (OH) radical [3]. Using detailed studies of electric-field-induced losses, we identify a novel Majorana-like loss mechanism that is inherently connected to the dipolar nature of the molecules. We discuss a newly developed trap design that will mitigate this loss to enable further evaporative cooling.

[1] M. Yeo, M. T. Hummon, A. L. Collopy, B. Yan, B. Hemmerling , E. Chae, J. M. Doyle, J. Ye, Physical Review Letters **114**, 223003 (2015).
[2] A. L. Collopy, M. T. Hummon, M. Yeo, B. Yan, J. Ye, New Journal of Physics **17**, 055008 (2015).
[3] B. K. Stuhl, M. T. Hummon, M. Yeo, G. Quéméner, J. L. Bohn, J. Ye, Nature **492**, 396 (2012).

# Laser slowing of CaF molecules to near the capture velocity of a molecular magneto-optical trap

Eunmi Chae[1, 2], Aakash Ravi[1, 2], Loic Anderegg[1, 2], Benjamin Augenbraun[1, 2], Boerge Hemmerling[1, 2], Garrett K. Drayna[1, 2], Nicholas R. Hutzler[1, 2], Alejandra L. Collopy[4], Jun Ye[4], Wolfgang Ketterle[2, 3], John M. Doyle[1, 2]

[1]Harvard University, Cambridge, Massachusetts 02138, USA

[2]Harvard-MIT Center for Ultracold Atoms, Cambridge, Massachusetts 02138, USA

[3]Massachusetts Institute of Technology, Cambridge, Massachusetts 02139, USA

[4]JILA, NIST and University of Colorado, and Department of Physics, University of Colorado, Boulder, Colorado 80309, USA

Due to their complex internal structures and strong long-range interactions, diatomic molecules are promising platforms for quantum simulations of strongly correlated systems, qubits in quantum information, precision measurements, and studies of quantum chemistry. Most of these experiments require ultracold temperatures, and to fully exploit the diversity of molecular structures requires a general method of cooling. Trapping molecules is a necessary first step. A magneto-optical trap is a good starting point with the temperatures in the mK range [1]. Molecular MOTs have small capture velocities of about 10 m/s, as the large number of states involved in the optical cycling reduces their scattering rates. To trap sufficient number of molecules in the MOT, one needs to prepare as many molecules as possible with near the trap's capture velocity.

Here, a two-stage buffer-gas beam source is used to generate a slow beam of CaF molecules with peak forward velocity $\geq$ 50 m/s. The molecules are then slowed down by scattering photons from three counter-propagating frequency-broadened lasers (Fig.1) [2]. About $6 \times 10^4$ molecules are slowed to a forward velocity of $10 \pm 3$ m/s in the detection region, where we plan to make an AC-MOT for CaF. The latest results and the prospects for co-trapping CaF and Li to explore the feasibility of sympathetic cooling will be presented.

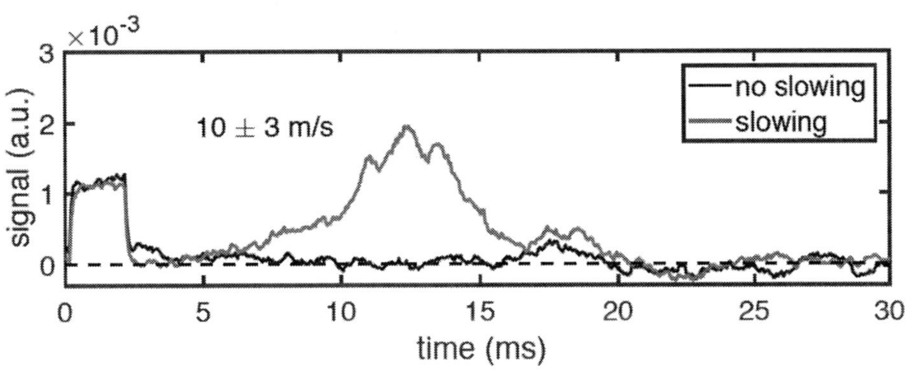


**Figure 1.** Slowed molecules with the forward velocity at $10 \pm 3$ m/s in the detection region.

[1] J. F. Barry et al., Nature **512**, 286 (2014).

[2] B. Hemmerling, E. Chae, et al., arXiv: 1603.02787 (2016), accepted to J. Phys. B.

# Towards Triplet Ground State NaLi molecules

Hyungmok Son[1,2], Timur M. Rvachov[1], Alan O. Jamison[1], Li Jing[1], Sepehr Ebadi[1], Yijun Jiang[1], Martin Zwierlein[1], Wolfgang Ketterle[1]

[1]MIT-Harvard Center for Ultracold Atoms, Research Laboratory of Electronics, and Department of Physics, Massachusetts Institute of Technology, Cambridge, MA 02139, USA

[2]Department of Physics, Harvard University, Cambridge, MA 02138, USA

Heteronuclear dimers can be a great tool for studying many-body dynamics and quantum information processing due to the long-range nature of the electric dipolar interactions. Ultracold collisions between the dimers can be used to explore controlled chemical reactions at the quantum level. For these purposes, the absolute, singlet ground state diatomic molecules have been created with various combinations of alkali atoms. In this conference, we present our recent results on the photoassociation spectroscopy of the NaLi triplet excited and ground state potentials. We also report progress in creating triplet ground state molecules through the Feshbach resonance and STIRAP. The fermionic NaLi molecules in the triplet ground state are expected to be stable with lifetime of >1 second and to have both non-zero magnetic and electric dipole moments. This provides us with opportunities to realize interesting spin lattice Hamiltonians and study molecule-molecule Feshbach resonances.

# A new $^{6}$Lithium-$^{53}$Chromium experiment

Andreas Trenkwalder[1,2,3], Elettra Neri[1,2,3], Massimo Inguscio[1,2,3], Matteo Zaccanti[1,2,3]

[1]Istituto Nazionale di Ottica del Consiglio Nazionale delle Ricerche (CNR), 50019 Sesto Fiorentino, Italy

[2]European Laboratory for Nonlinear Spectroscopy (LENS), 50019 Sesto Fiorentino, Italy

[3]Department of Physics and Astronomy, University of Florence, 50019 Sesto Fiorentino, Italy

The peculiar mass ratio of the $^{6}$Lithium-$^{53}$Chromium Fermi-Fermi mixture supports the existence of one Cr-Cr-Li bound state which allows a resonant control of three-body elastic interactions on top of the usual two-body ones [1,2]. This few-body effect, together with a mismatch between the Fermi surfaces of the heavy and the light components greatly enhances the possibility to realize new exotic superfluid phases. Additionally, our atomic mixture is extraordinary stable against recombination into two-body bound states in the regime of strong interspecies repulsion [3]. This might open a new route to explore magnetic phenomena in ultracold itinerant fermions. We report on the status and perspectives of the experiment we are currently setting up.

[1] O.I. Kartavstsev and A.V. Malykh, J. Phys. B **40**, 1429 (2007).
[2] J. Levinsen and D. Petrov, EPJD **65**, 67 (2011).
[3] D. Petrov, PRA **67**, 010703(R) (2003).

# Two impurities in a Bose-Einstein condensate: from Yukawa to Efimov attracted polarons

Pascal Naidon
Nishina Centre, RIKEN, Wakô 351-0198, Japan

The properties of two heavy impurities immersed in a dilute Bose-Einstein condensate are investigated using a non-perturbative variational approach. For a weak impurity-boson attractive interaction, the two impurities form two free polarons that interact through a weak Yukawa attraction mediated by a boson. For sufficiently strong interactions, the boson-mediated interaction turns into an Efimov attraction leading to the binding of the two impurities and one boson into an Efimov trimer. It is found that this in-medium Efimov trimer exists for weaker interactions than in vacuum.

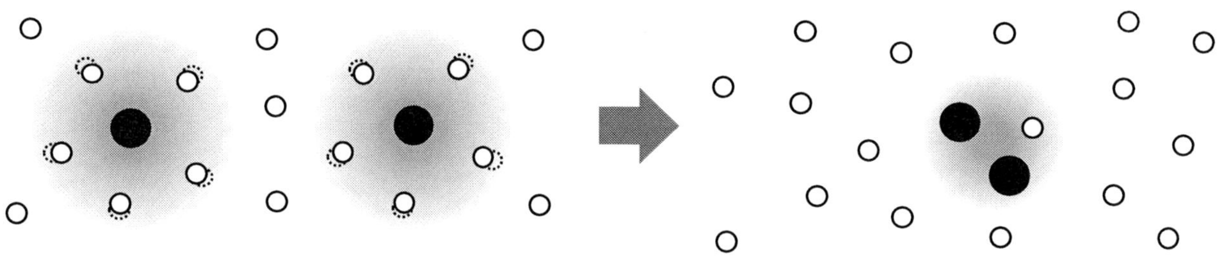

**Figure 1.** Schematic representation of two polarons (left side) created by two impurities (black dots) in a condensate of bosons (white dots) merging into an Efimov trimer (right side) made of two impurities and one boson.

# Heteronuclear Efimov scenario in an ultracold Bose-Fermi mixture with large mass imbalance

Stephan Häfner[1], Juris Ulmanis[1], Rico Pires[1], Eva D. Kuhnle[1], Matthias Weidemüller[1]
[1]Physics Institute, Heidelberg University, Im Neuenheimer Feld 226, 69120 Heidelberg, Germany

Universality in few-body systems is a prominent topic in fundamental quantum physics. One of it's hallmarks is the Efimov effect, where pairwise resonantly interacting particles can form an infinite geometric series of weakly-bound three-body states, the Efimov states, following a discrete scaling law. Due to the large mass imbalance, an ultracold Bose-Fermi mixture of $^{133}Cs$ and $^{6}Li$ atoms features a drastically reduced scaling factor, making it a prototypical system for the thorough investigation of the heteronuclear Efimov effect.

Here we present our recent measurements and analysis of three-body recombination spectra of Li+Cs+Cs collisions close to two broad Li-Cs Feshbach resonances [1]. For negative Cs-Cs scattering length a series of three Efimov resonances is observed [2,3]. The excited resonances are found to be in good agreement with the universal zero-range theory at finite temperature [4], whereas for the ground state deviations become apparent [5]. For positive Cs-Cs intraspecies scattering lengths two resonances are observed. The Efimov ground state transforms into the $Li+Cs_2$ scattering channel and the corresponding three-body resonance is missing. Both, the modification of the scaling factor via short-range effects and the absence of the ground-state resonance, are in agreement with the spinless van der Waals theory [6].

Additionally, we show that the three-body parameter and the scaling factor can be controlled not only by the van der Waals forces, but also by the scattering length between the two bosons, independent of short-range physics.

[1] J. Ulmanis et al., National Science Review (2016)
[2] R. Pires et al., Phys. Rev. Lett. 112, 250404 (2014)
[3] J. Ulmanis et al., NJP 17, 055009 (2015)
[4] D.S. Petrov and F. Werner, Phys. Rev. A 92, 022704 (2015)
[5] J. Ulmanis et al., Phys. Rev. A 93, 022707 (2016)
[6] Y. Wang et al., Phys. Rev. Lett. 109, 243201 (2012)

# $^{174}Yb$ and $^{6}Li$ mixtures: Rapid quantum degenerate gas production and large mass-mismatched Bose-Fermi dual superfluid studies

Richard Roy[1], Alaina Green[1], Ryan Bowler[1], Subhadeep Gupta[1]
[1]Department of Physics, University of Washington, Seattle, WA, United States

We present recent experimental work on few and many-body physics in an ultracold mixture of Yb and Li. By dynamically manipulating the shape of a time-averaged ("painted") optical potential, we achieve highly optimized evaporative cooling of $^{174}$Yb, and develop a theoretical model which successfully captures the evaporation dynamics over more than 3 orders of magnitude in phase space density. This generally applicable tool allows us to create nearly pure BECs of $(0.5 - 1) \times 10^5$ Yb atoms with a total cycle time of (1.6 - 1.8) s, as well as notably large Yb BECs of $1.2 \times 10^6$ atoms in 15 s [1]. By extending these techniques to experiments with Yb and Li, we create a large mass-mismatched ($m_{\mathrm{Yb}}/m_{\mathrm{Li}} = 29$) dual-species Bose-Fermi superfluid mixture, and investigate the elastic and inelastic effects of a Bose (Yb) impurity probe inside a strongly interacting Fermi (Li) gas. Finally, using photoassociation spectroscopy of excited electronic intermolecular potentials in a dual species MOT, we identify 4 rovibrational lines below the Yb($^1S$) + Li($^2P$) asymptote, which forms the first step towards realizing a 2-photon coherent transfer into heteronuclear YbLi ground state molecules [2].

[1] R. Roy, A. Green, R. Bowler, and S. Gupta, Phys. Rev. A **93**, 043403 (2016).
[2] R. Roy, R. Shrestha, A. Green, S. Gupta, M. Li, S. Kotochigova, A. Petrov, and C.H. Yuen, arXiv:1606.03120 (2016).

# Spectroscopy of ultracold $^6Li^{40}K$ molecules

Sambit B. Pal[1], Mark Lam[1], and Kai Dieckmann[1,2]
[1]Centre for Quantum Technologies, Singapore
[2]Department of Physics, National University of Singapore

With the creation of dipolar molecules in their ro-vibrational ground state, a long-standing scientific goal has been achieved [1-6]. At high phase-space density, ultracold heteronuclear molecular samples are promising tools for quantum simulation of a large class of many-body effects and for quantum bit manipulation due to their long-range anisotropic dipolar interactions.

$^6$Li-$^{40}$K molecules in its absolute ground state have a large dipole moment of 3.6 debye, which makes them a suitable candidate for investigating long range dipole-dipole interactions. Starting from Feshbach molecules, we plan to transfer them to the ground state using stimulated Raman adiabatic passage (STIRAP). We give an overview of our single-photon spectroscopic measurements of the $(1)^1\Pi$ and $(2)^1\Sigma$ potentials of $^6$Li-$^{40}$K molecules. We have discovered a variety of new lines and combined with existing literature data for different isotopologues nearly the complete range of the $(1)^1\Pi$ has been covered. From the data the long range behavior of the respective potentials is discussed as well as the suitability of the lines for a Raman transfer. We also present our ongoing work on the electronic ground state where we have identified the v=3 level of the $X^1\Sigma$ potential, by means of two-photon spectroscopy. Figure 1 shows a spectrum in the strong pumping regime with an observed Autler-Townes splitting of about 30 MHz.

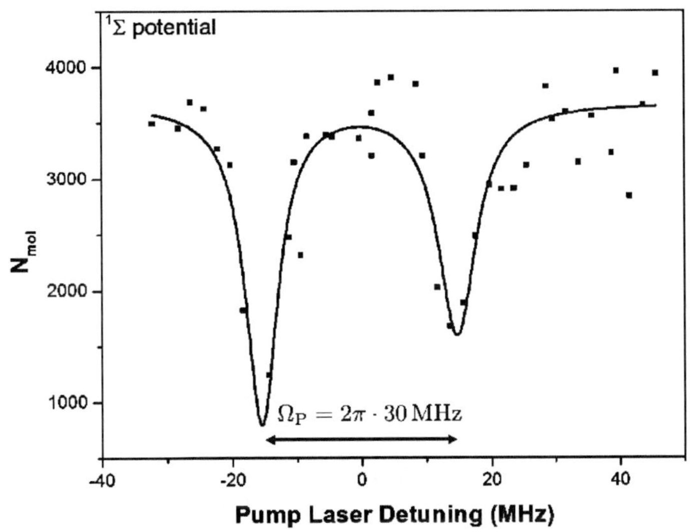


**Figure 1.** Autler-Townes splitting in two-photon spectroscopy of $^6Li^{40}K$.

[1] K. - K. Ni., et al., Science, **322**, 231 (2008).
[2] M. Debatin doctoral thesis (2013). T. Takakoshi, et al. Phys. Rev. Lett, **113**, 205301 (2014).
[3] Peter K. Molony, Philip D. Gregory, Zhonghua Ji, Bo Lu, Michael P. Köppinger, C. Ruth Le Sueur, Caroline L. Blackley, Jeremy M. Hutson, and Simon L. Cornish, Phys. Rev. Lett., **113**, 255301 (2014).
[4] J. W. Park, S. Will, M. W. Zwierlein, Phys. Rev. Lett., **114**, 205302, (2015).
[5] J. W. Park, S. Will, M. W. Zwierlein, New Journ. Phys., **17**, 075016, (2015).
[6] M.Y. Guo, *et al.*, Phys. Rev. Lett., **116**, 205303 (2016).

# Progress towards a MOT of CaF molecules

Moritz Hambach, Hannah Williams, Stefan Truppe, Noah Fitch, Thomas Wall, Ben Sauer, Mike Tarbutt and Ed Hinds

Centre for Cold Matter, Imperial College London, UK

Laser cooling is a powerful and widely used technique in atomic physics, but difficult to apply to molecules due to their complex energy level structure. Only recently the first molecules have been laser cooled and finally trapped in a magneto-optical trap (MOT) [1]. Ultra-cold molecules have applications in precision measurements, quantum information and cold chemistry.

In this poster I will present our recent progress towards a MOT of Calcium fluoride (CaF) molecules. We produce CaF via ablation in a cryogenic buffer gas cell. The molecules are cooled via collisions with Helium buffer gas at 4K and exit the cell as a bright, cold and slow beam (v=140 m/s). Using frequency-chirped radiation pressure slowing, we are able to slow most of the beam down to 15 m/s while simultaneously compressing its velocity width by factor 10, leading to a final longitudinal temperature below 100 mK (see figure)[2]. This method produces pulses of about $10^6$ CaF molecules passing through a typical MOT volume of 1 $cm^3$, 1 m from the source, below the expected capture velocity of a MOT. I will show our next steps for trapping molecules in a MOT, and discuss methods for reducing transverse divergence losses in the slowing process.

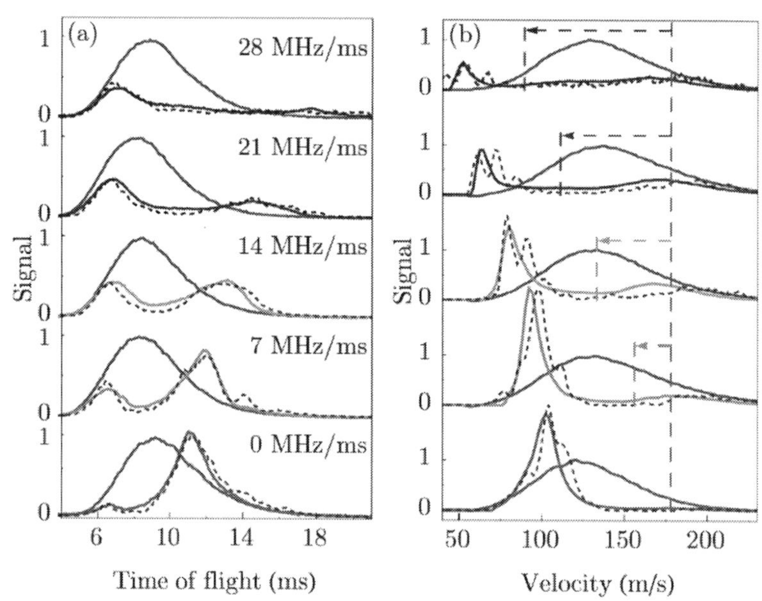


**Figure 1.** Time-of-flight profiles (a) and velocity distributions (b) for various laser chirp rates (coloured lines). Dashed lines are simulations, solid gray lines show the beam without slowing light.

[1] Barry et al, *Nature* **512**, 286-289 (2014).

[2] S.Truppe et al, *arXiv:160506055.*

# Attosecond lighthouse in highly ionizing medium

Kyung Taec Kim[1,2], Emeric Balogh[1], Chang Hee Nam[1,2]
[1]Center for Relativistic Laser Science, Institute for Basic Science (IBS), Gwangju, Korea
[2]Department of Physics and Photon Science, Gwangju Insitutue of Science and Technology (GIST), Gwangju, Korea

The generation and characterization of isolated attosecond pulses have been extensively studied for the time-resolved studies. Many different approaches were developed [1]. These are techniques have in common that they require a pulse with a very large bandwidth, or the combination of two pulses. Isolated attosecond pulse generation by such a field is based on at least one quickly varying parameter during the process, which can be the field's intensity, polarization, or the ionization of the medium. These temporal gating techniques microscopically apply for all atoms in the medium, thus individual atoms generate isolated attosecond pulses.

The attosecond lighthouse uses a macroscopic property - the wavefront rotation of a laser pulse - for separating attosecond pulses generated in each half optical cycle of the laser pulse [2,3]. Attosecond pulses are generated in every half cycle of the laser pulse in the medium. Due to the rotating wavefront of the laser pulse, the attosecond pulses propagate in different directions. The attosecond lighthouse provides multiple isolated attosecond pulses which can be used in attosecond pump and attosecond probe experiments. Also, it provides a direct experimental access to the ultrafast temporal dynamics in successive half cycles.

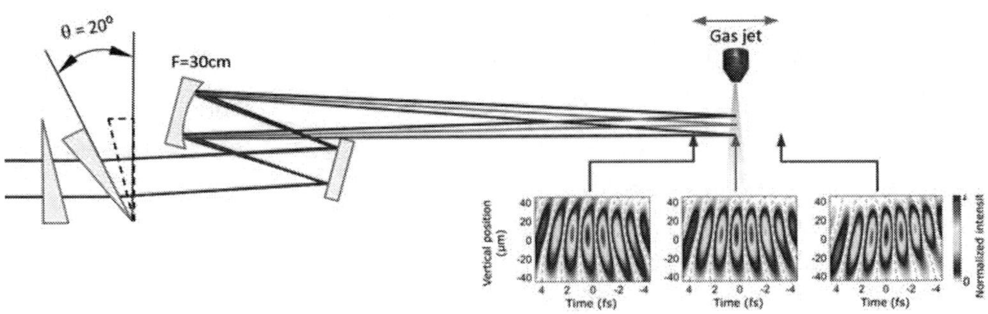


Fig. 1. Schematic diagram of an attosecond lighthouse experiment

In this work we show that propagation effects in a highly ionizing medium can dynamically increase the angular separation of individual attosecond pulses. We present experimental and numerical results of attosecond lighthouse experiments, and show that the choice of a generating medium, focusing geometry and laser intensity affects the angular separation of produced attosecond pulses. This dynamic wavefront rotation enables the separation of attosecond pulses even in the case when the conditions derived from a linear diffraction theory are not satisfied.

[1] M. Chini, K. Zhao, and Z. Chang, Nat. Photonics **8**, 178 (2014).
[2] J. A. Wheeler, A. Borot, S. Monchocé, H. Vincenti, A. Ricci, A. Malvache, R. Lopez-Martens, and F. Quéré, Nat. Photonics **6**, 829 (2012).
[3] K. T. Kim, C. Zhang, T. Ruchon, J.-F. Hergott, T. Auguste, D. M. Villeneuve, P. B. Corkum, and F. Quéré, Nat. Photonics **7**, 651 (2013).

# Wigner time delay near dipole photoionization Cooper minima: Effects of quadrupole channels

Ankur Mandal[1], Pranawa C Deshmukh[1], Anatoli Kheifets[2], Steven T Manson[3]

[1]Department of Physics, Indian Institute of Technology, Madras, Chennai, 600036, India

[2]Research School of Physics and Engineering, The Australian National University, Canberra ACT 0200, Australia

[3]Department of Physics and Astronomy, Georgia State University, Atlanta, 30303, USA

Attosecond time resolution of atomic processes has been achieved recently using ultrafast interferometric techniques [1, 2]. This has led to the possibility of complete photoionization experiments where the phases of the transition amplitudes are measured, in addition to the magnitudes. Since Wigner time delay [3, 4] is related to the energy derivative of the phase of the transition amplitude, this can be used to infer the phases. A number of theoretical studies of the time delay in the photoionization process have been carried out [5]; almost all of these calculations use only dipole approximation. However, in energy regions where the dipole matrix elements go through a Cooper minimum, the nondipole channels become important and can even dominate the photoionization process with a substantial change in the angular distribution of the photoelectrons as a result [6, 7]. In the present work we study the Wigner time delay in the photoionization of Mg 3s in the vicinity of its dipole Cooper minima [8] (~11.1eV). The energy and angular dependence of the photoionization time delay is studied in the relativistic random-phase approximation (RRPA) [9] for both dipole and quadrupole photoionization channels arising from the initial 3s subshell of atomic Mg. Results are presented for time delay for spin-polarized photoelectrons, as well as averages over the photoelectron spin.

[1] A. Baltuška, *et al*, Nature **421**, 611 (2003).

[2] M. Schultze, *et al*, Science **328**, 1658 (2010).

[3] E. P. Wigner, Phys. Rev. **98**, 145 (1955).

[4] F. T. Smith, Phys. Rev. **118**, 349 (1960).

[5] R. Pazourek, S. Nagele and J. Burgdörfer, Rev. Mod. Phys. **87**, 765 (2015).

[6] G. B. Pradhan, *et al*, J. Phys. B **44**, 201001 (2011).

[7] O. Hemmers, *et al*, J. Phys. B **30**, L727 (1997).

[8] J. W. Cooper, Phys. Rev. **128**, 681 (1962).

[9] W. R. Johnson and C. D. Lin, Phys. Rev. A **20**, 964 (1979).

# Gaussian wave packet dynamics with re-initialization in Gabor frame

Chang Woo Byun[1], Min-Ho Lee[1], Nark Nyul Choi[1], Dae-Soung Kim[2]

[1]School of Liberal Arts and Teacher Training, Kumoh National Institute of Technology, Gumi 39177, Republic of Korea

[2]Department of Mechanical Automation, Gyeonggi College of Science and Technology, Gyeonggigwagi-daero 269, Siheung City, Gyeonggi-do 15073, Republic of Korea

Gaussian Wave Packet Dynamics (GWPD) has great advantages as a tool for time-dependent approach to semiclassical dynamics: 1. the evolution of GWP is determined by classical equations of motion, 2. no boundary conditions are needed to be imposed at asymptotic regions, and 3. there is a possibility of avoiding the exponential increase of computational resource as the degrees of freedom of a system grow [1]. However, GWPD suffers from severe limitations: it is valid only for a limited short time, and it cannot describe the classical forbidden processes such as tunneling [2]. These are owing to the broadening of GWP as time goes on, which breaks the validity of GWPD based on quadratic approximation to the potential. As a solution to this problem, it has been suggested to re-initialize the GWPs, i.e., to transform the broadened GWPs to linear combinations of the initial GWPs with a narrow width after evolving in every short time-interval [3-5]. Following the suggestion in ref. 3, we develop a GWPD with re-initialization in a Gabor frame. Our method is shown to work well for a system with tunneling through a potential barrier as can be seen in Fig. 1. And we demonstrate the validity and advantages of the method as a time-dependent Schrődinger equation solver by applying to the above-threshold ionization spectra of the one-dimensional atomic model exposed to ultrafast strong laser pulses [6].

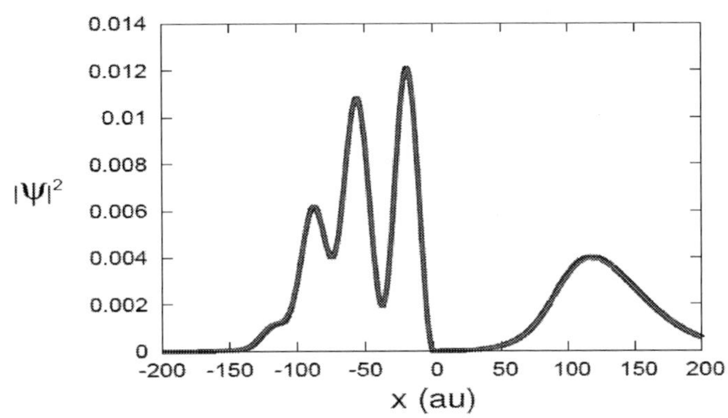


**Figure 1.** Wave packet at time t = 80 a.u. evolved from an initial Gaussian wave packet, whose central position and momentum are respectively q=-7 and p=1, under the potential barrier $V(x) = \exp(-x^2)$. The result obtained by GWPD with re-initialization in a Gabor frame (red line) agrees well with that by the exact quantum method (blue line).

[1] E. J. Heller, J. Chem. Phys. **62**, 1544 (1975).
[2] D. Huber and E. J. Heller, **87**, 5302 (1987).
[3] L. M. Andersson, J. Chem. Phys. **115**, 1158 (2001)
[4] S. Jin, P. Markowich, C. Sparber, Acta Numer. **20**, 121 (2011).
[5] X. Kong, A. Markmann, V. S. Batista, J. Phys. Chem. **120**, 3260 (2016).
[6] Q. Su and J. H. Eberly, Phys. Rev. A **44**, 5997 (1991).

# Signatures of intramolecular quantum interference in photoelectron momentum distributions under strong-field molecular ionization

Vladimir Usachenko[1], Pavel Pyak[2]

[1]Institute of Applied Physics, National University of Uzbekistan, Tashkent, 100174, Uzbekistan

[2]Physics Department, National University of Uzbekistan, Tashkent, 100174, Uzbekistan

The strong-field phenomenon of multiphoton above-threshold ionization (ATI) in laser-irradiated diatomic molecular species ($H_2$, $N_2$ and $O_2$) is addressed theoretically within the *velocity-gauge* formulation of *strong-field approximation* (SFA) under conditions of recent experiments with a mid-infrared wavelength ($\lambda \geq 800$ *nm*) laser field [1-4]. In particular, under the currently applied SFA-based approach [5] only the *direct* ATI process is suggested as quite a sufficient underlying physical mechanism solely contributing to the phenomenon under study. Moreover, the *density functional theory* (DFT) is also essentially exploited here for numerical composition of initial (laser-free) atomic/molecular state using the routines of GAUSSIAN-03 code [6]. Our DFT-SFA calculation results (see Figure 1) demonstrate and reproduce clear signatures of *intramolecular* quantum interference phenomenon manifested in molecular photoelectron momentum distributions as well-established and pronounced *interference* minima, whose form and location proved to be sensitive to the incident laser field wavelength λ and internuclear separation $R_0$ in diatomic molecule.

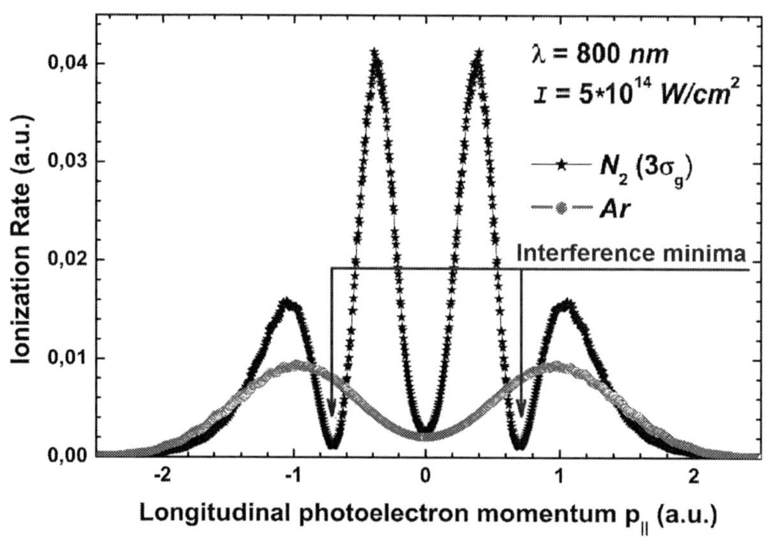


**Figure 1.** Photoelectron momentum distributions calculated for $N_2$ molecule vs *Ar* atom ionization in strong laser field of fixed wavelength λ and peak intensity *I* (as indicated in the picture inset).

[1] C. I. Blaga et al., Nat. Phys. **5**, 335 (2009).

[2] W. Quan et al., Phys. Rev. Lett. **103**, 093001 (2009).

[3] A. Rudenko et al., J. Phys. B **37**, L407 (2004).

[4] R. Moshammer et al., Phys. Rev. Lett. **91**, 113002 (2003).

[5] V. I. Usachenko, P. E Pyak, V. V. Kim, Phys. Rev. A **79**, 023415 (2009).

[6] M. J. Frisch, J. A. Pople, Gaussian-03, Revision A.1 (2003, Gaussian Inc., Pittsburgh, PA).

# Ultrafast cold electron bunches from atomic gases

Andrew McCulloch, Rory Speirs, Ben Sparkes, Robert Scholten
School of Physics, University of Melbourne, Victoria, 3010, Australia

High brightness, ultrashort electron bunches are a core component of free electron lasers [1], particle colliders [2], inverse Compton scattering experiments [3], and ultrafast electron diffraction [4]. In many of these applications, the electron bunch brightness is a limiting factor, stimulating the development of new kinds of pulsed electron sources. Metal or semiconducting photocathodes are the current state-of-the-art in producing bright, ultrafast electron bunches, but ultimately the brightness is limited by the high initial temperature of the electrons produced, typically $10^3$ K to $10^4$ K.

Electron sources based on near-threshold photoionisation of laser-cooled atoms by ultrafast lasers offer potentially large increases in brightness by greatly reducing the temperature at which electrons are generated. These cold atom electron sources have been shown capable of producing electron bunches with temperatures as low as 10 K [5] and bunch charges up to 80 fC [6], which are approaching the values required to perform single-shot electron diffraction from biomolecular microcrystals [7].

We have investigated the temporal structure of the excitation and ionization processes involved in the generation of ultrafast cold electron bunches from laser cooled rubidium. The duration of generated electron bunches was measured with a streaking technique, and pulse durations down to 130 ps were observed using a two-color multiphoton excitation process. We found that pulse broadening processes can increase the bunch duration by up to a factor of 106, due to slow photoexcitation pathways and ionisation by below-threshold tunnelling. Verification that ultrafast cold electron bunches can be generated by photoionization of cold atoms is a step towards these sources being used in applications requiring the highest brightness ultrafast electron bunches, such as ultrafast electron diffraction or injectors for particle accelerators.

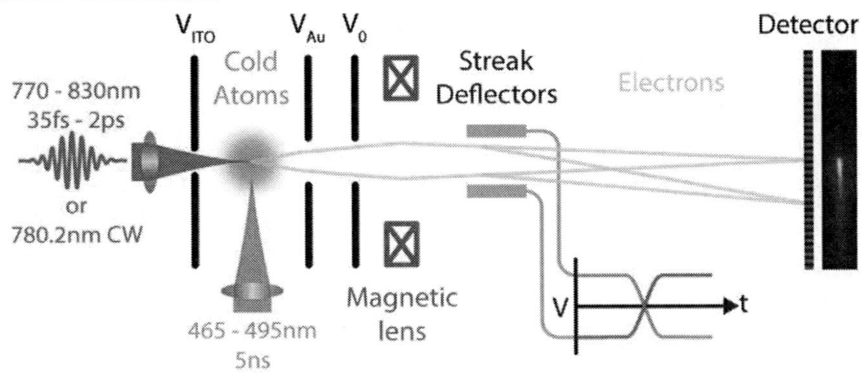


**Figure 1.** Electron bunches are produced by photoionization of laser cooled rubidium gas. The temporal bunch length is determined by applying a time varying deflection to the bunch while it is drifting, and measuring the length of the resulting streak on the detector.

[1] P. Emma *et al.*, Nature Photon. **4**, 641-647 (2010).
[2] C. Hernandez-Garcia *et al.*, Physics Today **61**, 44-49 (2008).
[3] R. W. Schoenlein *et al.*, Science **274**, 236-238 (1996)
[4] G. Scianini and R. J. D. Miller, Rep. Prog. Phys. **74**, 096101 (2011)
[5] A. McCulloch *et al.*, Nat. Phys. **7**, 785-788 (2011)
[6] R. Speirs *et al.*, J. Phys. B **48**, 213002 (2015)
[7] T. van Oudheusden, J. Appl. Phys. **102** (2007)

# Time-dependent multiphoton ionization of Mg atoms in strong laser pulses

Dae-Soung Kim[1], Min-Ho Lee[2], Chang Woo Byun[2], Nark Nyul Choi[2]

[1]Department of Mechanical Automation, Gyeonggi College of Science and Technology, Gyeonggigwagi-dearo 269, Siheung City, Gyeonggi-do 15073, Republic of Korea.

[2]School of Liberal Arts and Teacher Training, Kumoh National Institute of Technology, Gumi 39177, Republic of Korea

A recently developed general non-perturbative method to solve the time dependent Schrodinger equation (TDSE) for the interaction of a strong attosecond laser pulse with a general atom [1] is applied to multiphoton ionization of Mg atoms. The field-free Hamiltonian and the dipole matrices are generated using a flexible B-spline R-matrix(BSR) method [2] combined with MCHF method [3], which allows for the use of non-orthogonal sets of one-electron orbitals. The solution of the TDSE is propagated in time using the Arnoldi- Lanczos method. We are to focus on understanding the feature of two- and three-photon ionization rate as a function of photon energy as shown in Fig. 1.

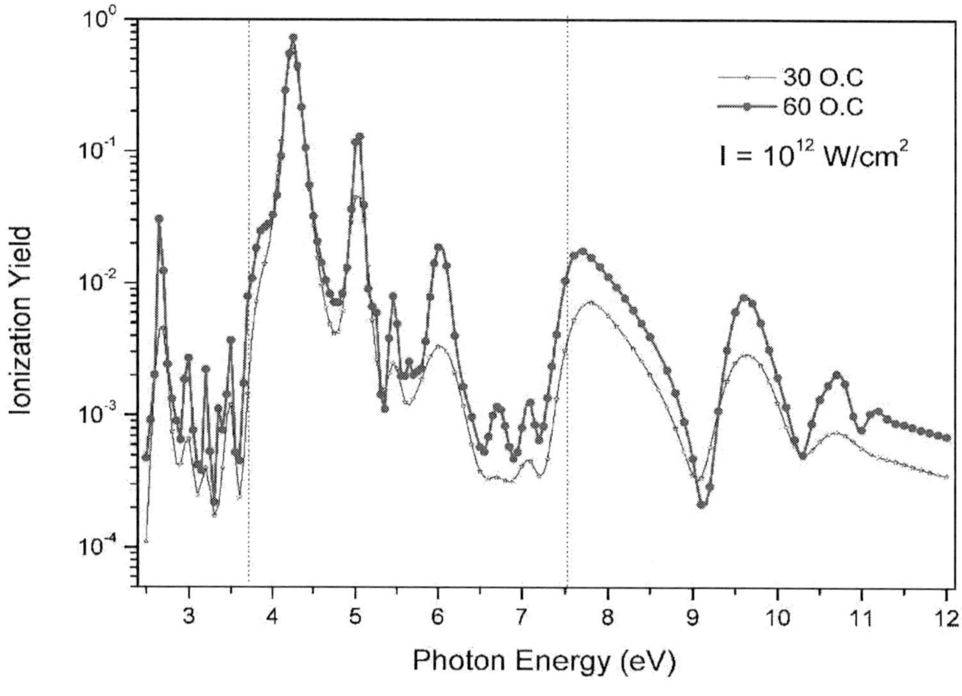


**Figure 1.** Ionization yield as a function of photon energy in 30-cycle and 60-cycle laser pulses of peak intensity $10^{12}$ W/cm$^2$.

[1] X. Guan *et al.*, Phys. Rev. A **76**, 053411 (2007).

[2] O. Zatsarinny, Comput. Phys. Commun. **174**, 273 (2006).

[3] C. F. Fischer *et al.*, Computational Atomic Structure: A MCHF Approach (Institute of Physics Publishing, Bristol, 1997).

# Coherence property of femtosecond X-ray FEL pulses and its applications to ultrafast lenseless imaging

Dohyung Cho[1], Sangsoo Kim[2], Daewoong Nam[1], Changyong Song[1]
[1]Department of Physics, POSTECH, Pohang 790-784, S. Korea
[2]Pohang Accelerator Laboratory, Pohang 790-784, S. Korea

Femtosecond X-ray laser pulses are now available through Self Amplified Spontaneous Emission (SASE) free electron lasers (FELs). This uncommon SASE lasing process makes statistical characteristics of the X-ray photons from such FELs distinct, which can facilitate various types of new applications. We have analyzed the first-order coherence properties of 5.5 keV X-ray photons from SACLA XFEL. Near complete spatial coherence is attained, whilst the temporal coherence is poor due to the native SASE process. The intense femtosecond X-ray laser pulses with the pulse energy higher than $10^{18}$ W/cm$^2$ are well suited for single-pulse imaging with several nm image resolution surpassing the radiation damage limit. Applications of XFEL pulses to nanoscale imaging of nano- & bio-systems are made to realize 3D imaging of catalytic nanoparticles, quantitative imaging of nano-bio complex systems of RNAi microsponges developed for drug-targets. This work introduces new aspects of photon characteristics from XFELs and the applications to coherent diffraction imaging at the spatial and temporal resolution beyond the present limit.

This work is supported by NRF through SRC (2015R1A5A1009962) and NRF-2016R1A2B3010980.

[1] M. Gallagher-Jones, *et al.*, Nature Communications. **5**:3798 (2014).
[2] R. Xu, *et al.*, Nature Communications. **5**:4061 (2014).

# Calculations of long-range three-body interactions for S-S-P system

Pei-Gen Yan[1,2], Li-Yan Tang[2], Zong-Chao Yan[1,2,3], and James F Babb[4]

[1]Department of Physics, University of New Brunswick, Fredericton, New Brunswick, E3B5A3, Canada

[2]State Key Laboratory of Magnetic Resonance and Atomic and Molecular Physics, Chinese Academy of Sciences, Wuhan 430071, China

[3]Center for Cold Atom Physics, Chinese Academy of Sciences, Wuhan 430071, China

[4]Harvard-Smithsonian Center for Astrophysics, ITAMP, Cambridge, Massachusetts 02138, USA

General formulas for calculating the long-range interactions among three identical atoms where two atoms are in identical S states and the other atom is in a P state are obtained using perturbation theory up to second order. The first order (dipolar) interaction coefficients already show a dependence on the geometrical configuration of the three atoms and nonadditive dispersion terms start to appear at the second order, in contrast to the case where all three atoms are in identical S states for which nonadditive dispersion terms appear at the third order. The formalism is demonstrated by the calculation of the coefficients for the Li($2^2$S)-Li($2^2$S)-Li($2^2$P) system using variationally generated atomic lithium wave functions in Hylleraas coordinates. These additive and nonadditive long-range coefficients may be useful in constructing precise potential energy surfaces for this three lithium atom system.

# Orbital Feshbach Resonance in Alkali-Earth Atoms

Ren Zhang[1], Yanting Cheng[1], Hui Zhai[1], and Peng Zhang[2]
[1]Institute for Advanced Study, Tsinghua University
[2]Department of Physics, Renmin University of China

For a mixture of alkali-earth atomic gas in the long-lived excited state $^3P_0$ and the ground state $^1S_0$, in addition to nuclear spin, another ``orbital'' index is introduced to distinguish these two internal states. In this letter we propose a mechanism to induce Feshbach resonance between two atoms with different orbital and nuclear spin quantum numbers. Two essential ingredients are inter-orbital spin-exchanging process and orbital dependence of the Land\'e g-factors. Here the orbital degrees of freedom plays similar role as electron spin degree of freedom in magnetic Feshbach resonance in alkali-metal atoms. This resonance is particularly accessible for $^{173}$Yb system. The BCS-BEC crossover in this system requires two fermion pairing order parameters, and displays significant difference comparing to that in alkali-metal system.

[1] Ren Zhang, Yanting Cheng, Hui Zhai and Peng Zhang, Phys. Rev. Lett. **115**,135301(2015).

[2] Junjun Xu, Ren Zhang, Yanting Cheng, Peng Zhang, Ran Qi and Hui Zhai, arXiv:1602.06 513

# Angular resolved measurements of collisions between ultracold $^{40}$K and $^{87}$Rb

Ryan Thomas[1], Amita Deb[1], and Niels Kjærgaard[1]

[1]Department of Physics, QSO—Centre for Quantum Science and Dodd-Walls Centre for Photonic and Quantum Technologies, University of Otago, 730 Cumberland Street, Dunedin 9016, New Zealand

The scattering experiment is a paradigm of modern physics, as the distribution of scattered waves/particles allows one, in principle, to reconstruct the interaction between the projectile and the target during their collision. It is thus curious that in ultracold atomic physics, a field which is deeply concerned with the interactions between atoms, that most information about the scattering properties of pairs of atoms is inferred through indirect measurements of cross-dimensional rethermalization or the loss of atoms through three-body recombination; almost all information about the angular distribution of scattered atoms is lost or ignored. In the case of Feshbach resonances, these indirect measurements can be time-consuming and require significant care to obtain values that can be reliably compared with theoretical predictions. Here, we present angular resolved measurements of the scattering halos produced from collisions between bosonic $^{87}$Rb and fermionic $^{40}$K. Using an all-optical atom collider[1] built using time-averaged crossed-beam dipole traps, we can prepare specific states of the two species in separate optical traps and accelerate them towards each other at energies up to a temperature equivalent of 350 μK. In the absence of a magnetic field, direct analysis of absorption images of the scattering halo yields the s and p-wave scattering phase shifts: the first measurement of this type for K-Rb collisions. We also demonstrate measurements of a magnetically tunable s-wave Feshbach resonance at 299 G[2] through the numbers of potassium and rubidium atoms scattered by the collision. By analyzing the magnetic field dependence of the scattered number of atoms for a small number of collision energies, we can readily determine both the elastic width of the resonance and the residual magnetic moment of the bound state – measurements that are much more difficult to perform without the benefit of directly counting the number of scattered atoms.

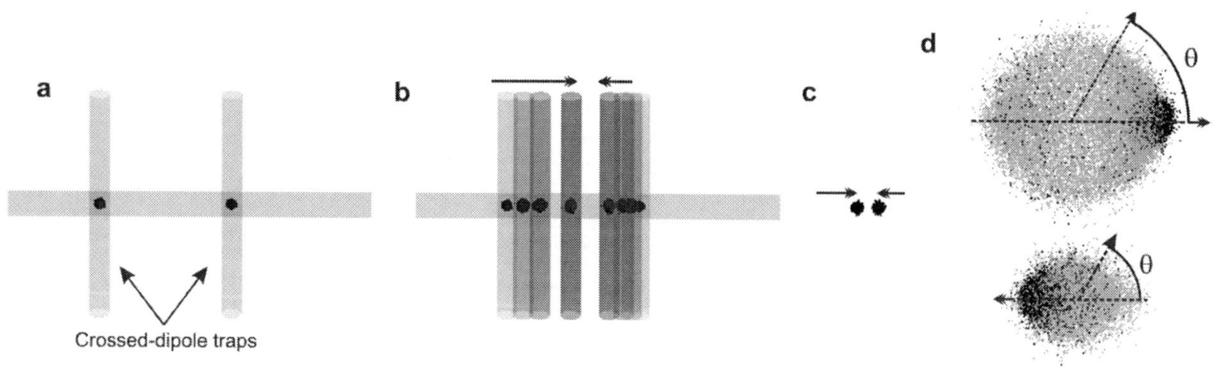


**Figure 1**: Collision sequence. Clouds of Rb and K are held in separated optical traps, which are accelerated towards each other with zero total momentum. In **d,** the top panel is K atoms and bottom Rb atoms. Black dots represent atoms that have not scattered.

[1] A. Rakonjac, A.B. Deb, S. Hoinka, D. Hudson, B.J. Sawyer, and N. Kjærgaard, Laser based accelerator for ultracold atoms. Opt. Lett. **37** 1085-1087 (2012).

[2] F. Ferlaino, C. D'Errico, G. Roati, M. Zaccanti, M. Inguscio, G. Modugno, Feshbach spectroscopy of a K-Rb atomic mixture. PRA **73** 040702(R) (2006).

# Combining Single Cesium Atoms with a Rubidium BEC

Daniel Mayer [1,2], Michael Hohmann [1], Farina Kindermann [1], Tobias Lausch [1], Felix Schmidt [1,2], and Artur Widera [1,2]

[1]Department of Physics, University of Kaiserslautern
Erwin-Schrödinger-Straße 46, 67663 Kaiserslautern, Germany
[2]Graduate School Materials Science in Mainz
Gottlieb-Daimler-Straße 47, 67663 Kaiserslautern, Germany

We have constructed an ultracold quantum gas experiment aiming at combining single, tightly controlled neutral atoms with a quantum many-body system by immersing a single Cesium atom into a Rubidium Bose-Einstein condensate [1]. While the Rubidium atoms provide a bath for the impurity atom, the impurity itself can serve as a probe for this bath.

The interaction dynamics between both species is revealed by space and time resolved fluorescence imaging of the impurity in a species-selective optical lattice [2]. Additionally, a release-recapture-technique is used to access the corresponding impurity energy distribution when immersed into the bath.

On my poster, I will present recent results on using single atoms as local temperature probes for an ultracold gas [3] as well as the current status of the experiment and future perspectives.

[1] M. Hohmann et. al., EPJ Quantum Technology **2**:23 (2015).

[2] F. Schmidt et. al., Phys. Rev. A **93**, 022507 (2016).

[3] M. Hohmann et. al., Phys. Rev. A **93**, 043607 (2016).

# Modulation of association efficiency in a resonant magneto-association of dimers

Lev Khaykovich
Department of Physics and Institute for Nanotechnology and Advanced Materials, Bar Ilan University, Ramat Gan 5290009, Israel

We report on experimental investigation of dimer association in an ultracold thermal gas of bosonic lithium atoms by means of a resonant magnetic field nodulation [1,2]. The modulation is performed in the vicinity of a Feshbach resonance, and we explore this technique in a novel regime in which the modulation amplitudes are strong. Typically, we observe atom-dimer production efficiency as large as 50% within as short time as few ms. The dimers, due to their deep binding energies, can be easily distinguished from free atoms and we monitor the short time dynamics of atom-dimer conversion efficiency. Surprisingly, at low temperatures we observe high contrast coherent oscillations between free atoms and dimer. We characterize these oscillations and examine the question of their origin.

[1] S.T. Thompson, E. Hodby, C.E. Wieman, Phys. Rev. Lett. **95** 190404 (2005).
[2] O. Machtey, Z. Shotan, N. Gross, L. Khaykovich, Phys. Rev. Lett. **108** 210406 (2012).

# Experimental Investigation on the Ionization Cross Sections of $CH_4$ and $CO_2$ Molecules by Electron Impact at Intermediate Energies

Mevlut Dogan, Zehra Nur Özer, Murat Yavuz
Afyon Kocatepe University, Department of Physics, e-COL Laboratory, Afyonkarahisar, Turkey

Cross section data for electron impact ionization of molecules are very important for modeling and obtaining information of electron impact ionization processes at the molecular level. Due to the challenges in the electron interactions with larger molecules in literature, there are currently limited data of targets of especially three or more atomic molecules. $CH_4$ and $CO_2$ are important molecules in gas discharges and also have atmospheric significance. $CO_2$ have been studied in experiments [1-2], and several theoretical models [3-4]. The aim of this work is to present experimental ionization cross sections meaning double and triple differential cross sections of $CH_4$ and $CO_2$ molecules by electron impact. The apparatus and measuring technique have been described in detail previously in [5]. Briefly, the experimental set-up consists of a conventional 7 lens element electron gun, a Faraday cup and two hemispherical electrostatic electron energy analyzers with channel electron multipliers mounted on them. Double differential cross sections (DDCSs) are measured for low detected electron energies (up to 50 eV) whereas triple differential cross sections (TDCSs) of $CH_4$ and $CO_2$ at 250 eV electron impact and at 37 eV ejection energy [6-8]. In low detected energy region <50 eV ionization is very sensitive to different processes such as post collisional interactions so this energy regime has comparable importance on accurate modeling of the reaction. In this manner, $CO_2$ being a linear triatomic molecule is a crossing point for understanding the ionization dynamics of polyatomic molecules. In the conference, we will discuss the theoretical and experimental progress of ionization cross sections on $CH_4$ and $CO_2$ molecules by electron impact in addition to experiments carried out in e-COL Laboratory.

This work was supported by BAPK through Grant No. 15.HIZ.DES.130 and 15.FENED.03.

[1] M. J. Hussey and A. J. Murray, J. Phys. B: At. Mol. Opt. Phys. 38, 2965 (2005).
[2] A. Lahmam-Bennani, E. M. Staicu Casagrande and A. Naja , J. Phys. B: At. Mol. Opt. Phys. 42, 235205 (2009).
[3] O. Chuluunbaatar and B. Joulakian, J. Phys. B: At. Mol. Opt. Phys. 43, 155201 (2010).
[4] O. Alwan, O. Chuluunbaatar, X. Assfeld, A. Naja and B. Joulakian. J. Phys.B: At. Mol. Opt. Phys. 47, 225201 (2014).
[5] M. Dogan, M. Ulu, Z. N. Ozer, M. Yavuz, and G. Bozkurt, Journal of Spectroscopy, (2013).
[6] Z. N. Ozer, E. Ali, M. Dogan, M. Yavuz, O. Alwan, A. Naja, O. Chuluunbaatar, B.B. Joulakian, C. Ning, J. Colgan, Don Madison, Physical Review A, accepted (2016).
[7] N Işık, M Dogan and S Bahçeli, Triple differential cross section measurements for the outer valence molecular orbitals (1t2) of a methane molecule at 250eV electron impact, J. Phys. B: At. Mol. Opt. Phys. 49 (2016) 065203 (5pp)
[8] Murat Yavuz, Zehra Nur Ozer, Melike Ulu, Christophe Champion and Mevlut Dogan, Experimental and theoretical double differential cross sections for electron impact ionization of methane, The Journal of Chemical Physics 144, 164305 (2016)

# Design and simulation of mass analyzer for electron-ion coincidence measurements

Mevlut Dogan, Murat Yavuz, Zehra Nur Ozer
Afyon Kocatepe University, Physics Department, e-COL Laboratory, Afyon, Turkey

Coincidence measurement of energy-selected electrons and mass-selected ions (electron–ion coincidence spectroscopy) is a powerful technique to clarify the behavior of interacting many-body systems and to measure all types of cross sections. Electron-ion coincidence measurements and related experimental apparatus are well studied in the literature [1]. Most of the research based on time-of-flight analysis of fragment ions after photon-molecule interaction.

We have developed a new electron-ion coincidence apparatus, which consists of an electron gun, hemispherical deflector electron energy analyzer and double focusing mass spectrometer (DFMS) [2]. The DFMS consists of an electric sector for kinetic energy focusing, magnetic sector for momentum focusing and focal plane position sensitive detector. Using the new experimental setup electrons recorded simultaneously in a coincidence with all mass selected positive ions produced in the experiment. By using SIMION particle trajectory simulation software, we simulated and optimized ion trajectories of fragment ions for different acceleration potentials and magnetic field strengths.

This work was supported by BAPK through Grant No. 15.HIZ.DES.130 and 15.FENED.03.

[1] T. Baer, International Journal of Mass Spectrometry 200, 443–457 (2000).

[2] M. Yavuz, Ozer ZN., Dogan M., The design and construction of hybride mass spectrometer and importance of vacuum, IV YUKPOP- International Vacuum Workshop, 28-30 March, Afyon, Turkey (2016)

# Tuesday Posters

# Compensation of Coriolis effect in differential measurement of gravity by a dual-species atom interferometer

Shi-Tong Long[1,2,3], Wei-Tao Duan[1,2,3], Chuan He[1,2,3], Wen-Dong Zhang[1,2,3], Biao Tang[1,2], Lin Zhou[1,2], Jin Wang[1,2], Ming-Sheng Zhan[1,2]

[1]State Key Laboratory of Magnetic Resonance and Atomic and Molecular Physics, Wuhan Institute of Physics and Mathematics, Chinese Academy of Sciences, Wuhan 430071, China

[2]Center for Cold Atom Physics, Chinese Academy of Sciences, Wuhan 430071, China

[3]School of Physics, University of Chinese Academy of Sciences, Beijing 100049, China

In our previous work [1], we performed a test of the weak equivalence principle (WEP) by an $^{85}$Rb-$^{87}$Rb double-diffraction Raman atom interferometer with a resolution of $8\times10^{-9}$ after an integration time of 3200 s, and the uncertainty is $3\times10^{-8}$ which is mainly limited by Coriolis effect due to the Earth's rotation. Now, the dual-species atom fountain source has been improved, the resolution is increased to $3\times10^{-9}$ after 3200 s of integration time. To further reduce the uncertainty, the Coriolis effect should be compensated [2]. Firstly, we optimize the geometry of magneto-optical trap and power balance of six cooling and trapping laser beams, in order to reduce biased horizontal velocity of atom clouds and enhance overlapping of two isotopes. Secondly, we perform the compensation of Coriolis effect by rotating the mirror reflecting Raman beams. As the result of the compensation, the amplitudes of interference fringes for $^{85}$Rb and $^{87}$Rb are modulated simultaneously, and the interference phases for $^{85}$Rb and $^{87}$Rb are also modulated. The dependence of differential gravity measured by $^{85}$Rb and $^{87}$Rb on the rotation rate of the mirror is investigated. Experimental data show that Coriolis effect is efficiently suppressed, the fractional uncertainty due to residual rotation effect is $3\times10^{-9}$. This work is a part of efforts towards the WEP test with atoms at level of $10^{-9}$ or better.

[1] L. Zhou, S. T. Long, B. Tang, X. Chen, F. Gao, W. C. Peng, W. T. Duan, J. Q. Zhong, Z. Y. Xiong, J. Wang, Y. Z. Zhang, and M. S. Zhan, Phys. Rev. Lett. **115**, 013004 (2015).

[2] S. Y. Lan, P. C. Kuan, B. Estey, P. Haslinger, and H. Müller, Phys. Rev. Lett. **108**, 090402 (2012).

# Reciprocity and time-reversal symmetry in quantum reflection of atomic matter-waves from a diffraction grating

Byung Gwun Jin[1], Lee Yeong Kim[2], Wieland Schöllkopf[3], Bum Suk Zhao (조범석)[1,2]

[1]Department of Chemistry, Ulsan National Institute of Science and Technology, Ulsan 44919, Korea

[2]Department of Physics, Ulsan National Institute of Science and Technology, Ulsan 44919, Korea

[3]Fritz-Haber-Institut der Max-Planck-Gesellschaft, Faradayweg 4-6, 14195 Berlin, Germany

The reciprocity theorem in wave optics states that the amplitude and phase of a signal does not change if source and detector are interchanged in a given setup. We validate the reciprocity theorem in quantum reflection of helium atoms from a blazed diffraction grating. At grazing incidence conditions the helium atoms are quantum reflected tens of nanometer above the grating surface. We observe identical diffraction efficiencies of the $n$-th order diffraction peaks at standard grating mount and of the $-n$-th orders at reversed geometry when plotted as a function of deflection angle, thereby confirming reciprocity. For purely elastic scattering as in quantum reflection, observation of reciprocity implies time-reversal symmetry. Considering time-reversal invariance we discuss details of the quantum reflection process such as the distances above the surface where quantum reflection occurs for diffraction beams of different order $n$.

# Using atomic interactions for rotation sensing approaching the Heisenberg Limit in ultra-cold gases

Stephen Ragole[1,2], Jacob M. Taylor[1,2,3]

[1]Joint Quantum Institute, University of Maryland, College Park, MD 20742, USA

[2]Joint Center for Quantum Information and Computer Science, University of Maryland, College Park, MD 20742, USA

[3]National Institute of Standards and Technology, Gaithersburg, MD 20899, USA

Recent experiments have realized ring shaped traps for ultracold atoms. We consider the one-dimensional limit of these ring systems with a moving weak barrier induced, e.g., by a blue-detuned laser beam. Focusing on the strongly interaction regime approaching the Tonks-Girardeau limit, we use Luttinger liquid theory to find that the system is analogous to the phase-slip qubit. We propose a method for the creation and interference of many-body superpositions. These superpositions can be used for rotation sensing and we demonstrate improvement over non-interacting schemes of the same system

# Detection method for topological band defects and supersymmetric quantum mechanics

Ludwig Mathey, Marlon Nuske, Martin Lahrz, Nick Fläschner, Benno Rem, Christof Weitenberg, Klaus Sengstock

Zentrum für Optische Quantentechnologien and Institut für Laserphysik, Universität Hamburg, 22761 Hamburg, Germany

On this poster, we present two recent projects. In the first one, we develop and implement a versatile detection scheme to measure topological defects of Bloch bands in optical lattices. Such defects arise in systems with a two-atomic basis due to the momentum-dependent coupling of the two sublattice states, which gives rise to a pseudo-spin texture. A measurement of azimuthal phase windings of the pseudo-spin directly reveals the topological defects. Here, we propose a method to map out this phase using time-of-flight images after off-resonant lattice modulation. We experimentally demonstrate the method in a hexagonal optical lattice and illustrate its power for the characterization of the merging transition of the two Dirac points induced by beam imbalance. Our method is easily implemented in existing setups and generally applicable to two-band systems in optical lattices.

In the second project, we propose an ultra-cold atom experiment to realize and detect supersymmetric quantum mechanics. We provide a pair of potentials that operate on two internal states of an atom and that are supersymmetric partners. These potentials are easily realized with current technology. We describe how the atomic dynamics in the two potential wells is initialized, how it evolves, and how it is brought to interference for detection. We demonstrate that the contrast of the two wave functions that emerge after the time evolution give a clear signal for the supersymmetry of the two potentials.

# Universal diffraction of atoms and molecules quantum reflected from a plane ruled grating

Bong Jun Kim[1], Ju Hyeon Lee[1], Hye Ah Kim[1], Byung Gwun Jin[1], Lee Yeong Kim[2], Bum Suk Zhao (조범석)[1,2], Weiqing Zhang[3], Wieland Schöllkopf[3]

[1]Department of Chemistry, National Institute of Science and Technology, Ulsan 44919, Korea

[2]Department of Physics, Ulsan National Institute of Science and Technology, Ulsan 44919, Korea

[3]Fritz-Haber-Institut der Max-Planck-Gesellschaft, 14195 Berlin, Germany

The observation of "matter-wave" diffraction of electrons [1] and atoms [2] confirmed de Broglie's wavelength formula and provided unambiguous evidence for the wave-particle duality, which is one of the key aspects of quantum mechanics. Since those seminal experiments diffraction, interference, and further phenomena known from classical optics have been observed with matter waves of atoms, molecules, and even clusters and macromolecules [3,4]. While diffraction experiments with atoms and molecules share the basic features with electromagnetic-wave diffraction, there are also differences arising from atomic and molecular properties that are absent in photons. These include a complex internal structure and the long-range interaction with the optical element. For instance, for atoms and molecules diffracted from a crystal surface the particle-solid interaction affects the effective surface corrugation. As a result, different diffraction patterns are observed for He and $D_2$ at the same de Broglie wavelength [5]. Most recently our group demonstrated that quantum reflection of different atomic and molecular species (He, $He_2$, and $D_2$) from a ruled grating leads to identical diffraction patterns if the particles' de Broglie wavelengths are identical [6]. This behavior, termed universal diffraction, is surprising because particles that are quantum reflected probe the strongly species-dependent attractive branch of the surface-particle interaction potential. Universal diffraction was traced back to the principles of quantum reflection. Here, we present a quantitative experimental investigation of the equality of diffraction efficiencies observed with He and $D_2$ quantum reflected from a blazed ruled diffraction grating. The data confirms universal diffraction behavior.

[1] C. Davisson, L. H. Germer, Phys. Rev. **30**, 705-740 (1927).

[2] I. Estermann, O. Stern, Z. Phys. **61**, 95-125 (1930).

[3] A. D. Cronin, J. Schmiedmayer, D. E. Pritchard, Rev. Mod. Phys. **81**, 1051-1129 (2009).

[4] K. Hornberger, S. Gerlich, P. Haslinger, S. Nimmrichter, M. Arndt, Rev. Mod. Phys. **84**, 157-173 (2012).

[5] M. F. Bertino, F. Hofmann, J. P. Toennies, J. Chem. Phys. **106**, 4327-4338 (1997).

[6] B. S. Zhao, W. Zhang, W. Schöllkopf, Sci. Adv. **2**, e1500901 (2016).

# Atomic multiwave interferometer for Aharonov-Casher-phase measurements

Ke Zhang[1], Min-kang Zhou[1], Zhong-kun Hu[1]

[1]MOE Key Laboratory of Fundamental Physical Quantities Measurement, School of Physics, Huazhong University of Science and Technology, Wuhan, People's Republic of China

We present an atomic multiwave interferometer with magnetic sublevels to precisely determine the Aharonov-Casher (AC) geometric phase [1]. Simulations show that this interferometer has sharper fringes than a normal two-wave interferometer, which means a higher phase resolution can be achieved. Moreover, atoms evolving in a single hyperfine structure state make the interferometer insensitive to the dc Stark phase shift. This dc Stark shift is one of the main noise sources in AC phase measurements. The constraint of the photon rest mass is also discussed when using this atomic interferometer to measure the Aharonov-Casher phase.

[1] M. K. Zhou, K. Zhang, X. C. Duan, Y. Ke, C. G. Shao, and Z. K. Hu, Phys. Rev. A **94**, 023641 (2016).

# Very long baseline atom interferometry

Dennis Schlippert[1], Étienne Wodey[1], Christian Meiners[1], Sven Abend[1], Holger Ahlers[1], Henning Albers[1], Naceur Gaaloul[1], Hendrik Heine[1], Dipankar Nath[1], Logan L. Richardson[1], Christian Schubert[1], Wolfgang Ertmer[1], and Ernst M. Rasel[1]

[1]Institut für Quantenoptik, Leibniz Universität Hannover, Welfengarten 1, 30167 Hannover, Germany

Modern physics stands on two pillars: quantum field theory and Einstein's theory of general relativity. While both theories have been confirmed independently to highest accuracy, no framework unifying all four fundamental interactions consistent over all energy scales exists to date. Certain theories of "quantum gravity" enable a reconciliation of quantum field theory and general relativity through violations of the universality of free-fall (UFF), the key component of Einstein's equivalence principle. The UFF states that all bodies at the same point in space-time experience the same acceleration in a gravitational field independently of their composition. It has been tested extensively, mostly with macroscopic test masses, with the best inaccuracies reaching down to a few parts in ten trillion [1, 2].

Matter-wave interferometers are a novel method for testing the UFF. By using genuine quantum test masses they differ considerably from the experiments described in [1, 2]. We report on a quantum test of the UFF with 100 ppb uncertainty using two different chemical elements, $^{39}$K and $^{87}$Rb [3] and ongoing improvements of the experiment aiming towards test with ppb inaccuracy including correlated noise suppression.

In the future, Very Long Baseline Atom Interferometers (VLBAI) [4, 5, 6] represent a new class of matter-wave sensors which extend the baseline from tens of centimeters to several meters, enabling free-fall times on the order of seconds and a corresponding increase in the sensitivity to the square of the free-fall time. This does not only establish a complementary method to state of the art superconducting gravimeters when operated as single species gravimeters but also aims at quantum tests of the UFF at levels comparable to the best classical tests and beyond [7].

We present our strategies and recent progress on the construction of a very-long-baseline-atom-interferometry facility in Hannover. We focus in particular on the use of ultra-cold mixtures of ytterbium and rubidium for quantum tests of the UFF and the development of our source of ultra-cold ytterbium atoms.

The very-long-baseline-atom-interferometry facility is a major research equipment funded by the German Research Foundation (DFG). We also acknowledge support from DFG through SFB 1128 (geo-Q) and SFB 1227 (DQ-mat), the Research Training Group 1729 and from the German Aerospace Center (DLR) with funds provided by the Federal Ministry of Economics and Technology (BMWi) due to an enactment of the German Bundestag under Grant No. DLR50WM1142 (project PRIMUS-II).

[1] S. Schlamminger *et al.,* Phys. Rev. Lett. **100,** 041101 (2008)
[2] J. Müller *et al.,* Class. Quant. Grav. **29,** 184006(2012)
[3] D. Schlippert *et al.,* Phys. Rev. Lett. **112,** 203002 (2014)
[4] R. Geiger *et al.,* Proc. 50th Rencontres de Moriond, La Thuile, Italy (2015)
[5] L. Zhou *et al.,* Gen. Rel. Grav. **43,** 1931 (2011)
[6] T. Kovachy *et al.,* Nature **528,** 530 (2015)
[7] J. Hartwig *et al.,* New J. Phys. **17,** 035011 (2015)

# Atom Interferometry on Sounding Rockets

Stephan Tobias Seidel[1,2], Ernst Maria Rasel[1] for the MAIUS-Team
[1]Institut für Quantenoptik, Leibniz Universität Hannover, Germany, Welfengarten 1, 30167 Hannover, Germany
[2]Deutsches Zentrum für Luft und Raumfahrt e.V.,(DLR/MORABA), Germany

The possibility of precise measurements of inertial forces using atom interferometry has led to a multitude of proposals for future satellite missions. These include missions aimed at geodetic measurements like a characterization of earth's gravitational field gradient and fundamental physics like a test of the universality of free fall or the detection of gravitational waves.

Current ground based atom interferometer setups are not suitable for the use on a satellite mission. Therefore a series of new technological as well as experimental techniques have to be implemented and qualified. This creates the necessity for pathfinder missions to test atom interferometer setups in relevant environments.

As a step towards the transfer of such a system in space three sounding rocket missions with atom interferometers are currently being prepared. The launch of the first mission, aimed at the first demonstration of a Bose-Einstein-condensate in space and using this quantum degenerate matter as a source for atom interferometry is planned for the end of 2016 from ESRANGE, Sweden. As a launch vehicle a VSB-30 sounding rocket was chosen with a flight ticket provided by the German Aerospace Center (DLR). It will be followed by two more missions that extend the scientific goals to the creation of degenerate mixtures in space and simultaneous atom interferometry with two atomic species. Their success would mark a major advancement towards precise measurements with space-born atom interferometers.

In this poster an outline of the missions, with a focus on the experimental realization of atom interferometers on sounding rockets, is given and the current status is presented.

This research is funded by the German Space Agency DLR under grant number DLR 50 1131-37.

# Atom chip based guided atom interferometer for rotation sensing

S. Bade[1], W. Yan[1], M.-A. Buchet[1], J.-M. Martin[1,2], A. Landragin[1], and C. L. Garrido Alzar[1]

[1]SYRTE, Observatoire de Paris, PSL Research University, CNRS, Sorbonne Universités, UPMC Univ. Paris 06, LNE, 61, avenue de l'Observatoire, 75014 Paris, France

[2]Department of Physics and Astronomy, University of Sussex, Falmer, Brighton, BN1 9QH, UK

The physical aspects as well as the experimental progress towards the realization of a rotation sensor using cold atoms magnetically guided on an atom chip are presented. The design and derivation of the magnetic guiding potential, the expected sensitivity, and the study of a highly efficient matter-wave beam splitter are in detail analyzed. This device is designed taking into account the stringent requirements of inertial navigation. Besides the usual constraints imposed on the physical dimensions and power consumption for the aforementioned application, we also investigate here the on-chip incorporation of keys elements needed in the realization of a cold atom interferometer. In particular, we discuss different strategies to overcome the fundamental limitations of guided [1] and free falling atom interferometer inertial sensors [2]: wire roughness induced decoherence, cloud fragmentation, interrogation time and quantum projection noise.

The working principle of this inertial sensor is based on a magnetic polarized cloud of cold $^{87}$Rb atoms coherently split by a $\pi/2$ pulse that creates a superposition of two opposite wave-packet propagation modes. Both wave-packets will be constrained to propagate along a circular guide of a few millimeters diameter. At the output of the guide, the application of a second $\pi/2$ pulse produces an interference signal sensitive to rotation via the Sagnac effect, measured as an atom number imbalance. If, for example, we fix the interrogation time to 1s, and the atoms are launched via a Bragg process ($v/v_{recoil} = 2$) on a guide of 500μm radius then the expected sensitivity is $3\times10^{-8}$ rad.s$^{-1}$/ $\sqrt{Hz}$ ($6\times10^{-3}$ deg/hr/ $\sqrt{Hz}$ ) for $10^6$ atoms. Such a sensitivity is already close to the navigation grade bias stability requirement for a gyroscope which is on the order of $10^{-4}$ deg/hr.

[1] J.-B. Trebbia, C. L. Garrido Alzar, R. Cornelussen, C. I. Westbroock, and I. Bouchoule. Phys. Rev. Lett. **98**, 263201 (2007).

[2] I. Dutta, D. Savoie, B. Fang, B. Venon, C.L. Garrido Alzar, R. Geiger, and A. Landragin. Phys. Rev. Lett. **116**, 183003 (2016).

# Advances in ytterbium BEC contrast interferometry for a new determination of the fine structure constant α

Benjamin Plotkin-Swing[1], Katherine McAlpine[1], Daniel Gochnauer[1], Subhadeep Gupta[1]
[1]University of Washington, USA

The narrow momentum and position spread of a Bose-Einstein condensate (BEC) can help improve atom interferometric measurements. In earlier work, we demonstrated a contrast interferometer with ytterbium (Yb) BECs[1]. Here, we report progress towards implementing a second generation Yb BEC interferometer with the goal of measuring h/m, where h is Planck's constant and m is the mass of a Yb atom, in order to determine the fine structure constant α. The use of the non-magnetic Yb atom and the symmetric geometry of the interferometer make the measurement immune to several error sources. The precision of our measurement scales with $N^2$ where 2N is the number of photon recoils separating the interfering momentum states in the interferometer. We have produced Yb BECs in a new apparatus, and we have demonstrated the laser pulse atom-optics required for implementing our interferometry sequence with large N. We have achieved N=100 using Bloch oscillations (BO) with 99.5% transfer efficiency per two photon recoil. As an extension of our previous analysis[1], we will also discuss the role of diffraction phases in our interferometer due to the BO pulses.

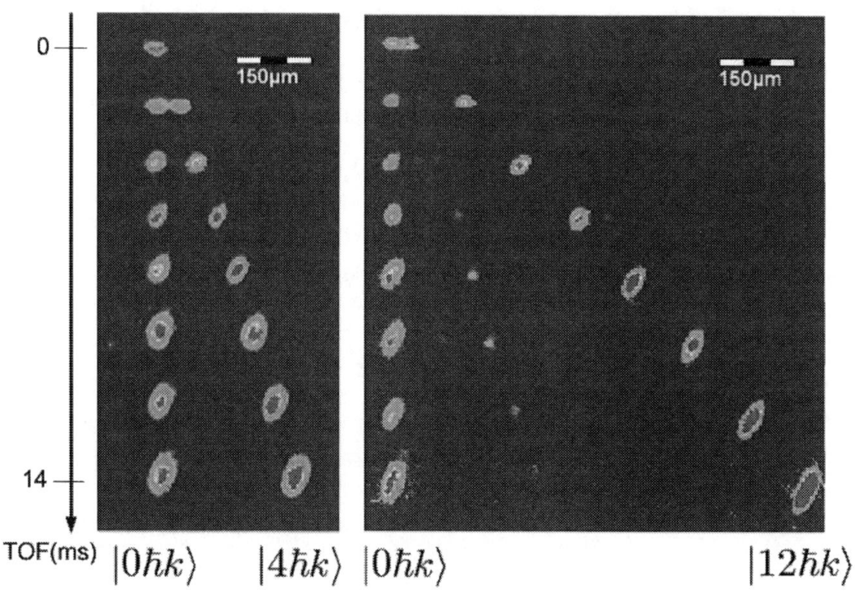


**Figure 1.** Acceleration using Bloch Oscillations (BO). In the left image, a 2nd order Bragg π/2 pulse evenly splits the atoms into the 0 and 4ħk momentum states. In the right image, the splitting pulse is followed by an N=4 BO pulse which imparts a further 8ħk of recoil momentum to the atoms. These images show that it is possible to accelerate the 4ħk state without affecting the 0ħk state, which is crucial for creating the contrast interferometer geometry. We have achieved up to N=100 BO with 99.5% transfer efficiency per two photon recoil.

[1] A.O. Jamison, B. Plotkin-Swing, S. Gupta, Phys. Rev. A **90**, 063606 (2014).

# Our cold atom gravimeter join the comparison of the absolute gravity measurement

Zhaoying Wang[1], Zhijie Fu[1], Qiyu Wang[1], Bing Cheng[2], Bin Wu[2], Qiang Lin[2]
[1]Institute of Optics, Department of Physics, Zhejiang University, Hangzhou, China
[2]Center for Optics and Optoelectronics Research, College of Science, Zhejiang University of Technology, Hangzhou, China

Nowadays, the cold atom gravimeter offers the prospect of a new generation of inertial sensors for field applications. We have already accomplished a mobile cold atom gravimeter (CAG) with a portable and stable laser system [1]. The laser system is based on two distributed feedback (DFB) lasers diodes of 1560 nm. With power amplified firstly by the erbium-doped fiber amplifier (EDFA), and then frequency doubled by using a periodically poled lithium niobate(PPLN) crystal, we can realized 1 W laser output at 780 nm [2]. Finally, a gravity sensitivity of $1.4\times10^{-7}$g $Hz^{-1/2}$ has been achieved. Moreover, A continuous gravity monitoring of 80 h is carried out.

Recently, the first Asia-Pacific Comparison of Absolute Gravitymeters (APMP.M.G-K1) was organized by National Institute of Metrology (NIM) of China from December 21st, 2015 to March 25th, 2016 in Changping, Beijing. Our compact cold-atom gravimeter was transported about 1200km to participate in the comparison, and it is the only one atom gravimeter in this comparison. Three sites were measured as requested by the organizer. The accuracy budget is formulated, considering the significant terms such as Raman laser frequency uncertainty, two-photon light shift and more, with the total instrument uncertainty of 19μGal.

The successful participating of international gravity measurement campaign have proven the mobility and robust of our system. The sensitivity of our cold atom gravimeter is comparable with that of traditional mechanical dropping mirror method. The accuracy of the instrument needs more study and further improvement.

[1] B. Wu, Z. Y. Wang, B. Cheng, Q. Y. Wang, A. P. Xu, Q. Lin, Metrologia, **55**, 452 (2014).
[2] Q.Y. Wang, Z. Y. Wang, Z. J. Fu, W. Y. Liu, Q. Lin, Opt. Commun., **358**, 82 (2016).

# A proposal on improving the accuracy of gravity gradient measurement by modulating magnetic field

Lei Zhu[1,2,3], Yu-Ping Wang[1,2,3], Jia-Qi Zhong[1,2], Jin Wang[1,2], Ming-Sheng Zhan[1,2] State Key Laboratory of Magnetic Resonance and Atomic and Molecular Physics, Wuhan

1

Institute of Physics and Mathematics, Chinese Academy of Sciences, Wuhan 430071, China
[2]Center for Cold Atom Physics, Chinese Academy of Sciences, Wuhan 430071, China
[3]University of Chinese Academy of Sciences, Beijing 100049, China

We present a new scheme for measuring the differential phase in dual atom interferometers. The magnetic field is modulated in one interferometer, and the differential phase can be extracted without measuring the amplitude of magnetic field by combining the ellipse and linear fitting methods. As the system error caused by magnetic field is an important item for gravity gradient measurement, it can be completely eliminated by modulating the pulse duration of magnetic field in our proposal where the phase extracted from the intercept of the fitted line does not contain the system error caused by the magnetic field. This method will make a significant contribution to improve the accuracy of gravity gradient measurement[1].

[1]Yu-PingWang, Jia-Qi Zhong, Xi Chen, Run-Bing Li, Da-Wei Li, Lei Zhu, Hong-Wei Song, Jin Wang, and Ming-Sheng Zhan, Opt. Commun. **375,** 34-37 (2016).

# Cold atom interferometer in earthquake monitoring

Bin Wu[1], Bing Cheng[1], Zhijie Fu[2], Zhaoying Wang[2], Qiang Lin[1]
[1]Center for Optics and Optoelectronics Research, College of Science, Zhejiang University of Technology, Hangzhou 300023, China
[2]Institute of Optics, Department of Physics, Zhejiang University, Hangzhou 310027, China

The performance of cold atom interferometer[1] has been improved substantially in recent years. The atomic gravimeter[2] based on cold atom interferometer has potential practical application in metrology, geodesy, inertial navigation, and earthquake monitoring[3]. For such field applications, the precision and repetition of the cold atom gravimeter should be enough high, and it is better to make the gravimeter enough compact and robust. We present a cold atom gravimeter to monitor the earthquake. The gravimeter has a sensitivity of about $1.0\times10^{-7}\,g/Hz^{1/2}$ and a repetition of 2.2 Hz. Based on this atom gravimeter, we have observed lots of earthquakes. On 28 September 2013, a whole seismic wave of about 1 hour that occurred in Pakistan has been recorded. The results coincide with that recorded by a traditional seismic detector very well including shear waves, compressional waves and surface waves, which can be shown in Figure 1. Recently, the earthquake of relatively small magnitude has been investigated. We have measured the three earthquakes occurred in Taiwan during several hours (2016-04-27/28), the magnitude is about 5.6, 4.9, and 5.2 respectively. The abnormality of the interference fringes has been analyzed.

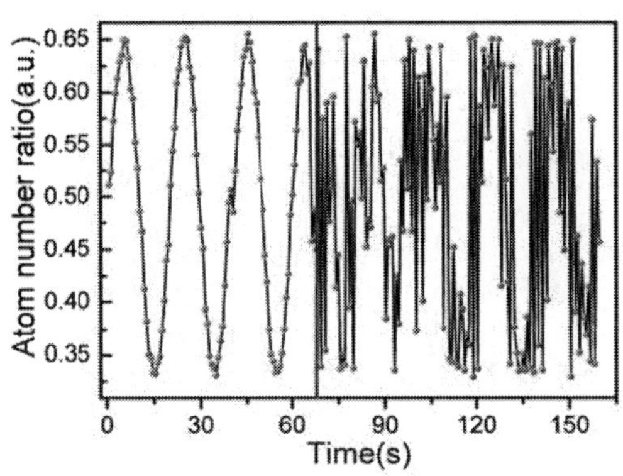


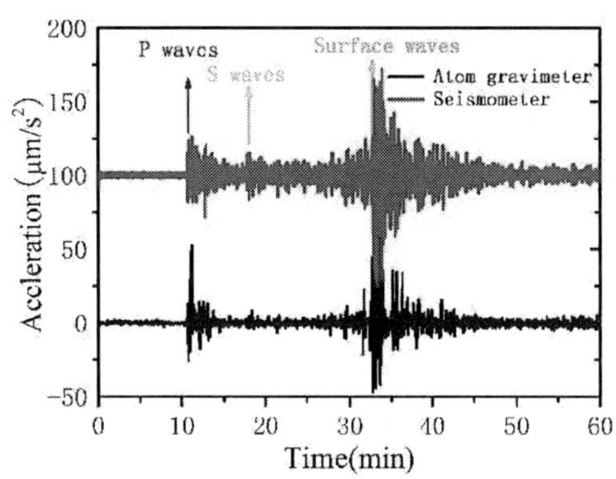


**Figure 1.** Cold atom interferometer in earthquake monitoring. left: The abnormality of atomic interference fringes; right: The comparison between our atomic gravimeter and the seismometer nearby.

[1] M. Kasevich and S. Chu, Phys. Rev. Lett **67**, 181 (1991).
[2] M. Kasevich and S. Chu, Appl. Phys. B, **54**, 321 (1992).
[3] Bin Wu, Zhaoying Wang, Bing Cheng, Qiyu Wang, Aopeng Xu and Qiang Lin, Metrologia, 51, 452 (2014).

# Atom Interferometry: Single-shot, optical-phase-insensitive interferometry using Bose-Einstein condensates

Billy I. Robertson, Andrew R. MacKellar, James Halket, Jonathan D. Pritchard, Aidan S. Arnold, Erling Riis, Paul F. Griffin.

Department of Physics, University of Strathclyde, Glasgow, United Kingdom

Atom interferometry provides a method of precision measurement through the detection of differential phase shifts induced in the atomic wavefunction, and is potentially orders of magnitude more sensitive than its optical analogue. Typically the phase of the wavefunction is readout against the spatial phase of an optical standing wave in the reference frame of the lab. This method is a leading limitation for practical measurement as it requires long stability of the optical phase. We have constructed a general purpose atom interferometer that is inherently insensitive to phase noise in the readout system. We present our experimental approach, including the development of a high-fidelity atom-optics toolbox comprising composite Kapitza-Dirac and Bragg pulses of light which we use to coherently manipulate the momentum of a Bose-Einstein condensate of $^{87}$Rb. An atomic homodyne detection system transfers the atomic phase to a temporal beat-note and the entire interferometric signal can be obtained in a single shot. This method provides a high contrast signal from which we can obtain the recoil frequency of the atomic species as well as infer information about magnetic and gravitational field gradients.

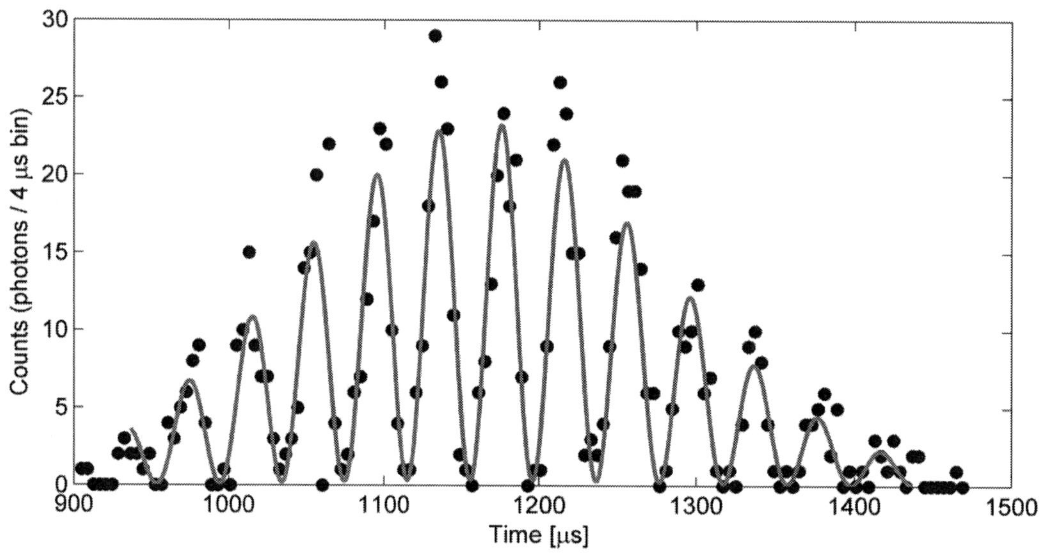


**Figure 1.** Single-shot readout from our interferometer.

# Study on nuclear magnetic resonance gyroscopes based on $^{133}Cs$-$^{129}Xe$/$^{131}Xe$

Zhengyi Xu, Dawei Zhang, Min Zhou, Xinye Xu

State Key Laboratory of Precision Spectroscopy and Department of Physics, East China Normal University, Shanghai 200062, China

We report the experiments on the frequency stabilization of the Distributed Bragg Reflector lasers against the $D_2$ transitions of cesium at 852 nm using the modulation transfer spectroscopy method [1]. For measuring the instability of the laser frequency, we built two independent 852-nm laser systems and detected the beat signal of them. Proper pump light power and modulation frequency are studied to optimize the frequency stabilization. The frequency drift of the stabilized laser is no more than 700 kHz in 2.5 hours. These frequency stabilized 852-nm lasers will be employed to the experiment on the nuclear magnetic resonance gyroscopes (NMRGs) based on $^{133}Cs$-$^{129}Xe$/$^{131}Xe$ [2-5]. In addition, for NMRGs we have theoretically investigated the optimal pressure of the quenching gas $N_2$ and corresponding pump laser intensity; the relationship between the detected signal and the probe laser frequency is analyzed; the optimal pump and probe light powers are also been studied.

[1] F. Bertinetto, P. Cordiale, and G. Galzerano, IEEE Trans. Instrum. Meas. **50**, 490 (2001).

[2] K. Liu, J. Micromech, Microeng **19**, 113001 (2009).

[3] J. H. Simpson, J. T. Fraser, and I. A. Greenwood, IEEE Trans. Aerosp. **1**, 1107 (1963).

[4] E. A. Donley, In Proceedings of the 2010 IEEE Sensors Conference (Kona, HI, USA) 17 (2010).

[5] S. J. Seltzer, Developments in alkali-metal atomic magnetometry Ph.D Thesis, Princeton University, USA (2008).

# Ramsey-Bordé interferometry for high performance laser stabilization

Judith B. Olson[1,2], R. W. Fox[1], C.W. Oates[1], A. D. Ludlow[1]
[1]National Institute of Standards and Technology, Boulder, CO, USA
[2]University of Colorado Boulder, Boulder, CO, USA

Ramsey-Bordé (RB) interferometry is a powerful spectroscopic tool for the interrogation of narrow optical resonances. Even for atomic systems with broad velocity distributions, spectral features free from first-order Doppler and transit-time broadening can be resolved using two counter-propagating pairs of co-propagating laser beams (Fig.1). In our system, a high-flux thermal calcium beam is excited on the $^1S_0$ to $^3P_1$ intercombination line at 657 nm [1]. The high spectral resolution afforded by RB interferometry allows exploration of spectral features approaching the transition's natural linewidth, 400 Hz. Together with the large atom number from the continuously fed thermal beam, this optical frequency reference has considerable potential for a compact frequency standard with extremely low instability. We previously observed fractional frequency instability of $5.5\times10^{-15}$ at 1s using this technique [2].

The addition of an excited state shelving detection scheme on the strong 431 nm cycling transition between $^3P_1$ and the doubly excited $^3P_0$ state reduces the photon shot noise limit. Consequently, using a second laser we have observed fractional frequency instability of $7.5\times10^{-16}$ at 1s. Performance was limited by technical fluctuations in the interferometer caused by ambient atmosphere. With our newly designed experimental system utilizing a low-expansion glass assembly in vacuum, the potential exists to achieve frequency stability below $10^{-16}$ at short times. We explore the implementation of this system and future enhancements to further improve the standard's short- and long-term performance.

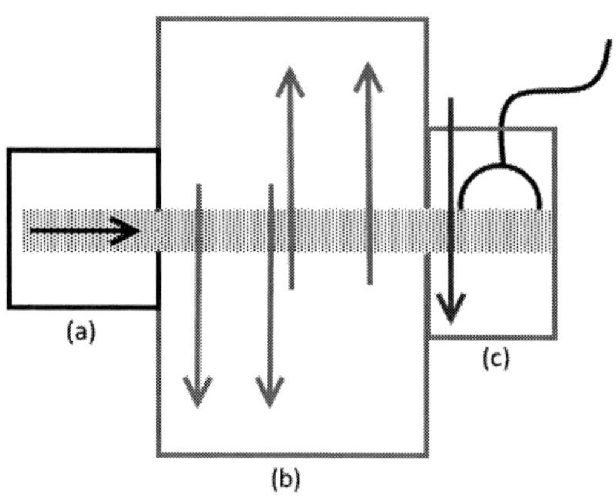


**Figure 1.** Schematic of experimental basis. (a) An oven and apertures generate a well-collimated calcium atomic beam. (b) The atomic beam is traversed in four zones by two anti-parallel pairs of parallel laser beams. (c) The excited state fluoresce is observed. Signal-to-noise is improved through a fast cycling transition from the excited state.

[1] R. L. Barger, J. C. Bergquist, T. C. English, and D. J. Glaze, Appl. Phys. Lett. **34**, 850 (1979).
[2] R. W. Fox *et al.*, Proceedings 2012 IEEE IFCS, 1 (2012).

# Inertial sensing using atom interferometry in a noisy environment

X. Cheng, J. Stammers, I. Dutta and E. Hinds
Centre for Cold Matter, Blackett Laboratory, Imperial College London, Prince Consort Road, London SW7 2AZ, United Kingdom

Precision navigation currently relies on the global navigation satellite system (GNSS). In environments of degraded GNSS signal, the alternative is to use conventional accelerometers and gyroscopes, but the bias drift of these limits their usefulness. The long-term stability of the atom presents an attractive solution to this problem. Atom interferometers also benefit from significantly higher sensitivity than conventional sensors. Here, we present our work towards a hybrid 3-axis cold atom accelerometer using Rubidium atoms, which combines the large operating range of a high-bandwidth MEMS accelerometer with the high sensitivity of the cold atom interferometer [1]. A compact sensing head with a volume of 10 liter is separated from a 19" rack that houses the lasers and electronics [2]. To approximate continuous operation, we aim to achieve a repetition rate of 10 Hz. Currently we are testing an interferometer to measure horizontal acceleration with a sensitivity of 100 ng/Hz$^{1/2}$ and a dynamic range of +/- 0.3g. This will eventually be integrated into an inertial navigation unit.

[1] J. Lautier, L. Volodimer, T. Hardin, S. Merlet, M. Lours, F. Pereira Dos Santos, and A. Landragin, Appl. Phys. Lett. **105**, 144102 (2014).
[2] B. Battelier, B. Barrett, L. Fouché, L. Chichet, L. Antoni-Micollier, H. Porte, F. Napolitano, J. Lautier, A. Landragin, P. Bouyer, Proc. SPIE **9900**, Quantum Optics, 990004 (2016).

# Atom interferometer gyroscope using cold atom beam

Jehyun Lee[1], Ji-Hyoun Kim[2], Seokchan Yoon[1], WonSang Yoon[1], Tai-Hyun Yoon[1], D. Cho[1]
[1]Department of Physics, Korea University, Seoul
[2]Department of Physics, Seoul National University, Seoul

We report on progress toward our construction of an atom interferometer gyroscope using cold $^{87}$Rb atoms from a low-velocity intense source (LVIS). The beam machine consists of an LVIS, transverse-cooling and optical-pumping stages followed by three regions for stimulated Raman transitions of $\pi/2-\pi-\pi/2$ configuration to form an interferometer and finally a detection area. The LVIS is operated in a pulsed mode at 3-4 Hz. The transverse-cooling region has a magnetic shield placed inside the chamber and cools the atoms to 25 μK. Optical pumping puts more than 85% of the atoms to the $|5S_{1/2}, F=1, m_F=0\rangle$ state. Counter-propagating Raman beams drive a clock transition to the $|5S_{1/2}, F=2, m_F=0\rangle$ state. The Raman beams are provided by a pair of optically phase locked laser beams with 6.8-GHz frequency offset. With a π pulse, more than 80% of the optically pumped atoms make a Raman transition at resonance.

In order to observe an interferometer signal, in addition to eliminate sources of dephasing as much as possible, we plan to image fluorescence from the detection area and analyze signal on each pixel of a camera. In order to measure rotational velocity precisely with a large dynamic range, we have frequency-shifting acousto-optic modulators at both upstream and downstream $\pi/2$ regions of the interferometer to track the velocity.

# Measurement progress of the high-sensitivity atom accelerometer to dark energy candidates

D. Sabulsky[1], I. Dutta[1], E. A. Hinds[1]

[1]Centre for Cold Matter, Department of Physics, Blackett Laboratory, Imperial College London, Prince Consort Road, London SW7 2AZ, United Kingdom

Theories of dark energy usually invoke a screening mechanism to explain why their scalar fields do not produce observable long-range fifth forces; a primary example of this mechanism is the so-called chameleon field. However, it is now known [1,2] that individual atoms are not massive enough and large enough to screen the chameleon field inside a large vacuum chamber under UHV. We present the experimental progress on our atom accelerometer experiment that will place strong new constraints on the chameleon [1], symmetron [2] and dilaton [3] scalar fields.

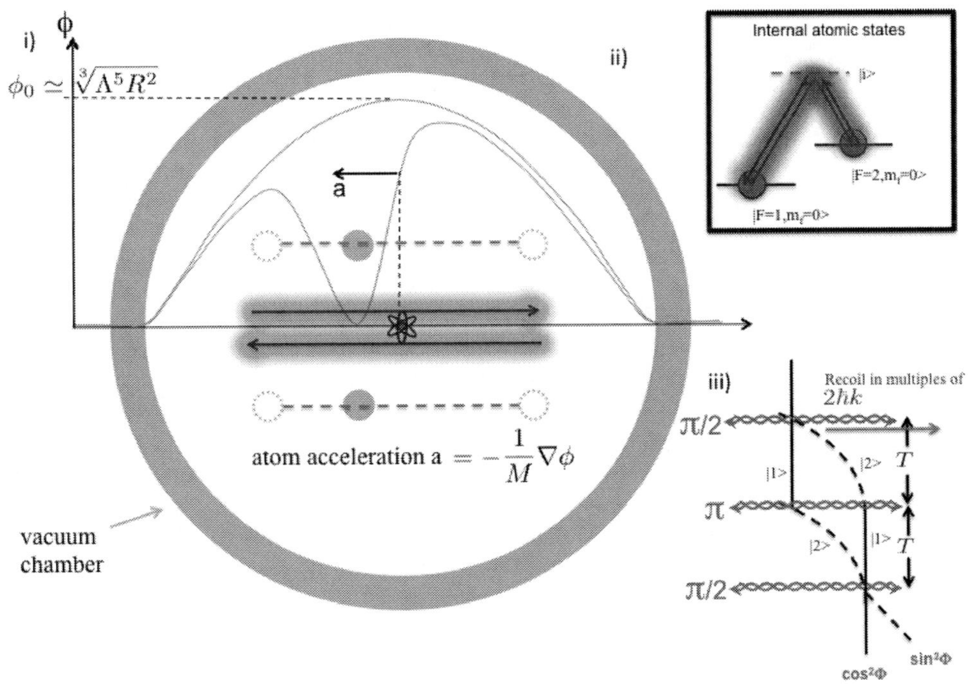


**Figure 1.** Overview of a general atom interferometer scheme sensitive to dark energy candidates. i) The schematic diagram showing the chameleon field inside of a vacuum chamber and how an atom interferometer can detect the field. ii). The internal atomic states diagram. iii) A rough look at how we apply the pulses to the atoms.

[1] C. Burrage, E. J. Copeland & E. A. Hinds, JCAP **03**, 042 (2015).
[2] P. Hamilton *et al.*, Science **349**, 849-851 (2015).
[3] K. A. Olive and M. Pospelov, Phys. Rev. D **77**, 043524 (2008).
[4] M. Pietroni, Phys. Rev. D **72**, 043535 (2005).
[5] P. Brax, C. van de Bruck, A.-C. Davis, B. Li, and D. J. Shaw, Phys. Rev. D **83**, 104026 (2011).

# Expansion of excited condensation into a ring trap

H. Yu[1], S. J. Kim[2], S. T. Kang[2], S. M. Jung[2], J. B. Kim[2]
[1]Hanwha Corporation Defense R&D center, Dae-Jeon, Rep. of Korea
[2] Physics Education Korea National University of Education, Chung-Buk, Rep. of Korea

We have constructed a ring potential on an atom chip by using RF dressed potential. Dimple potential added to the ring trap constructs an inhomogeneous potential called dimple-ring potential. We prepared Bose-Einstein condensation in the dimple-ring trap [1]. In here, we suggest to use odd excited state of the dimple-ring trap as an initial state to build an interferometer in a ring wave guide. We calculated the time evolution of the excited condensation when the dimple-ring trap was released into a ring trap. Optimal control theory helps to find a way to excite the condensation by tilting the dimple potential. Our scheme can be adopted to Sagnac interferometer in a ring wave guide.

[1] S. J. Kim, H. Yu, S. T. Kang, D. Z. Anderson, and J. B. Kim, Phys. Rev. A. **93**, 033612 (2016).

# Magnetically guided Cesium atom interferometer and progress towards a gravity gradiometer for space application

Lu Qi[1,2], Zhaohui Hu[1], Wei Quan[1], Tristan Valenzuela[2], Nick Waltham[2], Jiancheng Fang[1]

[1]School of Instrumentation Science and Opto-electronics Engineering, Beihang University, Beijing 100191, China

[2]Space Department, Rutherford Appleton Laboratory, Didcot OX11 0QX, United Kingdom

In this presentation, we in the first time report the realization of a magnetically guided Cesium(Cs) atom interferometer. As featured in compact volume and suspension atoms from gravity, atom interferometer based on guided atoms are deemed to be a competitive candidate for practical application such as precision measurements and inertial sensing. In our system, about $5\times10^{7}$ Cs atoms are loaded from a mirror MOT to a nearly perfect two dimensional magnetic waveguide generated by two pairs of macroscopic coils, in which atoms are confined in the radial directions(y and z) and are coherently manipulated within the Kapitza-Dirac regime along the longitudinal direction(x) as shown in Fig 1. Various interference loops, including the 3-pulse and elongated 4-pulse, are realized. A measurement uncertainty of $7\times10^{-5}$ m/s$^{2}$ on horizontal acceleration is achieved. The performance of our interferometer is now mainly limited by the decoherence mechanism induced by the magnetic guide, which is characterized and discussed with numerical simulations and agreeing experiment results. According to the knowledge of the decoherence mechanism, a fringe contrast increasing method based on atom optic kicked rotor(AOKR) technique is introduced and a signal amplitude enhancement up to 1.9 is observed.

We also report experimental work on a gravity gradiometer, which is built as an on-earth, technology test platform for future space application. Latest progress towards the gravity gradiometer and experimental results will be shown and discussed.

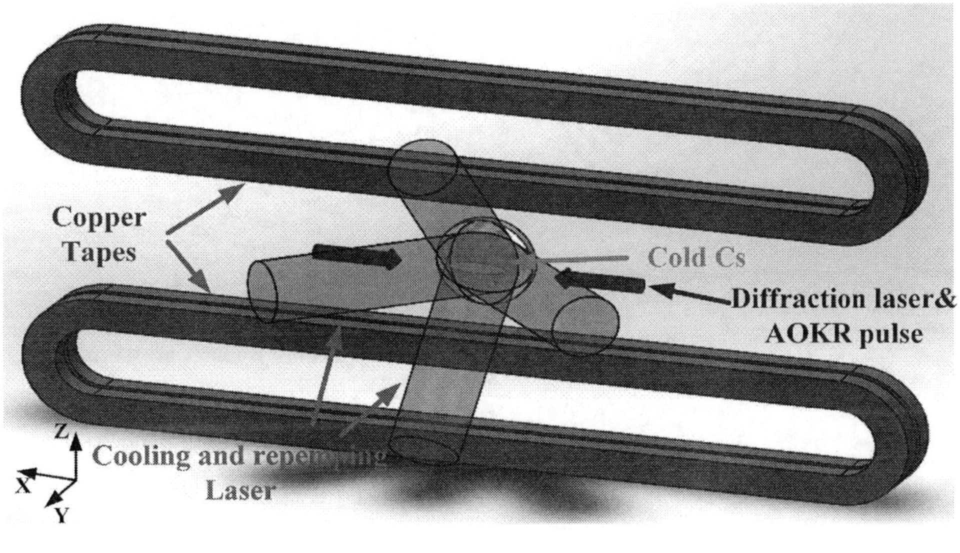


**Figure 1.** Scheme of magnetically guided Cesium atom interferometer .

# New geometric phase of a dipole under a nonlocal magnetic field

Kang-Ho Lee and Kicheon Kang
Chonnam National University, Korea

We predict a geometric phase of an electric dipole interacting with an external magnetic field under the condition that the dipole is moving in the field-free region. Based on the Lorentz-covariant field interaction approach [1], this phase shift is shown to be geometric and non-topological, and is distinct from the topological He-McKellar-Wilkens phase generated by a direct interaction with the external field [2]. We discuss experimental feasibility of detecting this phase with atomic interferometry. Detecting this phase would imply a deeper understanding the nature of electromagnetic interaction, in particular, concerning the issue of the Aharonov-Bohm type nonlocality [3].

[1] K. Kang, arXiv:1303.7019; K. Kang, Phys. Rev. A **91**, 052116 (2015).

[2] X.-G. He and H. J. McKellar, Phys. Rev. A **47**, 4 (1993); M. Wilkens, Phys. Rev. Lett. **72**, 5 (1994).

[3] K. Kang, arXiv:1512.02765.

# Long time performance of a cold atom gravimeter

Min-Kang Zhou, Le-Le Chen, Qin Luo, Xiao-Chun Duan, Zhong-Kun Hu
MOE Key Laboratory of Fundamental Physical Quantities Measurement, School of physics, Huazhong university of Science and technology, Wuhan 430074, China

Atom gravimeters based on atom interferometry have excited world-wide interests in fundamental research and practical applications. For high-precision gravity measurements and gravitational experiments, a cold-atom interferometry gravimeter with the aimed resolution of sub-micro Gal is being built in our cave laboratory. An atomic fountain based absolute gravimeter with an sensitivity of $4.2\times10^{-9}$ g/Hz$^{1/2}$ has been demonstrated after dramatically suppression of the technical noise. The main noise sources including vibration noise and detection noise are analyzed, and a sensitivity calibration experiment is performed. The mechanism of light-shift-related uncertainty in this atom-interferometry gravimeter was investigated systematically, which arises from the inhomogeneous intensity of Raman beams and the Gaussian distribution of an atomic cloud. This systematic error was verified by directly modulating the integration time T and the intensity ratio of Raman laser beams. The local gravity was measured with our atom gravimeter in several months, and the long time performance shows that the gravimeter has a repeatability of several micro-Gal.

[1] M. K. Zhou, L. L. Chen, Q. Luo, K. Zhang, X. C. Duan, and Z. K. Hu, Phys. Rev. A **93**, 053615 (2016).

[2] M. K. Zhou, Q. Luo, L. L. Chen, X. C. Duan, and Z. K. Hu, Phys. Rev. A **93**, 043610 (2016).

[3] M. K. Zhou, Z. K. Hu, X. C. Duan, B. L. Sun, L. L. Chen, Q. Z. Zhang, and J. Luo, Phys. Rev. A **86**, 043630 (2012).

[4] Z. K. Hu, B. L. Sun, X. C. Duan, M. K. Zhou, L. L. Chen, S. Zhan, Q. Z. Zhang, and J. Luo, Phys. Rev. A **88**, 043610 (2013).

# Optimal transport and magnetic lensing of degenerate matter waves in free fall

Waldemar Herr[1], Ernst M. Rasel[1] and the QUANTUS-Team[1,2,3,4,5,6,7]

[1]Institut für Quantenoptik, Leibniz Universität Hannover, Germany

[2]Institut für Physik, Humboldt-Universität zu Berlin, Germany

[3]Ferdinand-Braun-Institut, Leibniz Institut für Höchstfrequenztechnik Berlin, Germany

[4]ZARM, Universität Bremen, Germany

[5]Institut für Physik, Johannes Gutenberg Universität Mainz, Germany

[6]Institut für Quantenphysik, Universität Ulm, Germany

[7]Institut für angewandte Physik, TU Darmstadt, Germany

Quantum degenerate matter waves promise to be ideal input states for highly accurate atom interferometry, with the scope of testing fundamental physics, such as a test of the Einstein equivalence principle [1]. These measurements benefit greatly from long free propagation times of the atomic test body. Especially space platforms promise a leap forward in sensitivity [2], but also set high requirements to such experiments regarding compactness and robustness [3]. As a next step towards space the 2nd generation high-flux BEC experiment QUANTUS-2 [4] is being operated at the ZARM drop tower in Bremen, Germany. Microgravity times up to 9.2 s can be achieved by catapulting the whole experimental setup into a free fall parabola inside a 110 m high vacuum tube. To optimally benefit from the extended time of free evolution, a high degree of control of the atomic ensemble in its inner and outer degrees of freedom is crucial.

On this poster we report on our results on magnetically transporting the atomic ensemble by 1.3 mm within 150 ms away from the atom chip's surface, applying reverse engineering methods to minimize center of mass excitations of the wave packet. After transport and a free propagation phase of 80 ms a magnetic lens [5] is applied collimating two directions of the wave packet. The expansion rate of the third dimension of the wave packet is controlled via induced breathing during transport, leading to a three-dimensional collimation of the atomic ensemble into the pK-regime. After preparation of the magnetically insensitive sub-state, time of flight pictures up to 2000 ms free propagation of the lensed atomic ensemble were obtained.

Having demonstrated the necessary control of the inner and outer degrees of freedom of the atomic ensemble, double-Bragg interferometry [6] will be investigated in the next step.

[1] D. Schlippert et al., PRL **112**, 203002 (2014)

[2] T. Schuldt et al., Exp Astron **39**, 167–206 (2015).

[3] T. van Zoest et al., Science, **329**, 1540 (2010).

[4] J. Rudolph et al., New J. Phys., **17**, 065001 (2015).

[5] H. Müntinga et al., PRL, **110**, 093602 (2013).

[6] H. Ahlers et al., PRL, **116**, 173601 (2016).

# Atom-chip based BEC interferometry

Ernst M. Rasel[1]

[1] Institut für Quantenoptik, Leibniz Universität Hannover, Welfengarten 1, D-30167 Hannover, Germany

Today's inertially sensitive atom interferometry devices operate mostly with sources of laser cooled atoms. The finite temperature and size of these sources limit the efficiency of employed beam splitters and the analysis of systematic uncertainties. These limits can be overcome by the use of ultracold sources such as a delta-kick collimated Bose-Einstein condensate (BEC) with extremely narrow velocity distribution [1,2]. Atom-chip technologies offer the possibility to generate a BEC and perform delta-kick collimation in a fast and reliable away. We show a combination of such an ensemble generated in a miniaturized atom-chip setup with the application of Bragg beam splitting to perform inertially sensitive measurements. A specialty of this setup is that the beam splitting light field, which forms a Mach-Zehnder interferometer (MZI), is reflected at the atom chip itself. In this way, the chip is serving as the inertial reference inside the vacuum, allowing for a very compact prototype of a quantum gravimeter that is capable of determining the local gravitational acceleration g to $5 \cdot 10^{-5}$ limited by seismic noise. Using ultracold atoms gives us the possibility to implement new methods of coherent manipulation at high fidelity. With the help of an optical lattice in combination with a double Bragg diffraction pulse [3,4] we developed a novel coherent relaunch technique in a retro-reflected optical lattice and implemented a fountain geometry where the atomic ensemble is coherently relaunched on a parabolic trajectory. This allows us to increase the intrinsic sensitivity of the employed MZI by a factor of twenty, while keeping the region, in which the whole sequence is realized, as compact as a cube with one centimeter side length below the atom chip. In the framework of the Collaborative Research Center geo-Q a next generation quantum gravimeter is set-up in collaboration with the Institut für Erdmessung in Hannover, which targets an accuracy below μGal for geodetic field-application.

[1] T. van Zoest et al., Bose-Einstein Condensation in Microgravity, Science, **328**, 1540-1543, (2010)

[2] H. Müntinga et al., Interferometry with Bose Einstein Condensates in Microgravity, Phys. Rev. Lett., **110**, 093602, (2013)

[3] E. Giese et al., Double Bragg diffraction: A tool for atom optics, Phys. Rev. A, **88**, 053608, (2013)

[4] H. Ahlers et al., Double Bragg Interferometry, Phys. Rev. Lett., **116**, 173601, (2016)

# Towards an atom interferometer with guided matter waves in an optical fiber

Mingjie Xin, Wui Seng Leong, Zilong Chen, Shau-Yu Lan
Division of Physics and Applied Physics, School of Physical and Mathematical Sciences, Nanyang Technological University, Singapore

Precision measurement with light-pulse grating atom interferometers in free space have been used in the study of fundamental physics and applications in inertial sensing [1]. However, due to the diffraction nature of light, this approach requires large-diameter laser beams to maximize its sensitivity while minimizing systematic effects and therefore complex the experimental setup. Recent development of photonic crystal fibers allow light for travelling in hollow region while preserving its fundamental Gaussian mode [2]. Optically guided matter waves inside a hollow-core photonic crystal fiber can mitigate this diffraction limit problem and has the potential to bring the research in the field of atomic sensing and precision measurement to the next level of compactness and accuracy. Here, we will show our experimental progress towards an atom interferometer in an optical fiber. We demonstrate an atom trapping scheme inside a hollow-core photonic crystal fiber to create an optically guided matter waves system. Inside the fiber, the atoms are confined by a red-detuned optical dipole trap and probed by a resonant beam. The efficient trapping procedure results in loading of $\sim 10^5$ rubidium atoms, which corresponds to an atomic cloud with optical depth ~10 inside the fiber. This guiding system will provide a platform for optically guided matter-wave interferometer.

[1] A. D. Cronin, J. Schmiedmayer, D. E. Pritchard, Rev. Mod. Phys. **81**, 1051–1129 (2009).
[2] R. F. Cregan, B. J. Mangan, J. C. Knight, T. A. Birks, P. St. J. Russell, P. J. Roberts, D. C. Allan, Science **285**, 1537 (1999).

# Self-spin-squeezing in a spinor Bose-Einstein condensate

Theo Laudat[1], Vincent Dugrain[2], Carlos Garrido Alzar[1], Jakob Reichel[2], Peter Rosenbusch[1]
[1]SYRTE, Observatoire de Paris, PSL Research University, CNRS, Sorbonne Université, UPMC, LNE, 61 Avenue de l'Observatoire, 75014 Paris, France
[2]Laboratoire Kastler Brossel, ENS, UPMC, CNRS, 24 rue Lhomond, 75005 Paris, France

State-of-the-art microwave-frequency standards have now reached the quantum projection noise limit [1], and using spin-squeezing is an appealing approach to overcome this fundamental boundary. Spin-squeezed states, a particular kind of entangled states, can be generated in spin ½ systems through a non-linear interaction known as one-axis-twisting Hamiltonian [2]. Recently much effort has been devoted to generating such states through state-dependent microwave potentials [3], or Feshbach resonances [4]. Both methods artificially reduce the interstate atomic interaction $g_{\uparrow\downarrow}$. We present a new method, where the system reduces $g_{\uparrow\downarrow}$ by itself without user intervention. In [5] the two states of a $^{87}$Rb hyperfine spinor condensate exhibit large spatial separation and recombination due to the different s-wave scattering lengths $a_{\downarrow\downarrow}$ and $a_{\uparrow\uparrow}$. We experimentally show that when the two components spatially separate, the twisting interaction strongly increases and shears the spin noise distribution. Its small axis can end-up below quantum projection noise.

In our atom-chip-generated cigar-shaped magnetic trap, we observe state separation and recombination of our $^{87}$Rb spinor condensate (fig. 1) as well as contrast oscillations in Ramsey spectroscopy. We also demonstrate an elliptic spin noise distribution probed at the first recombination time (fig. 2). We measure a squeezing parameter [6] of $\xi^2 = -1.7 \pm 0.3$ dB for our Ramsey contrast of 96%, testifying of spontaneous and metrologically useful spin squeezing. Optimum parameters are being investigated. This result can be of particular interest in the context of BEC interferometry and many-particle entanglement with Bose Einstein condensates [7].

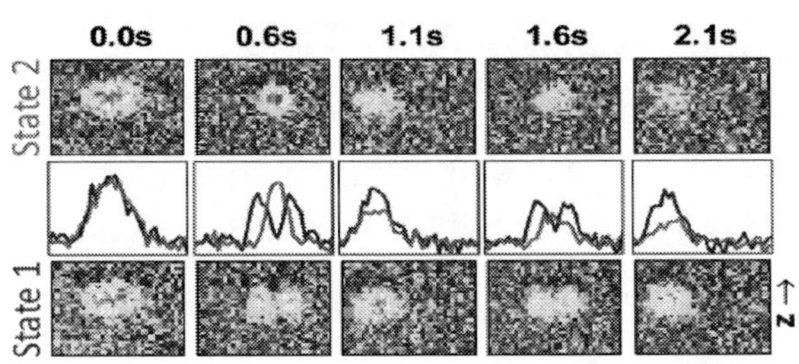


**Figure 1:** State sensitive absorption images of our $^{87}$Rb spinor BEC. The trap holding time after the first π/2 pulse is indicated. Demixing and remixing is observed.

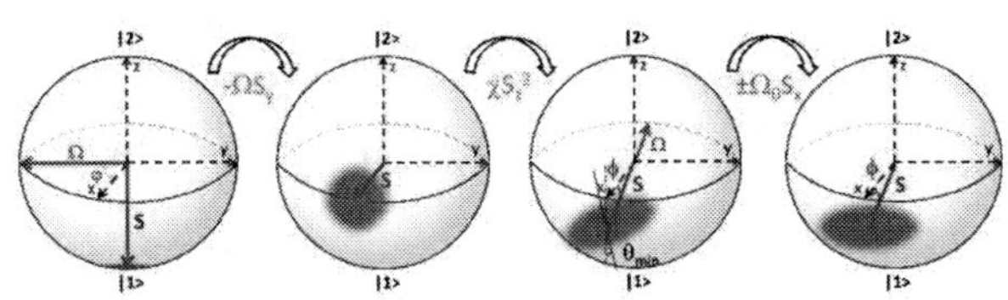

**Figure 2**: Experimental sequence represented on the Bloch sphere. After an initial π/2 pulse, atomic interactions shear the spin noise distribution into a squeezed state. A second pulse rotates the spin noise distribution around itself by an angle θ, followed by detection of the population difference $S_z$.

[1] G. Santarelli, *et al*, PRL **82**, 4619-4622 (1999)
[2] M. Kitagawa, M. Ueda, PRA **47**, 5138-5143 (1993)
[3] M. F. Riedel, P. Böhi, Y. Li, T. W. Hänsch, A. Sinatra, P. Treutlein, Nature **464**, 1170-1173 (2010)
[4] C. Gross, T. Zibold, E. Nicklas, J. Estève, M. K. Oberthaler, Nature **464**, 1165-1169 (2010)
[5] R. P. Anderson, C. Ticknor, A. I. Sidorov, B. V. Hall, PRA **80**, 02360, (2009)
[6] D. J. Wineland, J. J. Bollinger, W. M. Itano, and D. J. Heinzen, PRA **50**, 67-88 (1994)
[7] A. Sorensen, L. M. Duan, J. I. Cirac, P. Zoller, Nature **409**, 63-66 (2001)

# Long-time nonlinear dynamical evolution for P-band ultracold atoms in an optical lattice

Xiaoji Zhou, Linxiao Niu, Dong Hu, Xuzong Chen
School of Electronics Engineering and Computer Science, Peking University, Beijing 100871, China

It is important to study the physics on excited band in an optical lattice. Based on a non-adi abatic shortcut method we can load a Bose-Einstein condensate (BEC) fast and efficiently into the excited band at zero quasi-momentum [1, 2]. On the P-band, after the non-adiabatic load ing process we observe the long-time nonlinear dynamical evolution of ultracold atomic gases. For the first one and half milliseconds, this momentum states undergoes oscillations due to c oherent superposition of different bands, which are followed by oscillations up to 60ms of a much longer period. Our analysis shows the dephasing from the nonlinear interaction is very conducive to the long-period oscillations induced by the variable force due to the harmonic c onfinement.
During the experiment we observe both a short-period oscillation due to coherent superpositio n of different bands and a long period oscillation reflecting the random relative phase betwee n neighboring lattice sites. The present experiment paves the way to study the long time dyn amical evolution of the high orbital physics for other novel quantum state such as fermionic superfluid, molecular condensate and condensate with special configuration of optical lattice.

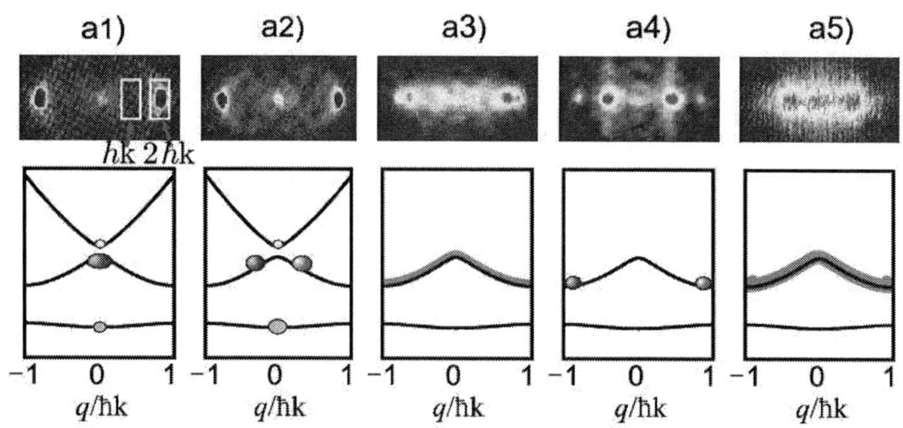


**Figure 1.** The first row: the measured momentum distributions of the BEC in the P-band at holding different times (0ms, 2ms, 5ms, 7ms, 30ms), the second row: schematic illustration of the corresponding population distributions in the Bloch band.

[1] Dong Hu, Linxiao Niu, Baoguo Yang, Xuzong Chen, Biao Wu, Hongwei Xiong, and Xiao ji Zhou, Phys. Rev. A **92**, 043614 (2015).
[2] Yueyang Zhai, Xuguang Yue, Yanjiang Wu, Xuzong Chen, Peng Zhang, and Xiaoji Zhou, Phys. Rev. A **87**, 063638 (2013).

# Interacting dynamics of half-quantized vortices in two-component Bose-Einstein condensates

Kenichi Kasamatsu[1], Minoru Eto[2], Muneto Nitta[3]
[1]Department of Physics, Kindai University, Higashi-Osaka, 577-8502, Japan
[2]Department of Physics, Yamagata University, Yamagata 990-8560, Japan
[3]Department of Physics, and Research and Education Center for Natural Sciences, Keio University, Hiyoshi 4-1-1, Yokohama, Kanagawa 223-8521, Japan

We study the interaction and dynamics of two half-quantized vortices in two-component Bose-Einstein condensates. Using the Pade approximation for the vortex core profile, we calculate the intervortex potential, whose asymptotic form for a large distance has been derived in Ref.[1]. Through numerical simulations of the two-dimensional Gross- Pitaevskii equations, we reveal different kinds of dynamical trajectories of the vortices depending on the combinations of signs of circulations and the intercomponent density coupling [2]. Under the adiabatic limit, we derive the equations of motion for the vortex coordinates, in which the motion is caused by the balance between Magnus force and the intervortex forces. The initial velocity of the vortex motion can be explained quantitatively by this point vortex approximation, but understanding the long-time behavior of the dynamics needs more consideration beyond our model.

[1] M. Eto, K. Kasamatsu, M. Nitta, H. Takeuchi, and M. Tsubota, Phys. Rev. A **83**, 063603 (2011).
[2] K. Kasamatsu, M. Eto, and M. Nitta, Phys. Rev. A **93**, 013615 (2016).

# Weak wave turbulence in spin-1 spinor Bose-Einstein condensates

Kazuya Fujimoto[1], Makoto Tsubota[2]

[1]Department of Physics, University of Tokyo, Bunkyo-ku 113-0033, Japan

[2]Department of Physics, Osaka City University, Sumiyoshi-ku, Osaka 558-8585, Japan

Recently, non-equilibrium dynamics is actively studied in cold atomic gases from various perspectives, e.g. thermalization in isolated quantum systems [1], dynamical phase transition [2], quantum turbulence (QT) [3] and so on. Such active researches for non-equilibrium phenomena are attributed to the high controllability of experiments, the long relaxation time, and high cleanness in the atomic gases.

These properties in the system give us a suitable stage for investigating weak wave turbulence (WWT) where the weakly interacting waves are dominant. Until now, most of studies for turbulence in the atomic gas focus on hydrodynamic turbulence dominated by quantized vortices, investigating the superfluid velocity field or the vortex distribution. On the other hand, WWT addresses not only the velocity field but also the density or spin profiles which have not been regarded as important observables, giving us a new insight into the research of QT in the atomic gas, e.g. importance of the density or spin correlation functions. At present, any experimental studies for WWT have not been done yet, but the non-equilibrium phenomena for spin wave is studied and the characteristic dynamics is observed [4].

On this background, we theoretically and numerically investigate WWT in the spin-1 ferromagnetic spinor Bose-Einstein condensate by using the spin-1 spinor Gross-Pitaevskii (GP) equation [5]. In the analytical calculation, the conventional WWT theory is applied to the GP equation, being found to give the two power laws in the transverse spin correlation function. One is the -7/3 power law in the direct cascade, and the other is the -5/3 power law in the inverse one. To confirm these power laws numerically, we perform the numerical calculations of the GP equation in the two-dimensional system, obtaining the time development of the transverse spin correlation function as shown in Fig. 1.

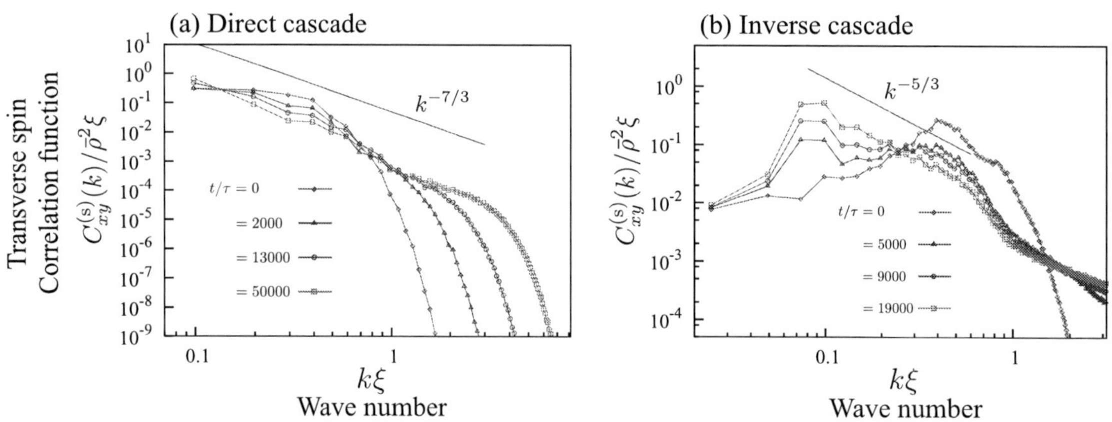


FIG. 1. Time development of the transverse spin correlation functions. The results are obtained by the numerical calculations of the spin-1 ferromagnetic spinor GP equation in the two-dimensional system. The (a) left and (b) right figures correspond to the direct and inverse cascade cases, respectively.

[1] M. Gring, et al., Science **337**, 1318 (2012).
[2] N. Navon, A. L. Gaunt, R. P. Smith and Z. Hadzibabic, Science **347**, 167 (2015).
[3] W. J. Kwon, G. Moon, J. Choi, S. W. Seo, and Y. Shin, Phys, Rev. A 90, 063627 (2014).
[4] F. Fang, R. Olf, S. Wu, H. Kadau, and D. M. Stamper-Kurn, Phys. Rev. Lett. **116**, 095301 (2016).
[5] K. Fujimoto and M. Tsubota, Phys. Rev. A **93**, 033620 (2016).

# Dirac monopoles with polar-core vortex induced by spin-orbit coupling in spinor Bose-Einstein condensates

Ji Li, Wu-Ming Liu

Beijing National Laboratory for Condensed Matter Physics, Institute of Physics,

Chinese Academy of Sciences, Beijing 100190, China

Dirac monopoles describe the isolated north or south magnetic poles, which are also predicted in the grand unified and superstring theories. These elusive elementary particles have fuelled a rich field of study, ranging from particle physics to condensed matter. Recently the study of monopole in spinor Bose-Einstein condensates is an intriguing topic. Although several types of monopoles have been found, convincing evidence of the existence of the stable monopoles in non-Abelian gauge field is still lacking. In particular, the recent realization of spin-orbit coupling in Bose-Einstein condensates, owing to coupling between the spin and the centre-of-mass motion of the atom, raises the question of whether exotic monopole structures can be created in quantum system. Hitherto it has remained elusive. We report Dirac monopoles with polar-core vortex induced by spin-orbit coupling in ferromagnetic Bose-Einstein condensates, which are stable in the time scale of the experiment and can be detected through directly imaging vortex lines. When the strength of spin-orbit coupling increases, Dirac monopoles with the vortex can be transformed into those with the square lattice. In the presence of spin-orbit coupling, increasing the strength of interaction can induce a cyclic phase transition from Dirac monopoles with polar-core vortex to those with Mermin-Ho vortex. The spin-orbit coupling not only is an unprecedented opportunity to produce exotic monopole structures, but also provides a unique insight into monopole in gauge fields. Our findings open new directions in the studies of topological defects and the corresponding stability in quantum systems [1].

[1] Ji Li, Yan-Mei Yu, Wu-Ming Liu. Spin-orbit coupling induced Dirac monopoles with polar-core vortex in spinor Bose-Einstein condensates. arXiv:1602.00189

# Thermally activated phase slips of one-dimensional Bose gases in shallow optical lattices

Masaya Kunimi and Ippei Danshita
Yukawa Institute for Theoretical Physics, Kyoto University, Kyoto 606-8502, Japan

A phase slip (PS) is one of the decay mechanisms of superflow, which is caused by thermal or quantum fluctuations. In low-dimensional systems, the PSs are enhanced due to their strong fluctuations. Recently, the LENS group has reported the observation of the crossover from the regime of pure quantum PSs to that of thermally assisted quantum PSs in a one-dimensional Bose gas flowing along a shallow optical lattice [1]. They measured a damping rate of the dipole oscillation, which is related to the nucleation rate of the PS [2].

Since the experiment has been performed at finite temperatures, some of their experimental results may include not only the contribution of the quantum PSs but also that of the thermally activated PSs. In this poster, to quantify the contribution of the quantum PSs to the observed damping rate, we theoretically investigate the effect of thermally activated PSs. Using the Kramers formula [3] and the Gross-Pitaevskii and Bogoliubov equations, we calculate the nucleation rate of the thermally activated PSs. In order to evaluate the damping rate that includes the effects of the harmonic trap, we extend the formula relating the nucleation rate with the damping rate by using the local density approximation. We compare our results with the experimental results to elucidate which gives the dominant contribution to the observed damping rate, thermal or quantum PSs.

[1] L. Tanzi *et al.*, Scientific Reports **6**, 25965 (2016).
[2] I. Danshita, Phys. Rev. Lett. **111**, 025303 (2013).
[3] J. S. Langer, Ann. Phys. **54**, 258 (1969).

# Quantum degenerate mixtures of bosonic potassium.

P. Cheiney[1], C. Cabrera[1], Julio Sanz[1], Luca Tanzi[1], Leticia Tarruell[1]
[1]ICFO- The Institute of Photonic Sciences,
Mediterranean Technology Park, 08860 Castelldefels, Spain

Quantum mixtures are versatile systems for the study of a broad range of phenomena, including both few- and many-body physics. Already at the mean field level, mixtures display rich physics. With repulsive interaction between all constituents, a miscible-immiscible phase transition is observed. With attractive interspecies interactions a collapse can occur. However, liquid-like droplets have recently been observed in dipolar gases [1] where the collapse is avoided by quantum fluctuations. It has also been predicted that a stabilization mechanism should occur and lead to the formation of liquid droplets in a Bose-Bose mixture with repulsive intraspecies and attractive interspecies interactions [2]. These droplets would be remarkable as their very own existence originates from quantum fluctuations which are usually a small correction to the mean field description.

I will present a new experimental setup that produces Bose-Bose mixtures of potassium isotopes with favorable Feshbach resonances to address these questions. Our apparatus allows producing large Bose-Einstein condensates (BEC) of $^{41}$K by evaporative cooling in a hybrid trap and dual BECs of $^{39}$K and $^{41}$K by sympathetic cooling. I will present our characterization of the Feshbach resonances in the new $^{39}$K-$^{41}$K mixture and of interspin resonances in $^{39}$K using RF association of Feshbach molecules close to the resonance (see Fig. 1). These results are compared with a simple asymptotic bound-state model and a full coupled channels calculation and are a prerequisite for the study of the miscible-immiscible phase transition and of the Bose-Bose liquid droplets.

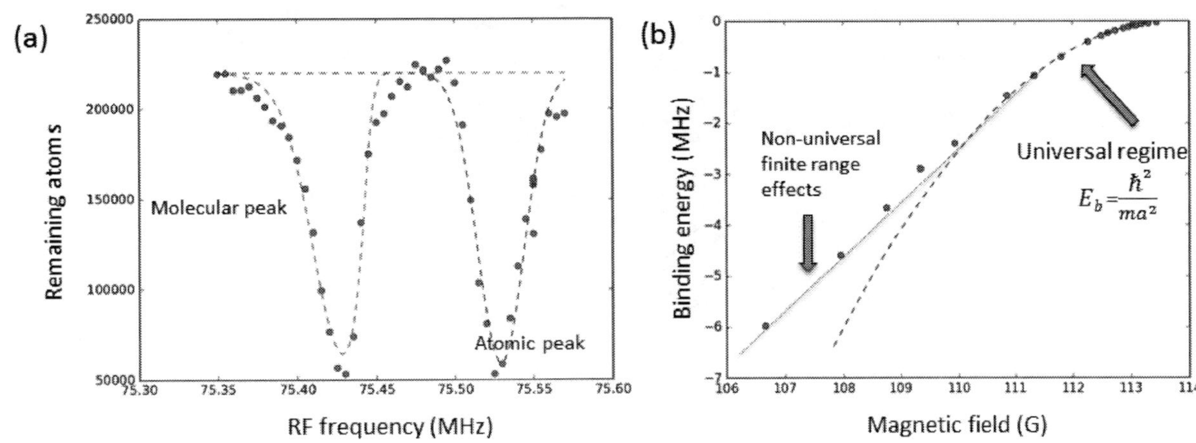


**Figure 1**. (a) Remaining population after an RF-pulse. The two peaks correspond to the transition to two free atoms (right peak) and to the Feshbach molecule (left). The frequency difference corresponds to the molecule binding energy. (b) Binding energy close to a Feshbach resonance.

[1] I. Ferrier-Barbut, H. Kadau, M. Schmitt, M. Wenzel, and T. Pfau, *PRL* **116**, 215301 (2016)
[2] D. S. Petrov, *PRL* **115**, 155302 (2015)

# Faraday Excitations in a Bose-Einstein Condensate

L. Liao[2], A. Groot[1], J. Smits[1], P.C. Bons[1], H.T.C. Stoof[2], and P. van der Straten[1]
[1]Debye Institute for Nanomaterials, Utrecht, The Netherlands
[2]Institute for Theoretical Physics, Utrecht, The Netherlands

Parametric excitations of a Bose-Einstein condensate can be induced by modulating the radial trap frequency of the trap. By shortly pulsing the radial trap frequency we induce quadrupole oscillations of the cloud in both directions and after some time we observe the emergence of a Faraday pattern. The period of the Faraday pattern depends strongly on the aspect ratio of the trap, as can be seen in Fig. 1. Since the Faraday pattern is a standing wave sound mode, it allows us to determine with a large accuracy the speed of sound in the condensate. The dynamics of the Faraday pattern is studied using phase-contrast imaging, which allows for the detection of the pattern of a single cloud for 100 consecutive instances. We study theoretically the interplay between the Faraday mode and the radial quadrupole mode of the condensate. In the model we want to establish the condition for which the Faraday pattern occurs.

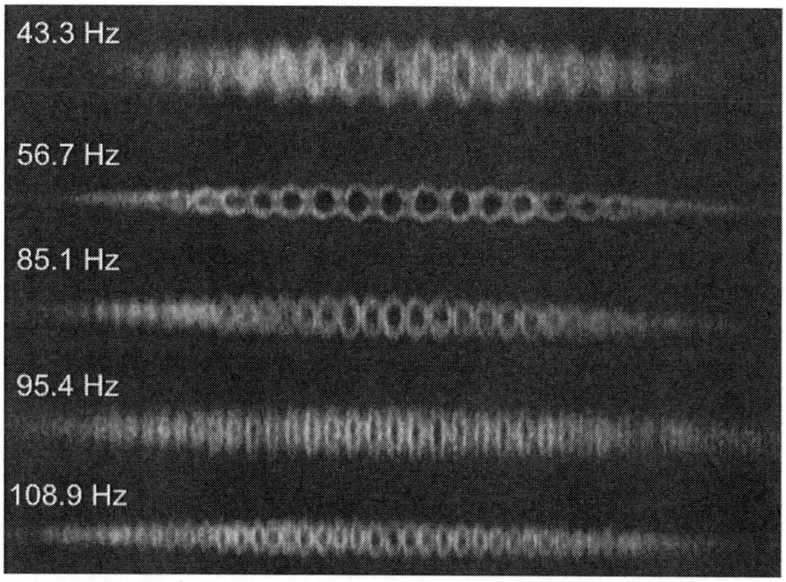


**Figure 1.** In-situ images of a Bose-Einstein condensate containing Faraday patterns. By changing the radial trap frequency, as shown in the upper left corner of the images, the pattern can be observed for different aspect ratios of the trap.

# Interspin Interaction Effects with Bose-Einstein Condensates in Optical Lattices

Colin Kennedy[1], William Cody Burton[1], Woo Chang Chung[1], Samarth Vadia[1], Wenlan Chen[1], Wolfgang Ketterle[1]

[1]Center for Ultracold Atoms, Research Laboratory of Electronics, MIT Department of Physics, 77 Massachusetts Ave, Cambridge, MA 02139, USA

We report on several experiments studying the effects of interactions in two spin component Bose-Einstein condensates. First, we show how interspin interactions can effect the decay of coherent Bloch oscillations, and we demonstrate the ability to increase the coherence time beyond the shot noise limit set by mean field fluctuations. Additionally, we show how these interspin interactions can be manipulated using a spin-dependent lattice to give direct control over the strength of the interaction. This control is then shown to give direct access to a ground state phase of the spin-1 Heisenberg model. Experimental progress towards realization is discussed along with plans for a new quantum gas microscope.

# Two Species Superfluid to Study Quantum Turbulence and Vortices: NaK System and the Production of Na BEC

Kilvia M. Farias[1], Edwin Pedrozo-Peñafiel[1], Patricia C. M. Castilho[1], Emmanuel Mercado-Gutierrez[1], Franklin A. J. Vivanco[1], Giacomo Roati[2], Daniel V. Magalhães[1,3], Vandereli S. Bagnato[1]

[1]Instituto de Física de São Carlos - Universidade de São Paulo, Brazil

[2]INO-CNR and LENS, Universitá di Firenze, 50019 Sesto Fiorentino, Italy

[3]Departamento de Engenharia Mecânica, Escola de Engenharia de São Carlos, Universidade de São Paulo, São Carlos, Brazil

In this work, we present the state of the art of our experiment for a mixture of Na and K Bose-Einstein Condensates. We are going to investigate effects of transferring quantum excitations, vortices formation, and quantum turbulence, as well. Our observation of superfluid turbulence in $^{87}$Rb [1] introduced the experimental tools of atomic optics into the world of quantum fluid dynamics. Prior to this experimental realization, quantum turbulence research had been confined to cryogenic helium which requires a large and relatively inaccessible apparatus. Conversely, atomic optics experiments fit on a tabletop and posses unprecedented measurement precision and paralleled access to the experimental volume. Our experimental system is a composition of two independently atomic sources attached to the main chamber. We produce magneto-optical traps (MOT) of Na and K atoms from 2D MOTs, which utilize a novel Zeeman slower configuration [2]. A Quadrupolar Trap has been installed and is able to trap magnetically both Sodium and Potassium, separately. We recently implemented Dark Spot MOT [3], PLUG associate to a Quadrupole Magnetic Trap [4] and achieved the Bose-Einstein condensation of Sodium by Evaporative Cooling [5]. The experimental designs, apparatus specifications, and first results of the Two Species Machine will be detailed in this presentation.

[1] E. A. L. Henn, J. A. Seman, G. Roati, K. M. F. Magalhães, and V. S. Bagnato, Physical Review Letters **103**, 045301 (2009).

[2] G. Lamporesi, S. Donadello, S. Serafini, and G. Ferrari, Review of Scientific Instruments **84**, 063102 (2013).

[3] W. Ketterle, K. B. Davis, M. A. Joffe, M. Martin, and P. D. E., Physical Review Letters **70**, 2253 (1993).

[4] K. B. Davis, M.-O. Mewes, M. R. v. Andrews, N. J. Van Druten, D. Durfee, D. M. Kurn, and W. Ketterle, Physical Review Letters **75**, 3969 (1995).

[5] W. Ketterle and N. V. Druten, Elsevier 37, **181** (1996).

# Dynamics of Half-Quantum Vortices in a Spinor Bose-Einstein Condensate

Sang Won Seo[1], Seji Kang[1,2], Joon Hyun Kim[1,2], Bumsuk Ko[1,2], and Yong-il Shin[1,2]

[1]Department of Physics and Astronomy, and Institute of Applied Physics, Soul National University, Seoul 08826, Korea

[2]Center for Correlated Electron System, Institute for Basic Science, Seoul 08826, Korea

We present experimental results on the dynamics of half-quantum vortices (HQVs) in the easy-plane polar phase of an antiferromagnetic spinor Bose-Einstein condensate (BEC). In magnetization sensitive phase-contrast imaging, we observed that singly quantized vortices are dissociated into HQV pairs with opposite magnetization. This dissociation shows that the HQVs are fundamental topological excitations in a spinor BEC [1].

We also prepare a collision experiment between two HQV pairs by exploiting the orbit motion of a vortex dipole. The observed collisional dyanmics is consistent with the short-range HQV interaction. Finally, the relaxation of the HQVs demonstrate that the spin excitations are generated by the collisional dynamics of the HQVs [2].

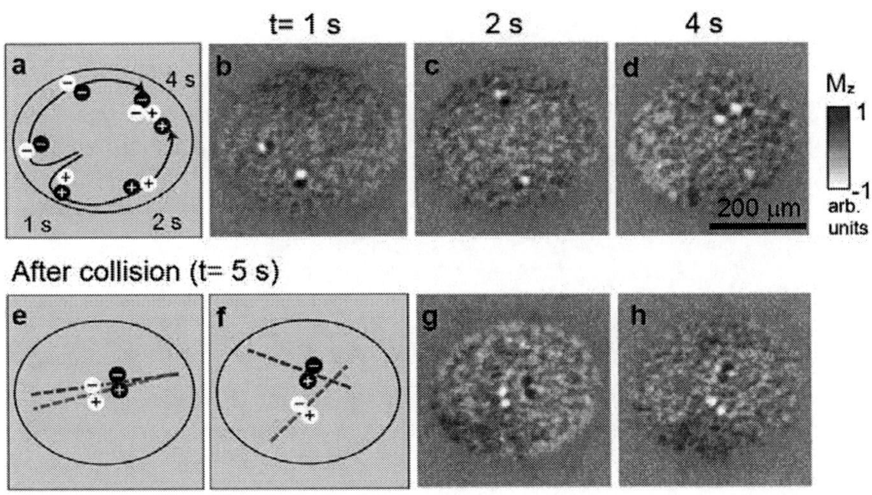


**Figure 1.** Collisional dyanmics of HQV pairs. We create a singly quantized vortex dipole and it becomes two HQV pairs with opposite magnetization. The HQV pairs follow original trajectory of the singly quantized vortex dipole and they collide. After the collision, they scatter into HQV dipoles with the same magnetization.

[1] S. W. Seo, S. Kang, W. J. Kwon, and Y. Shin, Physical Review Letters **115**, 015301 (2015).

[2] S. W. Seo, W. J. Kwon, S. Kang, and Y Shin, Physical Review Letters **116**, 185301 (2016).

# Thermal friction in the decaying turbulence of a highly oblate Bose-Einstein condensate

Joon Hyun Kim[1,2], Woo Jin Kwon[1,2], Yong-il Shin[1,2]

[1]Department of Physics and Astronomy, and Institute of Applied Physics, Seoul National University, Seoul 08826, Korea

[2]Center for Correlated Electron Systems, Institute for Basic Science, Seoul 08826, Korea

In this poster, we present our study on the effect of thermal friction on the decaying turbulence in a highly oblate Bose-Einstein condensate (BEC). In a recent experiment [1], the thermal friction coefficient, $\alpha$, was quantitatively measured from the dissipative evolution of a corotating vortex pair in a BEC at finite temperatures, and the decay rates of vortex number were also investigated for various sample conditions [2]. We extract the relationship between the vortex decay rate, $\Gamma$, and $\alpha$ from a correlation analysis of the two experimental measurements, which can be quantitatively accounted for our numerical simulations based on a dissipative point vortex model with implementing the vortex pair annihilation. Particularly, we observe that the friction much weaker than that of typical experiment can obstruct the vortex clustering behavior appearing in the relaxation process from our numerical simulation. These results point out the importance of the thermal dissipation on the relaxation of quantum turbulence in highly oblate BEC.

[1] W. J. Kwon, G. Moon, J. Choi, S. W. Seo, and Y. Shin, Physical Review A **92**, 051601(R) (2015).
[2] G. Moon, W. J. Kwon, H. Lee, and Y. Shin, Physical Review A **90**, 063627 (2014).

# The two-component Bose-Hubbard Model using $^7$Li in an Optical Lattice

Jesse Amato-Grill[1,2], P. Niklas Jepsen[2], Ivana Dimitrova[2], William Lunden[2], Wolfgang Ketterle[2]
[1]Department of Physics, Harvard University, USA
[2]Department of Physics, MIT, USA

The two-component Bose-Hubbard model with second-order tunneling can be mapped to spin *n/2* Heisenberg models (where *n* is the number of bosons per lattice site) [1]. The light mass of $^7$Li allows for maximal second-order tunneling ("super-exchange") rates among the Bosonic Alkalis, and its many Feshbach resonances additionally provide access to a wide, tunable variety of inter- and intra-component interactions, leading to predications of different magnetic orderings in the ground state. We report on progress towards spectroscopic identification, by use of interaction blockade effects [2], of the critical ratios of interactions producing a transition between two such states in the $n=2$ manifold, and also additional progress towards reaching the ground states.

[1] E. Altman, W. Hofstetter, E. Demler and M. D. Lukin. New Journal of Physics **5,** 113.1–113.19 (2003).
[2] P. Cheinet, S. Trotzky, M. Feld, U. Schnorrberger, M. Moreno-Cardoner, S. Fölling, and I. Bloch. Phys. Rev. Lett. **101**, 090404 (2008).

# Response of the Higgs amplitude mode in the 3D Bose-Hubbard model

Kazuma Nagao[1], Yoshiro Takahashi[2], Ippei Danshita[1]
[1]Yukawa Institute for Theoretical Physics, Kyoto University, Japan
[2]Department of Physics, Graduate School of Science, Kyoto University, Japan

To detect the Higgs mode (HM) of superfluid Bose gases in optical lattices, which is a collective excitation corresponding to amplitude fluctuations of the order parameter, is one of the important problems in the field of cold atoms. The experiment of 2D Bose gases in optical lattices [1] has observed the energy gap of the HM, which agrees with results predicted theoretically, by measuring a response to the temporal lattice-amplitude modulation as a function of the frequency. The observed response, however, exhibits a broad continuum above an onset frequency rather than a sharp peak. The existence of the HM as a sharp resonance has not yet been verified in the systems.

Reference [2] has shown that the HM in 2D systems is unstable due to the combined effects of the quantum fluctuations, thermal fluctuations, and spatial inhomogeneity. Motivated by Ref. [1, 2], we have studied the stability of the HM against quantum and thermal fluctuations in 3D systems [3]. We have calculated the damping rate of the HM at finite temperatures and shown that the HM in 3D systems is underdamped below a typical temperature in contrast to 2D. Although the damping rate characterizes the stability of the HM, this quantity is not directly observable in experiment.

In this work, to understand the detectability of the HM in 3D experimental systems, we calculate the response functions (RFs) of the 3D Bose-Hubbard model (BH) within the linear response theory. The previous study [4] has obtained the RFs to modulations of the hopping parameter and chemical potential at tree-level order, i.e. with no quantum correction. Using the effective model of the BH near the Mott-insulating transition [4], we analyze, in particular, the quantum corrections to the RFs to the hopping parameter and the onsite interaction within 1-loop order.

[1] M. Endres et al., Nature **487**, 454 (2012).
[2] L. Pollet and N. Prokof'ev, Phys. Rev. Lett. **109**, 010401 (2012).
[3] K. Nagao and I. Danshita, to be published in Prog. Theor. Exp. Phys. (2016).
[4] S. Huber et al., Phys. Rev. B **75**, 085106 (2007).

# Observation of self-organization to two crossed cavity modes

Julian Leonard[1], Andrea Morales[1], Philip Zupancic[1], Tilman Esslinger[1] and Tobias Donner[1]
[1]Institute for Quantum Electronics, ETH Zürich, Switzerland

Coupling a quantum gas to the field of a single high-finesse optical cavity gives rise to interactions of infinite range between the atoms, which can create a self-organized state when exceeding a critical strength [1]. It is desirable to tune range and directionality of these interactions in order to enable explorations of more complex self-organized states or quantum soft matter physics, such as superfluid glasses and associative memory. However, this requires extending the atom-photon interactions to multiple cavity modes [2].

We report on the realization of such an extended system, involving a Bose-Einstein condensate coupled to the modes of two optical cavities crossing at an angle of 60°. The interplay of the couplings between different momentum states and the cavity modes allows to spatially shape the interactions, leading to multiple new crystalline phases, e.g. with hexagonal, triangular or rectangular order. By independently varying the couplings to the cavity modes, we map out the associated phase diagram exhibiting multiple phase boundaries of first and second order with broken discrete and continuous symmetries. In addition, cavity-assisted Bragg spectroscopy gives us access to the excitations of the system.

[1] K. Baumann et al., Nature **464**, 1301 (2010).
[2] S. Gopalakrishnan et al., Nat. Phys. **5**, 845 (2009).

# Domain-size distribution and a fractal domain wall between two percolating domains in segregating Bose-Einstein condensates

Hiromitsu Takeuchi
Department of Physics, Osaka City University, 3-3-138 Sugimoto, Sumiyoshi-ku, Osaka 558-8585, Japan

Non-equilibrium dynamics of symmetry breaking phase transition can occur universally in different physical systems from condensed matter physics to cosmology and high energy physics. Superfluid system in ultra-cold atoms is now an ideal ground to study such phenomena thanks to the advanced experimental techniques, such as direct visualization of topological defects and realization of quasi-uniform system by trapping superfluids in a box potential. In this study, phase transition dynamics of $Z_2$ symmetry breaking, which is basic to systematic understanding of different kinds of phase transition dynamics, is explored on the basis of percolation theory and dynamic scaling hypothesis for segregating binary superfluids of atomic Bose-Einstein condensates (BECs).

Domain structures arising from the dynamic instability to segregation exhibit a universal scaling behavior in a dynamical scaling regime until the characteristic domain size becomes comparable to the system size. A domain structure is constructed of two percolating domains and a number of finite-sized domains. It is found that the time evolution of the percolating domains and an infinite domain wall existing between them [Figure (Left)] are universally characterized by their non-integer fractal dimensions [1]. The size distributions of finite-sized domains are well-described by the percolation analysis with the Fischer exponent, consistent with the dynamic scaling hypothesis [Figure (Right)] [2]. Our numerical experiments indicate the existence of a scaling regime unique to the quantum fluids in the size distribution function.

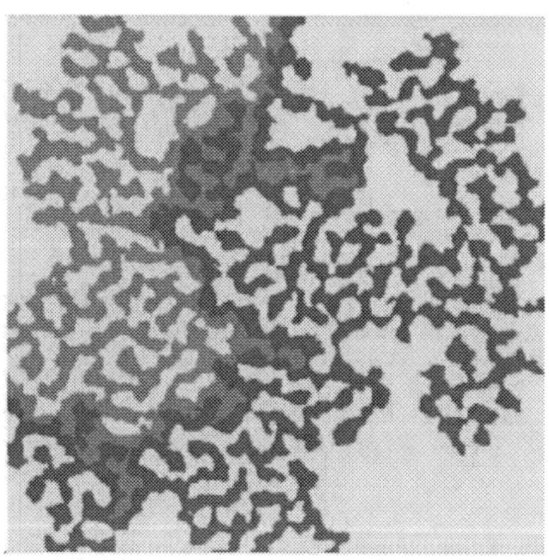

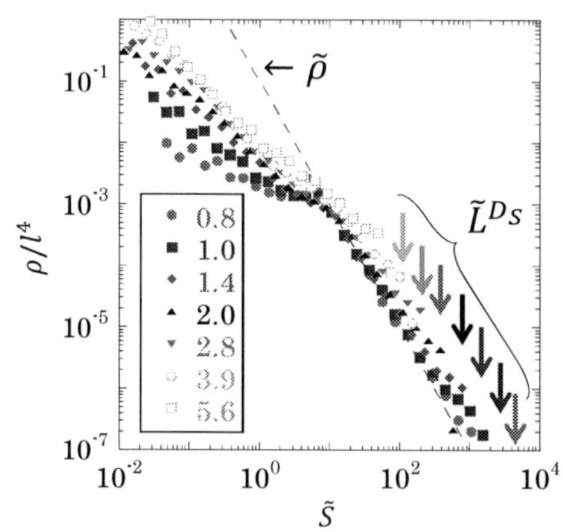


**Figure.** (Left) Two percolating domains in segregating BECs. There exists an infinite domain wall between the two domains. (Right) Dynamic scaling plot of domain-size distribution.

[1] Hiromitsu Takeuchi, Yumiko Mizuno, and Kentaro Dehara, Phys. Rev. A **92**, 043608 (2015).
[2] Hiromitsu Takeuchi, J. Low Temp. Phys. **183**, 169 (2016).

# Vortex quasicrystals in rotating Bose-Einstein condensates

Pekko Kuopanportti, Tapio Simula

School of Physics and Astronomy, Monash University, Victoria 3800, Australia

We study numerically the response of a disk-shaped Bose-Einstein condensate to a rotating optical Penrose quasicrystal potential [1,2]. The pinning effect of the potential is found to induce ordered but nonperiodic configurations of superfluid vortices in the rotating ground state of the condensate (Figure 1). When the numbers of superfluid and optical vortices are comparable, the superfluid vortices acquire a high degree of local tenfold orientational ordering. With fewer optical than superfluid vortices, the latter tend to form small clusters. In the opposite limit, we find numerical evidence for essentially degenerate rigidly rotating states which have genuinely different vortex configurations.

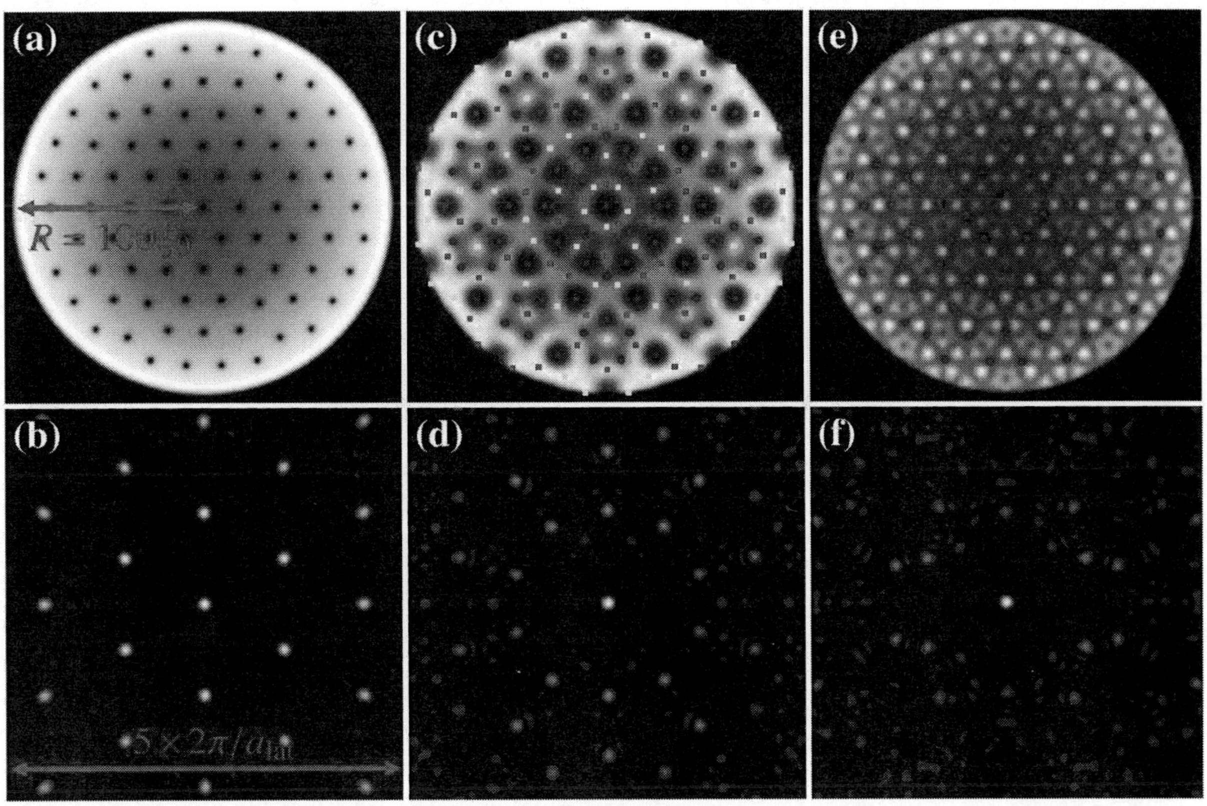


Figure 1. (a) Two-dimensional density profile of a rotating disk-shaped condensate in the absence of any pinning potentials; (b) the Fourier transform of its vortex distribution. Panels (c) and (d) are the corresponding images in the presence of a repulsive quasicrystal potential that has 1.6 optical vortices (squares) per superfluid vortex (circles). Panels (e) and (f) are for an attractive quasicrystal potential with 5.9 optical vortices (not shown) per superfluid vortex.

[1] L. Guidoni, B. Dépret, A. di Stefano, and P. Verkerk, Physical Review A **60**, R4233 (1999).

[2] L. Sanchez-Palencia and L. Santos, Physical Review A **72**, 053607 (2005).

# Phase ordering dynamics in ferromagnetic spin-1 Bose-Einstein condensates

Lewis Williamson[1], P. Blair Blakie[1]
[1]Dodd-Walls Center for Photonic and Quantum Technologies, Department of Physics, University of Otago, Dunedin 9016, New Zealand

Spinor Bose-Einstein condensates present rich phase diagrams for exploring phase transitions between states with different symmetry properties. In this work we simulate the approach to equilibrium of a spin-1 condensate quenched from an unmagnetized phase to three different ferromagnetic phases. The three ferromagnetic phases have $Z_2$, SO(2) and SO(3) symmetries respectively and possess different conservation laws. Following the quench, domains of magnetization form, with each domain making an independent choice of the symmetry breaking order parameter. These domains grow and compete for the global equilibrium state. We find that this growth follows universal scaling laws and identify the dynamic universality class for each of the three quenches. We also examine thermalization following the quenches. Contrary to what is expected, we find that in the SO(2) ferromagnetic phase the annihilation of vortices is not the slowest time scale. Even for long times after the vortices have disappeared, some long wavelength excitations are much hotter than predicted from an analytic calculation. We compare this microcanonical evolution to evolution in the canonical regime, and find that in the canonical evolution the long wavelength excitations are able to thermalize.

[1] L. A. Williamson and P. B. Blakie, Physical Review Letters 116, 025301 (2016).
[2] L. A. Williamson and P. B. Blakie, arxiv:1605.04016 (2016).

# Formation of Stable Soliton and Quantum Magnetism in Bose-Einstein Condensates

Daniel Benedicto-Orenes[1], Daryl Paul[1], Anna Kowalckzyk[1], Harry Proud[1], Kai Bongs[1], Nadine Meyer[1], Giovanni Barontini[1]

[1]University of Birmingham B15 2TT, Edgbaston, UK

We will present the first experimental realization of form-stable excitations in a 3d Bose-Einstein condensate, which have been predicted by Jones and Roberts in 1982 [1], but so far remained elusive. To create these Jones-Roberts solitons we imprint a non-trivial phase structure using a spatial light modulator (SLM). We find a pronounced self-stabilization effect, which assists the formation and enhances the lifetime of the Jones-Roberts solitons. In the future we intend to exploit the SLM to implement a cold atom simulator for twisted van der Waals heterostructures [2].

We will also present our progress in the construction of a "zero magnetic field" apparatus for Bose-Einstein condensates, featuring a multilayer mu-metal shield aiming at reducing the background fields to the nT regime. Exotic spinor dynamics and quantum phases are predicted in this novel experimental regime [3]. We will further exploit the unique features of our apparatus to generate different classes of multi-particle entangled states with the aim of performing precise quantum enhanced magnetometry.

[1] C. Jones & P.H. Roberts, J. Phys. A. Math. Gen. 15, 2599 (1982)

[2] M. T. Greenaway, E. E. Vdovin, A. Mishchenko, O. Makarovsky, A. Patanè, J. R. Wallbank, Y. Cao, A. V. Kretinin, M. J. Zhu, S. V. Morozov, V. I. Fal'ko, K. S. Novoselov, A. K. Geim, T. M. Fromhold and L. Eaves, Nat Phys. 11, 1057 (2015)

[3] S. Yi and H. Pu, arXiv:0804.0191v1 (2008)

# Vortex reconnections and bounces in trapped atomic Bose-Einstein condensates

S. Serafini[1], T. Bienaimé[1], E. Iseni[1], R. Bisset[1], L. Galantucci[2], C. F. Barenghi[2], F. Dalfovo[1], G. Lamporesi[1], G. Ferrari[1]
[1]INO-CNR BEC Center and Dipartimento di Fisica, Università di Trento
[2]Joint Quantum Centre and School of Mathematics and Statistics, Newcastle University

Vortex dynamics and reconnections play a fundamental role in quantum turbulent flows. In particular, quantum vortex reconnections are responsible for the redistribution of energy amongst scales and represent the ultimate process for incompressible kinetic energy dissipation into acoustic modes. Experimentally, direct visualization of the vortex–line dynamics in elongated Bose-Einstein Condensates (BECs) has been recently achieved in Trento [1], suggesting the possibility of exploring the effects of vortex-vortex interaction.

We will present a novel imaging technique suitable to follow the 3-dimensional dynamics of vortex–lines in harmonically trapped BECs, hence opening to the direct comparison between experimental observations and 3D numerical simulations based on the solution of the Gross–Pitaevskii equation. For the two-vortex interaction we find different regimes depending on the relative orientation and velocity of the vortices, their orbits and their initial distance: unperturbed precession around the center of the condensate, bounce dynamics, double and single reconnection events.

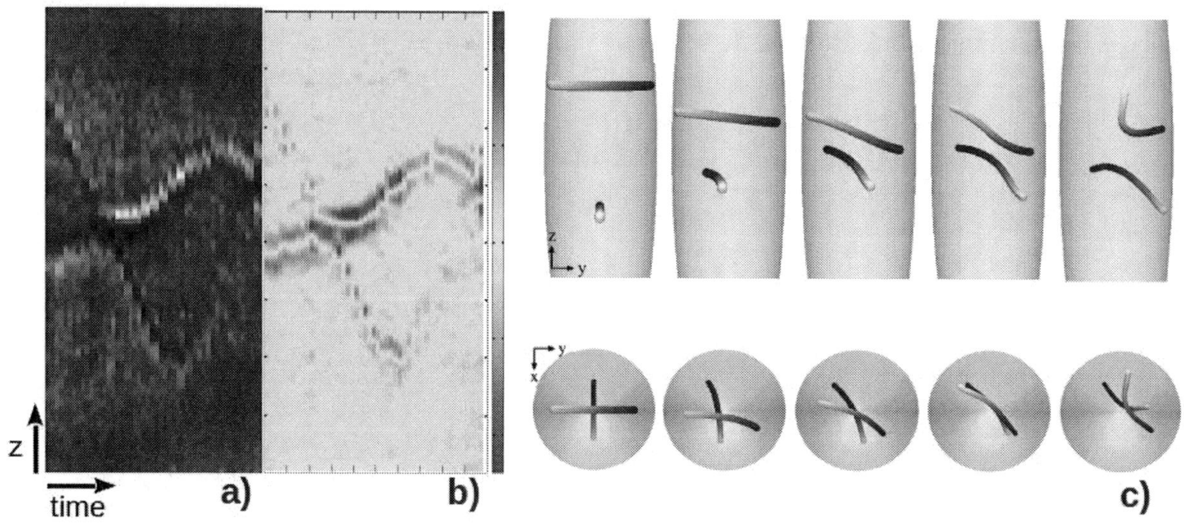


**Figure 1.** Bounce between vortices evolving in a BEC. a) The evolution of the axial position directly results from the residuals of the density profile. b) The orientation of the vortex line from the vertical direction appears from the asymmetry of the residuals signal. c) 3D numerical simulations of a bounce between two vortices.

[1] S. Serafini, M. Barbiero, M. Debortoli, S. Donadello, F. Larcher, F. Dalfovo, G. Lamporesi, and G. Ferrari, Phys. Rev. Lett. **115**, 170402 (2015).

# Phase fluctuations of Bose gases in the one- to three-dimensional crossover

Jorge Ferreras, Tom Barrett, Fedja Oručević and Peter Krüger
School of Physics and Astronomy, University of Nottingham, United Kingdom

The coherence properties of Bose-Einstein Condensates (BEC) have been of great interest throughout the last two decades. Phase coherence is studied for applications such as BEC interferometry. Dimensionality significantly affects the system's properties in this regard [1]. We experimentally demonstrate the appearance and characterize the properties of phase fluctuations across the dimensional crossover from 3d to 1d in BECs. Our atom chip-based trapping allows us to widely scan many relevant parameters, including atom number, temperature and aspect ratio of the trap. Furthermore, we are able to rapidly change those parameters in order to track the time evolution of phase fluctuations after a quench, e.g. in temperature, effective interaction strength, trap aspect ratio or even dimensionality of the system.

[1] B. Fang et al. Momentum-Space Correlations of a One-Dimensional Bose Gas, Phys. Rev. Lett., **116**:050402 (2016).

# Dipolar Quantum Droplets

Igor Ferrier-Barbut, Matthias Schmitt, Matthias Wenzel, Fabian Böttcher, Holger Kadau and Tilman Pfau

5. Physikalisches Institut and Center for Integrated Quantum Science and Technology, Universität Stuttgart, Pfaffenwaldring 57, 70550 Stuttgart, Germany

We present experimental observations of quantum droplets forming in strongly dipolar Bose-Einstein condensates. These droplets arise thanks to the combination of two effects. First, the mean-field contributions of the dipole-dipole and short-range contact interaction nearly screen each other resulting in an effective weak attraction. Second, the presence of these two interactions amplifies the quantum-mechanical correction to the mean-field energy. As a result, beyond mean-field effects balance the attraction, stabilizing a liquid-like state identified by the formation of quantum droplets. We have experimentally characterized the droplet density and its scaling with interaction parameters, in good agreement with theory. We are furthermore able to locate the critical point for the gaseous BEC to liquid-like transition. A strong signature is also found in collective modes of the BEC when approaching the transition. While we have observed here these droplets in the case of dipolar BECs, they should be found for other bosonic systems under competing inter-particle interactions.

[1] D. Petrov, Phys. Rev. Lett. **115,** 155302 (2015)
[2] H. Kadau *et al.,* Nature **532**, 194 (2016).
[3] I. Ferrier-Barbut *et al.,* Phys. Rev. Lett. **116,** 21530 (2016)

# Non-Abelian turbulence in spinor Bose – Einstein condensates

Thomas Mawson[1], Gary Ruben[2], Timothy Petersen[1], Tapio Simula[1]
[1]School of Physics and Astronomy, Monash University, Victoria 3800, Australia
[2]CSIRO Manufacturing Flagship, Clayton, Victoria 3168, Australia

We have studied computationally collisions of three spin-2 condensates and the resulting dynamics. These collisions produce vortex-antivortex lattices of fractional-charge vortices, which later evolve to quantum turbulence. We have shown that the initial state vortex lattices emerge as a result of multi-wave interference [1]. Furthermore, we find that three-source interference can be controlled to systematically create vortex lattices containing fractional-charge vortices whose collision dynamics may be non-Abelian [2]. We have phase imprinted initial states of vortex lattices populated with either Abelian or non-Abelian vortices, shown in Fig.1, and have studied the characteristics of two-dimensional quantum turbulence involving both spin and mass currents. Energy spectra of the incompressible kinetic energy and the spin interaction energy for such turbulent states are found to be characterized by scaling laws associated with classical and spin quantum turbulence [3]. We will also present our progress on characterizing three-dimensional non-Abelian quantum turbulence.

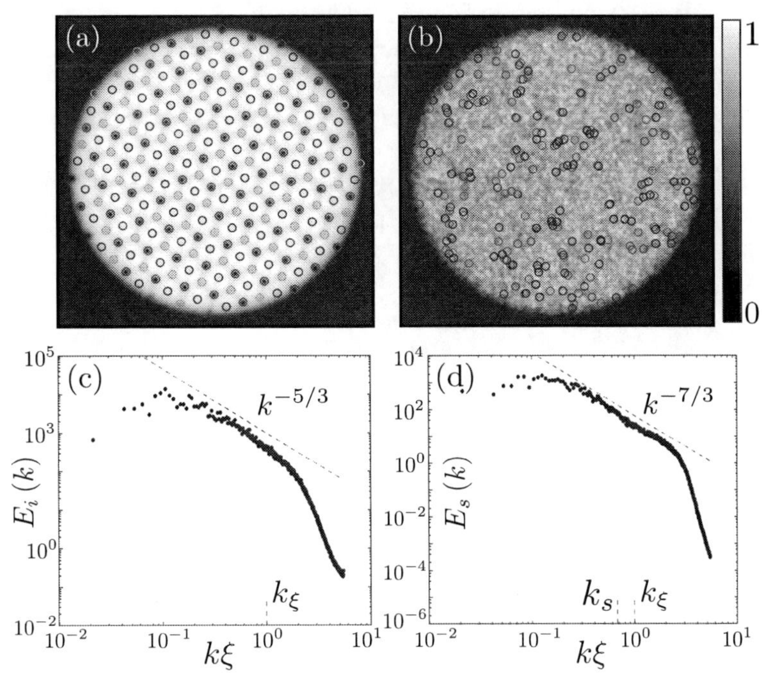


Figure 1: Two-dimensional quantum turbulence in a spin-2 Bose—Einstein condensate. Shown are the condensate density for an initial vortex lattice state of fractional-charged vortices (a) and a late-time quantum turbulent state (b). Power spectra of the incompressible kinetic energy (c) and spin interaction energy (d) are calculated for the state shown in (b).

[1] T. Mawson, G. Ruben, T. Simula, Physical Review A 91, 063630, (2015).
[2] M. Kobayashi, Y. Kawaguchi, M. Nitta, M. Ueda, Physical Review Letters 103, 115301, (2009).
[3] K. Fujimoto, M. Tsubota, Physical Review A 85, 053641, (2012).

# Many-body physics of out-of-equilibrium periodically modulated ultracold Bose atomic gases

Marios C. Tsatsos

Sao Carlos Institute of Physics, University of Sao Paulo, PO Box 369, 13560-970, Sao Carlos, SP, Brazil

Quantum fluctuations are in the heart of quantum theory; they determine how far from the expected value a particular single measurement of quantum system will be found. In a recent joint Rice-USP experiment, where a trapped BEC is periodically perturbed through external magnetic fields, novel granular phases appear. In this granulated phase the measured properties of the gas are not explained by a mean-field theory, such as the Gross-Pitaevskii model. We apply the MultiConfigurational Time-Dependent Hartree for Bosons (MCTDHB) [1,2] and find that quantum fluctuations are large and important in order to obtain faithful and realistic simulations of the experiment.

[1] Alexej I. Streltsov, Ofir E. Alon, and Lorenz S. Cederbaum, Role of Excited States in the Splitting of a Trapped Interacting Bose-Einstein Condensate by a Time-Dependent Barrier, Phys. Rev. Lett. 99, 030402 (2007); Ofir E. Alon, Alexej I. Streltsov, and Lorenz S. Cederbaum, Multiconfigurational time-dependent Hartree method for bosons: Many-body dynamics of bosonic systems, Phys. Rev. A 77, 033613 (2008).

[2] A. U. J. Lode, M. C. Tsatsos and E. Fasshauer, MCTDH-X: The time-dependent multiconfigurational Hartree for indistinguishable particles software (2016), http://ultracold.org.

# Quench Dynamics of Antiferromagnetic Spin-1 Condensates

Luke Symes[1] and Blair Blakie[1]

[1] Department of Physics, Centre for Quantum Science,
and Dodd-Walls Centre for Photonic and Quantum Technologies,
University of Otago, Dunedin, New Zealand

We study the quench dynamics of a spin-1 condensate with antiferromagnetic interactions. In this system the condensate prefers to adopt uni-axial nematic spin-ordering [1]. We consider the case of a quench in the quadratic Zeeman energy from positive value, where the nematic director is aligned with the field direction, to negative value, where the director prefers to be transverse to the field. The quench thus takes the spinor condensate across a phase transition to a phase that spontaneously breaks a continuous symmetry of the Hamiltonian.

We discuss the relevant order parameter and present results for the short- and long-time dynamics of the system post-quench. Our results show that half-quantum vortices play an important role in the dynamics as the topological defects of the order parameter. We also show that the long-time coarsening dynamics [2] of the antiferromagnetic system are distinct in behavior from the ferromagnetic case [3] and that this system is a rich playground for studying the interplay of spin and superfluid order. Recent experiments by the Shin group have realized a system where our predictions might be explored and have already made beautiful observations of half-quantum vortex dynamics [4].

[1] T. Ohmi and K. Machida, Journal of the Physical Society of Japan **67**, 1822 (1998); T.-L. Ho, Physical Review Letters **81**, 742 (1998); T. Zibold, V. Corre, C. Frapolli, A. Invernizzi, J. Dalibard, and F. Gerbier, Physical Review A **93**, 023614 (2016).

[2] A. J. Bray, Advances in Physics **43**, 357 (1994).

[3] L. A. Williamson and P. B. Blakie, Physical Review Letters **116**, 025301 (2016).

[4] S. W. Seo, S. Kang, W. J. Kwon, and Y.-i. Shin, Physical Review Letters **115**, 015301 (2015); S. W. Seo, W. J. Kwon, S. Kang,, and Y.-i. Shin, Physical Review Letters **116**, 185301 (2016).

# One-dimensional spinor Bose gases: Quantum Phase Transitions and Elementary Excitations

Karina Jiménez-García, Camille Frapolli, Andrea Invernizzi,
Bertrand Evrard, Jean Dalibard, Fabrice Gerbier
Laboratoire Kastler Brossel, Collège de France, CNRS, ENS-PSL Research University,
UPMC-Sorbonne Universités, 11 place Marcelin Berthelot, 75005 Paris, France

Ultracold atom systems are excellent candidates to investigate the underlying physical mechanisms of spatial organization in quantum phases of matter. Spatial order typically arises when the system undergoes a phase transition at sufficiently low temperatures, and eventually determines the macroscopic properties of materials.

Ultracold *spinor* gases (systems with an internal *spin* degree of freedom) exhibit a rich landscape of quantum phases. In such systems, the competition between the spin dependent interaction energies and the hyperfine Zeeman energies, gives rise to a quantum phase diagram of generalized magnetic phases [1]. Spinor Bose gases in reduced dimensions are particularly interesting to study non-equilibrium phenomena across such magnetic quantum phase transition.

Here we present the experimental study of one-dimensional spin-1 Bose gases of Na atoms across a magnetic quantum phase transition, and the experimental progress toward the observation of their non-equilibrium dynamics. By quenching the system we expect to observe defects, manifested as axial spin domains, arising from the Kibble-Zurek mechanism. The introduction of an external periodic potential will further allow us to investigate magnetically ordered phases in a lattice.

Our experiments also include the characterization of scalar and vectorial excitations such as dark and magnetic solitons[2].

[1] D. Jacob, *et al.*, Phys. Rev. A **86**, 061601(R) (2012).
[2] C. Qu, *et al.*, Phys. Rev. Lett. **116**, 160402 (2016).

# Observation of Atomic Four-Wave-Mixing Processes of Multiple Diffraction Orders

Michael Keller[1,2], Mateusz Kotyrba[1,2], Mario Rusev[1,2], Mariusz Semczuk[2], Anton Zeilinger[1,2]

[1]Vienna Center for Quantum Science and Technology (VCQ), Faculty of Physics, University of Vienna, Boltzmanngasse 5, A-1090 Vienna, Austria

[2]Institute for Quantum Optics and Quantum Information (IQOQI), Boltzmanngasse 3, A-1090 Vienna, Austria

Bragg diffraction of ultracold atoms on an optical lattice has been shown to be a versatile tool for coherently transferring momentum to atoms over recent years. It not only allows to put atoms from a Bose-Einstein condensate in a state with well-defined momentum but also allows for transferring the atoms in a superposition of momenta 0, $\pm\hbar k$, $\pm 2\hbar k$, ... (where k is the wave vector of the optical lattice). We prepare condensates of metastable helium (He*) atoms and put them in such superpositions of momenta by pulsing an optical lattice. Subsequent s-wave scattering processes between atoms of different diffraction orders lead to a very rich scattering state which can be seen on a single-atom resolving detector (see figure).

The multiple shell structure of this state offers the potential to create more atom pairs with anticorrelated momentum in one collision process while still keeping the production rate at a level that allows for identification of the scattering partners, which is essential for experiments like the ones proposed in [1].

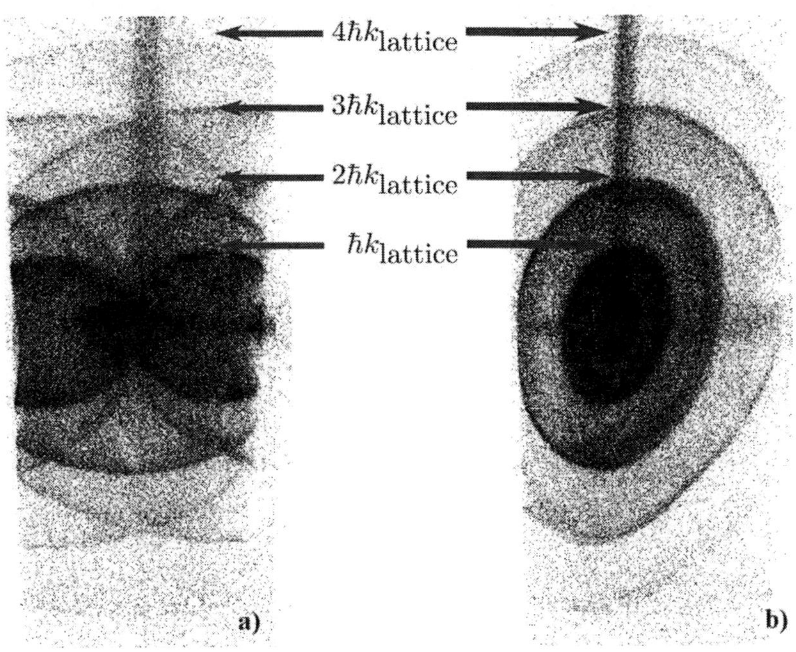


**Figure 1.** Observed shell structure after s-wave scattering of atoms from various momentum states. The total momentum is always an integer multiple of the wave vector of the optical lattice which the atoms are originally diffracted from. Shown are views perpendicular to the direction of the lattice (a) and along the direction of the lattice (b).

[1] J. Kofler, M. Singh, M. Ebner, M. Keller, M. Kotyrba, A. Zeilinger, Phys. Rev. A 86, 032115 (2012).

# Dynamics of vortex lines in three-dimensional Bose-Einstein condensates

Shan Zou[1], Tao Yang[2,*]
[1]School of Physics, Northwest University, Xi'an, 710069, P. R. China
[2]Institute of Modern Physics, Northwest University, Xi'an, 710069, P. R. China

We investigate the dynamics of two vortex lines carrying opposite charges in a pancake-shaped Bose-Einstein condensate. The real-time dynamics of the vortex lines shows different characteristics from a dipole structure of two-dimensional vortex cores for some ranges of initial separations. We find that the vortex lines could connect to form a vortex ring structure in the high density region of the condensate. The shape and length of vortex rings vary during the movement depending on the initial configuration of the system and the position of the vortex ring in the condensate. The vortex ring can also break into two new bent vortex lines in the direction with tight trapping frequency, and finally decay on the surface of the condensate. In addition, for large initial separations the vortex lines will annihilate and create solitary waves which expand fast out of the condensate.

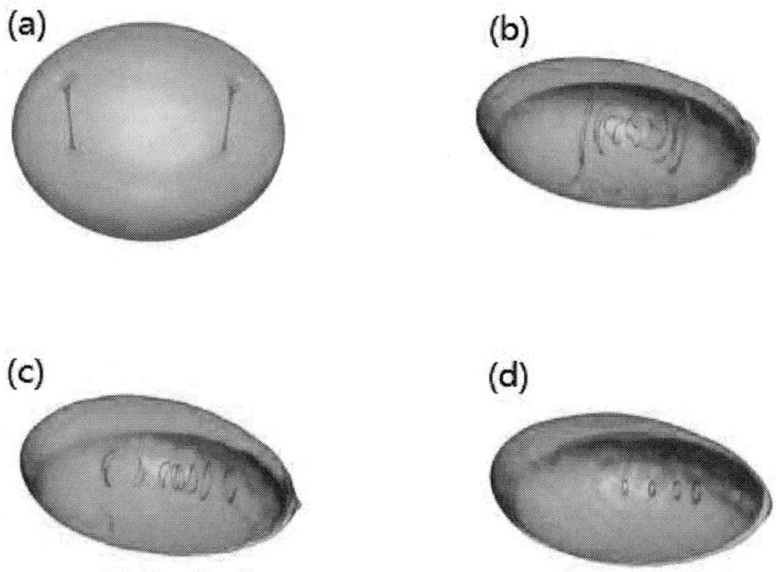


**Figure** The snapshot images of the isosurface plots. (a) One of the initial states of a pancake-shaped condensate with two vortex lines. (b)-(d) Dynamical processes of vortex rings with different initial separations of dipole structure.

[1] P. Rosenbusch, V. Bretin, and J. Dalibard, Physical Review Letters **89**, 200403 (2002).
[2] Weibin Li, Masudul Haque, and Stavros Komineas, Physical Review A **77**, 053610 (2008).
[3] S. Middelkamp, P. J. Torres, P. G. Kevrekidis, D. J. Frantzeskakis, R.Carretero-Gonzalez, P. Schmelcher, Physical Review A **84**, 011605 (2011).
[4] S. Serafini, M. Barbiero, M. Debortoli, S. Donadello, F. Larcher, F. Dalfovo,G. Lamporesi, and G. Ferrari, Physical Review Letters **115**, 170402 (2015).

# A quantum gas microscope of Yb atoms

Eunmi Chae, Ryuta Yamamoto, Jun Kobayashi, Shuya Yamanaka, Yuto Sakura, Takuma Kuno, Kohei Kato, Yoshiro Takahashi

Department of Physics, Graduate School of Science, Kyoto University, Kyoto 606-8502, Japan

A quantum gas microscope (QGM) has been used as a new tool to detect each atom in an optical lattice site *in situ*, probing quantum correlations in both bosonic and fermionic systems in real space [1]. So far, most of the QGMs are for alkali atoms and expanding the scope of species is desired to explore rich physics available with other atomic classes such as two-electron atoms. Ytterbium (Yb) atoms have long-lived metastable states and many isotopes including both bosonic and fermionic species, offering possibilities for quantum simulations of various systems such as one with SU(N) symmetry and for quantum information exploiting the internal states. A QGM enables observation of features in these systems that would be averaged out with typical ensemble detections, and also enables site-resolved control of the atoms.

Here, we report on the realization of the QGM of Yb atoms using two methods – fluorescence imaging [2] and Faraday imaging [3], which is equivalent to a polarization-phase-contrast imaging [4]. A strong transition of $^1S_1$-$^1P_1$ is used for probing as in the previous report [5], while an intercombination transition of $^1S_1$-$^3P_1$ is applied for cooling during the detection in both cases. Fidelity of 87% (84%) is achieved to resolve the atoms in the 266 nm lattice sites with the fluorescence (Faraday) imaging. By using Faraday interference, heating from the probe laser can be suppressed with a large detuning of the laser. Since the Faraday imaging does not rely on the spontaneous emission but utilized a Quantum–Non–demolition type dispersive interaction, this new technique can be implemented for quantum information processing and quantum feedback of single atoms in the optical lattices. Detailed descriptions of the experimental setup including current progress to improve the resolution and the fidelity, along with the planned experiments using the QGM will be presented.

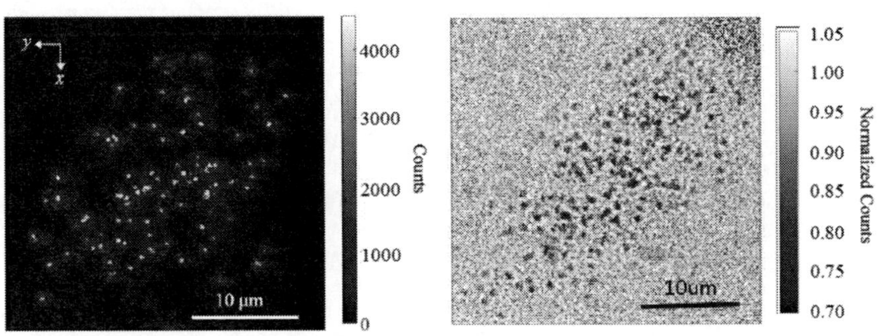


**Figure 1.** Site resolved image of ultracold Yb atoms. (Left) Fluorescence image. (Right) Faraday image.

[1] R. Islam et al., Nature 528, 77 (2015); M. F. Parsons et al., arXiv: 1605.02704 (2015); M. Boll et al., arXiv: 1605.05661 (2016); A. Omran et al., Phys. Rev. Lett. **115**, 263001 (2015); L. W. Cheuk et al., Phys. Rev. Lett. **116**, 235301 (2016); E. Haller et al., Nat. Phys. **11**, 738 (2015);

[2] R. Yamamoto et al., New J. of Phys. **18**, 031001 (2016).

[3] R. Yamamoto et al., in preparation.

[4] C. A. Sachett et al., Appl. Phys. B **65**, 433 (1997).

[5] M. Miranda et al., Phys. Rev. A **91**, 063414 (2015).

# Dissipative Bose-Hubbard system with cold atoms in an optical lattice

Takafumi Tomita, Shuta Nakajima, Yosuke Takasu, Yoshiro Takahashi
Department of Physics, Graduate School of Science, Kyoto University, Kyoto 606-8502, Japan

A cold atom system is often regarded as an ideal closed quantum system. In real experiments, however, some dissipation processes such as photon scattering and particle losses exist. In some situations the dissipation, including a measurement process, could play an essential role in quantum physics and it is important to study the effect of the dissipation on quantum many-body phenomena such as a quantum phase transition [1].

We experimentally study a Bose-Hubbard system with the dissipation of the particle losses using ultracold atoms in an optical lattice. Firstly, we report an observation of atom loss suppression induced by strong dissipation using ytterbium (Yb) atoms in a three-dimensional (3D) optical lattice. Yb atoms in the metastable $^3P_2$ state have large inelastic collision rate [2], inducing strong dissipation. We measure the two-body loss rate of $^3P_2$ atoms in a 3D optical lattice and observe significant reduction of the loss rate compared to the tunneling rate (Fig. 1(a)). Also we irradiate a photoassociation (PA) laser beam to the ground state ($^1S_0$) of Yb atoms in order to induce two-body loss artificially with high controllability on the dissipation strength. We investigate the atom loss behavior in a wide range of the dissipation strength and observe a non-monotonic dependence of the two-body loss rate on the dissipation strength (Fig. 1(b)). These results can be interpreted as the continuous quantum Zeno effect [3]: strong dissipation suppresses the coherent dynamics of the tunneling and reduces the atom loss.

We also report the effect of the dissipation on the quantum phase transition between a Mott insulator and a superfluid state. We observe the momentum distribution of the atoms using a time-of-flight method and investigate the phase coherence of the atoms in a shallow lattice with dissipation. Latest results and detailed analysis will be presented.

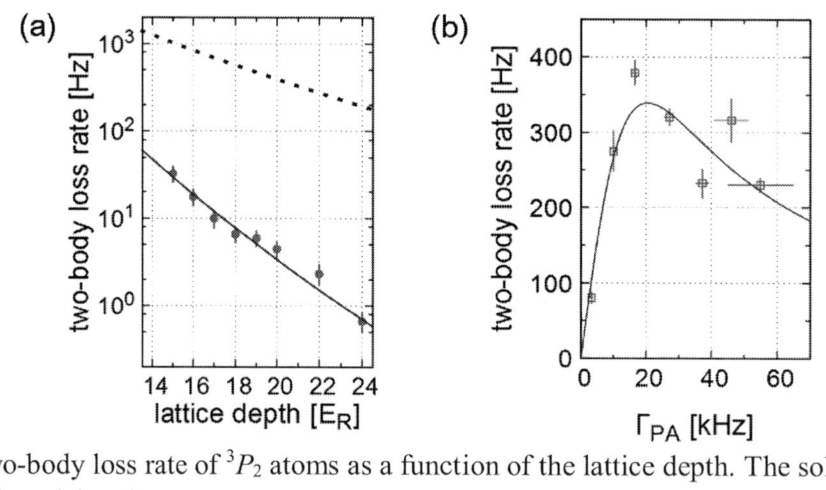


Figure 1: (a) Two-body loss rate of $^3P_2$ atoms as a function of the lattice depth. The solid line shows a fit of a theoretical model to the experimental data and the dashed line shows the tunneling rate. (b)Two-body loss rate of $^1S_0$ atoms as a function of the dissipation strength of PA. The lattice depth is set to 8 $E_R$. The solid line is a fit of a theoretical model to the experimental data.

[1] Y. Ashida *et al.*, arXiv:1605.06899v1 (2016).
[2] S. Uetake *et al.*, Phys Rev. A **86**, 032712 (2012).
[3] N. Syassen *et al.*, Science **320**, 1329 (2008).

# The dynamics of $^{87}$Rb BECs configured at the microscale

Tyler W. Neely[1, 2], Guillaume Gauthier[1, 2], Isaac Lenton[2], Mark Baker[2], Stuart Szigeti[1, 2], Matthew Davis[2], Halina Rubinsztein-Dunlop[1, 2]

[1]ARC Centre of Excellence for Engineered Quantum Systems (EQuS), School of Mathematics and Physics, University of Queensland, Brisbane, Australia 4075

[2]School of Mathematics and Physics, University of Queensland, Brisbane, Australia 4075

The implementation of optical tweezing techniques in quantum gas experiments has allowed the realization of ever more complex trapping geometries. Our laboratory pursues configurable traps based on the direct (nearly diffraction limited) imaging of a digital micromirror device (DMD).

Our optical potentials are achieved using commercially available microscope objectives external to our vacuum chamber (see Fig. 1(a)) that directly project the DMD to the atom plane, producing optical traps over an area of 130 μm × 200 μm, with a resolution of 630(10) nm full width at half maximum (FWHM) at 532 nm illumination [1]. We combine these potentials with a horizontally propagating $TEM_{00}$ or $TEM_{01}$ Hermite-Gaussian optical sheet that provides vertical confinement. The combined potentials enable near arbitrary control of the condensate density in a 2D plane, as shown in Fig. 1(b). The DMD is additionally highly dynamic, with a maximum full-frame rate of 20 KHz and on-board storage of >13,000 frames. Using the dynamic control enabled by the DMD, we study the transport of atoms prepared at different temperatures and condensate fractions. We utilize a double-reservoir geometry as shown in Fig. 1(c). Subsequent imaging of the BECs with 960(80) nm FWHM resolution allows the in situ imaging of vortices excited by the superfluid transport.

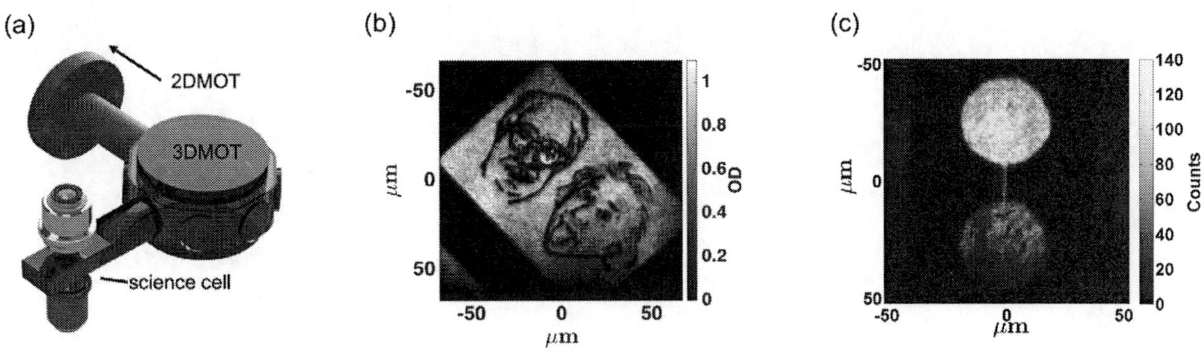


**Figure 1.** (a) Experimental apparatus used to microscopically confine BECs. A horizontally propagating sheet (not shown) provides vertical confinement, while a DMD-based potential is imaged with the top microscope objective. The lower microscope objective is used to image the BECs at high resolution. (b) Example of a highly configured BEC, onto which the portraits of Bose and Einstein have been imprinted. (c) Flow between reservoirs across a temperature gradient, with a vortex stream imaged in situ.

[1] G. Gauthier, I. Lenton, N. McKay Parry, M. Baker, M. J. Davis, H. Rubinsztein-Dunlop, and T. W. Neely, arXiv preprint arXiv:1605.04928 (2016).

# Probing the phase coherence and excitation spectrum of a toroidal BEC

Thomas Bell[1, 2], Jake Glidden[1, 2], Mark Baker[2], Tyler W. Neely[1, 2], Matthew Davis[2], Halina Rubinsztein-Dunlop[1, 2]

[1]ARC Centre of Excellence for Engineered Quantum Systems (EQuS), School of Mathematics and Physics, University of Queensland, Brisbane, Australia 4075

[2]School of Mathematics and Physics, University of Queensland, Brisbane, Australia 4075

We present details of our $^{87}$Rb BEC apparatus, which makes use of configurable optical potentials using time-averaged optical dipole traps. These are produced by rapidly scanning a far-detuned laser beam using a 2D acousto-optical deflector (2D-AOD). By using an intensity correction system, which uses the local atom density as a measure of trap imperfections, we correct for variation in the trap depth to achieve smoothness <10% of the chemical potential [1]. With this technique, we demonstrate smooth toroidal traps up to up to 300 μm diameter, large atom number (N= $2.5 \times 10^6$ atoms) with typical trap frequencies $\omega_z=2\pi$ (140) Hz, $\omega_\rho=2\pi$ (60) Hz.

We are using a number of methods and experimental signatures to probe the phase coherence properties of these extended toroidal traps for various ring diameter, temperature, and atom number. These including observing the self-interference of the BEC in time-of flight imaging [2], as well as utilizing Bragg spectroscopy and interference [3]. We present results on studying the Bogoliubov excitation spectrum, examining both extant excitations at finite temperature and those introduced through resonant driving.

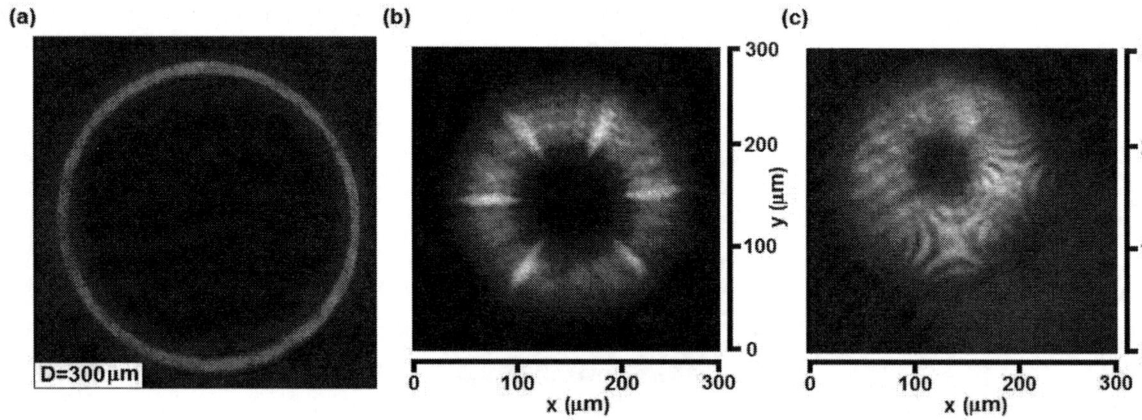


**Figure 1.** (a) BEC confined in a 300 μm toroidal optical trap (b) Resonant modulation of a 150 μm diameter ring BEC at 17 Hz, corresponding to m=6 mode of standing phonon wave (20 ms time of flight) (c) Matter wave interference of ring BEC(120 μm diameter) in time-of-flight through Bragg π/2 pulse and subsequent recombination π/2 pulse 12 ms later.

[1] T. A. Bell, J. A. Glidden, L. Humbert, M. W. Bromley, S. A. Haine, M. J. Davis, T. W. Neely, M. Baker, and H. Rubinsztein-Dunlop, New Journal of Physics **18** 035003 (2016).

[2] L. Mathey, A. Ramanathan, K.C. Wright, S.R. Muniz, W.D. Phillips, C.W. Clark, Phys. Rev. A **82** , 033607 (2010).

[3] D. Hellweg, L. Cacciapuoti, M. Kottke, T. Schulte, K. Sengstock, W. Ertmer, J.J. Arlt,, Phys. Rev. Lett. **91**, 010406 (2006).

# Density-adjustable all-optical Bose-Einstein condensate and its thermal equilibrium properties from 1D gases to 3D gases

Chen Li[1], Tianwei Zhou[1], Jinggang Xiang[1], Qi Huang[1], Xuzong Chen[1]
[1]School of Electronics Engineering and Computer Science, Peking University, Beijing, China

We create the BEC of $^{87}Rb$ through all optical evaporation cooling directly in a gravity-titled optical dipole trap. The tilting which increases the final trap frequency brings us not only the fast cooling, but also the relatively high temperature as well as the large and inhomogeneous filling factor in optical lattices. So we develop a method to levitate the atoms with a magnetic field gradient and a frequency-reduced dipole trap in the lattices loading sequence to lower the filling factor. We also investigate the thermal equilibrium properties in the transition from the arrays of 1D Bose gases to the 3D Bose gases. By dividing the atoms into two parts with opposite momentums and arranging an additional confining potential along the longitudinal direction of the 1D Bose gases, we observe the atoms oscillate and collide with each other until thermal equilibrium. The relaxation dynamics during the equilibrium are greatly changed in the transition from the 1D gases to the 3D gases.

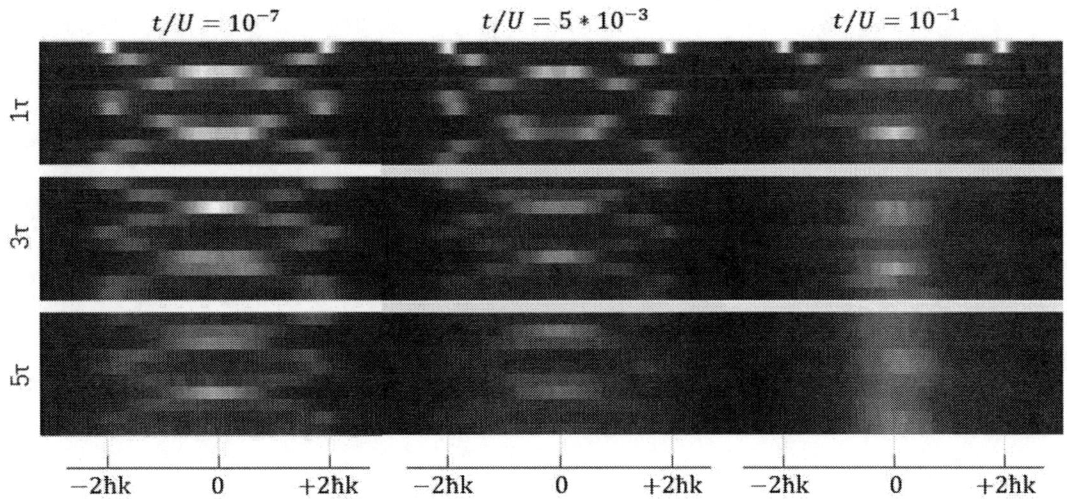


**Figure 1.** A 2D optical lattice is used to divide the atom cloud into an array of tubes. We investigate how is the relaxation process changed in the lattices with different lattice depth, corresponding to the transition from 1D gases to 3D gases. The three pictures in a row show the oscillation in the first cycle, the third cycle, and the fifth cycle respectively.

# H-wire trap to improve atomic number in BEC state

Donghoon Min, Sugtae Kang, SeungJin Kim, Ashish Kumar Sharma, Yoonseong Jeon, Sangmin Jung, Dongjin Lee and Jungbog Kim

Physics Education, Korea National University of Education, Chung-Bok, Republic of Korea

We have made Bose-Einstein condensate (BEC) on the atom chip with dimple wire or T-wire. As shown in figure 1, dimple wire (a) uses main guide wire fully, but T wire (b) uses only half of it. The trap behaviors are very similar; however, T-wire has spatial advantage[1]. But transfer rate in T-wire trap is smaller due to shift of magnetic field minimum position when transferring atoms from external coil trap to T-wire trap. In order to increase transfer rate, we add H-wire trap process between external coil trap and T-wire trap. We compare the atom number in BEC state when transferring atoms in external coil to T-wire trap, and to T-wire via H-wire trap.

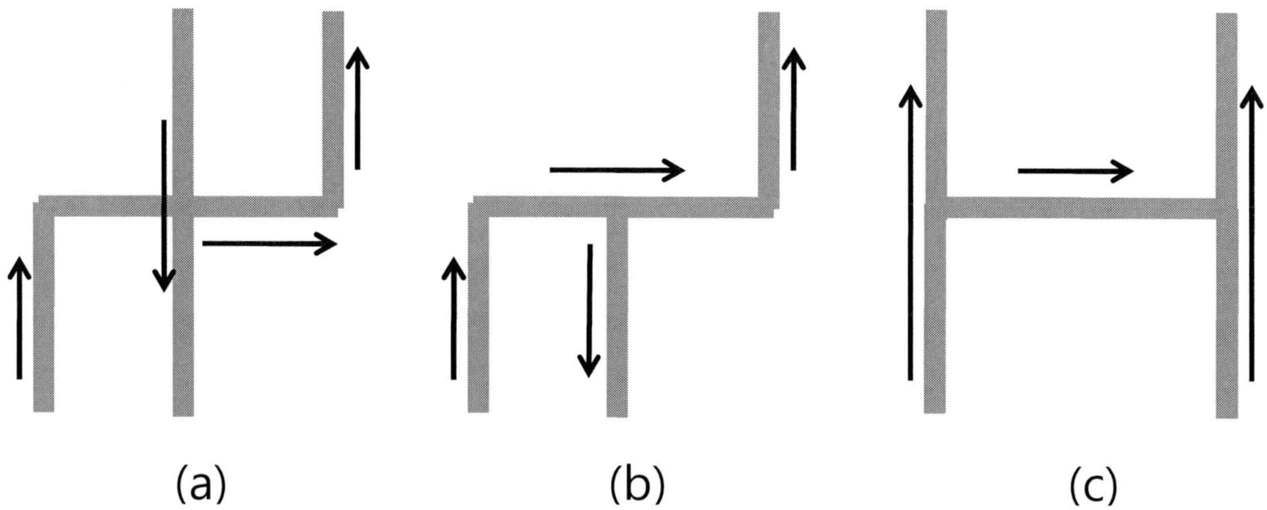


**Figure 1.** Three type of chip traps used for BEC experiments (1) Dimple trap, (b) T-wire trap, (c) H-wire trap.

[1] Evan Ali Salim "Ultracold matter systems and atomtronics instrumentation", University of Colorado, PhD Thesis (2011)

# Superfluidity of light in hot atomic vapor

Stefano Pierini[1], Dhruv Sharma[1], Alberto Bramati[1], Quentin Glorieux[1]
[1]Laboratoire Kastler Brossel, UPMC-Sorbonne Universités, CNRS, ENS-PSL, FRANCE

**Quantum fluid of light in propagating geometries.**

We present here, recent experiments using a propagative geometry to study quantum fluids of light. As an alternative approach to the confined light in cavities, we use a monochromatic light field propagating in a non-linear $\chi(3)$ medium (a near-resonant rubidium vapor) showing an intensity dependent refractive index.[1] Under the paraxial approximation, the transverse amplitude of the field can be described using an analogy with the Gross-Pitaevskii equation (GPE) for the macroscopic wave-function of a BEC.[2] However, as the GPE describes a time evolution, the non-linear wave equation refers to an evolution in space along the propagation direction and a space-time mapping must be made when discussing fluids of light in the propagating geometry. As a consequence, we can study the time evolution of a 2D Bose gas in this configuration. The spatial evolution is observed in the transverse direction and the temporal evolution is mapped along the propagation axis. Simple imaging techniques give a access to the temporal evolution of the fluid in position and momentum.

**Preliminary results on supefluidity.**

The basic idea of the preliminary measurement is to use a strong pump beam that initiates the nonlinearity in the near resonance atomic vapor and thus defines the photon fluid properties, i.e. speed of sound, and use a weak probe beam to seed a sound wave of desired frequency (Interference pattern) and measure its phase velocity.[3] The dispersion of these small amplitude excitations is given by the Bogoliubov dispersion relation. By changing the flow velocity of the Bogoliubov waves (incident angle) we can observe a transition from a sonic dispersion to a parabolic dispersion demonstrating the possibility of studying superfluidity of light in this configuration.

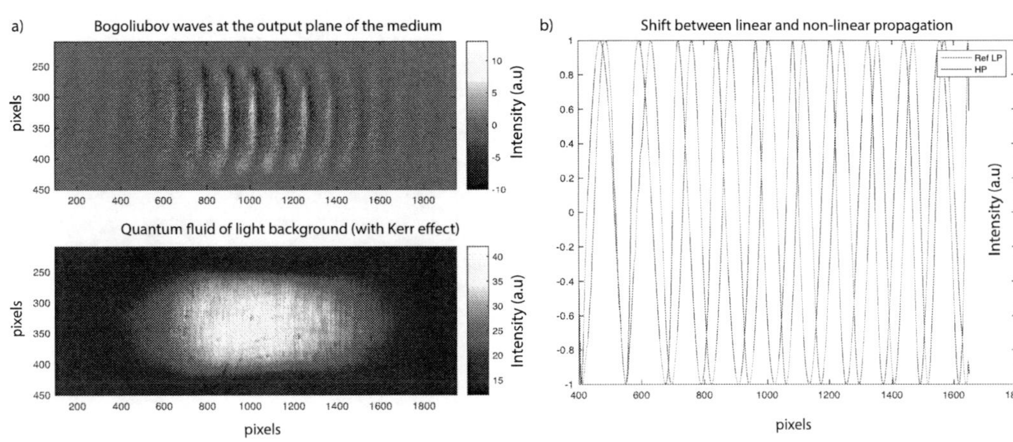


**Figure 1.** a) Image at the ouptut of the non-linear medium. B) 1D profile of the waves in the transverse plane for the linear and non-linear case.

## REFERENCES

1. P.E. Larré and I. Carusotto, Phys. Rev. A **92** 043802 (2015)
2. R.Y. Chiao and J. Boyce, Phys. Rev. A **60**, 4114 (1999)
3. D. Vocke, T. Roger, F. Marino, E. Wright, I. Carusotto, M. Clerici, and D. Faccio. Optica **2**, 484

# Exploring the many-body localization transition in two dimensions

Jae-yoon Choi[1], Sebastian Hild[1], Johannes Zeiher[1], Peter Schauß[1], Antonio Rubio-Abadal[1], Tarik Yefsah[1], Vedika Khemani[3], David. A. Huse[3,4], Immanuel Bloch[1,2], Christian Gross[1]

[1]Max-Planck-Institut für Quantenoptik, 85748 Garching, Germany,

[2] Fakultät für Physik, Ludwig-Maximilians-Universität, 80799 München, Germany

[3] Physics Department, Princeton University, Princeton, 08544 New Jersey, USA

[4] Institute for Advanced Study, Princeton, 08540 New Jersey, USA

Thermalization of isolated quantum many-body systems under their own dynamics is a fundamental assumption in statistical physics. While general conditions for the thermalization of isolated quantum many-body system are not well established, the many-body localization has been considered as a general system that breaks the thermalization, which could extend our understandings of the connection between statistical physics and quantum mechanics. In this presentation we report on the observation of many-body localization transition in two-dimensional disordered optical lattice [1]. Using single-site resolved imaging system, we prepare a high energy density state far from equilibrium and trace its dynamical evolution under disorder potential. We observe the system does not thermalize above a nonzero disorder and find strong evidence for a diverging length scale when approaching the localization transition.

[1] J.-y. Choi *et al.*, Science **352**,1547 (2016).

# Josephson dynamics of a Bose-Einstein condensate in the RF-dressed double well potential

Sugtae Kang[1], SeungJin Kim[1], Ashish Kumar Sharma[1], Yoonseong Jeon[1], Donghoon Min[1], Sangmin Jung[1], Dongjin Lee[1] ,Jungbog Kim[1]

[1]Physics Education, Korea National University of Education, Chung-Bok, Republic of Korea

We have demonstrated the tunneling phenomena of Bose-Einstein Condensate (BEC) in the rf-dressed double well potential on the atom chip. The control ability of the initial population imbalance is important because the oscillation frequency of BEC only depends on the difference of chemical potential. Using independently controllable 3 rf-fields on the atom chip, the initial population imbalance can be easily achieved without shift of minimum potential position [1]. We measured the population imbalance along the tunneling time after large distance splitting due to the image resolution limit. This allows us to demonstrate the Josephson oscillation and self-trapping phenomena along the initial population imbalance. This experimental result is described by simulation result of the nonlinear Gross-Pitaevskii equation.

[1] S. J. Kim, H. Yu, S. T. Kang, D. Z. Anderson, and J. B. Kim, Phys. Rev. A. **93**, 033612 (2016).

# Fate of Anderson localization of a one-dimensional Bose-Einstein condensate after long-time expansion

Stefan Donsa[1], Harald Hofstätter[1], Othmar Koch[2], Joachim Burgdörfer[1], Iva Březinová[1]
[1]Institute for Theoretical Physics, Vienna University of Technology, Vienna, Austria, EU
[2]Faculty of Mathematics, University of Vienna, Vienna, Austria, EU

Anderson localization is an interference effect of non-interacting particles in disordered systems which leads to suppression of transport [1]. It is still an open question whether Anderson localization occurs in interacting systems. In recent experiments Bose-Einstein condensates were used to investigate Anderson localization in a highly controllable setup [2]. A Bose-Einstein condensate initially trapped in a harmonic potential was released and its expansion in a weak disorder potential was monitored. The observed suppression of its expansion was explained assuming that the initially interacting system can be regarded as non-interacting after long expansion times [3]. We test this assumption and examine the long-time influence of pair interactions by calculating the expansion with and without inter-particle interactions. The dynamics is described on the mean-field level within the Gross-Pitaevskii equation (1)

$$i\hbar\frac{\partial}{\partial t}\Psi(x,t) = \left[-\frac{\hbar^2}{2m}\frac{\partial^2}{\partial x^2} + V(x) + g|\Psi(x,t)|^2\right]\Psi(x,t), \quad (1)$$

which assumes that the motion of all atoms is fully coherent at all times and can be described effectively by a single wave function with self-interaction (Eq. 1). We compare the scenario of non-interacting (NI) particles [2, 3] with a calculation in which the condensate is interacting at all times (GPE). For the first case (NI), we switch off interactions when most of the initial interaction energy is converted into kinetic energy. Monitoring the width $\Delta x$ of the Bose-Einstein condensate for the two scenarios, we observe a difference for long times ($t > 5$s, fig. 1). This is a time scale long compared to the experiment [2]. In contrast to the interpretation of the experiment that the condensate is sufficiently dilute during the propagation and can therefore be treated as non-interacting we show that the effect of interactions becomes increasingly visible after long propagation times. We propose to revisit this problem experimentally and are confident that the influence of interactions on Anderson localization at long expansion times can be observed.

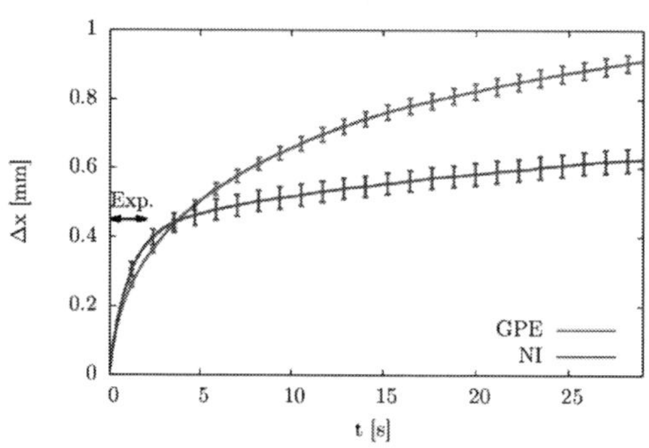


**Figure 1**. (a) Comparison of the width $\Delta x$ of a Bose-Einstein condensate that is interacting throughout the simulation (solid line - GPE) with the case where interactions are turned off after an initial expansion period where most of the interaction energy is converted into kinetic energy (dashed line - NI). We show the mean value and the standard deviation over 30 different realizations of the disorder potential. The arrow denotes the experimentally investigated time span [2].

[1] P. Anderson, Physical Review, **109**, 1492 (1958).

[2] J. Billy et al., Nature **453**, 891 (2008).

[3] L. Sanchez-Palencia et al., New Journal of Physics **10**, 045019 (2008).

# Exact finite-temperature dynamics of a harmonically quenched Tonks-Girardeau gas

Yasar Atas[1], Isabelle Bouchoule[2], Dimitri Gangardt[3], Karen Kheruntsyan[1]

[1]University of Queensland, School of Mathematics and Physics, Brisbane, Queensland, Australia

[2]Laboratoire Charles Fabry, Institut d'Optique, CNRS, Univesit´e Paris Sud 11, Palaiseau, France

[3]School of Physics and Astronomy, University of Birmingham, Edgbaston, Birmingham, UK

Ultracold quantum gases offer a fascinating and insightful playground for exploring and understanding out-of-equilibrium behaviour of quantum matter. Experimental systems whose non-equilibrium dynamics can be described by exactly solvable models of quantum many-body theory play a particularly important role. A paradigmatic example in this realm is the Tonks-Girardeau (TG) gas of strongly interacting bosons, previously treated only for zero-temperature dynamics. Here, we develop an exact finite-temperature dynamical theory of a harmonically trapped TG gas, and apply it to the problem of breathing-mode oscillations after a sudden confinement quench. We identify physical regimes for observing a phenomenon of frequency doubling in the oscillations of the momentum distribution of the gas, first discovered and treated theoretically in a weakly interacting quasicondensate regime [1,2]. The momentum-space frequency doubling relative to the breathing-mode oscillations of the real-space density distribution represents a novel example of a quantum many-body bounce problem due to a collective self-reflection mechanism, and can surprisingly well be captured by a finite-temperature hydrodynamic approach which we also develop here. Our exact dynamical theory opens the way to solve finite-temperature problems involving anharmonic traps, as well as arbitrary quench protocols of the trapping potential.

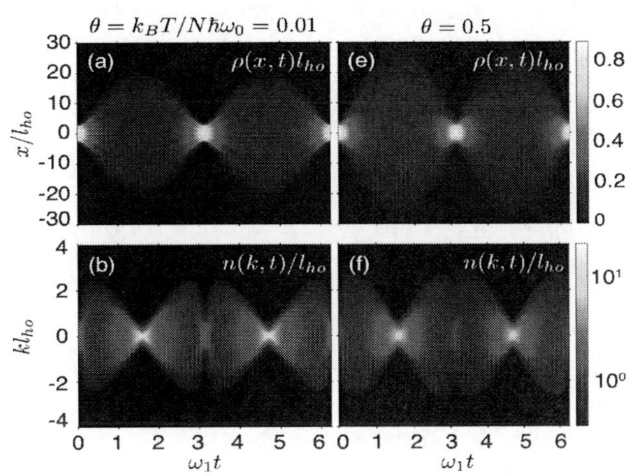


**Figure 1.** Breathing-mode dynamics of a harmonically trapped Tonks-Girardeau gas following a sudden quench of the trap frequency from $\omega_0$ to the final value $\omega_1$. Shown are the real space density (top row) and momentum-space (bottom row) distributions, for two different values (left and right columns) of the dimensionless temperature parameter $\theta$ (for $N$=5 atoms), with the initial harmonic oscillator length $l_{ho}$ used as the length scale.

[1] B. Feng, G. Carleo, A. Johnson, and I. Bouchoule, Phys. Rev. Lett. 113, 035301 (2014).

[2] I. Bouchoule, S. S. Szigeti, M. J. Davis, and K. V. Kheruntsyan, arXiv:1602.07889.

# 1D and 2D localization of excited states of Bose-Einstein Condensates in presence of disorder

Marisa Pons[1], Anna Sanpera[2,3]

[1]Applied Physics Department I, University of the Basque Country, UPV-EHU, Bilbao, Spain

[2]ICREA, Institució Catalana de Recerca i Estudis Avançats, 08010 Barcelona, Spain.

[3]Departament de Física, Universitat Autònoma de Barcelona, UAB, Bellaterra, Spain

We study the onset of localization of excited states of trapped Bose-Einstein Condensates expanding in presence of Gaussian uncorrelated random disorder. Quantum localization emerges as a consequence of the coherent back scattering that appears when a wave packet spreads in a disordered media [1]. To date, most studies of quantum localization in ultracold gases have exhaustively analyzed the onset of localization for the ground state of a trapped gas expanding freely in a disordered media, see e.g. [2] and references therein. We observe, in 1D systems, that for a fixed ratio between the disorder strength and the initial energy, excited states localize exponentially with a localization length that decreases as the energy of the initial state increases. Moreover, the localized state keeps the shape of the initial state wave function with an exponential tail [3].

In 2D, we analyze the interplay between vorticity and localization by examining the dispersion of a state containing a vortex on it in a disordered media [3]. We observe that if interactions are absent the wave function clearly localizes even in the presence of the vortex. In addition, we recover the vortex as soon as we turn the disorder off and let the system evolve freely. If the interactions are included, even if they are very small, localization is no longer possible at least in our a model of Gaussian uncorrelated disorder. The study of dispersion of a bosonic condensate with vorticity bears similarities to the stability of topological excitations in 2D p-wave fermionic superfluids.

[1] P.W. Anderson, Phys. Rev. **109**, 1492 (1958)

[2] M. Lewenstein, A. Sanpera, V. Ahufinger, Ultracold Atoms in Optical Lattices: mimicking condensed matter and beyond, Oxford University Press, (2012)

[3] M. Pons, A. Sanpera, arXiv:1511.08588

# Calorimetry and coherence of a photon Bose-Einstein condensate in a dye microcavity

Julian Schmitt[1], T. Damm[1], D. Dung[1], C. Wahl[1], F. Vewinger[1], J. Klaers[1,2], and M. Weitz[1]
[1]University of Bonn, Institut für Angewandte Physik, Wegelerstr. 8, 53115 Bonn, Germany
[2]ETH Zurich, Institute for Quantum Electronics, Auguste-Piccard-Hof 1, 8049 Zürich, Switzerland

Bose-Einstein condensation has been studied in different systems over the last years. One of them are photons trapped in a high-finesse microcavity, where thermalization is achieved by thermal contact to a dye medium [1]. Here, the dye molecules serve both as a heat bath and a particle reservoir, and grand-canonical statistical conditions can be realized experimentally in this system. Owing to the effective particle exchange with the reservoir, statistical physics predicts unusually large number fluctuations of the condensate [2], which for the first time can be unveiled in our experiments [3].

As the photons are close to a non-interacting quantum gas, our system can serve as a model for the ideal, trapped Bose gas in two dimensions. We have measured the thermodynamic properties, and found the characteristic cusp singularity for the specific heat at criticality (Fig. 1), which strikingly resembles the well-known λ-transition in superfluid liquid helium [4]. Our measurements follow the theory, and can serve as a textbook example for the ideal Bose gas.

Following this, we have studied the temporal interference of a Bose-Einstein condensate of photons in a dye microcavity with a phase stable optical reference source, observing the phase evolution and the emergence of temporal coherence of the photon condensate [5]. In a Hanbury Brown-Twiss experiment, we identify a regime with large statistical intensity fluctuations, which are a consequence of grand-canonical ensemble conditions [3]. For small condensate sizes, we observe phase jumps of the condensate following statistical condensate fluctuations to small photon numbers. Subsequently, upon reemergence of the condensate a new and random macroscopic phase develops, demonstrating spontaneous symmetry breaking at the phase transition. For large composite condensate-reservoir-systems, our experimental data shows phase coherence even in the presence of statistical number fluctuations [5].

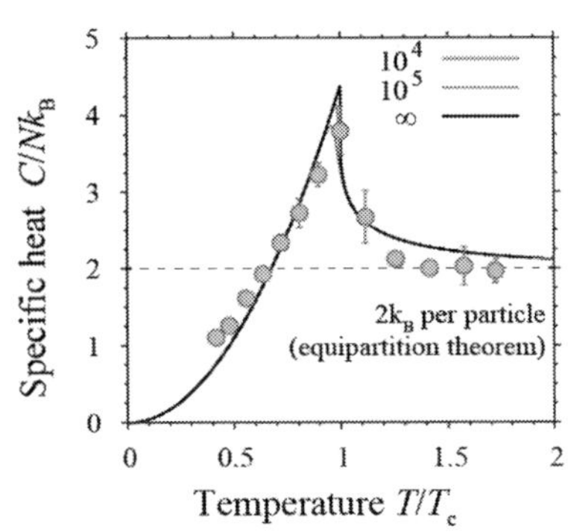


**Figure 1.** Specific heat of the photons as a function of the reduced temperature $T/T_c$. At the critical temperature, a cusp singularity indicates the phase transition from a classical gas at high temperatures ($T > T_c$) to a Bose-Einstein condensate of "cold" photons ($T < T_c$).

[1] J. Klaers, J. Schmitt, F. Vewinger, M. Weitz, Nature **468**, 545–548 (2010).
[2] See e.g.: R. M. Ziff, G. E. Uhlenbeck, M. Kac, Phys. Rep. **32**, 169–248 (1977).
[3] J. Schmitt, T. Damm, D. Dung, F. Vewinger, J. Klaers, M. Weitz, PRL **112**, 030401 (2014).
[4] T. Damm *et al.*, Nat. Comm. **7**, 11340 (2016)
[5] J. Schmitt *et al.*, PRL **116**, 033604 (2016).

# Atom-cavity physics with a Bose-Einstein condensate in an ultra-narrow band resonator

Hans Keßler[1], Jens Klinder[1], Christoph Georges[1], José Vargas[1], Andreas Hemmerich[1,2]
[1]Institut für Laser-Physik, Universität Hamburg, Luruper Chaussee 149, 22761 Hamburg, Germany
[2]Wilczek Quantum Center, Zhejiang University of Technology, Hangzhou 310023, China

Bose-condensed atoms are trapped in a cavity operating in the ultimate quantum regime, when the Purcell factor exceeds unity and the frequency shift associated with scattering of a single photon exceeds the cavity bandwidth. We explore topics such as recoil resolved cavity sideband cooling [5], non-linear Bloch oscillations [4], matter wave superradiance [3], physics of the Dicke model [2] and the bosonic Hubbard model with infinite range retarded interactions [1].

[1] J. Klinder *et al.*, PRL **115**, 230403 (2015).
[2] J. Klinder *et al.*, PNAS **112**, 3290 (2015).
[3] H. Keßler *et al.*, PRL **113**, 070404 (2014).
[4] H. Keßler *et al.*, New Journal of Physics **16**, 053008 (2014).
[5] M. Wolke *et al.*, Science **337**, 85-87 (2012).

# Quantum Critical Behavior Influenced by Measurement Backaction in Ultracold Gases

Yuto Ashida[1], Shunsuke Furukawa[1], Masahito Ueda[1,2]

[1]Department of Physics, University of Tokyo, 7-3-1 Hongo, Bunkyo-ku, Tokyo 113-0033, Japan

[2]RIKEN Center for Emergent Mater Science (CEMS), Wako, Saitama 351-0198, Japan

Recent realization of quantum gas microscopy [1] has realized the single-site detection of ultracold atoms trapped in an optical lattice and emerged as a powerful tool for studying strongly-correlated phenomena. Rapid developments of this technique are now approaching the level at which we can continuously monitor the quantum many-body dynamics at the single-particle level [2,3]. We expect that at such an ultimate situation the measurement backaction of continuous observation significantly alters the many-body phenomena. Of particular interest is the question of how the quantum critical behavior is influenced by the backaction of high precision measurement.

We study ultracold atomic systems subject to continuous observation and investigate unique roles played by the measurement backaction in quantum critical regimes [4]. By analyzing effective non-Hermitian Hamiltonians of interacting bosons, we show that the influence of the measurement backaction on the eigenenergies and eigenstates of a system significantly changes the quantum critical point and yields a critical phase unique to continuous observation. Specifically, by performing the mean-field analysis and the exact diagonalization of the effective Hamiltonian, we show that the superfluid-Mott transition point shifts toward the superfluid phase due to the backaction of quantum measurements (Figure 1(a)). We also study a one-dimensional Bose gas by analyzing the low-energy effective Tomonaga-Luttinger liquid (TLL) Hamiltonian. Under continuous observation, the two critical exponents emerge (Figure 1(b)). This indicates a unique critical behavior beyond the realm of the standard TLL universality class, where the single TLL parameter governs the critical behavior.

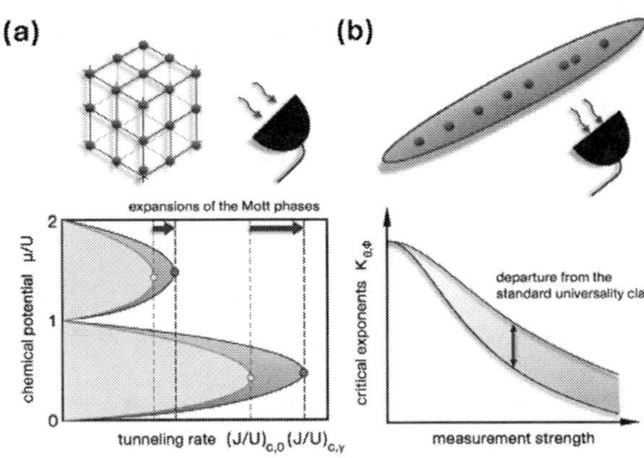


**Figure 1.** Schematic illustration of continuous observation of ultracold atoms trapped in (a) an optical lattice and (b) a one-dimensional trap, leading to (a) expansions of the Mott lobes and (b) a unique critical phase beyond the TLL class.

[1] W. S. Bakr *et al.*, Nature **462,** 74-77 (2009).

[2] Y. S. Patil *et al.*, Phys. Rev. Lett. **115,** 140402 (2015).

[3] Y. Ashida and M. Ueda, Phys. Rev. Lett. **115,** 095301 (2015).

[4] Y. Ashida, S. Furukawa, and M. Ueda, arXiv:1605.06899.

# Impurities immersed in a BEC. Quantum simulator of the polaron?

Luis A. Peña Ardila[1,2], Thomas Pohl[1], Stefano Giorgini[2]
[1] Max Plank Institute for the Physics of Complex Systems, Dresden, Germany
[2] INO-CNR BEC Center and Dipartimento di Fisica, Università di Trento, Povo, Italy.

We investigate the properties of an impurity immersed in a Bose gas at zero temperature using both analytical and Quantum-Monte Carlo methods. The interaction between bosons are modeled by a hard-sphere potential with scattering length **a**, whereas the impurity-boson interaction is modeled by a short-range attractive square-well potential, where both the sign and the strength of the scattering length **b** can varied by adjusting the well depth. We characterize the repulsive and attractive [Fig. 1] polaron branch by calculating the binding energy and the effective mass **[1]**. Furthermore, we study the structure of the bosonic bath such as the boson-boson correlation function and the density profile around the impurity. For resonant interactions between the impurity and the bosonic bath, the Ground state properties are also investigated as well as Efimov Effects **[1]**. The implication for the phase diagram of binary Bose-Bose mixtures is also discussed **[1]**. We also discuss more complicated interactions between the impurity and the bosonic bath. For this case we consider a Quasi-2D Dipolar Bose gas at zero temperature. Furthermore, the impurity-Bose interaction is dipolar. Using perturbation theory, the Ground-state properties are investigated based on the low-energy Fröhlich Hamiltonian. We studied the role of the long-range interaction for this model in contrast with contact interactions previously investigated. The polaron in this model is formed by the impurity dressed by the Low-energy modes of the Dipolar Bose gas. The dynamics of the low-energy modes is also discussed **[3]**.

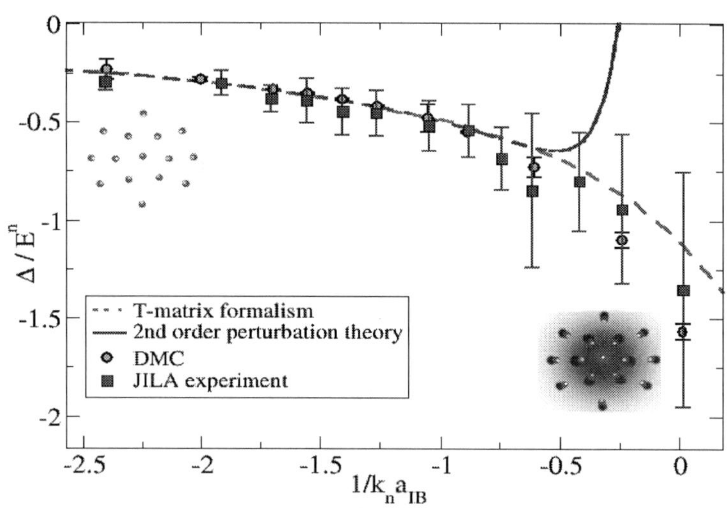


**Figure 1.** Attractive branch for an impurity immersed in a BEC. Binding energy as a function of the impurity-boson coupling strength characterized by the gas parameter of the gas $\mathbf{na^3}$ and the ratio $a_{IB}$= **b/a. [1], [2].**

[1] LAP Ardila, S. Giorgini, Phys. Rev. A **92**, 033612 (2015).

[2] M-G. Hu, M. J. Van de Graff, D. Kedar, J. P. Corson, E A. Cornell, D. S Jin. ArXiv:1605.00729

[3] LAP Ardila, W. Casteels and Thomas Pohl, In preparation (2016).

# Criticality and scaling properties during BEC transition obtained from global thermal expansion

Emmanuel Mercado-Gutiérrez[1], Freddy J. Poveda-Cuevas[1,2], Patrícia C. M. Castilho[1], Kilvia M. Farias[1] and Vanderlei S. Bagnato[1]

[1]Instituto de Física de São Carlos, Universidade de São Paulo, C.P. 369, 13560-970, São Carlos, SP, Brazil.

[2]Instituto de Física, Universidad Nacional Autónoma de México, Apartado Postal 20-364, 01000 México, D. F. México.

In this work, we have studied the thermodynamics susceptibilities as isothermal compressibility and the thermal expansion coefficient from a Bose-Einstein condensate of $^{87}Rb$. The critical phenomena and the critical exponent across the transition can be obtained through the behavior of the isothermal compressibility and the thermal expansion coefficient near the critical temperature. With the use of the global thermodynamics variables, we present a statistical treatment of a Bose gas in a 3D harmonic potential and comparing the results in this treatment we find the most appropriate state variables to describe the system. These variables are the volume parameterand the pressure parameter Π, they are experimental variables related with the confinement frequencies of the hybrid trap and the density distribution of the atoms in which the condensate is reached. We apply this approach to define a set of new thermodynamic variables on the BEC, and to construct the isobaric $V-T$ and thus, extract the compressibility and the thermal expansion coefficient and respectively. With the behavior of the thermal expansion coefficient around the critical point as with we obtain a critical exponent, where is the reduced temperature of the system. This treatment allows us to determine the dimensionality of the system with the help of the scaling theory, relating the critical exponents with the dimensionality. Finally, we find the dimensionality of the system d~3 which correspond to the real dimension of the system.

# Equation of state for a trapped quantum gas: remnant of zero point energy effects

Patrícia C. M. Castilho[1], Freddy J. Poveda-Cuevas[1,2], Jorge A. Seman[2], Rodrigo F. Shiozaki[1,3], Giacomo Roati[4], Sergio R. Muniz[1], Daniel V. Magalhães[1], Vanderlei S. Bagnato[1]

[1]Instituto de Física de São Carlos, Universidade de São Paulo, C.P. 369, 13560-970, São Carlos, SP, Brazil.

[2]HisInstituto de Física, Universidad Nacional Autónoma de México, Apartado Postal 20-364, 01000 México, D. F. México.

[3]Departamento de Física, Universidade Federal de São Carlos

[4]INO-CNR and LENS, Università di Firenze, 50019 Sesto Fiorentino, Italy.

[5]Escola de Engenharia de São Carlos, Universidade de São Paulo, 13566-590, São Carlos, Brazil.

The study of thermodynamic properties of trapped gases have attracted great attention during the last years. The majority of these studies apply the Local Density Approximation (LDA) in an attempt to deal with the inhomogeneous density profile of such clouds losing its global thermodynamic description. In a series of recent papers [1,2,3], our group has been characterizing the thermodynamic properties of trapped clouds by means of non-standard global thermodynamic variables and proving its capability on describing such systems. Besides that, these global variables allow to explore limits in which the standard Thermodynamics is not defined. In this work [4], we obtain an equation of state for a harmonically trapped Bose-Einstein condensate in the limit of $T \rightarrow 0$ and $N \rightarrow 1$ by means of the new global thermodynamic variables. Our results qualitatively show the existence of a zero point energy for harmonic potentials in which the determination of conjugated variables is limited by the quantum nature of the system.

[1] V. Romero-Rochìn, R. F. Shiozaki, M. Caracanhas, E. A. L. Henn, K. M. F. Magalhães, G. Roati and V. S. Baganto, Physical Review A **85**, 023632 (2012).

[2] R. F. Shiozaki, G. D. Telles, P. Castilho, F. J. Poveda-Cuevas, S. R. Muniz, G. Roati, V. Romero-Rochìn and V. S. Bagnato, Physical Review A **90**, 043640 (2014).

[3] F. J. Podeva-Cuevas, P. C. M. Castilho, E. D. Mercado-Gutierrez, A. R. Fritsch, S. R. Muniz, E. Lucioni, G. Roati and V. S. Bagnato, Physical Review A **92**, 013638 (2015).

[4] P. C. M. Castilho, F. J. Poveda-Cuevas, J. A. Seman, R. F. Shiozaki, G. Roati, S. R. Muniz, D. V. Magalhães and V. S. Bagnato, New Journal of Physics **18**, 023014 (2016).

# From quantum turbulence to statistical atom optics: new perspectives in speckle matter wave

Pedro E. S. Tavares[1], Amilson R. Frisch[1], Franklin A. J. Vivanco[1], Arnol D. G. Orozco[1], Gustavo D. Telles[1], François Impens[2], Robin Kaiser[3], Emanuel A. L. Henn[1], Vanderlei S. Bagnato[1]

[1]Instituto de Física de São Carlos, Universidade de São Paulo, Brazil

[2]Instituto de Física, Universidade Federal do Rio de Janeiro, Brazil

[3]Université Nice Sophia Antipolis, Institut Non-Linéaire de Nice, France

The properties of a turbulent trapped atomic superfluid were investigated by taking into account its matter wave nature. The quantum turbulent regime is achieved by perturbing a $^{87}$Rb Bose-Einstein condensate (BEC), produced in a QUIC trap, with a time oscillating magnetic field [1, 2]. The randomness of amplitude and phase observed in free-expanding turbulent BEC (disordered matter wave) were found to have strikingly similar properties to those of a traveling speckle light field (disordered optical wave), Fig. 1(a)-(d). More specifically, we observed that the divergence of both propagating waves present strong similarities, and diffraction is at the origin of the wave divergence. In addition, the second-order (spatial) correlation function was evaluated, Fig. 1(e) and (f), again resulting in essentially the same qualitative behavior for both waves [3]. The analogy in between the light and the matter waves may create new perspectives to study disordered quantum matter states. Finally, the study of the propagation of a three-dimensional matter wave speckle might bring some new insights to better understand the light speckle phenomenon in higher dimensions (beyond 2D), expanding the horizons of this phenomena.

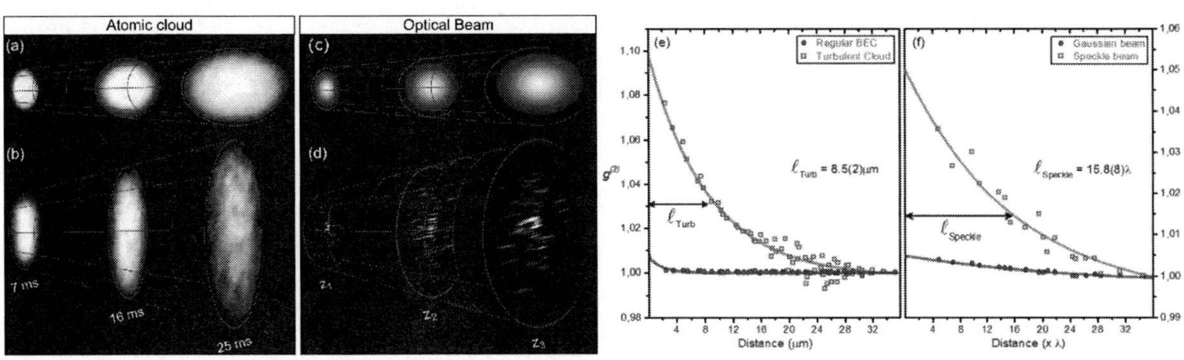


**Figure 1.** Atomic cloud expansion, optical beam propagation and Second-order correlation function. (a) Time-of-flight expansion of a regular BEC and (b) turbulent cloud. (c) Propagation of a Gaussian and (d) Speckle beam. (e) Second-order correlation function for an atomic cloud (matter wave) and (f) optical beam.

[1] E. A. L. Henn, J. A. Seman, G. Roati *et al.*, Phys. Rev. Lett. **103**, 045301 (2009).

[2] A. Bahrami, P. E. S. Tavares, A. R. Fritsch *et al.*, J. Low Temp. Phys. **180**, 126 (2015).

[3] P. E. S. Tavares, A. R. Fritsch, G. D. Telles *et al.*, arXiv:1606.01589 (2016).

# Observation of chiral superfluid order in orbital optical lattices

Robert Büchner[1,2], Raphael Eichberger[1], Carl Hippler[1], Thorge Kock[1], Andreas Hemmerich[1,2]

[1]Institut für Laserphysik, Universität Hamburg Luruper Chausee 149, Gebäude 69, 22761 Hamburg, Germany

[2]The Hamburg Center for Ultrafast Imaging, Luruper Chausee 149, 22761 Hamburg, Germany

We investigate ultra-cold bosonic quantum gases in excited bands of an optical lattice. We use a bipartite interferometric lattice allowing for dynamically changing the lattice geometry, which means in particular that the potential depth of adjacent lattice sites can be varied [1]. Among the atoms, we observe formation of an orbital superfluid in the second Bloch band of the lattice, arising from an interplay between the contact interaction of the atoms on each lattice site and the degeneracy of the p-orbitals in the second Bloch band. Here, a periodic pattern of locally alternating orbital currents and circular currents arises in the lattice, time reversal symmetry being spontaneously broken. By using the interference of the two independent samples of the superfluid order, we could recently directly observe the phase properties of the superfluid order parameter[2].

[1] T. Kock, C. Hippler, A. Ewerbeck, and Andreas Hemmerich, Journal of Physics B, Volume **49**, Number 4 (2016).

[2] T. Kock, M. Ölschläger, A. Ewerbeck, W.-M. Huang, L. Mathey, and A. Hemmerich, Phys. Rev. Lett. **114**, 115301 (2015).

# Vortex Clustering Transition in a Confined 2D Quantum Fluid

Xiaoquan Yu[1], Thomas P. Billam[2], Jun Nian[3], Matthew T. Reeves[1], Ashton S. Bradley[1]

[1]Department of Physics, Centre for Quantum Science, and Dodd-Walls Centre for Photonic and Quantum Technologies, University of Otago, Dunedin, New Zealand.

[2] Joint Quantum Centre (JQC) Durham-Newcastle, School of Mathematics and Statistics, Newcastle University, Newcastle upon Tyne, NE1 7RU, United Kingdom.

[3] Institut des Hautes E´dutes Scientifiques, Bures-sur-Yvette, France.

Clustering of like-sign vortices in a planar bounded domain is known to occur at negative temperature, a phenomenon that Onsager demonstrated to be a consequence of bounded phase space. In a confined superfluid, quantized vortices can support such an ordered phase, provided they evolve as an almost isolated subsystem containing sufficient energy. Here we develop an analytical theory of clustering transition in a neutral system of quantum vortices confined to a two-dimensional disc geometry, within the microcanonical ensemble. As the system energy increases above a critical value, a clustered vortex state emerges characterized by the continuous growth of the system dipole moment, resembling a continuous phase transition. The macroscopic dipole structure emerges from the homogeneous phase of vortices by spontaneously breaking the internal $Z_2$ symmetry, and the rotational SO(2) symmetry due to the disc geometry. The critical temperature of the transition, and the critical exponent associated with the dipole moment are obtained exactly within mean field theory. The cluster locations are found analytically, both near the transition, and in the high energy regime known as supercondensation. The microcanonical theory shows excellent agreement with Monte Carlo simulations, and signatures of the transition are apparent even for a modest system of 100 vortices, accessible in current Bose-Einstein condensate experiments.

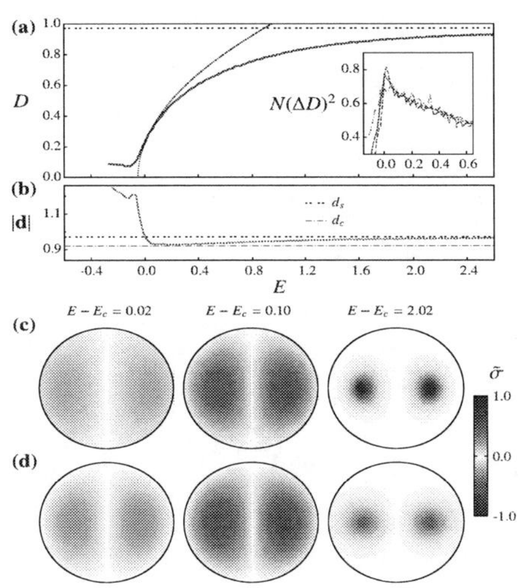


**Figure 1**. A comparison between microcanonical theory and MC sampling for N=100 point-vortices on the clustering transition. (a) shows the dipole moment D (solid red) for increasing energy E; also shown is the mean field prediction $D=(E-E_c)^{1/2}$(dotted black). The inset in (a) shows the variance of dipole moment for N=50 (dash dotted green), N=100 (solid red), and N=200 (dashed blue). (b) shows the separation between vorticity centres d(E); the numerical data (solid line) is shown alongside asymptotic mean field predictions for high energies $d_s$ and energies close to the transition $d_c$. (c) and (d) show, respectively, results from MC-sampling and mean field theory for the scaled vortex density difference.

[1] X.Yu, T. P. Billam, J. Nian, M. T. Reeves, A. S. Bradley , arxiv: 1512.05517, (2015).

# Controlled polarization of two-dimensional quantum turbulence in atomic Bose-Einstein condensates

A. Cidrim[1,2], F. E. A. dos Santos[3], L. Galantucci[2], V. S. Bagnato[1], C. F. Barenghi[2]

[1]Instituto de Física de São Carlos, Universidade de São Paulo, Caixa Postal 369, 13560-970, São Carlos, São Paulo, Brazil

[2]JQC (Joint Quantum Centre Durham-Newcastle) and School of Mathematics and Statistics, Newcastle University, Newcastle upon Tyne, NE1 7RU, United Kingdom

[3]Departamento de Física, Universidade Federal de São Carlos, 13565-905, São Carlos, SP, Brazil

We propose a scheme for generating two-dimensional turbulence in harmonically trapped atomic condensates with the novelty of controlling the polarization (net rotation) of the turbulence. Our scheme is based on an initial giant (multicharged) vortex which induces a large-scale circular flow. Two thin obstacles, created by blue-detuned laser beams, speed up the decay of the giant vortex into many singly-quantized vortices of the same circulation; at the same time, vortex-antivortex pairs are created by the decaying circular flow past the obstacles. Rotation of the obstacles against the circular flow controls the relative proportion of positive and negative vortices, from the limit of strongly anisotropic turbulence (almost all vortices having the same sign) to that of isotropic turbulence (equal number of vortices and antivortices). Using the new scheme, we numerically study the decay of 2D quantum turbulence as a function of the polarization. Finally, we present a model for the decay rate of the vortex number which fits our numerical experiment curves, with the novelty of taking into account polarization time-dependence.

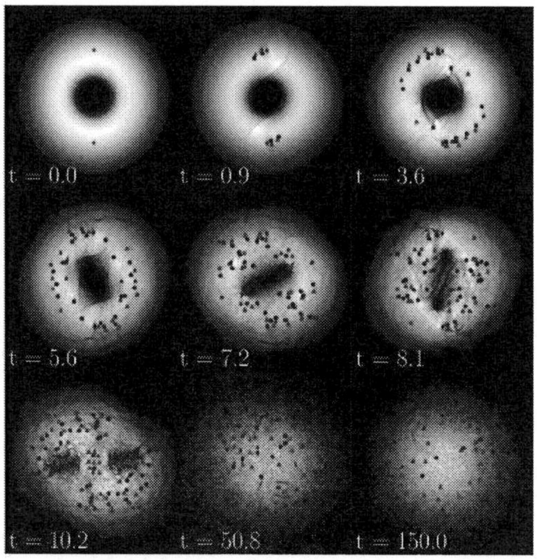


**Figure 1.** Density plots of the condensate at different times t for non-rotating obstacles. Regions of large/low density are displayed in white/black respectively. Red triangles and blue circles identify positive-charged and negative-charged vortices respectively. The giant vortex decays by injecting a large number of singly-quantized (positive) vortices into the condensate, whilst the pins generate vortex pairs, as can be clearly seen at time t = 0.9.

[1] A. Cidrim, F. E. A. dos Santos, L. Galantucci, V. S. Bagnato, C. F. Barenghi. Phys. Rev. A **93**, 033651 (2016).

# Bose-Einstein Condensates in Triple-well Superlattice

Woo Chang Chung[1], Colin Kennedy[1], William Cody Burton[1], Samarth Vadia[1], Wenlan Chen[1], Wolfgang Ketterle[1]

[1]MIT-Harvard Center for Ultracold Atoms, Research Laboratory of Electronics, Department of Physics, Massachusetts Institute of Technology, Cambridge, Massachusetts 02139, USA

Optical triple-well superlattice is a simple generalization of optical double-well superlattice, which has been utilized in cold neutral atom systems for realization of interesting quantum phenomena such as quantum charge pump [1, 2], spin-orbit coupling [3] and artificial gauge-field [4]. The presence of a third well can simplify some of the schemes used in the realizations of the aforementioned quantum phenomena in double-well systems; in particular, triple-well superlattice can be used to generate uniform artificial magnetic flux for neutral atoms without the need for rectification of staggered flux observed in double-well superlattice [5]. We report on the progress of the construction of a laser system for triple-well superlattice and discuss possible heating issues and their effects on the study of topological physics using triple-well superlattice.

[1] M. Lohse et. al., Nature Physics **12**, 350-354 (2016).
[2] S. Nakajima et. al., Nature Physics **12**, 296-300 (2016).
[3] J. Li et. al., arXiv: 1606.03514
[4] M. Aidelsburger et. al., Nature Physics **11**, 162-166 (2015).
[5] C. J. Kennedy et. al., Phys. Rev. Lett. **111**, 225301 (2013).

# Global thermodynamic variables: Determination of the first and second sound velocity of a Bose gas

A.R. Fritsch, P.E. Tavares, F.A. Vivanco, G.D. Telles, V.S. Bagnato, E.A.L. Henn
Instituto de Física de São Carlos, Universidade de São Paulo, 13560-970, São Carlos-SP, Brazil

We present a completely new approach for determining the sound velocity in atomic Bose-Einstein condensates (BECs). Here, the sound velocity measurement is based on thermodynamic global variables, which proved to be a valid alternative approach to determine some BEC properties, especially when the given conditions are not ideal or valid to the use of the standard local density approximation (LDA). Typically, the (first and second) sound velocity of trapped BECs are determined by introducing a density perturbation on the condensate and then recording its propagation as a function of time.

Our method provides another way to determine the sound velocities which does not require any density perturbation. Moreover, it is also possible to treat separately the thermal and the condensed fractions, allowing for the independent study of each fraction's contribution to the whole sound velocity. This makes our method more general, especially considering that density perturbations induced on thermal fractions are known to be difficult to track in the experiment. In Fig.1 below, we present, as the main result, the first and the second (global) sound velocities measured for a broad temperature range of temperatures for a $^{87}$Rb BEC. The dashed horizontal line, appearing in Fig.1, is the theoretical value predicted for the first sound velocity, at zero temperature, in very good agreement with our experimental results in the low temperature range. Note the good agreement also in the high temperature range, above $T_c$.

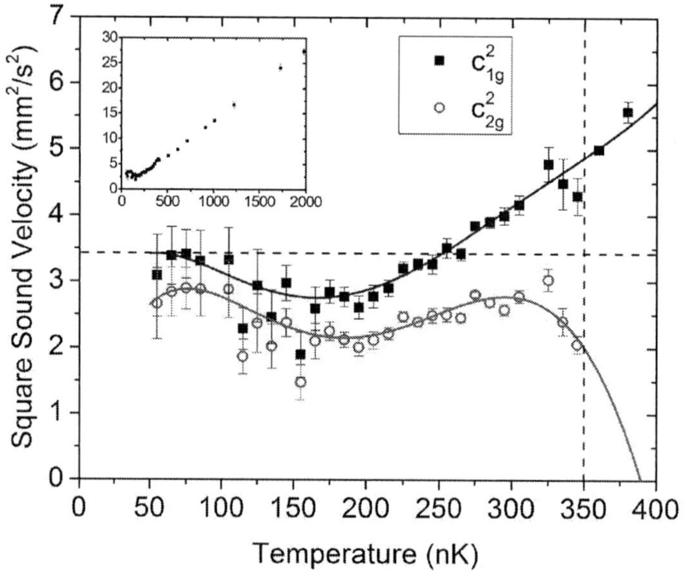


**Figure 1**: The squared sound velocity is plotted as a function of temperature; the solid squares are the first sound and open circles represent the second (global) sound velocities. The dashed horizontal line is the predicted theoretical value, calculated at zero temperature, for the first sound (Bogoliubov sound); the vertical dashed line shows the critical temperature, and the solid lines are eye guides only. The inset graph (upper left corner) shows the expected linear dependence at high temperatures..

# Twisted decay of quadruply quantized vortices in a Bose-Einstein condensate

G.D. Telles[1], P.E. Tavares[1], A. Cidrim[1,2], A.R. Fritsch[1], F.A. Vivanco[1], A.J. Allen[2], A.C. White[3], C.F. Barenghi[2], V.S. Bagnato[1]

[1]Instituto de Física de São Carlos, Universidade de São Paulo, 13560-970, São Carlos-SP, Brazil

[2]Joint Quantum Centre (JQC) Durham-Newcastle, School of Mathematics and Statistics, Newcastle University, Newcastle upon Tyne NE1 7RU, England, UK

[3]Quantum Systems Unit, Okinawa Institute of Science and Technology, Okinawa 904-0495, Japan

Quantized vortices, topologically imprinted in dilute Bose-Einstein condensates (BECs), have been intensively studied in the last decade [1], contributing to the early studies carried out in standard superfluids and superconductors [2]. In this work, we present results from quadruply quantized vortices, which were topologically imprinted in $|F=2, m_F=2>$ $^{87}$Rb Bose-Einstein condensates [1], produced and held in a QUIC trap. We investigated the vortex complex split decay process using 2-axis absorption imaging, and observed the spontaneous decay into four single charged vortices, shown in Fig.1. Moreover, we report the experimental observation of the twisted split decay of the quadruply charged vortices magnetically imprinted in the BEC. The observations were supported by numerical simulations showing that the process takes place in the shape of helical waves, finally splitting into separate singly-charged vortices as expected. We believe that the effect may be exploited to generate an almost isotropic state of turbulence, which we characterize in terms of the velocity statistics.

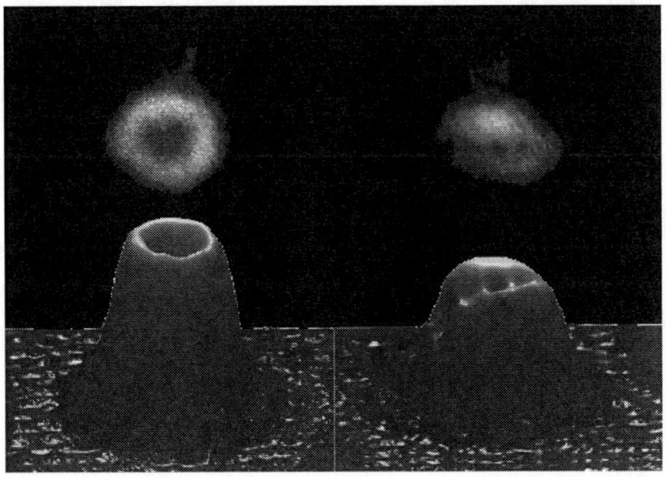

**Figure 1**: (Top) Experimental optical density (OD) images showing the quadruply charged vortex (left), and the 4-split singly charged vortices (right) in free fall, after 14 ms of time-of-flight. (Bottom) Corresponding 3D representations of the OD, emphasizing the central core(s) depletion.

[1] T. Isoshima *et al.*, Phys. Rev. A **61**, 063610 (2000); A. Leanhardt *et al.*, Phys. Rev. Lett. **89**, 190403 (2002); Y. Shin *et al.*, Phys. Rev. Lett. **93**, 160406 (2004); A. Mateo *et al.*, Phys. Rev. Lett. **97**, 180409 (2006); T. Isoshima *et al.*, Phys. Rev. Lett. **99**, 200403 (2007); T. Kuwamoto *et al.*, J. Phys. Soc. Jpn. **79**, 034004 (2010).

[2] R.J. Donnelly, *Quantized Vortices in Helium* II (Cambridge University Press, Oxford, 2003); M. Tinkham, *Introduction to Superconductivity* (McGraw–Hill, New York, 1996).

# Self-bound dipolar droplet: a localized matter-wave in free space

Danny Baillie[1], Ryan Wilson[2], Russell Bisset[3], Blair Blakie[1]

[1] Department of Physics, Centre for Quantum Science, and Dodd-Walls Centre for Photonic and Quantum Technologies, University of Otago, Dunedin, New Zealand

[2] Department of Physics, The United States Naval Academy, Annapolis, MD 21402, USA

[3] INO-CNR BEC Center and Dipartimento di Fisica, Università di Trento, Via Sommarive 14, I-38123 Povo, Italy

Recent experiments have demonstrated that dipolar condensates in the dipole-dominated regime are not necessarily unstable to mechanical collapse, but can instead form an array of dense droplets [1]. There is evidence that these droplets are stabilized by beyond-meanfield quantum fluctuations [2].

Based on these developments we demonstrate that a dipolar condensate can be prepared into a three-dimensional wavepacket that remains localized when released in free-space [3]. Such self-bound states arise from the interplay of the two-body interactions and quantum fluctuations. We develop a phase diagram for the parameter regimes where these self-bound states are stable, examine their properties, and demonstrate how they can be produced in current experiments.

[1] H. Kadau, M. Schmitt, M. Wenzel, C. Wink, T. Maier, I. Ferrier-Barbit, and T. Pfau, Nature **530** 194 (2016).

[2] I. Ferrier-Barbit, H. Kadau, M. Schmitt, M. Wenzel, and T. Pfau, Phys. Rev. Lett. **116**, 215301 (2016); H. Saiton, J. Phys. Soc. Jap. **85**, 053001 (2016); F. Wächtler and L. Santos arXiv:1601.04501 (*to appear Phys. Rev. A*); R. N. Bisset, R. M. Wilson, D. Baillie, P. B. Blakie, arXiv:1605.04964

[3] D. Baillie, R. M. Wilson, R. N. Bisset, and P. B. Blakie, arXiv:1606.00824

# Observation of attractive and repulsive polarons in a Bose-Einstein condensate

Nils B. Jørgensen[1], Lars Wacker[1], Kristoffer T. Skalmstang[1], Meera M. Parish[2], Jesper Levinsen[2], Rasmus S. Christensen[1], Georg M. Bruun[1], Jan J. Arlt[1]

[1]Institut for Fysik og Astronomi, Aarhus Universitet, 8000 Aarhus C, Denmark.

[2]School of Physics & Astronomy, Monash University, Victoria 3800, Australia

The problem of a mobile impurity particle immersed in a reservoir plays a fundamental role in physics. Studying the dressing of a particle by a bosonic medium is essential in understanding many physical systems such as technologically important semiconductors, Λ particles in nuclear matter, and high temperature superconductors. Even elementary particles acquire their mass by coupling to the bosonic Higgs particle. It is therefore highly desirable to study an impurity particle in a versatile environment such as offered by cold atomic gases. However, despite intense theoretical investigation, the canonical scenario of a mobile impurity immersed in a Bose-Einstein condensate (BEC) has never been realized. We use radio frequency spectroscopy of ultracold bosonic $^{39}K$ atoms to experimentally demonstrate the existence of a well-defined quasiparticle state of an impurity interacting with a BEC [1]. We measure the energy of the impurity both for attractive and repulsive interactions, and find excellent agreement with theories that incorporate three-body correlations as shown in Fig. 1. The spectral response consists of a well-defined quasiparticle peak at weak coupling, while for increasing interaction strength, the spectrum is strongly broadened and becomes dominated by the many-body continuum of excited states. Crucially, no significant effects of three-body decay are observed. Our results open up exciting prospects for studying mobile impurities in a bosonic environment and strongly interacting Bose systems in general.

[1] N. B. Jørgensen *et al.* arXiv:1604.07883

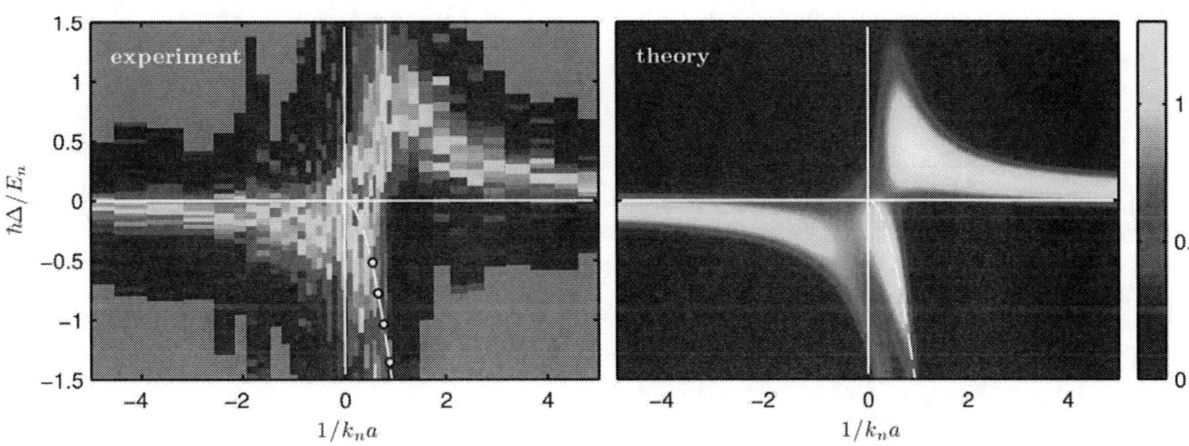


Fig. 1: The spectral response of an impurity in a BEC as a function of detuning Δ and interaction strength $1/k_n a$. Both experiment and theory show a clear shift in in spectral weight due to the interaction between the impurity and the BEC.

# 1D Bose gas in excited transverse motional state: optimal preparation and relaxation dynamics

Marie Bonneau, Sandrine van Frank, Mira Maiwöger, Raisa Trubko, Marine Pigneur, RuGway Wu, Thorsten Schumm, Jörg Schmiedmayer
VCQ, Atominstitut, TU Wien
Stadionallee 2, 1020 Vienna, Austria

Understanding the non-equilibrium dynamics of quantum many-body systems is an open problem, common to systems with very different length, time and energy scales, ranging from cosmology to ultracold gases. We consider here the relaxation dynamics of a 1D Bose gas prepared in a coherent superposition of the first two transverse vibrational levels of the trapping potential.

The system is initialized through optimal control of the displacement of the anharmonic transverse potential [1]. This method permits fast preparation of excited motional states with 99% fidelity and is robust towards atom number fluctuations [2]. Experimental characterization of the non-equilibrium evolution was performed. The system evolves towards a steady state, with a characteristic timescale strongly depending on the atom number of the Bose gas. We investigated the contribution to the relaxation of several effects: Local fluctuations of the vibrational levels' relative population arise during the system initialization, leading to prethermalization through propagation of relative density and relative phase excitations, in a similar fashion as observed for symmetrically split 1D gases [3]. Another source of longitudinal dephasing is the finite longitudinal coherence of the 1D gas, which results in fluctuations of the relative phase due to the asymmetry of the levels'sound velocities. Additionally, emission of atom pairs from the excited vibrational level [4] triggers many-body dephasing.

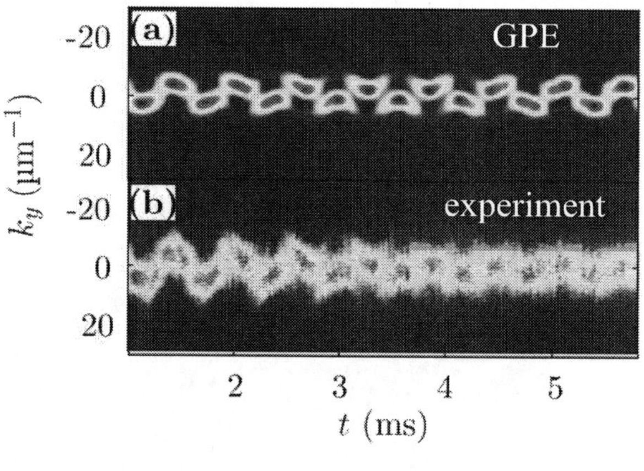


**Figure 1.** Comparison between simulated an experimental transverse momentum distributions for 700 atoms. (a) From a 1D (transverse) Gross-Pitaevskii simulation no relaxation is expected. (b) The experimental momentum distribution relaxes in a few ms.

[1] van Frank et al., Nat. Commun. **5**, 4009 (2014).
[2] van Frank et al., arXiv:1511.02247 (2015).
[3] Gring et al, Science **337**, 1318 (2012).
[4] Bücker et al., Nat. Phys. **7**, 608–611 (2011).

# Towards Cold Atom Experiments in Chinese Space Station

Xuzong Chen[1], Liang Liu[2], Weibiao Chen[2],Muming Liu[3], Xiaoji Zhou[1], Baolong Lu[4], Kaijun Jiang[4], Shuyu Zhou[2], Xinye Xu[5],Wei Xiong[1], Yin Zhang[1], Zhimin Liu[1], Zhaoyuan Ma[1], Bo Fan[1], Jingxing Sun[1], Qi Yu[1], Pingjun Wang[1], Tian Luan[1]

[1]Institute of Quantum Electronics, Peking University, Beijing 100871, China ,
[2]Shanghai Institute of Fine Optics and Fine Mechanics, CAS, Shanghai 201800, China
[3]Institute of Physics, CAS, Beijing100190, China
[4]Wuhan Institute of Physics and Mathematics, CAS, Wuhan 430071, China
[5]East China Normal University, Shanghai 200062, China

Chinese space station is due to launch in 2018, and the science module II is due to launch in 2022 with an Cold Atom Physics Rack (CAPR). The purpose for the cold atom physics rack is to achieve picokelvins ultra low temperature and implement six challenging experiments based on CAPR. There are three advantages for the quantum gas in space: (1) ultralow temperature: 10-12 K（pK）,three order lower than on the earth (nK), (2) longer observation time (20s) , three order longer than on the earth (20ms), (3) space uniform, no gravity gradient potential.

Four type of fundamental physics experiments based on quantum gas will be implemented in the Chinese space station:.

(1) Quantum State:（a）Quantum phase transition of Bose gas in zero gravity;（b）Topological quantum phase transition (c) Exotic material;（d） Ising model for quantum magnetism

(2) celestial body:（e） Acoustic black hole; Bose explosion

(3) Quantum Computation: （f） Topologic computation;

(4) Frontier Exploration:（g） test of quantum measurement theory;（h） relation between quantum mechanics and gravity;（i） test of standard model.

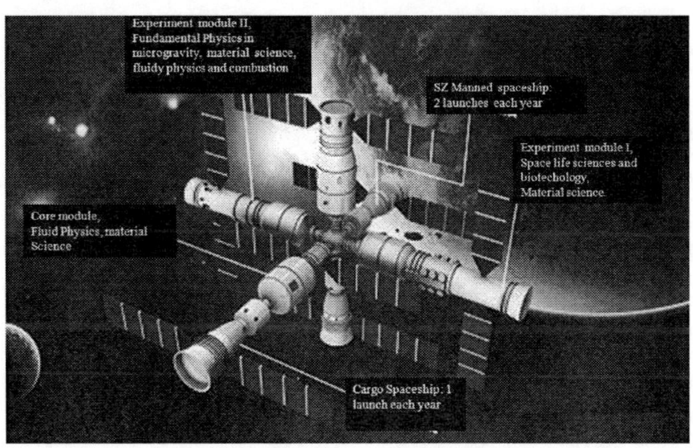


**Figure 1.** CAPR in the science module II of Chinese space station is due to launch in 2022.

References:

[1] Hepeng Yao a, Tian Luan, Chen Li, Yin Zhang, Zhaoyuan Ma, Xuzong Chen, Optics Communications, **359** ,123–128 (2016).

# Generation and decay of spin turbulence in a quenched antiferromagnetic spin-1 condensate

Seji Kang, Sang Won Seo, Joon Hyun Kim, Yong-il Shin
Department of Physics and Astronomy, Seoul National University

Depending on quadratic Zeeman energy, $q$, there are two distinctive ground states for an antiferromagnetic spinor Bose-Einstein condensate (BEC): an easy-axis polar (EAP) phase for $q>0$ and an easy-plane polar (EPP) phase for $q<0$. We investigate the generation and decay of spin turbulence through a rapid phase transition from an EAP phase to an EPP phase. Transverse magnon excitation causes initial instability to the phase transition [1], and then, spin wave arises, leading to spin turbulence. We observe the scaling behavior between the dynamic instability and the generation of spin turbulence with power law of $|q|^{1/2}$, which represents the controllability to generate spin turbulence. As spin turbulence decays, we can identify topological point defects, known as half-quantum vortices (HQVs) [2, 3].

[1] Y. Kawaguchi, and M. Ueda, Physics Reports 520, 253 (2012)
[2] S.W. Seo et al., Phys. Rev. Lett. 115, 015301 (2015)
[3] S.W. Seo et al., Phys. Rev. Lett. 116, 185301 (2016).

# Emergence of Metallic Quantum Solid Phase in a Rydberg-Dressed Fermi Gases

Wei-Han Li[1], Tzu-Chi Hsieh[1], Chung-Yu Mou[1,2], and Daw-Wei Wang[1,2]
[1]Department of Physics, National Tsing-Hua University, Hsinchu, Taiwan
[2]Physics Division, National Center for Theoretical Sciences, Hsinchu,Taiwan

We examine possible low-temperature phases of a repulsively Rydberg-dressed Fermi gas in a three-dimensional free space. It is shown that the collective density excitations develop a roton minimum, which is softened at a wavevector smaller than the Fermi wavevector when the particle density is above a critical value. The mean field calculation shows that unlike the insulating charge density waves states often observed in conventional condensed matters, a self-assembled metallic density wave state emerges at low temperatures. In particular, the density wave state supports a Fermi surface and a body-center-cubic crystal order at the same time with the estimated critical temperature being about one-tenth of the non-interacting Fermi energy. Our results suggest the emergency of a fermionic quantum solid that should be observable in current experimental setup [1].

[1] Wei-Han Li, Tzu-Chi Hsieh, Chung-Yu Mou, and Daw-Wei Wang, arXiv:1512.04079

# Probing Antiferromagnetic Ordering and Dynamics with the Fermi Gas Microscope

Anton Mazurenko[1], Maxwell F. Parsons[1], Christie S. Chiu[1], Geoffrey Ji[1], Daniel Greif[1], Markus Greiner[1]

[1]Harvard University Department of Physics, 17 Oxford Street, Lyman 137, Cambridge, MA, 02138, USA

Microscopy of Fermi gases in optical lattices has enabled studies of the Hubbard model at the single site level. In addition to site-resolved measurements of the charge degrees of freedom, we have implemented a scheme for measuring spin correlations between the constituent particles. We have observed beyond-nearest neighbor antiferromagnetic correlations in our system. This ability has enabled sensitive measurements of temperatures as low as $T/t = .39$ and allowed studying thermalization timescales within the system.

We also report on recent progress towards observing the dynamics of holes on a background of spins in two dimensions. The diffusive and ballistic character of the hole propagation is expected to be closely linked to the presence of magnetic ordering. In particular, arguments based on the Nagaoka theorem indicate that the hole propagation should locally increase ferromagnetic correlations.

We discuss recent progress towards implementing a single site addressing scheme using a digital micromirror device in the image plane of the microscope imaging system.

[1] D. Greif, M. Parsons, A. Mazurenko, C. Chiu, S. Blatt, F. Huber, G. Ji, M. Greiner, Science 351, 953 (2016).

# $d_{xy}$-density wave in fermion-fermion cold atom mixtures

Wen-Min Huang[1], Chen-Yen Lai[2], D. K. Campbell[3], and Shan-Wen Tsai[4]

[1]Department of Physics, National Chung-Hsing University, Taichung, 40227, Taiwan
[2]School of Natural Sciences, University of California, Merced, CA 95343, USA
[3]Department of Physics, Boston University, Boston, Massachusetts 02215, USA
[4]Department of Physics and Astronomy, University of California, Riverside, CA 92521, USA

We study density wave instabilities in a doubly-degenerate Fermi-Fermi mixture with SU(2) × SU(2) symmetry on a square lattice. For sufficiently large on-site inter-species repulsion, when the two species of fermions are both at half-filling, two conventional (s-wave) number density waves are formed with a $\pi$-phase difference between them to minimize the inter-species repulsion. Upon moving one species away from half-filling, an unconventional density wave with $d_{xy}$-wave symmetry emerges. When both species are away from the vicinity of half-filling, superconducting instabilities dominate. We present results of a functional renormalization-group calculation that maps out the phase diagram at weak couplings. Also, we provide a simple explanation for the emergence of the $d_{xy}$-density wave phase based on a four-patch model. We find a robust and general mechanism for $d_{xy}$-density-wave formation that is related to the shape and size of the Fermi surfaces. The density imbalance between the two species of fermions in the vicinity of half-filling leads to phase- space discrepancy for different inter-species Umklapp couplings. Using a phase space argument for leading corrections in the one-loop renormalization group approach to fermions, we show that the phase-space discrepancy in our system causes opposite flows for the two leading intra-species Umklapp couplings, and this triggers the $d_{xy}$-density-wave instability[1].

[1] C.-Y. Chen, W.-M. Huang, D.K. Campbell and S.-W. Tsai, Phys. Rev. A 90, 013610 (2014).

# Universal Properties in a Fermi Gas with a Resonant *p*-Wave Interaction

Shuhei M. Yoshida[1], Masahito Ueda[1,2]

[1] Department of Physics, University of Tokyo, Tokyo, Japan

[2] Center for Emergent Matter Science, RIKEN, Saitama, Japan

It has been known that, in the BCS-BEC crossover regime, the short- and long-distance physics are governed by universal laws, which are called the Tan relations; the short-range correlations obey universal power laws, and a single quantity, Tan's contact, gives both their coefficients and thermodynamic information of the gas. They hold quite generally in resonant atomic gases, including those in the strongly correlated regime, as long as the interaction is of the *s*-wave type.

Recently, such universal properties have been theoretically investigated in a Fermi gas with a resonant *p*-wave interaction [1, 2]. Also, an experimental evidence of the universal power-laws in a *p*-wave gas has been reported [3]. An essential difference between *s*-wave and *p*-wave interactions is that the latter can be anisotropic. In the presence of a magnetic field, the Feshbach resonance has only the axial symmetry around the magnetic field axis. Here, we discuss the universal relations of the *p*-wave Fermi gas and their rotational properties. We show that the momentum distribution has a high-momentum tail,

$$n_k \sim \frac{1}{k^2} \sum_{m,m'} C_{m,m'} Y_1^{m*}(\hat{\boldsymbol{k}}) Y_1^{m'}(\hat{\boldsymbol{k}})$$

in which we define the *p*-wave contact tensor $C_{m,m'}$. The *p*-wave contact tensor also has a direct relation to the thermodynamics through the adiabatic sweep theorem:

$$C_{m,m'} = \frac{M}{32\pi^2\hbar^2} \frac{\partial E}{\partial(-1/v_{m,m'})}$$

where $M$ is the mass of the fermions, $E$ is the total energy, and $v_{m,m'}$ is the *p*-wave scattering volume which characterizes the low-energy phase shift associated with the angular-momentum changing scattering. In contrast to the previous studies, it has nine components, which transform as $3 \otimes 3$ representation of SO(3). That the *p*-wave contact is a tensor means that the short-distance correlation may not have the axial symmetry which the *p*-wave Feshbach resonance possesses. This is in stark contrast to the *s*-wave case in which the short-distance correlation is always isotropic, as predicted by the Tan relations. We discuss two examples, the anisotropic *p*-wave superfluid and a Fermi gas in an anisotropic trap, in which the axial symmetry is broken spontaneously and externally, respectively. We show that, in these cases, the axially asymmetric components of $C_{m,m'}$ actually obtain finite values, which signal the breaking of the axial symmetry in the short-range correlation.

[1] S. M. Yoshida, and M. Ueda, Phys. Rev. Lett. **115**, 135303 (2015).

[2] Z. Yu, J. H. Thywissen, and S. Zhang, Phys. Rev. Lett. **115**, 135304 (2015).

[3] C. Luciuk, S. Trotzky, S. Smale, Z. Yu, S. Zhang, and J. H. Thywissen, Nat. Phys. (2016).

# Multiple Majorana Fermions in Bilayer Fermi gases with Spin-Orbit Coupling

Liang-Liang Wang, Wu-Ming Liu

Beijing National Laboratory for Condensed Matter Physics, Institute of Physics, Chinese Academy of Sciences, Beijing 100190, China

Majorana fermions, quantum particles that obey non-Abelian statistics have gained an increasing interest for their fundamental importance and potential application in fault-tolerant quantum information and computation. The experimental signatures of a pair of Majorana zero-energy states have been reported in a one-dimensional quantum wire with proximity-induced s-wave pairing. However, creating and manipulating multiple Majorana fermions, which could encode more qubits, remain an ongoing research topic. Here we propose an experimental set-up, one-dimensional trapped bilayer Fermi gases with the Raman coupling, for creating and manipulating multiple Majorana fermions. Two-dimensional bilayer Fermi gases with Rashba-type spin-orbit coupling are also discussed. The emergence of two diverse topological phases, induced by the inter-layer tunneling, supports the creation of multiple Majorana fermions by yielding more interfaces between different phases naturally formed in the trap. The number of resulting Majorana zero-energy modes, changing from eight to none, is equal to interfaces between distinct phases naturally formed in different areas of the trap, which are consistent with the results under analytical local-density approximation. This system is tunable and would be a significant step towards utilization of Majorana fermions in quantum computation [1].

[1] Liang-Liang Wang, Wu-Ming Liu. Topological Superfluidity of Spin-Orbit Coupled Bilayer Fermi Gases. arXiv:1601.03914

# Thermodynamic properties of a strongly interacting superfluid Fermi gas at zero temperature

Hiroyuki Tajima[1], Pieter van Wyk[1], Ryo Hanai[1], Daichi Kagamihara[1], Daisuke Inotani[1], Munekazu Horikoshi[2,3], Yoji Ohashi[1]

[1]Department of Physics, Keio University, 3-14-1, Hiyoshi, Kohoku-ku, Yokohama, Kanagawa 223-8522, Japan

[2]Institute for Photon Science and Technology, Graduate School of Science, The University of Tokyo, 7-3-1, Hongo, Bunkyo-ku, Tokyo 113-0033, Japan

[3]Photon Science Center, Graduate School of Engineering, The University of Tokyo, 2-11-16, Yayoi, Bunkyo-ku, Tokyo 113-8656, Japan

We theoretically investigate thermodynamic properties of a strongly interacting superfluid Fermi gas at $T$=0. Within the framework of an extended $T$-matrix approximation [1], we calculate various observable quantities, such as the Fermi chemical potential, isothermal compressibility. Our numerical results are shown to perfectly reproduce a recent experiment on a superfluid $^6$Li Fermi gas, measured from the weak-coupling BCS regime to the unitarity limit [2]. We also discuss how these thermodynamic quantities are affected by superfluid fluctuations in the whole BCS-BEC crossover region at T=0. Our result would be useful to understand many-body properties of strongly interacting Fermi superfluids.

[1] H. Tajima, R. Hanai, and Y. Ohashi, Phys. Rev. A, **93**, 013610 (2016).

[2] M. Horikoshi, H. Tajima, Y. Ohashi, M. Koashi, and M. Kuwata-Gonokami, in preparation.

# Gorkov--Melik-Barkhudarov correction to superfluid transition temperature of Rashba spin-orbit coupled Fermi gases in two dimensions

Juhee Lee, Dong-Hee Kim
Department of Physics and Photon Science, Gwangju Institute of Science and Technology, Gwangju 61005, Korea

We investigate the influence of the medium-induced interaction to the superfluid transition temperature of attractively interacting two-dimensional Fermi gases with Rashba spin-orbit coupling. The effective interaction is calculated to incorporate screening and anti-screening contributions induced from the medium, up to the second order of interaction strength within the Gorkov—Melik-Barkhudarov diagrammatic approach. We find that the SOC-dependence of the induced interaction effects shows distinctive behavior across the weak and strong SOC regimes in accordance with the undergoing Lifshitz-type transition of the Fermi surface topology. In computation of the induced interaction correction to the mean-field transition temperature and the Berezinskii-Kosterlitz-Thouless (BKT) transition temperature, we find that the correction factor generally decreases as the SOC strength increases, and in particular for the BKT transition temperature, the correction effect diminishes in the limit of strong SOC strength.

# Thermodynamic properties and effects of pairing fluctuations in a ultracold Fermi gas with a uniaxially anisotropic *p*-wave interaction

Daisuke Inotani, Pieter van Wyk, Yoji Ohashi
Department of Physics, Keio University, 3-14-1 Hiyoshi, Yokohama 223-8522, Japan

We investigate thermodynamic properties and strong coupling effects in an ultracold one component Fermi gas with a *p*-wave Feshbach resonance. In this system, it has been predicted that a *p*-wave interaction associated with a Feshbach resonance has a uniaxial anisotropy ($U_x \geq U_y = U_z$, where $U_i$ is the interaction strength of $p_i$-wave component.), due to the split of the *p*-wave Feshbach resonance. This splitting has experimentally been measured in a $^{40}$K Fermi gas[1]. In this presentation, including this uniaxial anisotropy, as well as associated pairing fluctuations within a strong coupling theory developed by Nozières and Schmitt-Rink, we calculate the specific heat at constant volume $C_V$ above the superfluid transition temperature $T_c$. At $T_c$, we show that, while $C_V$ in the weak coupling regime is almost independent on the anisotropy of the *p*-wave interaction (dashed line in Fig. 1), the anisotropy strongly affects $C_V$ in the strong-coupling regime (solid line in Fig. 1). We find that the enhancement of $C_V$ around $(\delta v p_F^3)^{-1} \sim 1$ in Fig. 1 originates from the thermal transition of molecular bosons from the $p_y$- and $p_z$-wave components to the $p_x$-wave component at low temperatures. We also show that, because of this phenomenon, the temperature dependence of $C_V$ is qualitatively different from that in the ordinary *s*-wave case[2]. Our results would help to understand how the anisotropic pairing fluctuations affect many-body properties of an ultracold Fermi gas with a *p*-wave Feshbach resonance.

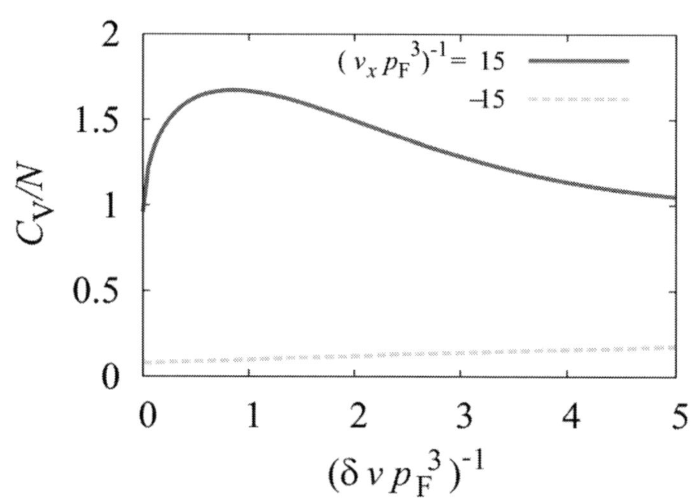


Fig. 1. Calculated specific heat at constant volume $C_V$ at $T_c$ as a function of the anisotropy parameter, $\delta v^{-1} = v_x^{-1} - v_y^{-1} = v_x^{-1} - v_z^{-1}$, where $v_i$ is the $p_i$-wave scattering volume. $p_F$ is the Fermi momentum, and $N$ is the number of the Fermi atoms.

[1] C. Ticknor, C. A. Regal, D. S. Jin, and J. L. Bohn, Phys. Rev. A **69**, 042712 (2004).
[2] P. van Wyk, H. Tajima, R. Hanai, Y. Ohashi, Phys. Rev. A **93**, 013621 (2016).

# Deterministic Preparation of Few-Fermion Systems in Multiple Potential Wells

Vincent M. Klinkhamer[1], Andrea Bergschneider[1], Jan-Hendrik Becher[1], Michael Dehabe[1], Philipp Preiß[1], Gerhard Zürn[1], Selim Jochim[1]
[1]Physikalisches Institut, Universität Heidelberg, Germany

In our experiment we aim to create mesoscopic quantum systems in multiple potential wells while maintaining full control over the wave function. Our vision is to study complex ordering and correlations from the bottom up with an increasing system size.

In our previous experiments with $^6$Li in a double-well potential, we were able to prepare two-particle systems in the lowest motional quantum state [1] by adiabatically transforming the single-well potential to a double-well potential. The two spin-½ particles, which are initialized as a spin singlet, stay in this spin configuration during the entire process. The next step is to create a product state of these singlets in multiple decoupled double wells and then to adiabatically introduce coupling, creating an entangled mesoscopic quantum state.

Our optical potential consists of optical tweezers which are controlled by an acousto-optic deflector (AOD). In order to create larger potential landscapes, we built a stabilization setup for the optical potential which directly produces and stabilizes potential landscapes by monitoring the potential on a camera in real time.

We implemented spin-resolved imaging by collecting the light of both hyperfine states sequentially on an EMCCD. Because of the short exposure time, we expect our atoms to diffuse less than the distance between the wells, which should also give us site-resolution.

Combined, this will enable us to study, e.g., one-dimensional Hubbard chains and four-site plaquettes, at half filling or with controlled amounts of doping.

[1] S. Murmann, *et al.*, Two Fermions in a double well: Exploring a fundamental building block of the Hubbard model, Phys. Rev. Lett. 114, 080402 (2015).

# Site-Resolved Observation of Charge and Spin Correlations in the 2D Fermi-Hubbard Model

Lawrence W. Cheuk*[,1], Matthew A. Nichols*[,1], Katherine R. Lawrence[1], Melih Okan[1], Hao Zhang[1], Ehsan Khatami[2], Nandini Trivedi[3], Thereza Paiva[4], Marcos Rigol[5], and Martin W. Zwierlein[1]

[1]Department of Physics, MIT-Harvard Center for Ultracold Atoms, and Research Laboratory of Electronics, MIT, Cambridge, MA 02139, USA

[2]Department of Physics and Astronomy, San José State University, San José, CA 95192, USA

[3]Department of Physics, The Ohio State University, Columbus, OH 43210, USA

[4]Instituto de Fisica, Universidade Federal do Rio de Janeiro, Cx.P. 68.528, 21941-972, Rio de Janeiro, RJ, Brazil

[5]Department of Physics, The Pennsylvania State University, University Park, PA 16802, USA

Strongly correlated fermionic systems can display many interesting phenomena, such as Mott-insulating behavior, magnetism, unconventional superconductivity, and anomalous transport. One simple paradigmatic model is the single-band 2D Fermi-Hubbard Hamiltonian. Despite its simplicity, this model is theoretically intractable, especially away from half-filling ($n = 1$), due to the fermion sign problem. In our experiment, we realize the Fermi-Hubbard model with ultracold $^{40}K$ atoms trapped in a square optical lattice. Using Raman sideband cooling, we perform quantum gas microscopy, site-resolving the occupation of individual lattice sites. This provides unprecedented access to spatial correlations in the system. We report on the site-resolved observation of the metallic, Mott insulating, and band insulating states of the Fermi-Hubbard model. Through the implementation of spin-sensitive imaging, we further study spin and charge correlations in the system as a function of filling. We find that antiferromagnetic correlations are strongest at half-filling, and decay monotonically as the filling is increased or decreased. In the charge sector however, the charge correlations (moment-moment correlations) are negative at low filling, but cross zero to become positive near half-filling. The negative charge correlations at low filling reveal the effects of Pauli blocking and strong repulsive interactions, which lead to a lower probability of finding two adjacent fermions. On the other hand, the positive correlations near half-filling indicate effective attraction between holes and doublons in the system. These findings agree well with novel numerical linked-cluster expansion (NLCE) and determinantal quantum Monte Carlo (DQMC) calculations that involve up to eight fermion operators. Further studies could shed light on the role of antiferromagnetic and doublon-hole correlations away from half-filling, a regime where intriguing phenomena, such as anomalous transport in a "strange metal" phase, and a pseudogap at lower temperatures, could appear.

* These authors contributed equally to this work.

# Exactly solvable model for a solitonic vortex in a compressible superfluid

L. A. Toikka, J. Brand
Dodd-Walls Centre for Photonic and Quantum Technologies, Centre for Theoretical Chemistry and Physics, and New Zealand Institute for Advanced Study, Massey University, Private Bag 102904 NSMC, Auckland 0745, New Zealand

An accurate theoretical understanding of particle interactions in strongly-correlated quantum liquids such as superfluid Helium, neutron matter, or the unitary Fermi gas is hampered by the lack of any known small expansion parameter, which is needed for precise perturbative analysis. We report an accurate perturbative approach to such non-perturbative fluids [1] by instead considering their response to a vortex excitation. By means of an exactly-solvable model of vortex dynamics in a slab geometry, we show how the underlying physical characteristics of the quantum fluid contribute at different orders of the naturally emerging expansion parameter to dynamical properties such as the vortex mass. Based on vortex properties that can be measured precisely, our results will enable the experimental determination of quantum pressure and vortex core filling and thus extend our knowledge of quantum liquids beyond the currently known hydrodynamic theory.

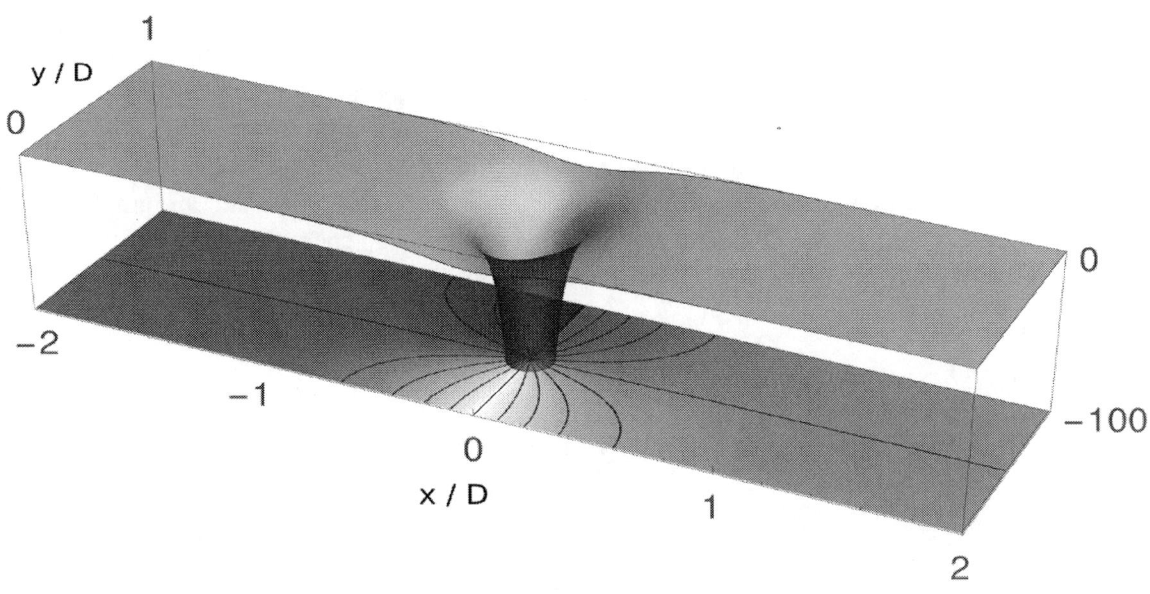


**Figure 1.** The first-order compressible density profile $n_1 = -q^2 n_0 \xi^2 |\nabla S_0|^2$ relative to the leading-order incompressible constant background $n_0$ together with $S_0$, the incompressible phase profile. Here $q$ is the vortex winding number, and $\xi$ is the healing length.

[1] L. A. Toikka, J. Brand, submitted (2016).

# Excitation spectra of a tunable Fermi superfluid

Sascha Hoinka[1], Paul Dyke[1], Cécile Carcy[1], Marcus G. Lingham[1],
Jami J. Kinnunen[2], Georg Bruun[3], and Chris J. Vale[1]

[1]Centre for Quantum and Optical Science, Swinburne University of Technology, Melbourne 3122, Australia

[2]COMP Centre of Excellence, Department of Applied Physics, Aalto University, FI-00076 Aalto, Finland

[3]Department of Physics and Astronomy, University of Aarhus, Aarhus C, Denmark

We present an experimental and theoretical study of the low-energy excitations of strongly interacting Fermi gases across the BCS to Bose-Einstein condensate (BEC) crossover using low-momentum Bragg spectroscopy. At long-wavelengths, Bragg scattering probes collective excitations, closely linked to the superfluid order parameter, as well as single-particle excitations [1]. In our setup, we shine two tightly focussed laser beams into the central, nearly uniform density region of an optically trapped cloud of lithium-6 atoms to directly measure the homogenous response. The Bragg laser beams intersect at a small angle giving momentum $\hbar k \simeq 0.5\hbar k_{\mathrm{F}}$ smaller than the Fermi momentum $\hbar k_{\mathrm{F}}$. Bragg spectra are obtained by evaluating the centre-of-mass displacement $\Delta X_{\mathrm{COM}}$ proportional to $\hbar k$ [2]. As shown in Figure 1, over a wide region of the crossover we observe a clear signature for macroscopic coherence in the form of a discrete collective mode peak in the Bragg spectrum at low excitation energies ($\hbar\omega$), which corresponds to the Goldstone mode, or Bogoliubov-Anderson (BA) phonon, providing a measure of the speed of sound. At higher energies, the onset of pair breaking excitations becomes visible close to unitarity, $1/(k_F a) = 0$, allowing a direct measure of the pairing (superfluid) gap. The measured results are in excellent agreement with our theoretical description based on the quasiparticle random-phase approximation. We also present progress towards measurement of the homogeneous contact using focussed beam Bragg spectroscopy.

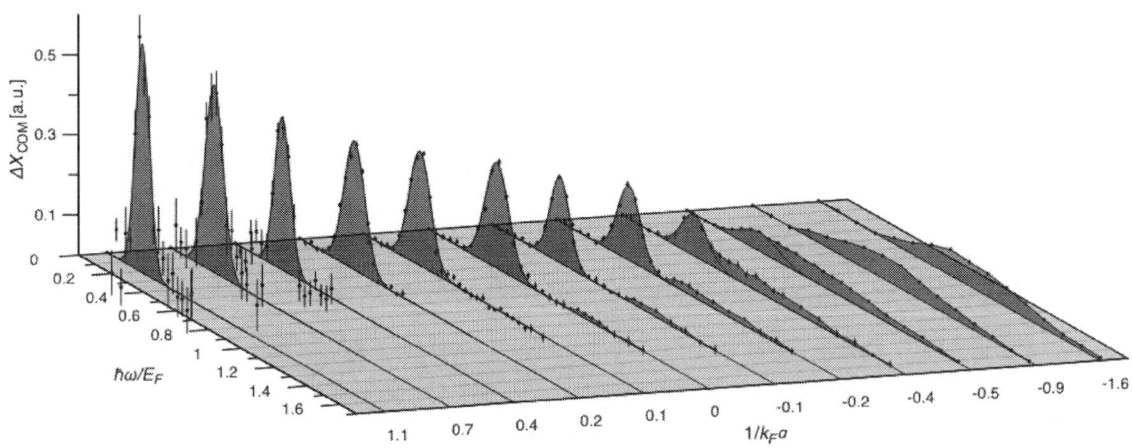


**Figure 1:** Bragg spectra for various interaction strengths $1/(k_F a)$ through the BCS-BEC crossover. Points are experimental data. At low excitation energies, the solid lines are Gaussian fits to the BA mode peak (shaded areas), while at higher $\hbar\omega$, the shaded regions indicate single particle excitations.

[1] D. Pines, P. Nozières, *The theory of quantum liquids, Vol. I*, Benjamin, New York (1966).

[2] S. Hoinka, M. Lingham, K. Fenech, H. Hu, C. J. Vale *et al.*, Phys. Rev. Lett. **110**, 055305 (2013).

# Phase separation and pair condensation in a quasi-2D Fermi gas

Debayan Mitra[1], Peter Brown[1], Peter Schauss[1], Stanimir Kondov[1], Waseem Bakr[1]
[1]Princeton University, Princeton, 08544 NJ, USA

We study the phase diagram of the strongly interacting spin-imbalanced Fermi gas in two dimensions, where the low dimensionality enhances correlations and phase fluctuations. Our interest is motivated by the connection of this system with superconductivity in the presence of a large Zeeman field. We observe pair condensation for a range of spin imbalance and interaction strengths. The measurement of the phase diagram opens the door for a detailed investigation of exotic phases such as the Sarma/broken pair phase and the elusive FFLO phase.

# Ferromagnetism of a repulsive Fermi gas

G. Valtolina[1], F. Scazza[1], A. Amico[1], A. Burchianti[1], A. Recati[2,3], T. Enss[4], M. Inguscio[1], M. Zaccanti[1], G. Roati[1]
[1]INO-CNR & LENS-University of Florence, Via Nello Carrara 1, 50019, Sesto Fiorentino, Italy
[2] INO-CNR BEC Center and Dipartimento di Fisica, Universita di Trento, 38123 Povo, Italy
[3] Technische Universitat Munchen, James-Franck-Straße 1, 85748 Garching, Germany
[4]Universitat Heidelberg, Philosophenweg 19, 69120 Heidelberg, Germany

Itinerant ferromagnetism represents one of the most spectacular manifestations of interactions within many-body fermion systems. In contrast to weak-coupling phenomena, it requires strong repulsion to develop, making a quantitative description of ferromagnetic materials notoriously difficult. In this frame, it is still debated whether the simplest case envisioned by Stoner of a homogeneous Fermi gas with short-range repulsive interactions can exhibit ferromagnetism at all.

Here, we will discuss our experimental route to tackle such an intriguing problem with ultracold Fermi gases.

We will describe a recent experiment in which we create an artificial ferromagnetic state by segregating degenerate spin mixtures of $^6$Li atoms into two initially disconnected reservoirs [1]. Starting from such a configuration, we characterize the behavior of the gas as a function of the interspecies interactions and temperature through two distinct but interconnected measurements of spin dynamics. First, we observe the softening of the spin-dipole mode, unambiguously connected to the diverging magnetic susceptibility approaching the ferromagnetic transition. Second, we investigate the diffusion dynamics of the two spin domains, whose temporary halt outlines the metastable character of the ferromagnetic state. Notwithstanding the limited lifetime of the repulsive Fermi gas due to the competing pairing instability, our findings allow us to draw in the temperature-interaction plane the critical boundaries for the ferromagnetic phase to exist, at least in a metastable sense.

[1] G. Valtolina et al., arxiv:1605.07850 (2016)

# Quench dynamics of a confined ultracold Fermi gas: Direct visibility of the Goldstone mode in the single-particle excitations

Peter Kettmann[1], Simon Hannibal[1], Mihail D. Croitoru[2], Vollrath Martin Axt[3], Tilmann Kuhn[1]

[1]Institute of Solid State Theory, Wilhelm-Klemm-Straße 10, 48149 Münster, Germany

[2]Condensed Matter Theory, Groenenborgerlaan 171, 2020 Antwerp, Belgium

[3]Theoretical Physics III, Universitätsstraße 30, 95440 Bayreuth, Germany

Ultracold Fermi gases are a convenient system to probe and study the properties of phases like the BEC and the BCS phase and the crossover in between these regimes. In particular, ultracold Fermi gases can be used as a test bed to study the two fundamental dynamical modes -the Higgs and the Goldstone mode- which result from spontaneous symmetry breaking in these phases. We investigate the Goldstone mode in the dynamics of a cigar-shaped ultracold $^6$Li gas after an interaction quench on the BCS side of the BCS-BEC crossover. To this end, we numerically solve Heisenberg's equations of motion for the Bogoliubov single-particle excitations in the framework of the Bogoliubov-de Gennes formalism making use of the Anderson approximation. In doing so, we find that the single-particle occupations oscillate in time with one dominant low-frequency component and several weak higher-frequency contributions. We show that the low frequency originates from the Goldstone mode of the BCS gap [1] and that it leads to a full inversion of the lowest-lying single-particle states. We investigate this low-frequency component over a wide range of parameters and show that the size-dependent superfluid resonances [2] have a strong impact on this mode. Furthermore, we find that the higher frequencies correspond to the energies of the single-particle states and thus to the fragments of the Higgs mode of the BCS gap [3].

[1] P. Kettmann, S. Hannibal, M.D.Croitoru, V.M.Axt, T.Kuhn, arXiv preprint arXiv:1511.04239 (2015).

[2] A.A. Shanenko, M.D. Croitoru, A.V. Vagov, V.M. Axt, A. Perali, F.M. Peeters, *PRA* **86** 033612 (2012).

[3] S. Hannibal, P. Kettmann, M.D. Croitoru, A. Vagov, V.M. Axt, T. Kuhn, *PRA* **91**, 043630 (2015).

# Fermi Gases in 2D

Tyson Peppler[1], Paul Dyke[1], Marta Zamarano[2], Sascha Hoinka[1], P.Hannaford[1], Chris Vale[1]
[1]Swinburne University of Technology, Melbourne, Australia
[2]École Normale Supérieure (ENS), Paris, France

Fermi gases cooled below the Fermi temperature with controllable interactions offer a rich experimental platform to study strongly interacting quantum systems. Constraining the dimensions to two changes fundamental properties of the system, including two-body scattering and the nature of the superfluid transition. Here we present our recent experimental results on the thermodynamics of such a gas[1] – wherein we analyze of a range of clouds fulfilling criteria to be kinematically 2D[2] at varying interaction strengths. We trap single 2D clouds in a blue-detuned $TEM_{01}$ mode optical trap, utilizing high power to achieve a large ratio (>200) of transverse to radial confinement, where radially the gas is trapped by residual magnetic field curvature. As the density of the gas trapped in a harmonic potential varies from zero in the outer wings to maximum in the central region, this thermodynamic analysis probes quite different regimes, from classical to quantum[3]. Results of the reduced density equation of state show a non-monotonic behavior, a result differing from that in 3D, and arising due to competition with an extra energy scale (that of a weakly bound dimer) ever-present for arbitrarily weak attractive interactions in 2D. As well as these results we present preliminary experimental data on collective oscillations of atom clouds in the 2D and near 2D regimes.

[1] K. Fenech *et al.*, Phys. Rev. Lett. **116.4**, 045302 (2016).
[2] P. Dyke *et al.*, Phys. Rev. A **93.1**, 011603(R) (2016) .
[3] M. Parish, Physics **9**, 10 (2016).

# Pseudogapped density of states in a two-dimensional ultracold Fermi gas near the observed BKT transition temperature

Morio Matsumoto, Ryo Hanai, Daisuke Inotani, Yoji Ohashi
Faculty of Science and Technology, Keio University, 3-14-1 Hiyoshi, Yokohama, 223-8522, Japan

We investigate single-particle excitations and strong-coupling effects in a two-dimensional Fermi gas near the observed Berezinskii-Kosterlitz-Thouless (BKT) transition temperature [1,2]. Including pairing fluctuations within the framework of a self-consistent T-matrix approximation, we clarify how the so-called pseudogap develops in the single-particle density of states in this temperature regio n (Fig.1). We also discuss to what extent the pseudogapped density of states reflects the bosoni c character of the system. Our results would contribute to further understanding of strong-cou pling properties of an ultracold Fermi gas in the crossover region.

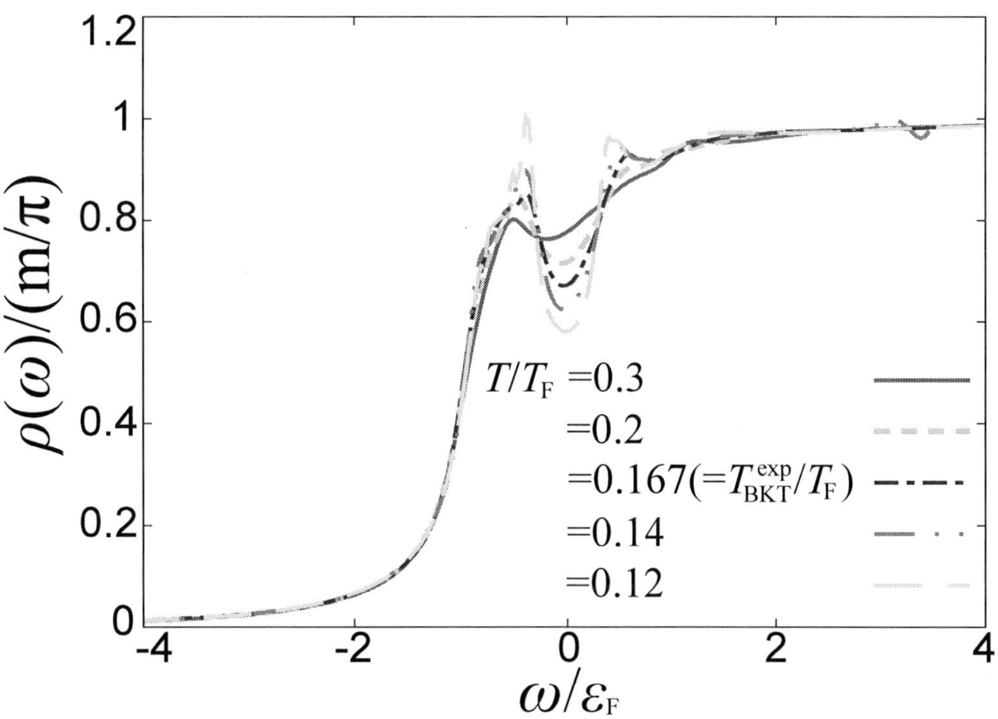


**Figure 1.** Calculated density of states near the observed BKT phase transition temperature $T_{\mathrm{BKT}}^{\mathrm{exp}} = 0.167T_{\mathrm{F}}$[1,2]. We set the interaction strength as $\ln(k_{\mathrm{F}}a_{2\mathrm{D}}) = 1.23$ (crossover regime), where $k_{\mathrm{F}}$ i s the Fermi momentum and $a_{2\mathrm{D}}$ is the two-dimensional s-wave scattering length.

[1] M. G. Ries, A. N. Wenz, G. Zürn, L. Bayha, I. Boettcher, D. Kedar, P. A. Murthy, M. Neidig, T. Lompe, and S. Jochim, Phys. Rev. Lett. **114**, 230401 (2015).

[2] P. A. Murthy, I. Boettcher, L. Bayha, M. Holzmann, D. Kedar, M. Neidig, M. G. Ries, A. N. Wenz, G. Zürn, and S. Jochim, Phys. Rev. Lett. **115**, 010401 (2015).

# Ratio of Shear Viscosity to Entropy Density of an Ultracold Fermi Gas in the BCS-BEC Crossover Region

Daichi Kagamihara, Yoji Ohashi
Department of Physics, Keio University, 3-14-1,
Hiyoshi, Yokohama, 223-8522, Japan

We theoretically investigate the shear viscosity $\eta$, as well as the entropy density $s$, in the BCS-BEC crossover regime of an ultracold Fermi gas above the superfluid phase transition temperature $T_c$. Including fluctuations in the Cooper channel within a T-matrix approximation (TMA)[1], we calculate these quantities, as well as $\eta/s$, to assess the conjecture by Kovtun, Son, Starinets (KSS)[2], stating the existence of the lower bound of this ratio.

We numerically evaluate $\eta$, within the neglect of the vertex correction. In the weak-coupling BCS side, the temperature dependence of $\eta$ is found to physically be divided into three regions (see Fig.1), that is, (1) high temperature classical regime, (2) low temperature Fermi degenerate regime, where the Fermi liquid like behavior of $\eta$ is seen, as well as (3) the region near $T_c$, where $\eta$ is remarkably affected by the so-called pseudogap phenomenon associated with pairing fluctuations. In the strong-coupling BEC side, the neglect of the vertex correction does not cannot correctly describe the behavior of $\eta$, because the contribution of tightly bound molecules (that dominate over the strong-coupling BEC regime) to $\eta$ is completely ignored.

We also calculate the entropy density $s$, within the framework of TMA. We show that $\eta$ is found to decrease as one passes through the BCS-BEC crossover region from the weak-coupling side, to become very close to the KSS bound, $\hbar/4\pi k_B$, slightly in the BEC side at $T_c$.

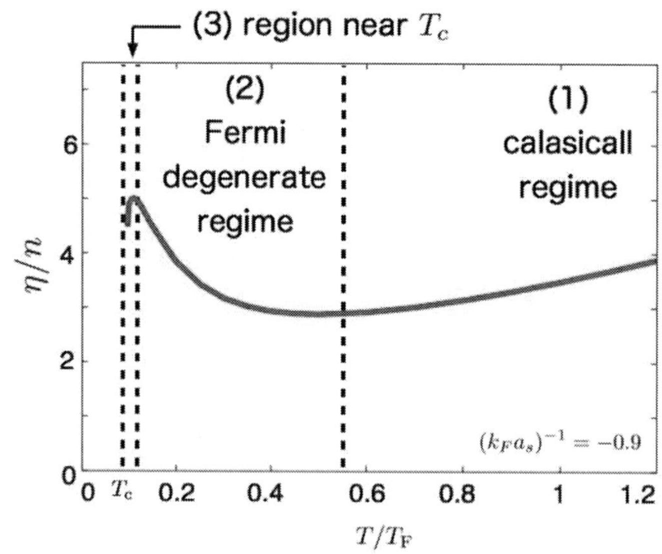


**Figure 1.** Calculated shear viscosity $\eta$ as a function of temperature. We set $(k_F a_s)^{-1} = -0.9$.

[1] S. Tsuchiya, R. Watanabe, and Y. Ohashi, Phys. Rev. A **80**, 033613 (2009).

[2] P. K. Kovtun, D. T. Son, and A. O. Starinets, Phys. Rev. Lett **94**, 111601 (2005).

# Non-equilibrium dynamics of finite one- and two-dimensional Hubbard systems

Stefan Donsa[1], Fabian Lackner[1], Joachim Burgdörfer[1], Iva Březinová[1]
[1]Institute for Theoretical Physics, Vienna University of Technology, Vienna, Austria, EU

The Hubbard model is one of the simplest models for correlation in solid state physics but still beyond a full theoretical description. Ultra-cold Fermi gases in optical lattices allow to investigate experimentally the Hubbard model for large systems under well-controlled conditions [1, 2]. So far, an adequate theoretical description has been missing. We present a new theoretical approach to simulate the dynamics of Hubbard systems in the regime of weak inter-particle correlations which is applicable to higher-dimensional systems. We use the time dependent two-particle reduced-density matrix (TD-2RDM) method [3] to simulate the dynamics of finite Hubbard systems following potential quenches. One of the strengths of the TD-2RDM method is that it includes two-particle correlations exactly. The time-evolution of the 2-RDM is coupled to two-particle and three-particle correlations via the interaction term of the underlying Hamiltonian. We reconstruct three-particle correlations partially using the Yasuda-Nakatusji approximation [4] and through an ansatz based on the contraction consistency of the 3-RDM to the 2-RDM [3]. To benchmark the theory we investigate small systems that are still accessible to an exact numerical solution. We show results for one-dimensional Hubbard systems far from equilibrium where we overcome stability problems observed in [5] by a procedure called purification [6] and achieve better results than non-equilibrium Green's function method calculations [7] (Fig. 1 (a)). We apply the TD-2RDM method to weak potential quenches and delimit its applicability. Finally, we present first results for the non-equilibrium dynamics of two-dimensional Hubbard nano-clusters (Fig. 1 (b)). We plan to study study two-dimensional Hubbard systems at sizes not treatable by exact methods and compare theoretical simulations to experimental results.

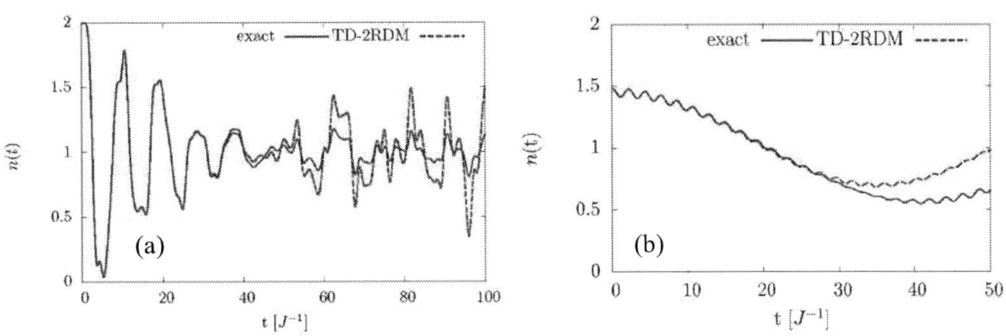


**Figure. 1** (a) Density fluctuations on the first site of an eight-site Hubbard model with interaction strength U = 0.1 J with far from equilibrium initial conditions as chosen in [7]. (b) Density fluctuations of the central site of a 3x3 Hubbard nano-cluster with interaction strength U = 0.1 J after an additional harmonic potential with strength $V_0$ = 0.3 J is turned off.

[1] M. Lewenstein et al., Advances in Physics **56:2**, 243 (2007).

[2] A. Omran et al., Phys. Rev. Lett. **115**, 263001 (2015).

[3] F. Lackner et al., Phys. Rev. A **91**, 023412 (2015).

[4] K. Yasuda, and H. Nakatsuji, Phys. Rev. Lett. **76**, 1039 (1996).

[5] A. Akbari et al., Phys. Rev. B **85**, 235121 (2012).

[6] D. Mazziotti, Phys. Rev. E **65**, 026704 (2002).

[7] S. Hermanns, N. Schlünzen, and M. Bonitz, Phys. Rev. B **90**, 125111 (2014).

# Manipulating the flow of ultracold atoms through a quantum point contact

Dominik Husmann[1], Martin Lebrat[1], Samuel Häusler[1], Sebastian Krinner[1], Jean-Philippe Brantut[1], Tilman Esslinger[1]

[1]Institute for Quantum Electronics, ETH Zürich, 8093 Zürich, Switzerland

Recent progress in the field of ultracold gases has made it possible to explore the transport properties of neutral atoms through mesoscopic structures engineered with optical potentials [1,2]. Here we report on conductance measurements through a precisely tailored quantum wire in the single mode regime.

Experimentally, we prepare a system of two reservoirs of fermionic Lithium 6 atoms connected by a narrow constriction, the quantum wire. To measure its transport properties, we impose chemical potential bias between the two reservoirs, driving current through the wire. From the current we directly infer the conductance of the wire via Ohm's law.

Extending previous experiments, we use a digital mirror device (DMD) to project holographically shaped optical potentials onto the wire. Using this technique, we can scan a small repulsive 'tip' (scanning gate) on the scale of the atomic Fermi wavelength over the constriction. This allows us to retrieve spatially resolved information about the response of the conductance to local perturbations, a technique called scanning gate microscopy and widely known in solid state physics. We achieve a spatial resolution close to the transverse wave function inside the contact. The clean geometry furthermore allows to directly compare the scanning gate pictures to exact simulations for non-interacting particles.

The fine control of the DMD laser beam can be extended to project more complex structures: by projecting several consecutive repulsive barriers, a lattice can be formed inside the quantum wire, one lattice site at a time. Upon scaling up the number of sites, we observe the emergence of a band gap, originating from quantum interferences among the scatterers. Further, the coherent character of transport can be investigated by measuring the visibility of the gap emergence as a function of temperature. We observe a washing out of the interference feature at temperatures higher than the gap energy, which we attribute to the loss of coherence. Our results agree to theory calculations in the Landauer-Büttiker formalism of transport with energy-dependent transmission coefficients obtained from numerical simulations using a Numerov algorithm.

This novel bottom-up approach to lattice physics opens the path to study metal-insulator physics in strongly correlated systems.

**Figure 1.** Artists view of two clouds of ultracold atoms connected by an optically shaped quantum wire.

[1] S. Krinner et al., Nature **517**, 64 (2015).

[2] D. Husmann et al., Science **350**, 6267 (2015)

# Spinor and orbital dynamics of trapped dipolar fermions

Ulrich Ebling[1], Masahito Ueda[1,2]

[1]Department of Physics, The University of Tokyo, 7-3-1 Hongo, Bunkyo-ku, Tokyo 113-0033, Japan

[2]RIKEN Center for Emergent Matter Science (CEMS), Wako, Saitama 351-0198, Japan

We study dynamcial properties of ultracold atoms with strong dipole-dipole interactions, such as rare-earth atoms like Erbium or Dysprosium. These atoms have a large number of spin states, and dipole-dipole interactions do not conserve total spin, in addition to being long-range and anisotropic. For the case of spinless Fermions, the anisotropic nature of dipole-dipole interactions leads to a deformation of the Fermi surface [1] recently observed experimentally for Erbium [2]. Here, we investigate the case of a large-spin system, where spinor dynamics is coupled to orbital dynamics by the dipole-dipole interactions. This can lead to the emergence of spatial structures associated with orbital angular momentum or nematic order, either in the total density or within individual spin components.

[1] T. Miyakawa, T. Sogo, H. Pu, Phys. Rev. A **77**, 061603R (2008)

[2] K. Aikawa, S. Baier, A. Frisch, M. Mark, C. Ravensbergen, F. Ferlaino, Science **345**, 1484 (2014).

# Beyond Gaussian Pair fluctuation theory in strongly interacting Fermi gases

Brendan C Mulkerin[1], Xia-Ji Liu[1], Hui Hu[1]

[1]Centre for Quantum and Optical Science, Swinburne University of Technology, Melbourne 3122, Australia.

Interacting Fermi systems in the strongly correlated regime play a fundamental role in many areas of physics and though weakly interacting fermions are understood, strongly correlated fermions are difficult to describe theoretically. Existing strong-coupling theories rely heavily on the so-called many-body *T*-matrix approximation that sums ladder-type Feynman diagrams. Here, by acknowledging the fact that the effective interparticle interaction (i.e., the vertex function) becomes smaller above three dimensions, we develop a theoretical framework for interacting Fermi gases beyond the ladder approximation. As an application, we solve the equation of state for three- and two-dimensional strongly interacting fermions and find excellent agreement with experimental and other theoretical results above temperatures of $0.5T_F$.

As an application of our beyond GPF theory, we determine the equation of state of above threshold strongly interacting Fermi gases in three dimensions, shown here in Figure 1.

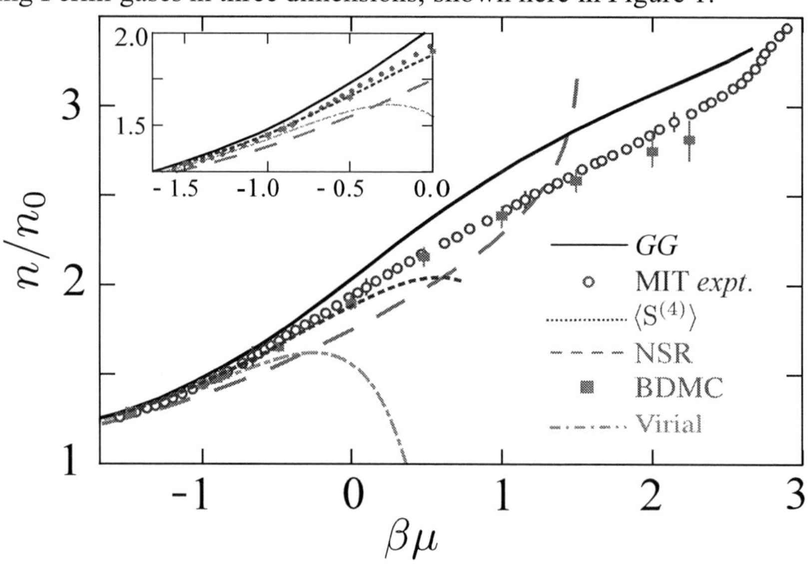


**Figure 1.** The density equation of state for 3D, $n/n_0$, plotted as a function of $\beta\mu$, normalized by the ideal gas result, $n_0$, at the same temperature $T$ and chemical potential $\mu$, found from the MIT experiment [1] (maroon circles with error bars), *GG* (black solid), NSR (red dashed), $\langle S^{(4)}\rangle$ (blue dotted), 3rd order virial (orange dot-dashed), and bold-diagrammatic Quantum Monte Carlo (green squares) [2]. The inset blows up the non-degenerate region for the same plot.

[1] M. J. H. Ku, A. T. Sommer, L. W. Cheuk, and M. W. Zwierlein, Science **355,** 563 (2012)
[2] K. V. Houcke, F. Werner, E. Kozik, N. Prokof'ev, B. Svistunov, M. J. H. Ku, A. T. Sommer, L. W. Cheuk, A. Schirotzek, and M. W. Zwierlein, Nature Phys. **8**, 366 (2012).

# Local Photoemission Spectra in the BCS-BEC Crossover Regime of an Ultracold Fermi Gas in a Harmonic Trap

Miki Ota[1], Hiroyuki Tajima[1], Ryo Hanai[1], Daisuke Inotani[1], Yoji Ohashi[1]
[1]Department of Physics, Keio University,
3-14-1, Hiyoshi, Yokohama, Japan

We theoretically investigate single-particle excitations in the BCS-BEC crossover regime of a trapped ultracold Fermi gas. Including strong paring fluctuations within the framework of a T-matrix approximation (TMA)[1], as well as effects of a harmonic trap in the local density approximation (LDA), we calculate the local photoemission spectrum (LPES) in the normal state above the superfluid phase transition temperature $T_c$. In the unitary regime, our result shown in Fig.1 agrees well with the recent LPES experiment on a $^{40}$K Fermi gas [2]. This experiment actually measures the spectrum, not at a spatial position, but around the trap center with a finite spatial volume. Thus, we also clarify to what extent the observed LPES is close to that in a homogeneous gas. Since LPES is a useful quantity to see the so-called pseudogap phenomenon in the BCS-BEC crossover regime of a trapped ultracold Fermi gas, our results would be useful for the detailed understanding of this strong interacting Fermi system.

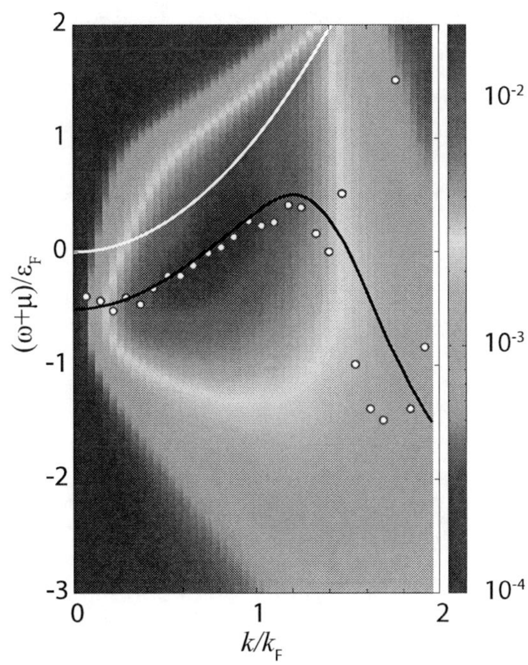


**Figure 1.** Calculated local photoemission spectrum (LPES) around the trap center in the unitary regime ($(k_F a_s)^{-1}$= 0.1) of an ultracold Fermi gas at $1.24T_c$. The solid line shows the peak positions of the spectrum, which is found to agree well with the recent experiment on a $^{40}$K Fermi gas (open circles)[2].

[1] S. Tsuchiya, R. Watanabe, and Y. Ohashi, Phys. Rev. A **80**, 033613 (2009).

[2] Y. Sagi, T. E. Drake, R. Paudel, R. Chapurin and D. S. Jin, Phys. Rev. Lett. **114**, 075301 (2015)

# A strongly correlated Fermi superfluid near an orbital Feshbach resonance: The stability, universal equation of state and Leggett mode

Liangyi He1, Jia Wang2, Shi-Guo Peng2, Xia-Ji Liu2, and Hui Hu2
[1]Tsinghua University, China
[2]Swinburne University of Technology, Australia

We theoretically study the superfluid phase of a strongly correlated 173Yb Fermi gas near orbital Feshbach resonances, by developing a quantitative fluctuation theory with the two-band model. We examine the low-energy density excitation spectrum of the system and determine a stability phase diagram (see the following figure 1). We find that the 173Yb Fermi gas is intrinsically metastable and has a new-type universal equation of state, due to the small but positive singlet scattering length. The massive Leggett mode, arising from the out-of-phase fluctuation of the relative phase of two pairing parameters, is severely damped. We discuss the parameter space where an undamped Leggett mode may exist.

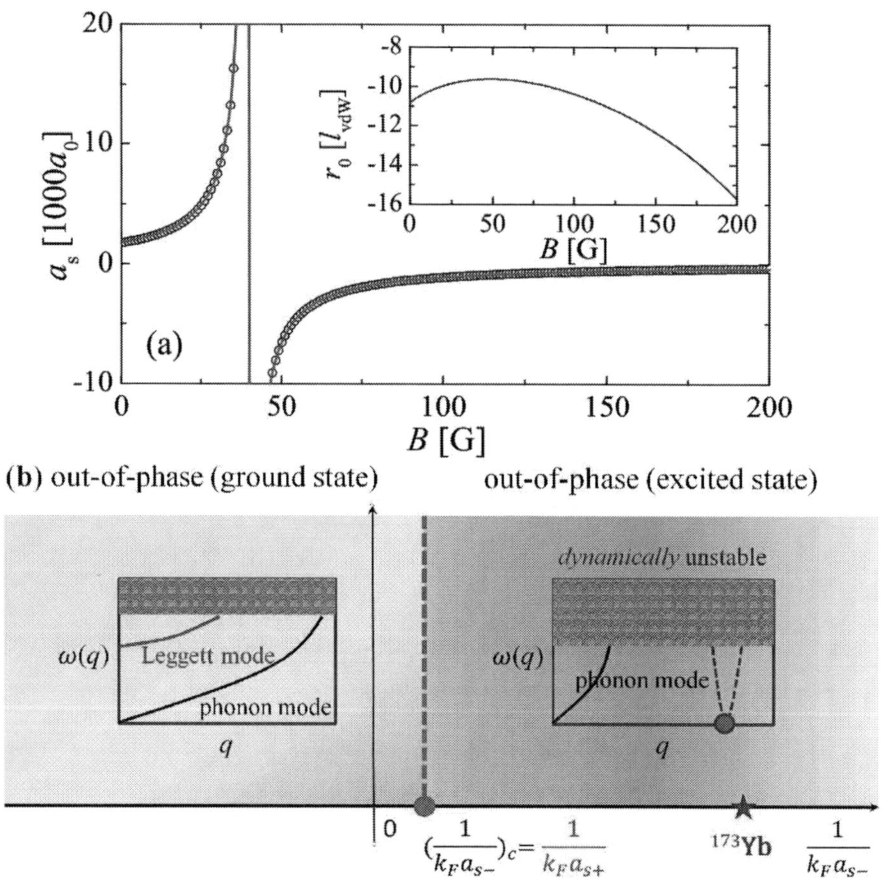


**Figure 1.** The stability phase diagram of orbital Feshbach resonance.

[1] Liangyi He, Jia Wang, Shi-Guo Peng, Xia-Ji Liu, and Hui Hu, arXiv: 1606.00188.

# Stoner ferromagnetism of a strongly interacting Fermi gas in the quasirepulsive regime

Liangyi He1, Xia-Ji Liu2, Xu-Guang Huang 3, and Hui Hu2
1Tsinghua University, China
2Swinburne University of Technology, Australia
3Fudan University, China

Recent advances in rapidly quenched ultracold atomic Fermi gases near a Feshbach resonance arise a number of interesting problems, in the context of observing the long-sought Stoner ferromagnetic phase transition. The possibility of experimentally obtaining a "quasirepulsive" regime in the upper branch of the energy spectrum due to the rapid quench is currently debated and theoretically, the Stoner transition has mainly been investigated by using perturbation theory or at high polarization, due to the limited theoretical approaches in the strongly repulsive regime. In this work, we present a nonperturbative theoretical approach to the quasirepulsive upper branch of a Fermi gas near a broad Feshbach resonance and determine the finite-temperature phase diagram for the Stoner instability (see the following figure 1). Our results agree well with the known quantum Monte-Carlo simulations at zero temperature and recover the known virial expansion prediction at high temperature for arbitrary interaction strengths. At resonance, we find that the Stoner transition temperature becomes of order of the Fermi temperature, around which the molecule formation rate becomes vanishingly small. This suggests a feasible way to observe Stoner ferromagnetism in the nondegenerate temperature regime.

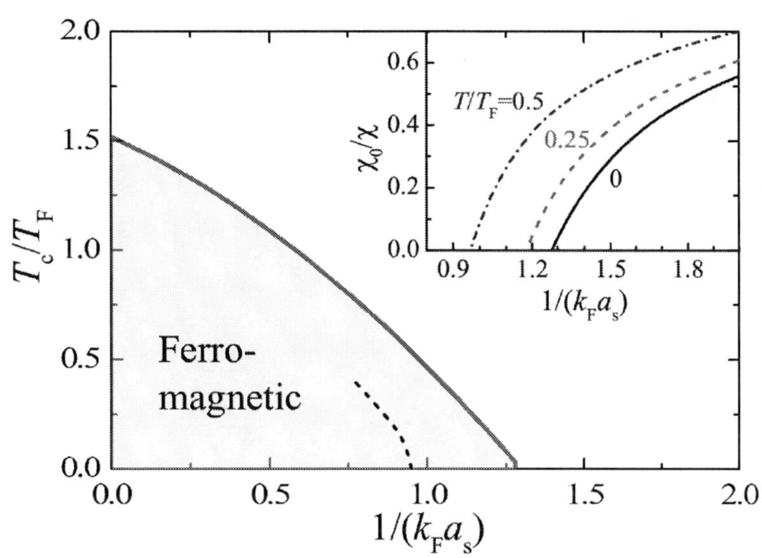


**Figure 1.** The finite-temperature phase diagram of a strongly interacting Fermi gas in its repulsive branch.

[1] Liangyi He, Xia-Ji Liu, Xu-Guang Huang, and Hui Hu, Phys. Rev. A, 93, 063629 (2016).

# Multi-photon spectroscopy and spin-orbit coupling of ytterbium degenerate Fermi gas

Moosong Lee[1], Jeong Ho Han[1], Jin Hyoun Kang[1], Yong-il Shin[1]
[1]Seoul National University, South Korea

We present our recent experimental results of multi-photon spectroscopy on 173 ytterbium degenerate Fermi gas. We inspect four photon resonance as well as two photon peaks and study their polarization dependency. With magnetic fields, we observe separation of resonance peaks which can be explained by Zeeman splitting of $^1S_0$ ground state. Among such transitions, spin-changing ones can be regarded as spin-orbit coupling. We find that at certain magnetic field, four photon and two photon spin-orbit coupling can be simultaneously coupled to the ground state. We also investigate dephasing of Rabi oscillation.

# Exact and semiclassical fermion dynamics in amplitude modulated lattices

Tomotake Yamakoshi[1], Shinichi Watanabe[2], Shun Ohgoda[2], Alexander P. Itin[3,4]
[1]University of Electro-Communications, Institute for Laser Science, Chofu-shi, Tokyo, Japan
[2]University of Electro-Communications, Dept. of Engineering Science, Chofu-shi, Tokyo, Japan
[3]Radboud University, Institute for Molecules and Materials (IMM), Nijmegen 6525 AJ, The Netherlands
[4]Space Research Institute, Russian Academy of Sciences, Moscow 117997, Russia

There have been continuing interest in manipulation and control of ultracold atoms in the combined optical lattice and parabolic trap potentials [1,2]. Recent Hamburg experiments [2] on quantum simulation of photoconductivity studied dynamics of fermions induced by an amplitude modulation of the optical lattice. The amplitude modulation excites atoms from the ground band, producing an excited wave packet in higher bands and leaving a hole in the particle distribution of the ground band. The experiments examined both the dynamics of the excited particles and holes, demonstrating nice correspondence with a nonlinear semiclassical theory developed by one of us [2]. Here, we provide a more detailed theoretical analysis of fascinating dynamics of fermonic holes, which undergo multiple collapses and revivals. In particular, we construct a semiclassical wave packet in the ground band more accurately, which allows us to predict the dynamics of holes on longer timescales. We study effects of temperature, dimensionality of the system, and anharmonicity of the parabolic trap. We previously investigated the excitation process and the dynamics of excited wave packets with non-interacting bosons [3] for the sake of a deeper understanding of the Aarhus experiment [1], developing an effective multi-level Rabi model for it, which remains useful for the current study. Accurate many-particle numerical simulations show that the hole has intrinsically long coherence time, in fact considerably longer than observed experimentally [4]. This work improves the early semiclassical approach [2] and provides useful information for future studies of interacting fermionic mixtures and dynamics of nonlinear fermionic waves in optical lattices.

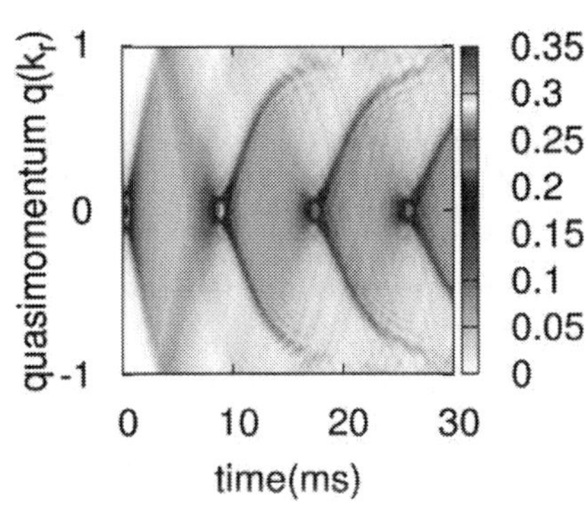


**Figure 1.** Density plot of the hole which serves as the counterpart to the excited wave packet in the higher bands. The time evolution of the hole is shown in the space of quasimomentum.

[1] P. L. Pedersen *et al.*, Phys. Rev. A **88**, 023620 (2013).
[2] J. Heinze *et al.*, Phys. Rev. Lett. **110**, 085302 (2013).
[3] T. Yamakoshi and S. Watanabe, Phys. Rev. A **91**, 063614 (2015).
[4] T. Yamakoshi, S. Watanabe, S. Ohgoda, and A.P. Itin, Phys. Rev. A **93**, 063637 (2016).

# Topological phase transitions in the repulsively interacting Haldane-Hubbard model

Tuomas I. Vanhala[1,2], Topi Siro[1], Long Liang[1], Matthias Troyer[2], Ari Harju[1], and Päivi Törmä[1,3]

[1]COMP Centre of Excellence, Department of Applied Physics, Aalto University, Helsinki, Finland

[2]Theoretische Physik, ETH Zurich, 8093 Zurich, Switzerland

[3]Institute for Quantum Electronics, ETH Zurich, 8093 Zurich, Switzerland

Using dynamical mean-field theory and exact diagonalization we study the phase diagram of the repulsive Haldane-Hubbard model, varying the interaction strength and the sublattice potential difference. In addition to the quantum Hall phase with Chern number C=2 and the band insulator with C=0 present already in the noninteracting model, the system also exhibits a C=0 Mott insulating phase, and a C=1 quantum Hall phase. We explain the latter phase by a spontaneous symmetry breaking where one of the spin-components is in the Hall state and the other in the band insulating state. [1] The noninteracting version of the model has recently been realized in ultracold gas experiments [2,3], and we discuss the possibilities for the experimental observation of the predicted phases.

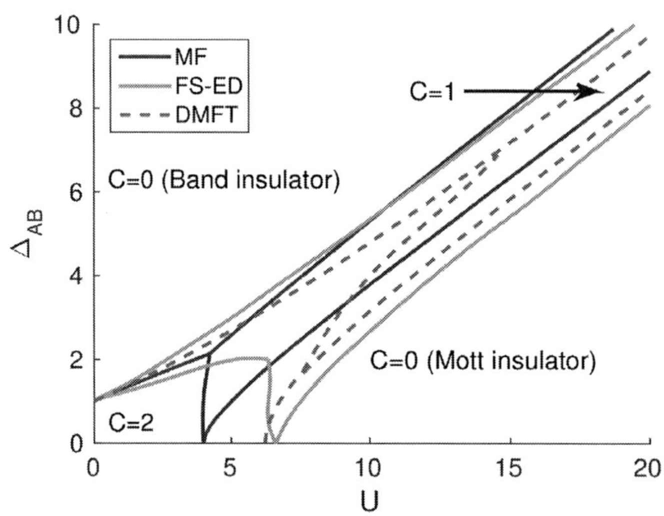


**Figure 1.** The figure shows the phase diagram of the model as a function of the Hubbard interaction strength U and the sublattice potential difference $\Delta_{AB}$ obtained within mean-field theory (MF), finite size exact diagonalization (FS-ED) and single-site dynamical mean-field theory (DMFT). The lines indicate the topological transitions where the Chern number C changes.

[1] T. I. Vanhala, T. Siro, L. Liang, M. Troyer, A. Harju, and P. Törmä, Phys. Rev. Lett. **116**, 225305 (2016)

[2] G. Jotzu, M. Messer, R. Desbuquois, M. Lebrat, T. Uehlinger, D. Greif, T. Esslinger, Nature **515**, 237-240 (2014)

[3] N. Fläschner, B. S. Rem, M. Tarnowski, D. Vogel, D.-S. Lühmann, K. Sengstock, C. Weitenberg, Science **352**, 1091-1094 (2016)

# Equation of state of a Superfluid Fermi atom gas and Neutron star interior: Similarity and Difference on the viewpoint of phase-shift corrections

Pieter van Wyk[1], Hiroyuki Tajima[1], Daisuke Inotani[1], Akira Ohnishi[2], Yoji Ohashi[1]
[1]Keio University, Japan
[2]Kyoto University, Japan

We investigate the equation of state (EOS) of a superfluid Fermi gas in the BCS-BEC crossover region. We go beyond the mean-field approach by including strong coupling effects, following the formalism developed by Nozières and Schmitt - Rink. We numerically calculate the internal energy far below the superfluid phase transition temperatu re, to compare our results with the recent EOS experiment in a superfluid $^6$Li Fermi atom gas. We th en apply this approach to pure neutron matter, expected to exist in the crust and outer core of neutron stars, by including effects of the detailed neutron-neutron phase shift. Our results are expected to make an important contribution in the understanding o f neutron star physics.

*This abstract modified in template by ICAP team*

# Observation of the Efimovian expansion in scale invariant Fermi gases

S. Deng[1], Z. Shi[2], P. Diao[1], Q. Yu[1], Hui Zhai[2], R. Qi[3], and Haibin Wu[1]
[1] State Key Laboratory of Precision Spectroscopy,
East China Normal University, Shanghai 200062, P. R. China
[2]Institute for Advanced Study, Tsinghua University, Beijing 100084, P. R. China
[3]Department of Physics, Renmin University of China, Beijing 100872, P. R. China

Scale invariance plays an important role in unitary Fermi gases. Discrete scaling symmetry manifests itself in quantum few-body systems such as the Efimov effect. In the poster we will present the theoretical prediction and experimental observation of a distinct type of expansion dynamics for scale invariant quantum gases. When the frequency of the harmonic trap holding the gas decreases continuously as the inverse of time t, the expansion of the cloud size exhibits a sequence of plateaus [1], as shown in the figure 1. Remarkably, the locations of these plateaus obey a discrete geometric scaling law with a controllable scale factor and the expansion dynamics is governed by a log-periodic function. This striking expansion shares the same scaling law and mathematical description as the Efimov effect. The results are universal and hold for all scale invariant quantum gases, including non-interacting gases, Tonks gas in one-dimension, weakly interacting gases in two-dimension and unitary Fermi gas in three-dimension.

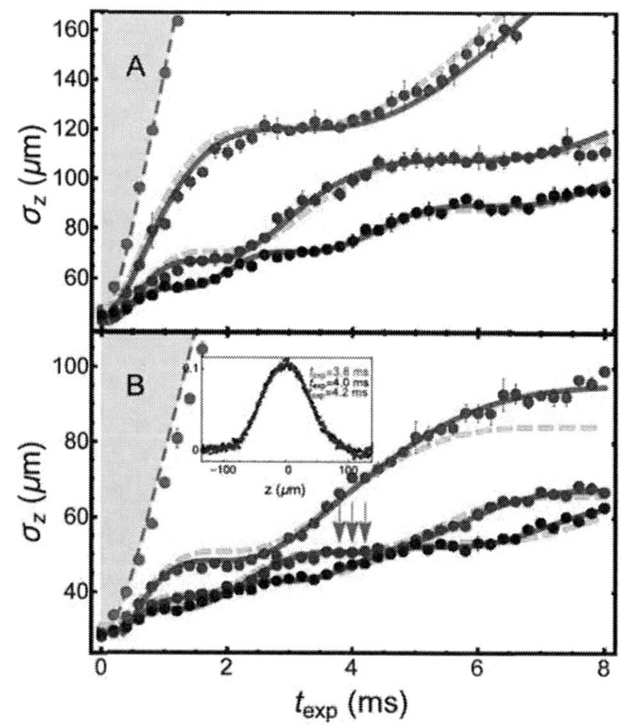


**Figure 1.** Experimental observation of Efimovian expansion. A, a non-interacting Fermi gas. B, a unitary Fermi gas.

[1] S. Deng, Z. Shi, P. Diao, Q. Yu, Hui Zhai, R. Qi, and Haibin Wu, arXiv:1512.02044.

# Scaling of a driven atomic gas from the weakly-dressed to the quantum critical regime

Stephan Helmrich[1], Alda Arias[1], Shannon Whitlock[1]
[1]Physics Institute, University of Heidelberg, Im Neuenheimer Feld 226, 69120 Heidelberg, Germany

The emergence of correlations in complex many-body systems can be accompanied by unexpectedly simple scaling laws which signal new physical regimes or universal relations between otherwise very different physical systems [1]. We demonstrate that non-equilibrium scaling laws can reveal the different regimes of strongly-interacting quantum systems driven to highly excited states [2]. For weak or far off-resonant driving we find that the dependence of the excitation rate on coupling strength is well described by power laws characteristic of the dissipative or weakly-dressed regimes, while for strong near-resonant driving we observe a crossover to the quantum critical regime. For intermediate detunings we discover superlinear intensity scaling in a new regime, indicative of cooperative excitation processes, which extends the domain where scale-invariant behavior can be found in driven quantum systems.

[1] R. Albert, A.-L. Barabási, Rev. Mod. Phys. **74**, 47 (2002).

[2] S. Helmrich, A. Arias, S. Whitlock, arXiv:1605.08609.

# Optical π Phase Shift Created with a Single-Photon Pulse

Daniel Tiarks[1], Steffen Schmidt[1], Gerhard Rempe[1],
Stephan Dürr[1]
[1]Max-Planck-Institut für Quantenoptik, Hans-Kopfermann-Str. 1
85748 Garching, Germany

A deterministic photon-photon quantum-logic gate is a long-standing goal. Building such a gate becomes possible if a light pulse containing only one photon imprints a phase shift of π onto another light field. Here we experimentally demonstrate the generation of such a π phase shift with a single-photon pulse [1]. This extends our previous work on a single-photon switch [2] and transistor [3]. A first light pulse containing less than one photon on average is stored in an atomic gas. Rydberg blockade combined with electromagnetically induced transparency creates a phase shift for a second light pulse which propagates through the medium. Postselected on the detection of a retrieved photon from the first pulse, we measure a π phase shift of the second pulse. This demonstrates a crucial step toward a photon-photon gate and offers a variety of applications in the field of quantum information processing.

[1] D. Tiarks, S. Schmidt, G. Rempe, and S. Dürr, arxiv:1512.05740.

[2] S. Baur, D. Tiarks, G. Rempe, and S. Dürr, Phys. Rev. Lett. 112, 073901 (2014).

[3] D. Tiarks, S. Baur, K. Schneider, S. Dürr, and G. Rempe, Phys. Rev. Lett. 113, 053602 (2014).

# Spectral shift and dephasing of Rydberg EIT

Thibault Vogt[1,2], JingShan Han[1], Wenhui Li[1,3]
[1]Centre for Quantum Technology, National University of Singapore, Singapore
[2]MajuLab, CNRS-UNS-NUS-NTU International Joint Research Unit UMI 3654, Singapore
[3]Department of Physics, National University of Singapore, Singapore

Electromagnetically induced transparency (EIT) involving a Rydberg state has advanced as a major means to obtain non linearity at the few photons level and tailor an effective interaction between photons. Rydberg EIT has also evolve as a fine tool to perform spectroscopic measurements of microwave fields and stray electric fields and is even used for the imaging of Rydberg atoms. However, we show in this communication that extreme care should be taken for the interpretation of the data. Indeed, not only can lensing modify dramatically the line shapes [1], but also the interaction between Rydberg atoms tends to shift the EIT frequency resonance [2]. All the measurements are made on cold atom systems and the data are systematically compared with relevant models.

[1] J. Han, T. Vogt, M. Manjappa, R. Guo, M.Kiffner, and W. Li, Lensing effect of electromagnetically induced transparency involving a Rydberg state. Phys. Rev. A 92, 063824 (2015)

[2] J. Han, T. Vogt and W. Li, Spectral shift and dephasing of electromagnetically induced transparency in an interacting Rydberg gas, arXiv:1603.00635 (2016)

# A High-Speed Single Ion Source using a Cold Atom Beam and Rydberg Blockade

B. M. Sparkes[1], D. Murphy[1], R. W. Speirs[1], A. J. McCulloch[1], R. E. Scholten[1]
[1]School of Physics, University of Melbourne, VIC 3010, Australia

We present a mechanism for generating single ions on demand based on Rydberg excitation from a cold atom beam. This source will have the ability to place ions into target surfaces with high precision due to their low temperature [1, 2], allowing for the possibility of new types of nanofabricated devices and processes in material sciences. For instance, it could be used for ion implantation in solid-state quantum computers, color-centers in diamond for the creation of single-photon sources, the manufacture of ever-shrinking transistors, as well as for high-resolution ion microscopy and lithography. The ideal single ion source would be fast, precise and fully deterministic, but current technologies are either slow or stochastic. Combining our second-generation cold atom-beam based ion source with the phenomena of Rydberg blockade [3] and stimulated Raman adiabatic passage (STIRAP) [4] will allow us to overcome these issues to create a source that is fast, heralded and quasi-deterministic (Fig. 1).

We will present simulations of the proposed single ion beam source as well as our latest experimental results demonstrating high-efficiency excitation using STIRAP in a MOT-based cold atom electron and ion source, with a peak efficiency of 82% (total efficiency of 60%) [5].

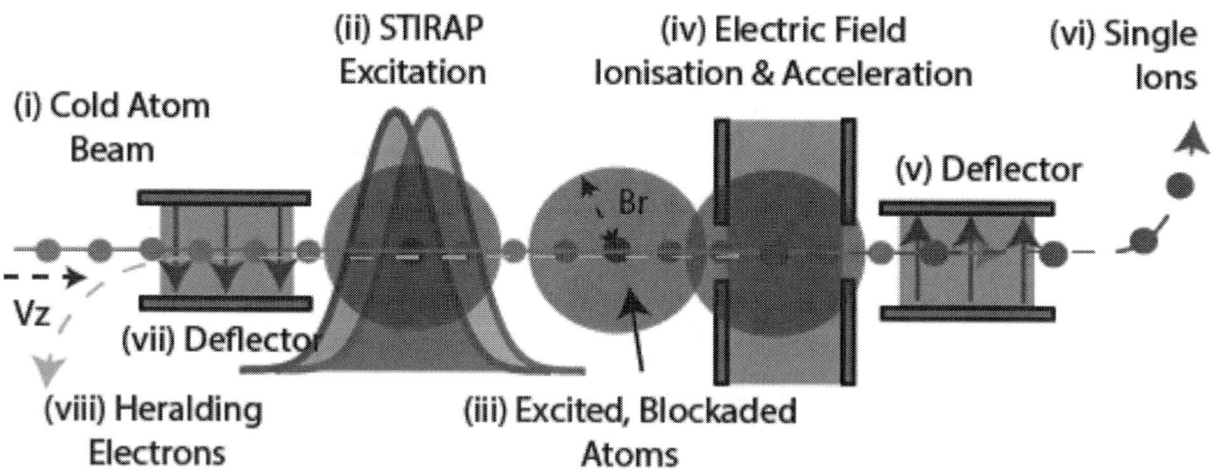


**Figure 1.** Main components of proposed single ion beam source.

[1] A. J. McCulloch *et al.*, Nature Physics **7**, 785 (2011).

[2] A. J.McCulloch *et al.*, Nature Communications **4**, 1692 (2013).

[3] R. Löw *et al.*, Journal of Physics B **45**, 113001 (2012)

[4] K. Bergmann *et al.*, Review of Modern Physics **70**, 1003 (1998).

[5] B. M. Sparkes *et al.*, arXiv:1602.07791 (2016).

# Progress towards Rydberg gate experiments with Non-destructive hyperfine state selective readout using EMCCD camera

Minho Kwon, Matthew Ebert, Thad G. Walker, Mark Saffman
Department of Physics, University of Wisconsin,
1150 University Avenue, Madison, Wisconsin 53706, USA

We report progress towards a hyperfine state-selective, non-destructive fluorescence detection using an EMCCD camera. State-selective methods have been developed however they work with single photon detectors either assisted by a cavity [1] or in free space [2,3]. We adopt and develop the technique [4] using an EMCCD camera, allowing multiple sites to be readout simultaneously. The method interrogates spatially separated ground state Rubidium atoms in parallel, while preserving their presence and hyperfine state. This improves our data rate for neutral atom quantum computing by an order of magnitude. We also present single and Rydberg-mediated gate experiments employing the readout scheme.

[1] Vuletic et al, Phys. Rev. Lett 109, 133603(2012)
[2] Gibbons et al, Phys. Rev. Lett 106, 133002(2011)
[3] Browaeys et al, Phys. Rev. Lett 106,133003(2011)
[4] Martinez-Dorantes, Ph.D Thesis, Univ. Bonn (2016)

# Study of electron source based on the ionization of colds atoms

Elias Moufarej[a], Collin Lopez[a], Melissa Vieille Grosjean[a], Yan Picard[a], Daniel Comparat[a]

[a] Laboratoire Aimé Cotton, CNRS, Université Paris-Sud, ENS Cachan, Bât. 505, 91405 Orsay, France.

*Author Email: elias.moufarej@u-psud.fr*

Initial energy dispersion of an atomic beam lies within the sub-meV range, what remains lower than standard electron sources energy spread (above 0.3 eV). Thus by photo-ionizing an atomic beam we expect to realize a bright and mono-energetic electron source.

We develop a new source of focused electron and ion beams, through ionization of a high flux effusive atomic beam. The atomic (cesium) beam is obtained from 2D Magneto-Optical Trap, or from an effusive oven. The very low transverse temperature (sub-K range) and the relative density of the starting atomic sample ensure excellent initial conditions in order to obtain a bright and monochromatic charge sources (Fig.1). We are also studying, and developing an oven where an intense laser can propagate longitudinally to the atomic beam in order to collimate it [1]. The collimation laser can have Gaussian or even donut shape [2]. This technique will allow us to increase the atomic density, and thus increase the brightness of the charged particles source.

We study the behavior of the produced electron beam using time of flight or velocity imaging methods. The main idea is to combine laser cooling with energy resolved electron microscope. This unique combination should allow us to focus a mono-energetic (~1 meV) and very low energy electron (1-10 eV) beam down to unprecedented nanometer spots [3]. On the other hand, this energy spread is limited by chromaticism created during the ionization process. To reduce those effects, we plan to use Rydberg atoms field ionization.

The exclusive injection and diagnostic microscope, with such a high spatial and energy resolution, will open the path toward control chemical reactions at the molecular level. Possible applications for new microscopes, lithography or implantation are foreseen.

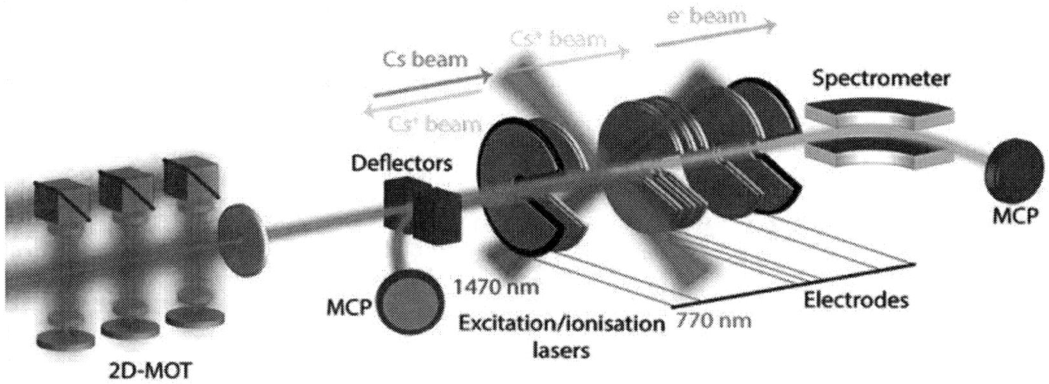


**Figure 1.** Electron and ion beams produced from cesium atoms beam cooled in a 2D MOT.

**[1]** Bruneau Y., *et al.* (2014). Guided and focused slow atomic beam from a 2 dimensional magneto optic al trap. *The European Physical Journal D, 68*(4), 1-6.

**[2]** Mestre, M., *et al.*(2010). Cold atom guidance by a holographically-generated Laguerre-Gaussian laser mode. *The European Physical Journal D, 57*(1), 87-94.

**[3]** Kime, L., *et al.(*2013). High-flux monochromatic ion and electron beams based on laser-cooled atom s. *Physical Review A, 88*(3), 033424.

# A compact apparatus of hybrid 2D-MOT and Zeeman slower for ultracold strontium Rydberg atoms

Peng Chen[1], L. Couturier[1], I. Nosskea[1], C. Qiao[1], F. Hua[1], Y. Jiang[1,2] and M. Weidemüller[1,3]

[1]University of Science and Technology of China, Shanghai Institute for Advanced Studies, Xiupu Road 99, 201315 Shanghai, China

[2]Shanghai Advanced Research Institute, Chinese Academy of Sciences, Shanghai, China

[3]Physikalisches Institut, Universität Heidelberg, Germany

We present a compact apparatus of hybrid 2D-MOT and Zeeman slower for ultracold strontium Rydberg atoms. We use four pieces of permanent magnets to create magnetic fields that can both use for 2D-MOT and Zeeman slower. The whole apparatus is relatively small, as shown in Fig. 1. We measure the strontium atomic flux by time-of-flight measurement and get a flux as high as $2\times10^{9}$/s at the temperature of 735 Kelvin.

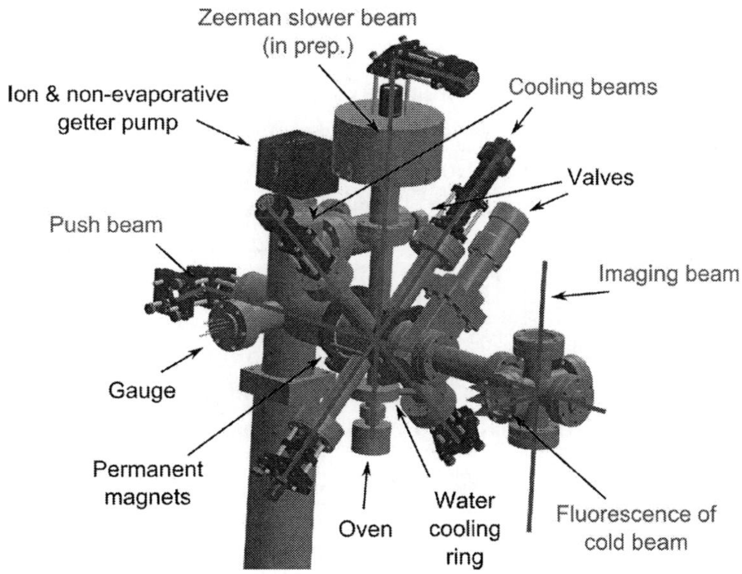


**Figure 1.** The compact apparatus of hybrid 2D-MOT and Zeeman slower for ultracold strontium Rydberg atoms.

[1] G. Lamporesi, S. Donadello, S. Serafini, and G. Ferrari, Rev. Sci. Instrum. **84**, 063102 (2013)
[2] S. Simon, R. Grimm, and F. Schreck, Physical Review A 87, 013611 (2013).
[3] C.S. Hofmann, et al., Frontiers of Physics **9,** 571(2014).

# Rydberg mediated long-range interactions between spatially separated photons

Hannes Busche[1], Paul Huillery[1], Simon W. Ball[1], Teodora V. Ilieva[1], Matthew P. A. Jones[1], and Charles S. Adams[1]

[1]Joint Quantum Centre (JQC) Durham-Newcastle, Department of Physics, Durham University, South Road, Durham, DH1 3LE, United Kingdom

While the lack of interaction between photons is an asset in applications like optical communications, it also makes all-optical processing notoriously difficult, even more so if interactions between spatially separated, non-overlapping optical modes are desired. Rydberg non-linear optics [1] has emerged as a viable approach to induce strong effective interactions between individual photons. Using electromagnetically induced transparency and storage of photons as Rydberg polaritons in cold such that the excitation blockade between Rydberg atoms limits the number of photons stored within a certain volume to a single one which is reflected in the retrieval of highly non-classical states of light. So far, Rydberg-induced photon interactions have been observed between photons propagating in overlapping optical modes inside the same atomic medium. However, thanks to the long-range dipolar character of the interactions, spatial overlap is no fundamental requirement, neither for the photons nor the media. Here, we demonstrate interactions between photons in individual optical modes stored in side-by-side atom clouds which are separated by distances on the order of 10 μm [2]. Analysing the statistics of the retrieved photons, we observe a strong suppression of simultaneous retrieval in both modes as the distance between them is reduced and the interaction strength increases (see figure below). Comparing our results to a Monte-Carlo model, we find that suppression already occurs at separations above the Rydberg blockade as a result of interaction induced dephasing of the polaritons [3]. The demonstration of long-range interactions between photons opens up perspectives amongst others for the implementation of all-optical quantum gates [4] and applications in quantum simulation.

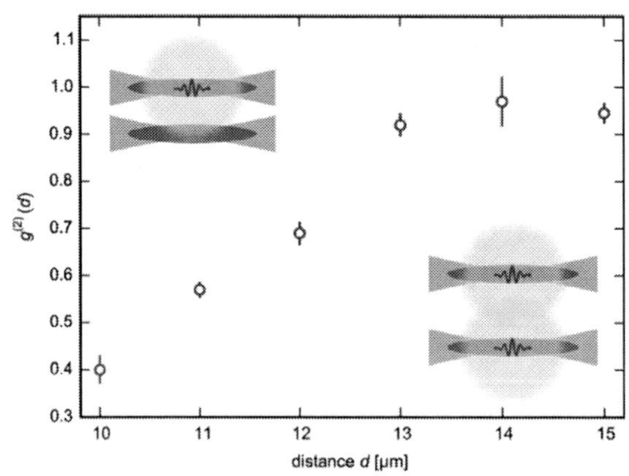


**Figure 1.** Observation of spatial correlations and suppression of simultaneous storage between photons stored as Rydberg polaritons in individual cold atom clouds as a result of Rydberg induced interactions.

[1] O. Firstenberg, C. S. Adams, and S. Hofferberth, arXiv:1602.06117 (2016).
[2] H. Busche *et al.*, in preparation.
[3] F. Bariani *et al.*, Phys. Rev. Lett. **108**, 030501 (2012).
[4] D. Paredes-Barato and C. S. Adams, Phys. Rev. Lett. **112**, 040501 (2014).

# Electromagnetically Induced Transparency in Rydberg gases above atom chips

Bo Zhang, Bo Yang, Zhibin Xu,

Wuhan University of Technology, 122 Ruoshi Road, Hubei, China

Electromagnetically induced transparency(EIT) with Rydberg atoms provides a promising way for quantum information processing and quantum optics. The interactions between Rydberg atoms can be translated into the observable atom-light interactions[1, 2]. We proposed here that EIT with cold Rydberg atoms can be used to detect the Casimir-Polder(CP) effect between Rydberg atoms and chip surfaces. We used kinetic Monte Carlo method[3] to model the Electromagnetically induced transparency in cold Rydberg gases with different atomic density at different heights above atom chips. Compared to the simulation with reduced density matrix, our approach may give the temporal evolution of Rydberg ensemble. We studied the impacts of van der Waals(vdW) interaction between Rydberg atoms and of the Casimir-Polder effect between Rydberg atoms and chip surfaces on the dissipative photon-photon interactions. We computed and plotted the probe beam absorption spectra for a Rubidium Rydberg ensemble with different densities in Fig.1. The black line(full circles) takes no account for either van der Waals interaction or Casimir-Polder effect, which corresponds to low Rydberg atomic density and large atom-surface distance. The Purple line(empty circles) considers only the Casimir-Polder effect; the blue line(triangles) consider only the van der Waals interaction; the red line(crosses) considers both. Compared those lines, we see that Rydberg-Rydberg atom interactions and Rydberg atom-surface interactions may shift the resonant frequency. When the Rydberg atomic density is low enough, the resonant shift is dominant by the Casimir-Polder effect.

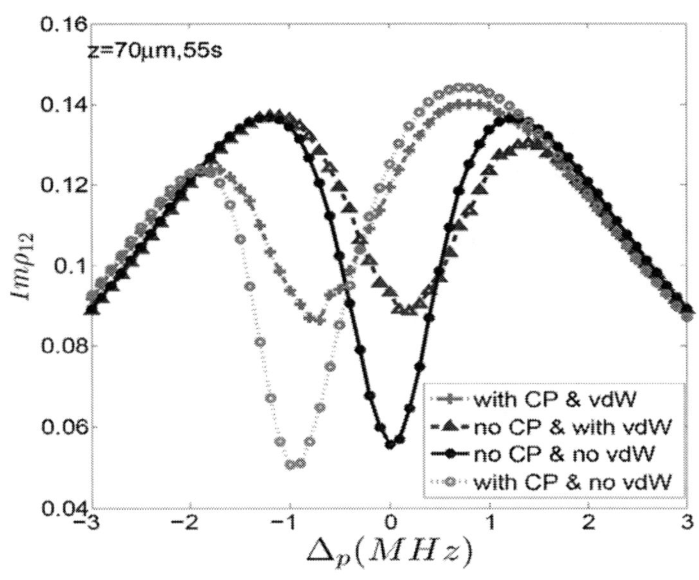


**Figure 1.** Absorption spectra of probe beam for a Rubidium Rydberg ensemble at different densities. The atoms are resonantly excited to 55S Rydberg state. The probe and coupling beams have Rabi frequencies of $\Omega_1$= 1 MHz and $\Omega_2$= 2 MHz, respectively. The line widths of both are 0.1MHz. The atom-surface distance is 70μm.

[1] C. Ates, S. Sevingli, T. Pohl, Phys. Rev. A, **83**, 041802 (2011).

[2] D. Petrosyan, J. Otterbach, M. Fleischhauer, Phys. Rev. Lett. **107**, 2013601 (2011).

[3]A. Chotia, M. Viteau, T. Vogt, D. Comparat, P. Pillet, New J. Phys. **10**, 045031(2008).

# Three-body resonant energy transfer between Rydberg atoms

Patrick Cheinet[1], Riccardo Faoro[1,2], Alexandre Zuliani[1], Bruno Pelle[1], Ennio Arimondo[2] and Pierre Pillet[1]

[1]Laboratoire Aimé Cotton, CNRS/ENS Cachan/Université Paris Sud, Orsay France

[2]INO-CNR, Universita di Pisa, Italy

I will present a generalization of Förster resonances in Rydberg atoms to few-body resonances that we have demonstrated in [1]. This generalization is valid for any atom and any system displaying 2-body Förster resonances as long as the starting state is close to degeneracy with other states. We find this condition in cesium atoms with the previously known Förster resonance:

$$2 \times np_{3/2} \rightarrow ns + (n+1)s \qquad \text{Eq. (1)}$$

where $n$ is the principal quantum number, $s$ and $p$ represent respectively the orbital momentum 0 and 1 and the index 3/2 explicits the total angular momentum J of the state. In an electric field, the degeneracy between $m_J$=1/2 and $m_J$=3/2 is lifted and one can find the 3-body resonance of the following equation 2 at a specific electric field, separated from the resonance field of equation 1.

$$2 \times np_{3/2}m_{3/2} \rightarrow ns + (n+1)s + np_{3/2}m_{1/2} \qquad \text{Eq. (2)}$$

Figure 1 displays one of our measurements on cesium atoms with $n$=35 where we can see the 2-body resonance at 3.6 V cm$^{-1}$ and the 3-body resonance at 3.8 V cm$^{-1}$. Such few-body processes are interesting for quantum information processing as they could lead to few-body quantum gates.

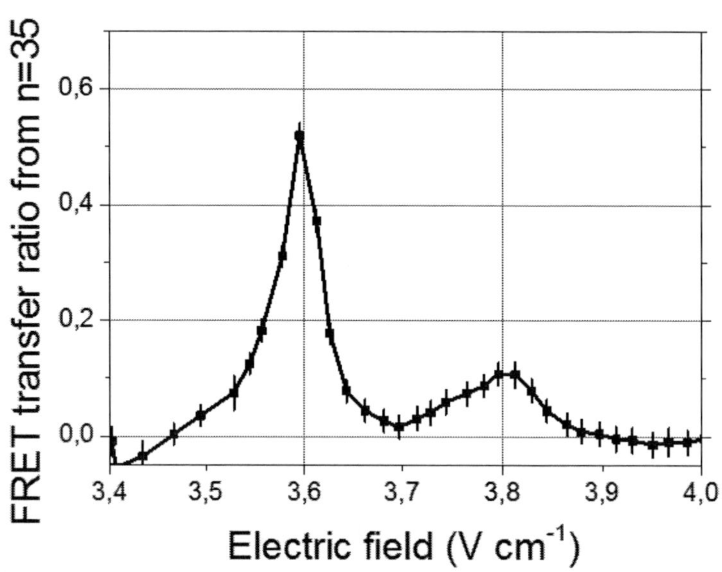


**Figure 1.** Observed transfer ratio with cesium atoms in the $35p_{3/2}$, m=3/2 state of cesium atoms to 35$s$ and 36$s$ states close to the 2-body resonance of eq. 1 at 3.6 V cm$^{-1}$. The 3-body resonance displays a well resolved peak at 3.8 V cm$^{-1}$.

[1] R. Faoro et al., Nature Communications **6**, 8173 (2015).

# Thursday Posters

# Elimination of Dick noise utilizing two cold-atom ensembles

N. Hinkley[1,2], M. Schioppo[1,2,3], R. C. Brown[1], W. F. McGrew[1,2], R. J. Fasano[1,2], K. Beloy[1], T. H. Yoon[1,4], G. Milani[1,5,6], D. Nicolodi[1], N. B. Phillips[1], C. W. Oates[1] and A. D. Ludlow[1]

[1]National Institute of Standards and Technology, 325 Broadway, Boulder, Colorado 80305, USA

[2]Department of Physics, University of Colorado, Boulder, CO 80309, USA.

[3]Institute für Experimentalphysik, Heinrich-Heine-Universität Düsseldorf, Germany

[4]Department of Physics, Korea University, 145 Anam-ro, Seongbuk-gu, Seoul 02841, Korea

[5]Instituto Nazionale di Ricerca Metrologica, Strada delle Cacce 91, 10135 Torino, Italy

[6]Politecnico di Torino, Corso duca degli Abruzzi 24, 10125 Torino, Italy

Optical clocks have emerged as the most stable timekeepers, yet remain far from their measurement potential. The key to realizing the full precision of these optical clocks, given by quantum projection noise from atomic state detection, is the mitigation of a technical noise source known as the Dick effect, arising from aliasing of laser frequency noise [1]. To eliminate Dick noise we implement a zero-dead-time clock architecture based on two optical lattice clocks providing quasi-continuous monitoring of the interrogation laser frequency. Zero-dead-time operation consists of engineering a unity spectroscopic sensitivity function, to eliminate the aliased noise at the heart of the Dick effect. We demonstrate this technique in an ytterbium optical lattice clock, leading to unprecedented clock instability of $6\text{x}10^{-17}/\sqrt{\tau}$ for an averaging time $\tau$ in seconds. With the Dick noise reduced, we assess other clock stability contributors from quantum projection noise, photon shot noise, and other technical noise stemming from measurement of the atomic system. Finally, we experimentally explore an alternative two-atomic-ensemble configuration where one clock provides short-term laser frequency corrections, enabling the second clock to interrogate a cold atom ensemble over longer time scales to realize higher clock quality factors and thus improved clock stability [2]. In a proof-of-concept demonstration of this technique, we observe persistent 120 mHz linewidth, Fourier-limited optical spectroscopic features.

[1] Dick, G. J. Proc. Precise Time and Time Interval Meeting 133-147 (1987).

[2] Kohlhaas, R., et al. Physical Review X **5**, 021011 (2015).

# Inner-shell magnetic dipole transition in Tm atom for optical lattice clock

N. Kolachevsky[1,2], D. Sukachev [1,2,3], S. Fedorov[4], I. Tolstikhina[1], E. Kalganova[1,2], G. Vishnyakova[1,2], K. Khabarova[1, 2], A. Golovizin[1, 2], V. Sorokin[1,2]

[1]P.N Lebedev Physical Institute, 119991, Moscow, Russia

[2]Russian Quantum Center, 143025, Skolkovo, Russia

[3]Harvard University, Physics Department, 17 Oxford str., Cambridge MA, 02138 USA

[4]Ecole polytechnique federale de Lausanne, Route Cantonale, 1015 Lausanne, Switzerland

We consider the possibility to use the inner-shell transition [Xe]$4f^{13}6s^2$ ($J$ = 7/2, $F$ = 4, $m$ = 0) → [Xe]$4f^{13}6s^2$ ($J$ = 5/2, $F$ = 3, $m$ = 0) in the Tm atom at $\lambda$ = 1.14 μm as a candidate for an optical lattice clock. The transition wavelengths and probabilities for two clock levels $J$ = 7/2 and $J$ = 5/2 in the spectral range 250 - 1200 nm are calculated using the COWAN package which allows deducing of the differential dynamic polarizability and suggests that the magic wavelength is at around 807 nm with an attractive optical potential.

Calculations show a reasonable correspondence with existing experimental data and significantly extend it to the UV and IR spectral ranges. The suggested clock transition demonstrates a low sensitivity to the BBR shift which provides a clock frequency instability at the low $10^{-18}$ level competing with the best known optical clocks. We also evaluated other feasible contributions to clock performance (magnetic interactions, light shifts, van der Waals, and quadrupole shifts) which, after reasonable assumptions, can be lowered to the $10^{-18}$ level. Together with the relative simplicity of laser cooling and trapping Tm atoms, our results demonstrate that Tm is a promising candidate for optical clock applications [1]. One of the disadvantages is the relatively low carrier frequency of only $2.6\times10^{14}$ Hz which requires longer integration time to reach the same instability as Sr and Yb lattice clocks.

Our experiments with direct excitation of the clock transition by spectrally narrow laser radiation at $\lambda$=1.14 μm set a lower limit for the upper clock level lifetime of 112 ms which corresponds to the natural linewidth of < 1.4 Hz. Experiments were done in a 1D optical lattice at 532 nm. Modulating the trap depth and analyzing the corresponding parametric resonances frequencies, we deduced the scalar polarizability of the Tm ground state at 532 nm which shows reasonable agreement with our calculations.

To experimentally study the magic wavelength and analyze systematic shifts, we plan to change the trapping wavelength to 806 - 807 nm using a tunable Ti:sapphire laser. This will also simplify our study of Feshbach resonances and may open a way to study and control dipole-dipole interactions using narrow band excitation of the clock transition at 1.14 nm.

[1] D. Sukachev *et al.*, arXiv:1605.09032

# Development of cryogenic Sr optical lattice clocks and their applications

Masao Takamoto[1,2,4], Ichiro Ushijima[1,2,4], Manoj Das[1,2,4], Tetsushi Takano[3,4], Noriaki Ohmae[1,3,4], Tomoya Akatsuka[1,3,4], Atsushi Yamaguchi[1,2,4], Hidetoshi Katori[1,2,3,4]

[1]Quantum Metrology Laboratory, RIKEN, Wako-shi, Saitama, Japan

[2]RIKEN Center for Advanced Photonics, RIKEN, Wako-shi, Saitama, Japan

[3]Department of Applied Physics, Graduate School of Engineering, The University of Tokyo, Bunkyo-ku, Tokyo, Japan

[4]Innovative Space-Time Project, ERATO, Japan Science and Technology Agency, Bunkyo-ku, Tokyo, Japan

The recent progress of optical lattice clocks has shown their ability to reach a fractional accuracy and stability at the $10^{-18}$ level, which surpasses the present realization of the SI second based on Cs clocks by a few orders of magnitude and has triggered the redefinition of the second. Towards achieving a fractional uncertainty of $10^{-18}$, a formidable challenge was how to address the uncertainty of frequency shift caused by the blackbody radiation, which is emitted from the ambient room-temperature environment. We have developed a cryogenic Sr optical lattice clock that operates in a low-temperature environment, and succeeded in reducing the fractional uncertainties of blackbody radiation shift down to $9 \times 10^{-19}$ [1]. Synchronous comparison method [2] precisely evaluates the frequency difference between two clocks in a short averaging time, while removing the degradation of the stability caused by the phase noise of the probe laser. By comparing two clocks precisely, we have investigated the blackbody radiation shift directly and verified the agreement of two clocks with an uncertainty of $10^{-18}$ [1].

Along with further improvement of the performance, it will be of importance to find applications of clocks even beyond frequency standard. High-accuracy clocks allow precision measurement of gravitational potential through the general relativity. The optical lattice clock, if available in the field rather than a laboratory, becomes a measurement tool of local gravitational potential. As a first step towards such applications, we have demonstrated precise measurement of gravitational potential difference by comparing remote clocks located 15 km apart. Using an optical fiber link between RIKEN and the University of Tokyo [3], we performed a synchronous comparison of cryo-clocks developed in both sites. The fractional frequency difference between the clocks were determined with the uncertainty of $5.9 \times 10^{-18}$, which corresponds to the measurement uncertainty for an elevation difference of 5 cm. As compared with spirit levelling method, chronometric levelling allows monitoring of geopotential variation in a shorter timescale, which may provide a new tool for geophysics.

The uncertainties of cryogenic Sr optical lattice clocks are now dominated by the uncertain ties related to the higher-order light shift of the lattice potential [4]. We will also present the progress on the evaluation of lattice-induced light shifts towards uncertainties below $10^{-18}$.

[1] I. Ushijima, T. Takamoto, M. Das, T. Ohkubo and H. Katori, Nature Photon. **9**, 185 (2015).

[2] M. Takamoto, T. Takano and H. Katori, Nature Photon. **5**, 288 (2011).

[3] T. Akatsuka, H. Ono, S. Hayashida, K. Araki, M. Takamoto, T. Takano and H. Katori, JJAP **53**, 032801 (2014).

[4] H. Katori, V. D. Ovsiannikov, S. I. Marmo and V. G. Palchikov, Phys. Rev. A **91**, 052503 (2015).

# Generation of a time scale steered by an optical clock

Tetsuya Ido, Hidekazu Hachisu, Fumimaru Nakagawa, and Yuko Hanado
National Institute of Information and Communications Technology
Koganei, Tokyo 184-8795, Japan

Realization of a time scale is one of the most important applications of optical atomic clocks, where stable oscillators without phase jumps are required. Considering this aspect, hydrogen masers (HM) are currently used as source oscillators for national time scales, and it is still difficult for optical oscillators to rival HMs in the balance of reliability and frequency stabilities. Optical clocks may play a role of the standards for the time being, to which the time scales refer in order to adjust their scale intervals. In this case, the benefit of using optical frequency standards would be the capability to evaluate the HM frequency more quickly. Then, we do not need to operate optical clocks continuously. One optical clock will be utilized not only for time keeping but also for other scientific applications from time to time.

To investigate such possibility of "optical" steering, we first evaluated the behavior of an HM over a few months with reference to a $^{87}$Sr lattice clock. The lattice clock was operated for 10000 − 24000 s per day. Three sets of five-day intermittent measurements and a one-day measurement were distributed in two months with a separation of ~10 days. The HM is a part of the Japan Standard Time (JST) system and is linked to the Universal Coordinated Time (UTC) via UTC(NICT). The evaluations referring to an optical clock clearly demonstrated a stable linear drift of HM frequency, indicating the capability of compensating the drift. This prospect was supported by a numerical simulation based on the record of HM − UTC(NICT) − UTC link. We assumed that the HM frequency and its linear drift rate are predicted relying on the optical calibration performed in past 45 days. Referring to the time differences of the HM from UTC(NICT), it was figured out that the difference between UTC and the "optically steered" time scale would be maintained below 3 ns, where the instability is $1.3\times10^{-15}$ for five days.

Following this promising simulation, we are currently continuing to intermittently operate the lattice clock for three months, and generated a real signal of a time scale steered by the calibrations provided by the lattice clock. The result of the comparison against other time scales such as UTC will be presented in the contribution.

# Manipulating cold $^{171}$Yb atoms in optical lattice clocks

Min Zhou, Xiaohang Zhang, Qi Gao, Chengyin Han, Yilin Xu, Peng Xu, Shangyan Li, Shuang Zhang, and Xinye Xu

State Key Laboratory of Precision Spectroscopy and Department of Physics, East China Normal University, Shanghai 200062, China

Optical lattice clocks embracing cold $^{171}$Yb atoms are promising frequency standards. Two $^{171}$Yb optical lattice clocks are being developed at East China Normal University (ECNU) [1-3]. Since the clock laser was locked to the atomic transition and initial measurements of its frequency stability were made, we have been working to reduce various systematic uncertainties and improve the clock stability by manipulating the cold $^{171}$Yb atoms. The frequency of one lattice laser at 759 nm is phase locked to another lattice laser which is pre-stabilized on a Fabry-Perot cavity, helping determine the magic frequency and reduce the lattice ac Stark shift uncertainty down to $10^{-16}$ level. Evaluation of the blackbody radiation shift is also realized. To suppress the collision shift, $^{171}$Yb atoms are efficiently spin polarized using a simple pumping method. Interleaved stabilization on two characteristic $\pi$ transitions, from which the self-comparison of each clock is examined, is performed to remove the first-order Zeeman shift and the residual vector Stark shift. To verify the improved clock stability, we finally implement the direct comparison of two $^{171}$Yb lattice clocks. This work paves the way to the absolute frequency measurement in the near future.

[1] X. Y. Xu, W. L. Wang, Q. H. Zhou, G. H. Li, H. L. Jiang, L. F. Chen, J. Ye, Z. H. Zhou, Y. Cai, H. Y. Tang and M. Zhou, Front. Phys. China, **4**, 160 (2009).

[2] N. Chen, M. Zhou, H. Q. Chen, F. Su, L. Y. Huang, X. H. Zhang, Q. Gao, Y. Y. Jiang, Z. Y. Bi, L. S. Ma and X. Y. Xu, Chin. Phys. B **22**, 090601 (2013).

[3] X. H. Zhang, M. Zhou, N. Chen, Q. Gao, C. Y. Han, Y. Yao, P. Xu, S. Y. Li, Y. L. Xu, Y. Y. Jiang, Z. Y. Bi, L. S. Ma and X. Y. Xu, Laser Phys. Lett. **12**, 025501 (2015).

# Strontium single-ion optical clock at NRC

Pierre Dubé[1], Bin Jian[1,2], Alan A. Madej[1,2]
[1]National Research Council Canada, Ottawa, ON, Canada
[2]Department of Physics and Astronomy, York University, Toronto, ON, Canada

The evaluated uncertainty of optical frequency standards in realizing the unperturbed line center of ultra-narrow resonances began to surpass that of the best realizations of the SI second about 10 years ago [1]. Reduction of this uncertainty in optical clocks has since continued to progress rapidly worldwide with a few optical clocks currently evaluated with fractional uncertainties of $10^{-17}$ to $10^{-18}$ [1-5], one to two orders of magnitude lower than state-of-the-art cesium fountain clocks. Optical clocks are considered serious candidates for an eventual re-definition of the SI second.

The optical clock developed at the National Research Council Canada (NRC) is based on the 445 THz 5s $^2S_{1/2}$-4d $^2D_{5/2}$ electric quadrupole transition of $^{88}Sr^+$. The reference is realized using a single trapped ion, laser cooled to milli-Kelvin kinetic temperatures and held within sub-micron dimensions using a radio-frequency (rf) trap field. During the past decade, the uncertainty in realizing the unperturbed *S-D* transition line center frequency has been reduced to a current level of $1.2\times10^{-17}$ [5].

For $^{88}Sr^+$, the main sources of uncertainty are the micromotion shifts, the electric quadrupole shift, and the blackbody radiation shift. Micromotion is the result of ion displacement from the trapping node where the electric fields vanish. Micromotion causes a second-order Doppler shift and Stark shifts. In our system, micromotion shifts are minimized using two methods. The first is with trim electrodes that center the ion in the trapping potential. The second relies on tuning the rf trap frequency to a special value where the scalar Stark shift and the second-order Doppler shift cancel each other. This property of the clock transition requires a negative value of the differential scalar polarizability, $\Delta\alpha_0$. An additional reduction of the shifts by a factor of 200 or more is obtained by this method, for a net micromotion shift evaluated at the $10^{-19}$ level. The second most important shift is the electric quadrupole shift (EQS), which is nominally at the $10^{-14}$ level. It is canceled by averaging the line centers of six Zeeman components that connect to the six *D*-state magnetic sublevels [6]. In addition, Zeeman averaging cancels tensor Stark shifts. Black body radiation (BBR) causes a scalar Stark shift of +0.2480 Hz at 300 K [5]. The BBR shift uncertainty depends on the accuracy of the $\Delta\alpha_0$ parameter and of the BBR field evaluation. We have measured $\Delta\alpha_0$ to high accuracy. Its contribution to the uncertainty budget is below the $10^{-18}$ level [5]. Evaluation of the BBR field contributes a fractional uncertainty of $\approx1.1\times10^{-17}$, an order of magnitude larger than the next leading uncertainty. Accurate evaluation of the BBR field should bring the total uncertainty of $^{88}Sr^+$ to the low $10^{-18}$ level. The presentation will include an overview of the $^{88}Sr^+$ ion clock developed at NRC, the control of the systematic shifts, as well as improvement of the clock stability to $3\times10^{-15}/\sqrt{\tau(s)}$.

[1] T. Rosenband *et al.*, Science **319**, 1808 (2008).
[2] A. D. Ludlow *et al.*, Rev. Mod. Phys. **87**, 637 (2015).
[3] C.W. Chou *et al.*, Phys. Rev. Lett. **104**, 070802 (2010).
[4] N. Huntemann *et al.*, Phys. Rev. Lett. **116**, 063001 (2016).
[5] P. Dubé *et al.*, Phys. Rev. Lett. **112**, 173002 (2014).
[6] P. Dubé *et al.*, Phys. Rev. Lett. **95**, 033001 (2005).

# Narrow-line laser cooling of cadmium towards a portable optical lattice clock

Atsushi Yamaguchi[1,2,3], Kurt Gibble[1,4], Hidetoshi Katori[1,2,3,5]

[1]Quantum Metrology Laboratory, RIKEN, Wako-shi, Saitama 351-0198, Japan

[2]Innovative Space-Time Project, ERATO, JST, Bunkyo-ku, Tokyo 113-8656, Japan

[3]RIKEN Center for Advanced Photonics, Wako-shi, Saitama 351-0198, Japan

[4]Department of Physics, The Pennsylvania State University, University Park, Pennsylvania 16802, USA

[5]Department of Applied Physics, Graduate School of Engineering, The University of Tokyo, Bunkyo-ku, Tokyo 113−8656, Japan

The light shift caused by the blackbody radiation (BBR) is a significant source of uncertainty for optical lattice clocks based on strontium and ytterbium atoms. A cryogenic environment is one approach to reduce this systematic frequency shift [1, 2]. Another is to identify and use an atomic species whose clock transition frequency is intrinsically insensitive to the BBR, such as mercury, magnesium and cadmium (Cd) [3-5]. These species avoid the complexity of a cryogenic apparatus, enabling more compact and portable optical lattice clocks.

As a first step towards a portable cadmium optical lattice clock, we demonstrate narrow-line laser cooling of Cd. We first trapped Cd atoms in a MOT using the $^1S_0 - {}^1P_1$ transition at 229 nm (Figure 1) [6]. These atoms were subsequently loaded into a narrow-line $^1S_0 - {}^3P_1$ MOT to further cool the atoms. We demonstrate two-stage cooling of both bosonic ($^{112}$Cd) and fermionic ($^{113}$Cd) isotopes to a temperature of 4.2(9) μK and 5.2(6) μK, respectively. We are preparing spectroscopy of the clock transition in an optical lattice to determine the magic wavelength for Cd. The experiment and future prospects will be discussed.

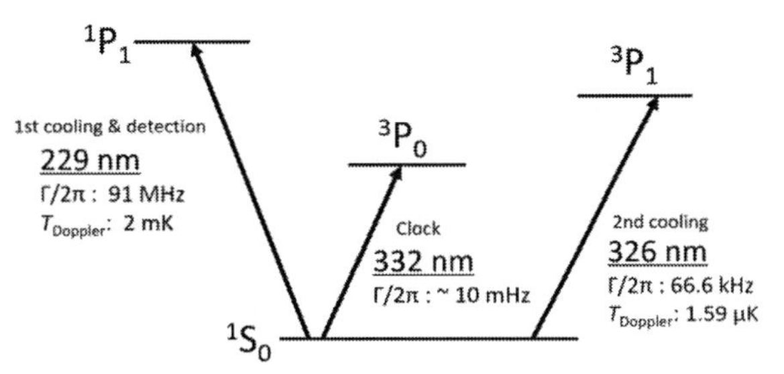


Figure 1 Energy levels of neutral cadmium.

[1] I. Ushijima *et al.*, Nat. Photonics **9**, 185 (2015).

[2] N. Nemitz *et al.*, Nat. Photonics **10**, 258 (2016).

[3] J. J. McFerran et al., Phys. Rev. Lett. **108**, 183004 (2012).

[4] K. Yamanaka *et al.*, Phys. Rev. Lett. **114**, 230801 (2015).

[5] A. P. Kulosa *et al.*, Phys. Rev. Lett. **115**, 240801 (2015).

[6] Y. Kaneda *et al.*, Opt. Lett. **41**, 705 (2016).

# A search for the low-energy $^{229}$Th nuclear isomeric transition Using nuclear resonant scattering

T. Masuda[1], H. Hara[1], T. Hiraki[1], Y. Kasamatsu[2], S. Kitao[3], Y. Kobayashi[3], K. Konashi[4], R. Masuda[3],Y. Miyamoto[1], K. Okai[1], S. Okubo[1], N. Sasao[1], T. Schumm[5], M. Seto[3], Y. Shigekawa[2], S. Stellmer[5], S. Uetake[1], M. Watanabe[4], A. Yamaguchi[6], Y, Yasuda[2], Y. Yoda[7], A. Yoshimi[1], K. Yoshimura[1], M. Yoshimura[1]

[1]Research Institute for Interdisciplinary Science, Okayama University, Okayama , Japan

[2]Department of Chemistry, Osaka University, Osaka, Japan

[3]Research Reactor Institute, Kyoto University, Osaka, Japan

[4]Institute for Materials Research, Tohoku University, Miyagi, Japan

[5]Institute for Atomic and Subatomic Physics, TU Wien, Vienna, Austria

[6]Quantum Metrology Laboratory, RIKEN, Saitama, Japan

[7]Japan Synchrotron Radiation Research Institute, Hyogo, Japan

The nuclear isomeric first excited state of Thorium-229 ($^{229m}$Th) has drawn attention from many fields. In general, nuclei are insensitive to ambient disturbance due to the shielding by their core electrons. In addition, the extraordinary-low energy level of $^{229m}$Th can be accessed by laser spectroscopy techniques. Since its insensitiveness may enable us to achieve longer decoherence time in comparison to atomic states, it attracts much interest from atomic clock, fundamental physics, and other fields. However large energy uncertainty of 0.5 eV makes the laser spectroscopy difficult. In spite of much effort, transition between the isomer state and the ground state has not been observed directly by optical methods.

Our experiment aims to determine the energy of the isomeric transition precisely by nuclear resonant scattering (NRS) with X-ray beam. Our basic strategy is as follows (see Fig. 1): (1) irradiating a $^{229}$Th target with 29.2-keV X-ray and exciting it to the rotational excitation level ($5/2^+$); (2) detecting the NRS signal of 29.2 keV with the estimated life time of ~0.2 ns to confirm the population transfer from ($5/2^+$) level to the low-energy isomeric state; (3) detecting the emitted 7.8-eV photons by photomultiplier tubes and a monochromator. An advantage of this method is that the wavelengths of the signal (7.8 eV) and excitation beam (29.2 keV) are quite different; therefore, the irradiated photons do not become background. We have performed several studies at a SPring-8 beam line, and attempted to detect the NRS signal and 7.8-eV signal. This poster presents the current status of the experiment.

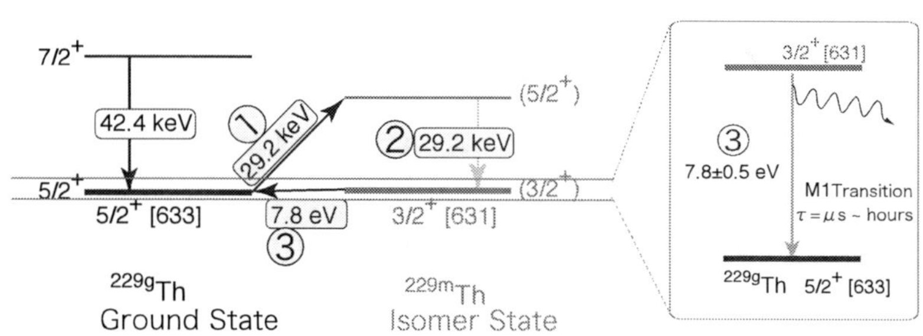


**Figure 1.** $^{229}$Th energy level and the transition scheme of the experiment.

# Optical frequency ratio between a stationary Yb lattice clock and a transportable Sr lattice clock

P. Thoumany[1], M. Pizzocaro[1], B. Rauf[1,2], F. Bregolin[1,2], J. Grotti[3], St. Vogt[3], S. Koller[3], A. Tampellini[1,2], P. Barbieri[1,2], C. Clivati[1], F. Levi[1], Ch. Lisdat[3], D. Calonico[1]

[1]Instituto Nationale di Ricerca Metrologica, strada delle cace 91, 10135 Turin, Italy
[2]Politecnico di Torino, Corso Duca degli Abruzzi 24, 10129 Turin, Italy
[3]Physikalisch-Technische Bundesanstalt (PTB), Bundesallee 100, 38116 Braunschweig, Germany

Optical clocks based on neutral atoms [1] and ions [2] are now reaching uncertainties below the most accurate microwave Cs primary standards [3] and frequency ratios between optical clocks [4] are now required to test and verify their consistency and overall accuracy. This is an essential step towards a redefinition of the second and testing fundamental law of physics. Optical lattice clocks based on ultracold neutral atoms trapped in an optical lattice operated at the Stark shift cancellation wavelength are among the most accurate clocks to date by combining the advantage of the Lamb-Dicke regime spectroscopy together with the possibility to probe a large number of quantum absorbers simultaneously. In the framework of the EMRP project ITOC, a transportable $^{87}$Sr optical lattice clock from PTB, Germany, has been moved to INRIM, Italy, where a stationary $^{171}$Yb optical lattice clock is operated. Here we report about the frequency ratio between the two clocks.

[1] H. Katori, Nature Photonics **5**, 203 (2011).
[2] C.W. Chou *et al.*, Phys. Rev. Lett. **104**, 070802 (2010).
[3] F. Levi *et al.*, Metrologia, vol. **51**, no.3, p. 270 (2014).
[4] N. Nemitz *et al.*, Nature Photonics **10**, 258 (2016).

# Absolute Frequency Measurement of the $^1S_0 \rightarrow {}^3P_0$ Transition in an $^{171}$Yb based Optical Lattice Clock

P. Thoumany[1], M. Pizzocaro[1], B. Rauf[1,2], F. Bregolin[1,2], G. Milani[1,2], M. Gozzelino[1,2], C. Clivati[1], G. A. Costanzo[1,2], F. Levi[1], D. Calonico[1]
[1]Instituto Nationale di Ricerca Metrologica, strada delle cace 91, 10135 Turin, Italy
[2]Politecnico di Torino, Corso Duca degli Abruzzi 24, 10129 Turin, Italy

Optical lattice clocks [1,2] based on ultracold neutral atoms trapped in an optical lattice operated at the Stark shift cancellation wavelength are among the most accurate clocks to date by combining the advantage of the Lamb-Dicke regime spectroscopy together with the possibility to probe a large number of quantum absorbers simultaneously and are regarded as promising candidates for a redefinition of the second and the test of fundamental physics with higher accuracy. Among the different atomic species considered for optical clocks, neutral Yb presents several favorable atomic properties and the narrow ($\Gamma$= 10 mHz), spin- and orbital-momentum forbidden $^1S_0 \rightarrow {}^3P_0$ atomic transition has been recognized by the BIPM as a secondary representation of the second [3]. We report on the full metrological characterization of our $^{171}$Yb optical lattice clock and a first absolute frequency measurement at the accuracy level of our cryogenic Cs fountain [4].

[1] H. Katori, Nature Photonics **5**, 203 (2011).

[2] N. Hinkley *et al.*, Science, vol. **341**, p.1215 (2013).

[3] CIPM Recommendation 1(CI-2013): Updates to the list of standard frequencies.

[4] F. Levi *et al.*, Metrologia, vol. **51**, no.3, p. 270 (2014).

# Single-ion spectroscopy of two quadrupole transitions in ytterbium ions towards search for temporal variation of the fine structure constant

Yasutaka Imai, Kazuhiko Sugiyama, Ren Irie, and Masao Kitano
Kyoto daigaku-katsura, Nishikyo-ku, Kyoto, Japan

We are developing optical clocks based on single $^{171}Yb^{+}$. $^{171}Yb^{+}$ has the $m_F$=0-$m_{F'}$=0 clock transitions, which are not disturbed by the 1st-order Zeeman shifts in low magnetic field. $^{171}Yb^{+}$ has relatively simple hyperfine structures owing to the nuclear spin of 1/2. $^{171}Yb^{+}$ has three clock transitions: the $^2S_{1/2}$-$^2D_{3/2}$ and the $^2S_{1/2}$-$^2D_{5/2}$ quadrupole transitions, and the $^2S_{1/2}$-$^2F_{7/2}$ octupole transition. Owing to large differences in the sensitivities to variation of the fine structure constant $\alpha$ between the octupole and the quadrupole transitions, one can search for a temporal variation of $\alpha$ by using only $^{171}Yb^{+}$ [1, 2]. The two quadrupole transitions have similar sensitivities, and therefore a potential for exploring other effects which may cause temporal variations.

We conducted spectroscopy of the $m_F$=0-$m_{F'}$=0 component in the $^2S_{1/2}(F=0)$-$^2D_{3/2}(F=2)$ transition in single $^{171}Yb^{+}$. After detection of resolved motional sidebands, we probed the carrier spectrum in higher resolution of 20-Hz intervals. In order to cancel the linear frequency drift of the clock laser, we detected the carrier spectrum twice by changing the sweep direction of the clock-laser frequency. After cancellation of the drift, we obtained the carrier spectrum of the full width at half maximum of 380 Hz. This is limited mainly by the linewidth of the clock laser. For the $^2S_{1/2}$-$^2D_{5/2}$ transition, we so far conducted single-ion spectroscopy by using single $^{174}Yb^{+}$. We resolved motional sidebands of the $m_j$=-1/2-$m_{j'}$=-5/2 component and probed the carrier spectrum as shown in Fig. 1. The full width at half maximum of the carrier spectrum was 2.5 kHz.

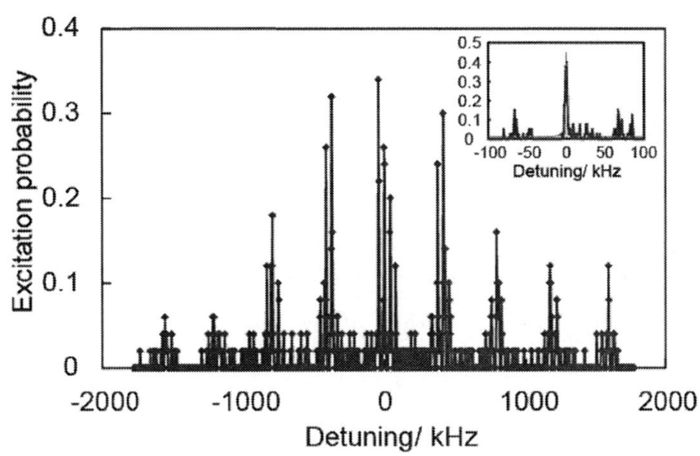


**Figure 1**. Spectrum of the $^2S_{1/2}(m_j=-1/2)$-$^2D_{5/2}(m_{j'}=-5/2)$ transition in single $^{174}Yb^{+}$. The interval of the frequency sweep is 4 kHz. Inset shows a carrier spectrum detected by 1-kHz intervals.

[1] R. M. Godun *et al.*, Phys. Rev. Lett. **113**, 210801 (2014).

[2] N. Huntemann *et al.*, Phys. Rev. Lett. **113**, 210802 (2014).

# Towards neutral mercury lattice clock in SIOM

Zhen Xu[1,2,3,4], Jian-Fang Sun[1,2], Kang-Kang Liu[1,2,3], Xaio-Hu Fu[1,2,3], Ru-Chen Zhao[1,2,3], Wei Gou[1,2,3], Wen-Jie Li[1,2,3], Yu-Zhen Wang[1,2,3,4]

[1]Key Laboratory of Quantum Optics, Chinese Academy of Sciences, Shanghai 201800, China

[2]Shanghai Institute of Optics and Fine Mechanics, Chinese Academy of Sciences, Shanghai 201800, China

[3]University of Chinese Academy of Sciences, Beijing 100049, China

[4]Center of Cold atom physics, Chinese Academy of Sciences, Shanghai 201800, China

Neutral mercury atom holds a large atomic number and it is insensitive to black body radiation in the room temperature [1], which is a great advantage for neutral mercury lattice clock [2]. Besides, Hg is also a good candidate in the search for variation of natural constants because of its high sensitivity to variations of the fine structure constant α [3]. However, one of the challenges for the neutral mercury lattice clock is the requirement of high power deep-UV laser sources for both the cooling laser and the lattice laser. Here we introduce our recent progress of neutral mercury lattice clock in SIOM.

An ultra-high vacuum system had been built to produce ultracold mercury atoms in a single chamber with a mercury source which is cooled to about -70°C. The frequency quadrupled semiconductor deep-UV laser is locked on $^1S_0$-$^3P_1$ transition (253.7 nm) by sub-Doppler frequency modulation spectroscopy. The ultracold mercury atoms were produced by a folded beam configuration magneto-optical trap (MOT) because of the shortage of laser power, and observed by fluorescence detection [4]. The temperature was measured by time of flight method. All of six rich abundant isotopes have been observed, and the atom number is about $1.7\times10^6$ for $^{202}$Hg and $1.5\times10^6$ for $^{199}$Hg. The temperature is about 170 μK for $^{202}$Hg and 50 μK for $^{199}$Hg.

To enhance the power of cooling laser, a room temperature 1014.8 nm fiber laser amplifier was developed [5]. After two stages of frequency doubling, about 75 mW of 253.7 nm UV laser were generated at infrared input power of 4 W. Recently, we have designed and constructed a homemade ECDL seed laser and two cascade efficient frequency doubling cavity to enhance the UV power. About 3 W at 507.4 nm and 500 mW at 253.7 nm laser power were generated. More power of UV cooling laser could help us to trap more atoms in the future.

These works laid a good foundation to realize the mercury lattice clock in SIOM. Now, an ultra-stable laser system with a linewidth of ~ 1 Hz is also under construction to further realize the clock laser at 265.6 nm.

[1] H. Hachisu, K. Miyagishi, S. G. Porsev, *et al.*, Phys. Rev. Lett. **100**, 053001 (2008).

[2] J. J. McFerran, L. Yi, S. Mejri, *et al.*, Phys. Rev. Lett. **108**, 183004 (2012).

[3] E. J, Angstmann, V. A. Dzuba, and V. V. Flambaum, Phy. Rev. A **70**, 014102 (2004).

[4] H.-L. Liu, S.-Q. Yin, K.-K. Liu, *et al.*, Chinese Physics B **22**, 043701 (2013).

[5] J.-M. Hu, L. Zhang, H.-L. Liu, *et al.*, Optics Express **21**, 30958 (2013).

# Improved limits on variations of the fine structure constant from the frequency ratio of a $^{171}Yb^{+}$ single ion clock and a $^{87}Sr$ lattice clock

Ekkehard Peik, Nils Huntemann, Burghard Lipphardt, Christian Sanner, Christian Tamm, Sören Dörscher, Ali Al-Masoudi, Stephan Falke, Nathan Lemke, Sebastian Häfner, Christian Grebing, Uwe Sterr, Christian Lisdat

Physikalisch-Technische Bundesanstalt, 38116 Braunschweig, Germany

We report on a direct comparison of two very different optical clocks. One is based on the $^2S_{1/2} - {}^2F_{7/2}$ electric octupole (E3) transition of a single $^{171}Yb^{+}$ ion stored in a radio-frequency Paul trap and the other clock realizes the $^1S_0 - {}^3P_0$ transition of many $^{87}Sr$ atoms confined in an optical lattice near the magic wavelength. The lattice clock achieves frequency instabilities smaller than $2 \times 10^{-16}/(\tau / s)^{1/2}$ [1] and a total systematic uncertainty of $2 \times 10^{-17}$ [2], while the ion clock is evaluated with a systematic uncertainty of $3.2 \times 10^{-18}$ [3]. More than eighty hours of acquired data result in a statistical uncertainty of $1.3 \times 10^{-17}$, and yield the frequency ratio of the two clocks with a total fractional uncertainty of $2.4 \times 10^{-17}$. This is the smallest uncertainty achieved for clocks of different type to date and significantly improves the knowledge of the ratio of these two secondary representations of the second. The large sensitivity of the E3 transition frequency to the value of the fine structure constant $\alpha$ makes the experiment suitable for a search for temporal variations of $\alpha$. With data from this recent measurement and a similar one performed two-and-a-half years earlier, we can provide the most stringent limit on a potential drift from laboratory experiments: $d(\ln \alpha)/dt = -6.5(4.7) \times 10^{-18}$/yr.

[1] A. Al-Masoudi, S. Dörscher, S. Häfner, U. Sterr, C. Lisdat, "Noise and instability of an optical lattice clock", Phys. Rev. A **92**, 063814 (2015).

[2] C. Lisdat, et al., "A clock network for geodesy and fundamental science", arXiv: 1511.07735, 2015.

[3] N. Huntemann, C. Sanner, B. Lipphardt, Chr. Tamm, E. Peik, "Single-Ion Atomic Clock with $3 \times 10^{-18}$ Systematic Uncertainty", Phys. Rev. Lett. **116**, 063001 (2016).

# Investigating lattice-light induced frequency shifts in an ytterbium optical lattice clock

Nils Nemitz[1], Asbjørn Jørgensen[2], Ryotatsu Yanagimoto[3], Hidetoshi Katori[1,2,3,4]
[1]RIKEN Center for Advanced Photonics, Wako, Japan
[2]Quantum Metrology Laboratory, RIKEN, Wako, Japan
[3]Department of Applied Physics, Graduate School of Engineering, The University of Tokyo, Japan
[4]Innovative Space-Time Project, ERATO, Japan Science and Technology Agency, Tokyo, Japan

In our previous measurement of the Yb/Sr frequency ratio [1], the overall uncertainty of $4.6\times10^{-17}$ was dominated by the residual lattice-light induced frequency shifts in the Yb clock. A further reduction of these uncertainties requires a better understanding of hyperpolarizability and multipolar effects [2]. We have recently equipped the Yb clock with additional optics that allow for a larger lattice intensity. In conjunction with control of the vibrational state through efficient sideband cooling, the increased range of trap depths will make these effects experimentally accessible. We will present the progress of our measurements that aim towards reducing the associated uncertainties to the $10^{-18}$ range.

**Table 1.** Current uncertainty budget (in fractional units of $10^{-18}$) for RIKEN's cryogenic Yb and Sr clocks. Addressing the lattice-light induced shifts in the Yb clock will provide a significant reduction of the overall uncertainty.

| | **$^{171}$Yb clock** | | **$^{87}$Sr clock** | |
|---|---|---|---|---|
| **Systematic effect** | **correction** | **uncert.** | **correction** | **uncert.** |
| Lattice light shift | 8.5 | 32.8 | 3.5 | 3.4 |
| Quadratic Zeeman shift | 67.7 | 9.8 | 54.2 | 0.9 |
| Collisional shift | 0.0 | 3.4 | 0.9 | 4.2 |
| Probe light shift | -0.8 | 3.2 | 0.09 | 0.05 |
| 1$^{st}$ order Doppler shift | 0.0 | 2.0 | 0.0 | 0.5 |
| AOM chirp | 0.0 | 1.1 | 0.0 | 0.2 |
| Servo error | 0.8 | 1.1 | 1.9 | 1.6 |
| BBR shift | 27.5 | 0.7 | 54.2 | 0.9 |
| **Total** | **103.7** | **34.7** | **177.6** | **5.8** |

[1] N. Nemitz, T. Ohkubo, M. Takamoto, I. Ushijima, M. Das, N. Ohmae, H. Katori, Nature Photon. **10**, 258 (2016).
[2] V. D. Ovsiannikov, S. I. Marmo, V. G. Palchikov and H. Katori, Phys. Rev. A **93**, 043420 (2016).

# Towards an optical characterization of the Th-229 nuclear isomer

Simon Stellmer[1], Matthias Schreitl[1], Georgy Kazakov[1], Koji Yoshimura[2], Thorsten Schumm[1]

[1]Vienna University of Technology and VCQ, Stadionallee 2, 1020 Vienna, Austria

[2]Okayama University, Tsushima-naka 3-1-1 Kita-ku, Okayama 700-8530, Japan

The iconic nucleus of the isotope Th-229 is infamous for two reasons: (1) It possesses a long-lived metastable state (called a nuclear isomer) with an exceptionally low energy of only a few eV, and (2) this isomer has resisted all attempts of a direct detection for over 40 years.

Among the 3000+ known isotopes, featuring some 180.000 charted nuclear levels, the Th-229 isomer stands out as a true freak of Nature. With a midget energy of only about 7 eV [1], compared to typical nuclear energies of a few MeV, and a lifetime of about an hour, it might be amenable to laser excitation. Transferring the well-established toolbox of optical laser spectroscopy to a nuclear system has sparked a wealth of ideas for future applications, among them development of optical clocks and gamma-ray lasers [2, 3]. The great enthusiasm to harness this nuclear two-level system is offset by the isomer's persistent habit to evade direct detection and a measurement of its energy or lifetime.

The Vienna group uses thorium-doped $CaF_2$ crystals as a platform for optical spectroscopy. Three different approaches are used to excite the isomer: direct excitation with undulator radiation in the VUV (experiments performed at the MLS synchrotron in Berlin, Germany) [4], "optical" pumping with synchrotron radiation at 29-keV (SPring-8 facility, Japan), and passive population of the isomer through alpha decay of U-233 [5]. In all of these experiments, we seek to measure the wavelength of the VUV gamma that is emitted as the isomer de-excites into the nuclear ground state. If successful, such a measurement will reduce the present uncertainty in the isomer energy (about 1 eV, corresponding to 20 nm) to about 0.1 nm, which would then allow for direct laser spectroscopy.

The nuClock project, established in summer 2015 as a pan-European initiative, aims to boost the development of a nuclear clock based on Th-229. We will introduce the consortium and highlight recent developments and invite you to visit www.nuClock.eu for more information.

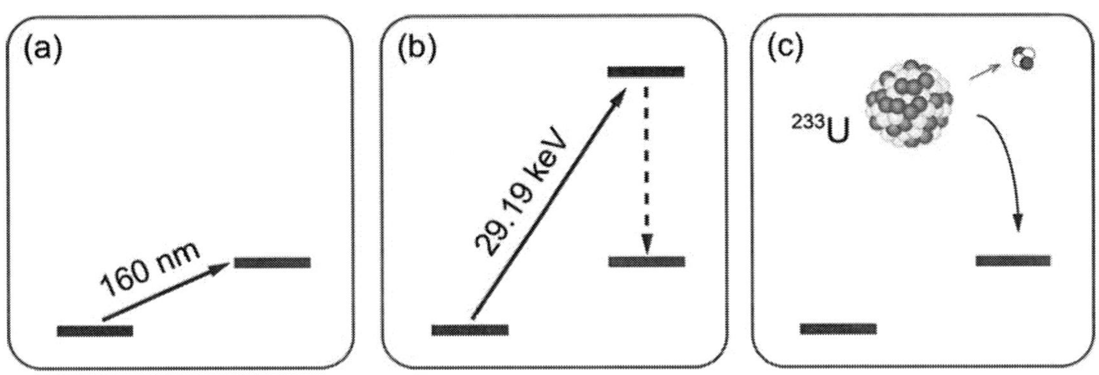


**Figure 1.** Three approaches to populate the isomeric state: (a) direct excitation around 160 nm, (b) X-ray pumping through higher-lying nuclear states, and (c) alpha decay of U-233.

[1] B. R. Beck *et al.*, Phys. Rev. Lett. **98**, 142501 (2007).
[2] E. Peik & C. Tamm, Europhys. Lett. **61**, 181 (2003).
[3] E. Peik & M. Okhapkin, Comptes Rendus Physique **16**, 516 (2015).
[4] S. Stellmer *et al*, Sci. Rep. **5**, 15580 (2015).
[5] S. Stellmer *et al*, arXiv:1511.07187.

# Spin squeezed $^{171}$Yb atomic clock beyond the standard quantum limit

Akio Kawasaki[1], Boris Braverman[1], Leonardo Salvi[1,2], Bojan Zlatković[1,3], and Vladan Vuletić[1]

[1]Department of Physics, MIT-Harvard Center for Ultracold Atoms and Research Laboratory of Electronics, Massachusetts Institute of Technology, Cambridge, Massachusetts 02139, USA

[2]Dipartimento di Fisica e Astronomia and LENS, Universita' di Firenze, INFN, Sezione di Firenze, Via Sansone 1, 50019 Sesto Fiorentino, Italy

[3]Institute of Physics, University of Belgrade, Pregrevica 118, 11080 Belgrade, Serbia

Optical lattice clocks have achieved the precision of $10^{-18}$ [1, 2], and are aiming to attain even higher precision. One of the limiting factors is the standard quantum limit (SQL) of the quantum projection noise occurring when state of the atoms is measured [3]. This limit has been overcome in a spin squeezed atomic clock on a microwave transition [4, 5]. We apply this technique to an optical lattice clock operating on an optical transition and try to exceed the SQL. In the poster, the experimental setup and the preliminary results of spin squeezing are presented.

[1] N. Hinkley, J. A. Sherman, N. B. Phillips, N. D. Lemke, K. Beloy, M. Pizzocaro, C. W. Oates, A. D. Ludlow, Science **341**, 1215 (2013).

[2] T. L. Nicholson, S. L. Campbell, R. B. Hutson, G. E. Marti, B. J. Bloom, R. L. McNally, W. Zhang, M. D. Barrett, M. S. Safronova, G. F. Strouse, W. L. Tew, J. Ye, Nat. Commun. **6**, 6896 (2015).

[3] T. L. Nicholson, M. J. Martin, J. R. Williams, B. J. Bloom, M. Bishof, M. D. Swallows, S. L. Campbell, J. Ye, Phys. Rev. Lett. **109**, 230801 (2012).

[4] Ian D. Leroux, Monika H. Schleier-Smith, and Vladan Vuletić, Phys. Rev. Lett. **104**, 250801 (2010)

[5] O. Hosten, N. J. Engelsen, R. Krishnakumar, M. A. Kasevich, Nature **529**, 505 (2016).

# New generation of a mobile primary frequency standard based on cold atoms

Rodrigo Duarte Pechoneri[1], Stella Torres Müller[1], Jair De Martin Jr[1], Vanderlei Salvador Bagnato[2], Daniel Varela Magalhães[1]
[1] São Carlos School of Engineering, Brazil
[2] São Carlos Institute of Physics, Brazil

Due to their exceptional accuracy and stability, atomic clocks have been finding applications in several fields. These fields ranges from tests of macroscopic scales theories like relativity or gravitation to the microscopic world, in atomic scale, thanks to quantum mechanics. Two of the main technological applications are navigation and telecommunication, since atomic references are in the heart of the most advanced navigation satellite systems (GNSS) and synchronize channels in high rate networks of data. On the other hand, most of the high performance frequency standards are extremely complex and their size are incompatible with some applications.

The compact frequency standard under development at USP São Carlos is a cold atoms system that works with a completely innovative principle, in which the different operation steps happen in one place: inside the microwave cavity. This kind of operation allows to obtain a standard much more compact than a conventional one, since different interactions occur in the same region of the apparatus. Currently, we are redefining all the instrumentation associated with the experiment. This work gives an overview of the topology we are adopting and the current status of the system.

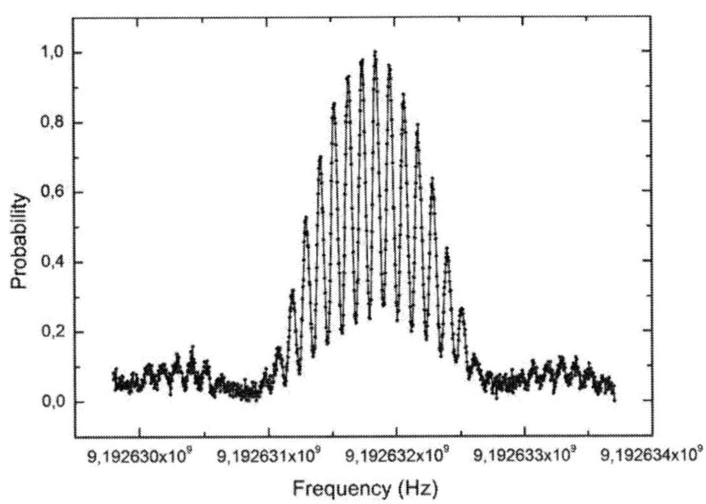


**Figure 1.** Ramsey Fringes obtained by the current version (2) of our compact frequency standard. The interrogation was performed in the microwave cavity by two microwave pulses of 1 ms each and separated by 8 ms. The linewidth of the central fringe is 47 Hz.

# Progress towards a Brazilian cesium atomic fountain second generation

Andrés Rodriguez Salas[1], Caio Bueno[2], Vanderlei Salvador Bagnato[1], Daniel Varela Magalhães[2]
São Carlos Institute of Physics, Brazil
São Carlos School of Engineering, Brazil

The goal of this work is to present the Brazilian Cesium fountain Br-CsF, which passed by several changes and currently is back to operation after improvements in its optical system. The main result for this new operation is Ramsey fringe showed in figure 1. The central fringe have a width of 1.5 Hz measure in FWHM. The points from Ramsey fringes were obtained without average, doing a cycle time of 2 s. This fringe is obtained loading ~$10^9$ atoms in a magneto optical trap (loading time of 1.2 s, red detuning $\Delta = -3\Gamma$, Doppler temperature T ~ 130 μK), with the laser beams at the maximum power in this phase. Immediately after the atoms are transferred in an optical molasses with $\Delta = -5\Gamma$ and laser beam intensity decreased to the half. This process is done in 60 ms and after the atoms are launched up using the moving molasses technique ($\Delta = 2\Gamma \pm \delta\upsilon$, where $\delta\upsilon = 2.4$ MHz). Once the cloud is launched, the atoms are cooled down until a temperature of T < 4 μK using sub-Doppler cooling stage. The detuning for this phase ramps down until $\Delta = -12\Gamma$. They then enter the interaction zone, where interact twice with an oscillatory magnetic field tuned to the transition of the ground state between F = 4, $m_f = 0$ and F = 3, $m_f = 0$, defined as 9192611770 Hz. Finally the atoms are detected with respect to the clock transition using the TOF technique (time of flight).

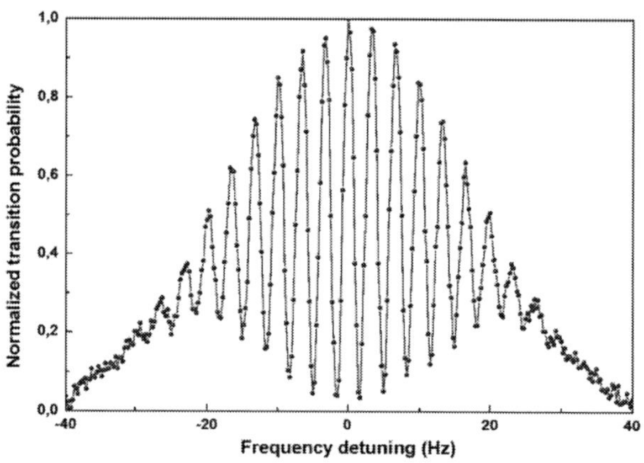


**Figure 1.** Ramsey fringes from the BR-CsF fountain clock, the central fringe is 1.5 Hz FWHM. The points acquired have no average and are taken from a single scan.

# Frequency ratio of a mercury optical lattice clock with optical and microwave frequency standards

Luigi De Sarlo[1], Rinat Tyumenev[1], Maxime Favier[1], Slawomir Bilicki[1], Eva Bookjans[1], Daniele Nicolodi[1], Michel Abgrall[1], Jocelyne Guéna[1], Yann Le Coq[1], Rodolphe Le Targat[1], Jérôme Lodewyck[1], Sébastien Bize[1]

[1]LNE-SYRTE, Observatoire de Paris, CNRS, PSL Research University, Sorbonne Universités, UPMC Univ. Paris 06, France

Neutral mercury is a promising candidate to build an optical lattice clock thanks to several favorable atomic properties. Its high vapor pressure at room temperature suppresses the need for an oven and thus reduces temperature gradients on the experimental setup. Furthermore, the 1S0 − 3P0 ultranarrow clock transition is very weakly coupled to static and thermal radiation fields easing the efforts needed to control the blackbody radiation shift, one of the limiting factors in almost all the other optical clocks. Additionally, $^{199}$Hg has a simple structure with spin ½ for which tensor component of light shift is null and vector component can be exactly canceled when alternatively interrogating the two spin states. Finally, the $^3P_1$ state used for laser cooling has a 1.3 MHz linewidth yielding Doppler limited temperatures as low as 30μK. This allows for direct loading in the magic wavelength optical lattice from a single stage MOT. In spite of these advantages, a big challenge lies in the need for adequate (narrow-linewidth, tunable…) cw laser sources in the UV region of the spectrum at 254, 362 and 266 nm respectively for cooling, trapping and probing the mercury atoms.

In this poster, after briefly presenting a new evaluation of the systematics of our Hg clock down to $1.6 \times 10^{-16}$ of relative uncertainty, we will report the first direct measurement of the Hg/Rb frequency ratio, and, to our knowledge, the best absolute frequency measurement of the Hg clock transition via comparison with FO2-Cs down to an uncertainty of $4 \times 10^{-16}$, close to the limit of the fountain and about 30 times better than the last measurement reported by our group [1]. Finally, we will report a direct optical to optical measurement of the Hg/Sr frequency ratio with an uncertainty of 1.8 × 10−16. Our value is in good agreement, within the stated 1σ uncertainties, with the value reported in [2]. To the best of our knowledge, no other frequency ratio has been measured in different labs with an uncertainty below that of the SI second. These kinds of comparisons are essential in assessing the reliability of optical frequency standards as candidates for a redefinition of the SI second, as well as for tests of the variation of fundamental constants.

[1] J. J. McFerran, L. Yi, S. Mejri, S. Di Manno, W. Zhang, J. Guéna, Y. Le Coq, and S. Bize, Phys. Rev. Lett., **108**, 183004 (2012).

[2] K. Yamanaka, N. Ohmae, I. Ushijima, M. Takamoto, and H. Katori, Phys. Rev. Lett., **114**, 230801 (2015).

# Quantum thermalization through entanglement in an isolated many-body system

Matthew Rispoli[1], Adam Kaufman[1], Eric Tai[1], Alex Lukin[1], Robert Schittko[1], Philipp Preiss[1], Markus Greiner[1]

[1]Department of Physics, Harvard University, Cambridge, Massachusetts, 02138, USA

Quantum and classical many-body systems appear to have disparate behavior due to the different mechanisms that govern their evolution. The dynamics of a classical many-body system equilibrate to maximally entropic states and quickly re-thermalize when perturbed. The assumptions of ergodicity and unbiased configurations lead to a successful framework of describing classical systems by a sampling of thermal ensembles that are blind to the system's microscopic details. By contrast, an isolated quantum many-body system is governed by unitary evolution: the system retains memory of past dynamics and constant global entropy. However, even with differing characteristics, the long-term behavior for local observables in quenched, non-integrable quantum systems are often well described by the same thermal framework. We explore the onset of this convergence in a many-body system of bosonic atoms in an optical lattice. Our system's finite size allows us to verify full state purity and measure local observables. We observe rapid growth and saturation of the entanglement entropy with constant global purity. The combination of global purity and thermalized local observables agree with the Eigenstate Thermalization Hypothesis in the presence of a near-volume law in the entanglement entropy [1].

[1] A. Kaufman, *et al.*, arXiv: 1603.04409 (2016)

# Parity symmetry breaking and hysteresis with BECs with tunable interactions in a double-well potential

Andreas Trenkwalder[1], Giacomo Spagnolli[2], Giulia Semeghini[2], Simon Coop[2,3], Manuele Landini[1], Patricia Castilho[1,4], Luca Pezzè[1,2,5], Giovanni Modugno[2], Massimo Inguscio[1,2], Augusto Smerzi[1,2,5], Marco Fattori[1,2]

[1]Istituto Nazionale di Ottica-CNR, 50019 Sesto Fiorentino, Italy

[2]LENS and Dipartimento di Fisica e Astronomia, Università di Firenze, 50019 Sesto Fiorentino, Italy

[3]ICFO-Institut de Ciences Fotoniques, 08860 Castelldefels (Barcelona), Spain

[4]Instituto de Fisica de Sao Carlos, Universidade de Sao Paulo, 369, 13560-970 Sao Carlos, Brazil

[5]Quantum Science and Technology in Arcetri, QSTAR, 50125 Firenze, Italy

In this work we report the experimental observation of the full phase diagram across a transition where the spatial parity symmetry is broken [1]. Our system consists of an ultra-cold gas of $^{39}K$ with tunable interactions trapped in a double-well potential. At a critical value of the interaction strength, we observe a continuous quantum phase transition where the gas localizes in one well or the other, thus breaking the underlying symmetry of the system. Furthermore, we show the robustness of the asymmetric state against controlled energy mismatch between the two wells. This is the result of hysteresis associated with an additional discontinuous quantum phase transition that we fully characterize. Our results pave the way to the study of quantum critical phenomena at finite temperature, the investigation of macroscopic quantum tunneling of the order parameter in the hysteretic regime and the production of quantum entangled states at critical points including Schroedinger cat states with macroscopic atom number.

[1] A. Trenkwalder, G. Spagnolli, G. Semeghini, S. Coop, M. Landini, P. Castilho, L. Pezzè, G. Modugno, M. Inguscio, A. Smerzi, M. Fattori, Nature Physics (2016) doi:10.1038/nphys3743

# Realization of Two-Dimensional Spin-orbit Coupling for Bose-Einstein Condensates

ZhanWu[1,2], Long Zhang[1,3],Wei Sun[1,2], Xiao-Tian Xu[1,2], Bao-ZongWang[1,3], Si-Cong Ji[1,2], Youjin Deng[1,2], Shuai Chen[1,2], Xiong-Jun Liu[3,4], and Jian-Wei Pan[1,2]

[1]Shanghai Branch, National Laboratory for Physical Sciences at Microscale and Department of Modern Physics, University of Science and Technology of China, Shanghai 201315, China

[2]Synergetic Innovation Center of Quantum Information and Quantum Physics, University of Science and Technology of China, Hefei, Anhui 230026, China

[3]International Center for Quantum Materials, School of Physics, Peking University, Beijing 100871, China

[4]Collaborative Innovation Center of Quantum Matter, Beijing 100871, China

Cold atoms with laser-induced spin-orbit (SO) interactions provide intriguing new platforms to explore novel quantum physics beyond natural conditions of solids. Recent experiments demonstrated the one-dimensional (1D) SO coupling for boson and fermion gases. However, realization of 2D SO interaction, a much more important task, remains very challenging. Here we propose and experimentally realize, for the first time, 2D SO coupling and topological band with 87Rb degenerate gas through a minimal optical Raman lattice scheme, without relying on phase locking or fine tuning of optical potentials. A controllable crossover between 2D and 1D SO couplings is studied, and the SO effects and nontrivial band topology are observed by measuring the atomic cloud distribution and spin texture in the momentum space. Our realization of 2D SO coupling with advantages of small heating and topological stability opens a broad avenue in cold atoms to study exotic quantum phases, including the highly-sought-after topological superfluid phases.

# Atomic quantum simulation of a three-dimensional U(1) gauge-Higgs model

Kenichi Kasamatsu[1], Yoshihito Kuno[2], Shinya Sakane[1], Ikuo Ichinose[2], Tetsuo Matsui[1]
[1]Department of Physics, Kindai University, Higashi-Osaka, 577-8502, Japan
[2]Department of Applied Physics, Nagoya Institute of Technology, Nagoya, 466-8555, Japan

We study theoretically atomic quantum simulations of a U(1) gauge-Higgs model on a three-dimensional (3D) optical lattice [1,2]. We start from an extended 3D Bose-Hubbard model with intersite repulsions and show that it can simulate a U(1) gauge-Higgs model with next nearest-neighbor Higgs couplings. Here the phase of the boson variable on each site of the optical lattice describes the vector potential on each link of the gauge-model lattice. To determine the phase diagram of the gauge-Higgs model at a zero temperature, we perform Monte-Carlo simulations of the corresponding 3+1-dimensional U(1) gauge-Higgs model, and obtain the three phases, i.e., the confinement, Coulomb and Higgs phases. To investigate the dynamical properties of the gauge-Higgs model, we apply the Gross-Pitaevskii equations to the extended Bose-Hubbard model. We simulate the time-evolution of an electric flux initially put on a straight line connecting two external point charges. We also calculate the potential energy between this pair of charges and obtain the string tension in the confinement phase. Finally, we propose a feasible experimental setup for the atomic simulations of this quantum gauge-Higgs model in the 3D optical lattice..

[1] K. Kasamatsu, I. Ichinose, T. Matsui, Phys. Rev. Lett. **111**, 115303 (2013).

[2] Y. Kuno, S. Sakane, K. Kasamatsu, I. Ichinose, T. Matsui, arXiv:1605.02502 (2016).

# Critical thermalization in a disordered dipolar spin system

Joonhee Choi[1], Georg Kucsko[1], Soonwon Choi[1], Peter Maurer[2], Junichi Isoya[3], Eugene Demler[1], Fedor Jelezko[4], Norman Yao[5], Mikhail Lukin[1]

[1]Department of Physics, Harvard University, Cambridge MA 02138, USA

[2]Department of Physics, Stanford University, Stanford, California 94305, USA

[3]Graduate School of Library, Information and Media Studies, University of Tsukuba, Japan

[4]Institut für Quantenoptik, Universität Ulm, 89081 Ulm

[5]Physics Department, University of California Berkeley, Berkeley, California 94720, USA

Ensembles of strongly interacting spins offer an attractive platform for the study of many-body quantum physics. We realize and study the critical dynamics in a dense ensemble of interacting electronic spins associated with Nitrogen Vacancy (NV) centers in diamond. In particular, we study thermalization behavior in a large, 3D, dipolar quantum system by preparing two independently addressable and oppositely polarized spin ensembles. By utilizing a spin-locking pulse sequence, we probe the spin decay dynamics as a function of on-site potential disorder and identify a critical regime of strongly interacting many-body dynamics. We find that the spin relaxation follows a slow, sub-exponential decay consistent with a power law with disorder-dependent exponents. In addition, the spin relaxation at long times reveals many-body effects as well as coupling to the environment.

# Extended Bose-Hubbard model and atomic quantum simulation of (1+1)D U(1) gauge-Higgs model

Yoshihito Kuno[1], Keima Kawaki[1], Shinya Sakane[2], Kenichi Kasamatsu[2], Ikuo Ichinose[1], Tetsuo Matsui[2]

[1] Department of Applied Physics, Nagoya Institute of Technology, Nagoya, 466-8555, Japan

[2] Department of Physics, Kindai University, Higashi-Osaka, 577-8502, Japan

Recently atomic quantum simulation for high-energy physics comes to become an active field in cold-atom physics society. In particular there are a plethora of proposals to build up a quantum simulator for the theoretical models of lattice gauge theory. In this conference we show theoretically an atomic quantum simulator of U(1) lattice gauge-Higgs model based on an extended Bose-Hubbard chain in 1D optical lattice. This quantum simulator of the lattice gauge theory is directly connected toward the Bose-Hubbard model with nearest-neighbor interactions. In the 1D system the most important ingredient, i.e., local gauge invariance can be implemented in a simple way. The interaction preserving this invariance is nothing but the nearest-neighbor interaction assumed to be dipole-dipole interactions in real experiments. Furthermore we show a global phase diagram, including some interesting phases, e.g., Haldane-insulator. This phase diagram is obtained by carrying out a stochastic series expansion method in quantum Monte-Carlo simulations. Also by using the correspondence between the Bose Hubbard model and the lattice gauge theory we interpret these phases from the view of the lattice gauge-theory. In this phase diagram we can detect the Higgs phase (Superfluid phase) and the confinement phase (Mott insulator phase). Consequently the 1D extended Bose-Hubbard model appears two gauge-theoretical phases in some parameter regimes. Next we study the non-equilibrium properties of the quantum simulator. We focus on simulating the dynamics of an electric flux (confinement string) in both Higgs and confinement phase. To detect this subject we use the Gross-Pitaevskii equation for the Higgs (Superfluid) regime and the semi-truncated Wigner method for shallow confinement regime. We certify that the electric flux spontaneously breaks in the Higgs phase, on the other hand in the shallow confinement regime the electric flux is dynamically stable. These results are expected to be measured in future real experiments. Moreover our numerical simulations find that the Schwinger-like mechanism can be observed in this quantum simulator; electric fields oscillate via a Higgs and anti-Higgs pair creation.

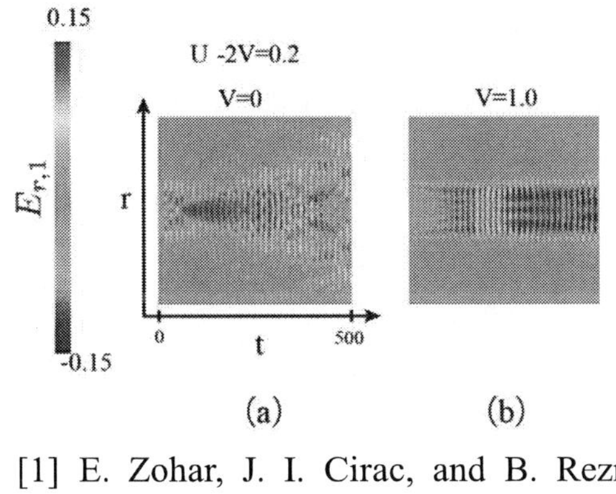


Fig: Real time-dynamics of a single electric flux. This figure shows electric field dynamics. (a) In Higgs regime, the initial electric flux gradually spreads out along the time evolution. (b) In confinement regime, the shape of the initial electric flux is not broken. The electric field oscillates changing their sign.This phenomenon is explained by the string-antistring oscillation that is also observed in the QED real-time simulation.

[1] E. Zohar, J. I. Cirac, and B. Reznik, Rep. Prog. Phys.**79**, 014401 (2016).

[2] Y. Kuno, K. Kasamatsu, Y. Takahashi, I. Ichinose, and T. Matsui, New J. Phys. **17**, 063005 (2015).

[3] Y. Kuno, K. Kawaki, S. Sakane, K. Kasamatsu, I. Ichinose, T. Matsui, arXiv:1605.00333 (2016).

# Spin and Charge Resolved Quantum Gas Microscopy of Antiferromagnetic Order in Hubbard Chains

Guillaume Salomon[1], Martin Boll[1],Timon Hilker[1], Ahmed Omran[1], Immanuel Bloch[1,2], Christian Gross[1]

[1]Max-Planck-Institut für Quantenoptik, 85748 Garching, Germany

[2]Fakultät für Physik, Ludwig-Maximilians-Universität, 80799 München, Germany

Ultracold fermions in optical lattices allow for the simulation of the Hubbard Hamiltonian with a unique control over kinetic energy, interactions and dimensionality. However the study of the onset of magnetic ordering at low temperature turns out to be challenging when looking at trap averaged quantities.We will present our recent experimental study of spin correlations in spin 1/2 Hubbard chains using a quantum gas microscope [1,2]. We revealed antiferromagnetic correlations at distances up to three sites corresponding to a local entropy well below ln2. Our ability to measure both charge and spin opens exciting perspectives for studying the interplay between magnetic ordering and doping in ultracold fermion systems.

[1] A. Omran *et al.*, Physical Review Letters **115**, 263001 (2015).

[2] M. Boll *et al.*, arXiv:1605.05661 (2016).

# Exciton-Polariton Quantum Simulators

Na Young Kim[1]

[1]Institute for Quantum Computing, Department of Electrical and Computer Engineering, University of Waterloo, 200 University Ave. West, Waterloo, ON N2L 3G1, Canada

Microcavity exciton-polaritons are hybrid light-matter quasi-particles arising from the mixed states between cavity photons and quantum well excitons. The inherent light-matter duality provides experimental advantages: the stimulated scattering among interacting particles and the small effective mass (~ $10^{-8}$ times the hydrogen atom) form coherent condensate states at high temperatures (e.g. 4 K in GaAs and room temperature in GaN materials). In addition, the dynamics of exciton-polaritons is accessed by capturing the leaked photons out of the cavity due to the short lifetime. Utilizing coherence and open-dissipative nature of exciton-polariton condensates, we engineer a two-dimensional (2D) polariton-lattice system for investigating exotic quantum phase order. Via micro-photoluminescence measurements in both real and momentum spaces, we have observed *d*-orbital condensate states, vortex-antivortex phase order, massless Dirac dispersions in 2D square, honeycomb, and triangular lattices respectively. These results demonstrate that the polariton-lattice systems will be promising solid-state quantum simulators in the quest for better understanding strongly correlated materials and in the development of novel optoelectronic devices.

# Disordered cold atoms in 2D

Donald White, Thomas Haase, Ivan Herrera-Benzaquen and Maarten Hoogerland
Dodd-Walls Centre for Photonic and Quantum Technologies,
Department of Physics, University of Auckland, New Zealand

Understanding the propagation of waves through disordered media is of fundamental importance to an extraordinary variety of systems, ranging from the propagation of seismic waves through fractured rock [1] to the conductivity behaviour of disordered materials [2]. Interference effects for coherent waves result in behaviour known as Anderson localization, in which disorder results in totally localized states suppressing propagation.

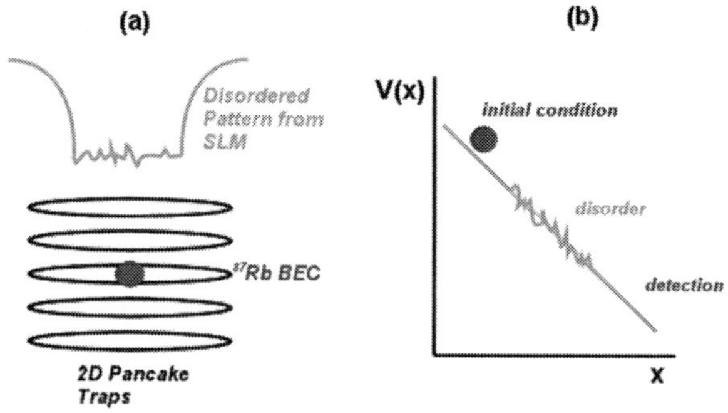


Figure 1: Experimental schematic. (a) A spatial light modulator imprints a pattern of controlled disorder onto a 532 nm laser beam. Our high resolution optical system of numerical aperture 0.5 permits structures on the order of 1 μm. The atoms are maintained in pancake-shaped traps (pitch 10 μm, radial trap frequency 1 Hz, vertical trap frequency 1 kHz) created by interfering two 1064nm laser beams. (b) A magnetic field gradient provides a 'voltage' for spin-polarised atoms, allowing a measurement of the resistance of the disordered sample.

We engineer an 'atomtronic' system in two dimensions. The components for this system are a $^{87}$Rb Bose–Einstein condensate; a two–dimensional trap created from the interference of laser beams; a disordered potential created via a spatial light modulator (SLM); and an external 'voltage' for spin-polarised atoms created by a magnetic field gradient, as shown in Figure 1.

Our system allows a direct quantum simulation of a disordered 2D system, similar to the experiment of Krinner et al. [3]. The use of a spatial light modulator allows us direct access to the Anderson localization regime, which is difficult to achieve with speckle in two dimensions [4]. The full–scale tunability of our system also allows us to vary system size, presenting an ability to investigate the scaling law of localization.

I will present our latest measurements on the 'resistance' of a disordered crystal of light in 2D using cold atoms, and our progress towards observing 2D Anderson localization.

[1] E. Larose, L. Margerin, B.A van Tiggelen, and M. Campillo, Phys. Rev. Lett. **93**, 048501. (2004)

[2] P. Lee, and T.V. Ramakrishnan, Rev. Mod. Phys. **57**, 287 (1985).

[3] S. Krinner, D. Stadler, J. Meineke, J.-P. Brantut, and T. Esslinger, Phys. Rev. Lett. **110**, 100601. (2013)

[4] W. Morong and B. DeMarco, Phys. Rev. A **92**, 023625 (2015).

# Imaging fluorescence from individual fermions in an optical lattice

Rhys Anderson[1], Graham Edge[1], Peihang Xu[1], Vijin Venu[1], Stefan Trotzky[1], Joseph Thywissen[1]

[1]Department of Physics, University of Toronto, Toronto, Ontario, Canada

We discuss the development of a quantum gas microscope for $^{40}K$. We load a degenerate Fermi gas into a cubic optical lattice of period 527nm, which is capable of simulating the Fermi-Hubbard model. The sample is prepared in UHV below a 200μm thick sapphire window, at the focus of a 5mm focal length objective located outside the chamber. To isolate a single plane for imaging, we perform spectroscopic selection in a 210G/cm gradient, which separates the hyperfine transition frequencies of adjacent vertical planes by 28kHz. The plane may be at any depth within the three-dimensional bulk, which allows access to surveys of three-dimensional physics in a tomographic fashion. A similar protocol allows for patterning of the distribution of spins within the lattice, which will be employed for preparation of initial states for dynamical studies, and for selective removal of high-entropy regions of the trap. EIT cooling on the 770.1nm D1 transition facilitates fluorescence imaging of our atoms with long exposures. Atoms remain pinned in a 200μK-deep lattice, with a 1/e lifetime of 67(9)s, while scattering $\sim 10^3$ photons per second. Collection of fluorescence photons onto an EMCCD via a 0.8NA objective results in a PSF of FHWM 600nm, and 94(2)% of atoms identified in the first frame remain pinned in successive frames, enabling reconstruction of the lattice-site occupancy. We present ongoing progress in obtaining lower entropy samples. Direct imaging of site-occupancy of fermions in the lattice will provide access to unambiguous signatures of strongly-correlated physics such as quantum magnetism, and will allow for the characterization of non-equilibrium dynamics for strongly interacting states of matter.

# Spin-Orbit Coupling and Spin Textures in Optical Superlattices

Junru Li, Wujie Huang, Boris Shteynas, Sean Burchesky, Furkan Cagri Top, Edward Su, Jeongwon Lee, Alan O. Jamison, and Wolfgang Ketterle

Research Laboratory of Electronics, MIT-Harvard Center for Ultracold Atoms, Department of Physics, Massachusetts Institute of Technology, Cambridge, Massachusetts 02139, USA

We propose and demonstrate a new approach for realizing spin orbit coupling with ultracold atoms [1]. A one-dimensional optical superlattice forms a stack of coherently coupled double wells, where the orbital levels are used as pseudospin states. Two-photon Raman transitions between left and right wells induce spin-orbit coupling. With single internal state of an atom and far-detuned beams being used, our scheme allows convenient generalization to wide range of atoms. The double well potential can be shaped to create a highly imbalanced inter- and intra- pseudospin interactions ($g_{\uparrow\downarrow}^2 < g_{\uparrow\uparrow}g_{\downarrow\downarrow}$), which provides a favorable parameter space for detection of the stripe phase [2]. We also show that the pseudo-spinor Bose-Einstein condensate spontaneously acquires an antiferromagnetic pseudospin texture, which breaks the lattice symmetry similar to a supersolid.

[1] J. Li, W. Huang, B. Shteynas, S. Burchesky, F. C. Top, E. Su, J. Lee, A. O. Jamison, and W. Ketterle, arXiv:1606.03514 (2016).

[2] Y. Li, L. P. Pitaevskii, and S. Stringari, Phys. Rev. Lett. **108**, 225301 (2012).

# Inter-sublattice dynamics of ultracold atoms in an optical Lieb lattice

Hideki Ozawa, Shintaro Taie, Tomohiro Ichinose, Shuta Nakajima, Yoshiro Takahashi
Department of Physics, Kyoto University, Japan

Recent advancements in optical lattices have brought about great success in engineering the band structures, tunneling amplitude and on-site interactions in the Hubbard regime. Furthermore, the highly tunable optical lattices realize unconventional band structures that are not easy to study in solid-state materials including flat bands. In special lattice structure such as Lieb lattice, kagome lattice and saw-tooth lattice, a flat band appears due to destructive interference of the hopping matrix element. For bosonic system, a fascinating question is considered whether a Bose-Einstein condensation (BEC) is possible in a flat band [1].

So far we have realized "Lieb lattice" configuration with an optical lattice and developed a phase-imprinting method to coherently transfer a BEC into a flat band of the optical Lieb lattice. In addition to measuring lifetime of a flat band population, we have investigated the inter-sublattice dynamics of the system by projecting the sublattice population onto the band population. This measurement clearly shows the formation of the localized state [2].

Our optical lattice potential has various degree of freedom, which enables to prepare the superposition of band-eigenstate of Lieb lattice and measure the band gap by extracting oscillation frequency of sublattice population. Using Bloch oscillation to transfer a BEC to desired quasimomenta, we reconstruct the band structure of an optical Lieb lattice (Fig. 1). The result indicates the distortion of a flat band due to on-site interaction.

A flat band in Lieb lattice is mathematically equivalent to "dark states" in three level systems with Λ-type transition. In this analogy, three sublattices, tunneling amplitudes and the energy difference correspond to the basis of three levels, laser-induced coupling and the detuning of laser, respectively. According to STIRAP in the three level system, we perform spatial adiabatic passage (SAP) [3] of fermionic atoms between sublattices in an optical Lieb lattice (Fig. 2). This is not only first demonstration of SAP in matter-wave, but also can be used to introduce fermionic atoms into a flat band.

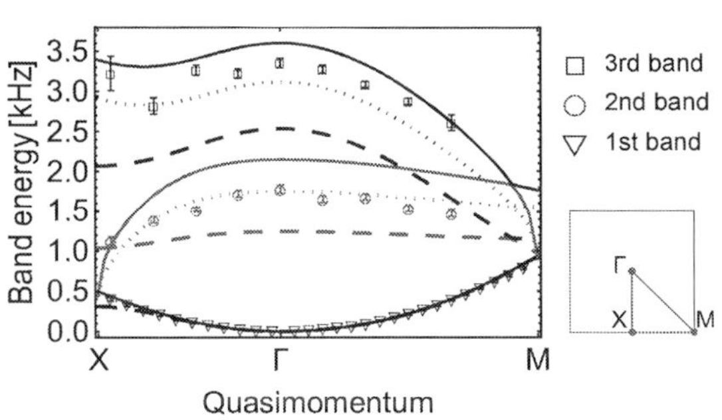


Figure 1: *The lowest three bands of optical Lieb lattice. Curves are numerical calculation results with different interaction strength. Right small figure shows 1st Brillouin zone.*

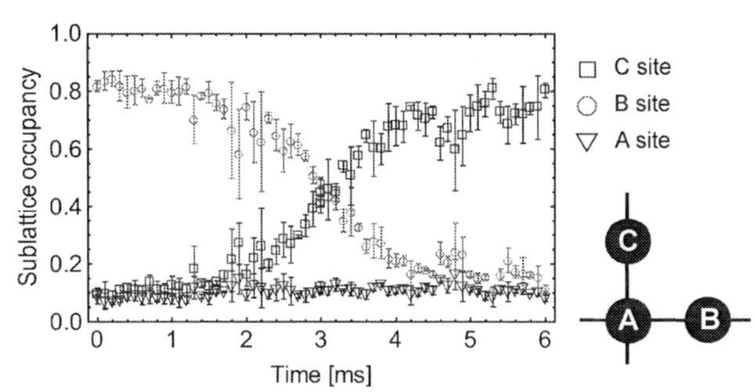


Figure 2: *Time evolution of sublattice occupancy during SAP from B site to C site. Occupancy of A site is kept low throughout SAP. Right small figure shows the lattice configuration.*

[1] Y.-Z. You *et al.*, Phys. Rev. Lett. **109**, 265302(2012).

[2] S. Taie *et al.*, Science Advances **1**, e1500854(2015).

[3] K. Eckert *et al.*, Phys. Rev. A **70**, 023606(2004).

# Experimental realization of a Bose-Hubbard model with cavity-mediated long-range interactions

Nishant Dogra[1], Renate Landig[1], Lorenz Hruby[1], Manuele Landini[1], Rafael Mottl[1], Ferdinand Brennecke[2], Sebastian Huber[3], Tobias Donner[1], Tilman Esslinger[1]

[1]Institute for Quantum Electronics, ETH Zurich, 8093 Zurich, Switzerland

[2]Physikalisches Institut, University of Bonn, Wegelerstrasse 8, 53115 Bonn, Germany

[3]Institute for Theoretical Physics, ETH Zurich, 8093 Zurich, Switzerland

We experimentally investigate an extended Bose-Hubbard model with cavity-mediated long-range interactions using an ultracold atomic gas. The long-range interactions are generated by coupling a BEC to the single mode of a high-finesse cavity and pumping it with a transverse laser-field. The competition among three energy scales- tunneling, short-range interactions and long-range interactions gives rise to a rich phase diagram consisting of four different phases - a superfluid, a supersolid, a Mott insulator and a charge density wave [1]. Moreover, we study the transition between the two insulating phases - charge density wave and Mott insulator - and observe a hysteretic behavior which signals towards a first order phase transition. We also investigate theoretically the various features of such an extended Bose-Hubbard model using different mean-field approaches.

[1] R. Landig, L. Hruby, N. Dogra, M. Landini, R. Mottl, T. Donner, and T. Esslinger, Nature **532**, 476 (2016).

# Realization of two-dimensional quantum walks of neutral atoms

Geol Moon, Stefan Brakhane, Volker Schilling, Carsten Robens, Gautam Ramola, Muhammad Sajid, Wolfgang Alt, Dieter Meschede, Andrea Alberti

Institute for Applied Physics, Bonn University

We report on the experimental realization of a spin-dependent square optical lattice to implement the two-dimensional discrete-time quantum walks of Cs atoms. We demonstrate high-resolution images of single atoms through an objective lens with very high numerical aperture (NA~0.92) which is installed inside the home-built ultra-low-birefringence dodecagonal vacuum glass cell [1]. Our system provides an ideal platform to study the topological features of 2D quantum walk as the simulator of topological phases, which can particularly expect to observe exotic matter wave flow at the boundary between different topological domains [2, 3]. Furthermore, manipulating the phase of single atoms accumulated through the movement from site to site on the 2D lattice make it possible to realize artificial gauge fields and to study the effect of magnetic fields on the 2D quantum walk [4].

[1] S. Brakhane, et al., *Ultra-low birefringence dodecagonal vacuum glass cell*, Rev. Sci. Instrum. **86**, 126108 (2015).

[2] T. Kitagawa, et al., *Exploring topological phases with quantum walks*, Phys. Rev. A **82**, 033429 (2010).

[3] Thorsten Groh, et al., *Robustness of topologically protected edge states in quantum walk experiments with neutral atoms*, arXiv:1605.03633.

[4] P. Arnault and F. Debbasch, *Quantum Walks and discrete Gauge Theories*, Phys. Rev. A **93**, 052301 (2016).

# Quantum annealing for the number-partitioning problem using a tunable spin glass of ions

T. Grass[1], D. Raventós[1], Bruno Juliá-Díaz[1,2], C. Gogolin[1,3], M. Lewenstein[1,4]

[1]ICFO-Institut de Ciències Fotòniques, The Barcelona Institute of Science and Technology, Castelldefels, Spain

[2]Dept. Física Quàntica i Astrofísica, Universitat de Barcelona, Spain

[3]Dept. Max-Planck-Institu für Quantenoptik, Garching, Germany

[4]ICREA-Institució Catalana de Recerca i Estudis Avançats, Barcelona, Spain

Exploiting quantum properties to outperform classical ways of information processing is an outstanding goal of modern physics. A promising route is quantum simulation, which aims at implementing relevant and computationally hard problems in controllable quantum systems [1]. Here we demonstrate that in a trapped ion setup [2], with present day technology, it is possible to realize a spin model of the Mattis-type that exhibits spin glass phases. Our method produces the glassy behaviour without the need for any disorder potential, just by controlling the detuning of the spin-phonon coupling [3]. Applying a transverse field, the system can be used to benchmark quantum annealing strategies which aim at reaching the ground state of the spin glass starting from the paramagnetic phase. In the vicinity of a phonon resonance, the problem maps onto number partitioning [4], and instances which are difficult to address classically can be implemented.

[1] R. P. Feynman, Int. J. Theor, Phys. **21**, 467 (1982), S. Lloyd, Science **23**, 273, 1073 (1996).

[2] D. Porras, and I. Cirac, Phys. Rev. Lett. **92**, 207901 (2004).

[3] T. Grass, D. Raventós, Bruno Juliá-Díaz,.C. Gogolin, M. Lewenstein, Nature Communications **7**, 11524 (2016).

[4] F. Barahona, J. Phys. A: Math. Gen. **15**, 3241 (1982); A. Lucas, Front. Phys. **2,** 5 (2014).

# Quantum gas microscopy of the interacting Harper-Hofstadter system

M. Eric Tai[1], Alex Lukin[1], Philipp M. Preiss[2], Matthew Rispoli[1], Robert Schittko[1], Adam Kaufman[1], Markus Greiner[1]

[1]Harvard University, 17 Oxford Street, Cambridge, MA, 02138, USA

[2]Physikalisches Institut, Heidelberg University, D-69120 Heidelberg, Germany

At the heart of many topological states is an underlying gauge field. Perhaps the most familiar example of a gauge field is the magnetic field which manifests in the Lorentz deflection of a moving charged particle. Gauge fields can also give rise to novel states of matter that are not characterized by order parameters and cannot be described in the framework of spontaneous symmetry breaking. Instead, these states, e.g. fractional quantum Hall states, are characterized by topological invariants, such as the Chern number, and often have rich entanglement properties. Theoretically, gauge fields can be incorporated into tight-binding Hamiltonians through the Peierls substitution. Experimentally, the Peierls substitution can be achieved in optical lattice emulators through laser-assisted tunneling[1,2,3].

In this poster, we report on experimental results upon introducing an artificial magnetic field into a system of strongly-interacting ultracold Rubidium-87 atoms confined to a two-dimensional optical lattice. With the single-site resolution afforded by our quantum gas microscope, we can prepare a fixed atom number and project hard walls. With an artificial gauge field, this quantum simulator realizes the Harper-Hofstadter Hamiltonian. We can independently control the tunneling strengths in our lattice as well as dynamically change the flux per unit cell. This flexibility enables studies of topological phenomena from many perspectives, e.g. site-resolved images of edge currents. With the strong on-site interactions possible in our system, these experiments will pave the way to observing chiral ground states.

[1] K. Jiménez-García, et. al., PRL **108**, 225303 (2012).
[2] H. Miyake, et. al., PRL **111**, 185302 (2013).
[3] M. Aidelsburger, et. al. PRL **111**, 185301 (2013).

# Evidence of orbital Feshbach resonances and chiral edge states in ytterbium Fermi gases

G. Cappellini[1], M. Mancini[1], L. F. Livi[2], G. Pagano[1], C. Sias[3,2], J. Catani[4,2], M. Inguscio[5,1,2], L. Fallani[1,2]

[1]Department of Physics and Astronomy, University of Florence, Via G. Sansone 1, I-50019 Sesto Fiorentino, Italy

[2]European Laboratory for Non-Linear Spectroscopy (LENS), Via N. Carrara 1, I-50019 Sesto Fiorentino, Italy

[3]Istituto Nazionale di Ricerca Metrologica (INRIM), Str. delle Cacce 91, I-10135 Torino, Italy

[4]Istituto Nazionale di Ottica del Consiglio Nazionale delle Ricerche (INO-CNR), Sezione di Sesto Fiorentino, Via N. Carrara 1, I-50019 Sesto Fiorentino, Italy

[5]Consiglio Nazionale delle Ricerche (CNR), Piazzale A. Moro 7, I-00185 Roma, Italy

The poster reports on recent results obtained in two different sets of experiments in which we exploit both nuclear spin and orbital degrees of freedom in a quantum degenerate gas of fermionic Ytterbium.

In the first set of experiments, we realize strongly interacting ultracold Fermi gases of Yb [1] by tuning the interaction strength between atoms in different electronic states, exploiting a novel kind of Feshbach resonance, the recently proposed Orbital Feshbach Resonance [2], based on the orbital spin-exchange mechanism recently demonstrated by us [3]. By observing the hydrodynamic expansion of the Fermi gas, we detect the emergence of the strongly interacting regime, determining the location of the resonance and observing its dependence on the nuclear spin mixture, which is a direct consequence of the SU(N) symmetry of interactions in Yb atoms. We finally estimate the lifetime of the strongly interacting regime in order to investigate the exploitability of this technique.

In a second set of experiments, we observe the emergence of chiral edge states in an atomic Hall ribbon [4], implemented by exploiting the concept of "synthetic dimension" envisioned in the controllable, internal nuclear spin degree of freedom coupled by a pair of Raman beams. Combining this technique with standard optical lattices, we create a 2D ribbon on which we overlay a strong artificial magnetic flux that imprints a topological phase onto the atoms tunneling around a "synthetic" plaquette. Similarly to the behavior of electrons in a Hall system, edge states are observed, as well as their "skipping" trajectories, cyclotron orbits naturally truncated by the geometrical edges of the atomic systems. Such edge states show a true chiral behavior, whereas inner bulk states of the systems are shown not to take part to the chiral transport.

[1] G. Pagano, et al., Phys. Rev. Lett. **115**, 265301 (2015).
[2] R. Zhang, et al., Phys. Rev. Lett. **115**, 135301 (2015).
[3] G. Cappellini et al., Phys. Rev. Lett. **113**, 120402 (2014).
[4] M. Mancini, et al., Science **349**, 1510 (2015).

# Verification of quantum non-equilibrium work relation in the presence of decoherence

Yao Lu[1], Andrew Smith[2], Shuoming An[1], Xiang Zhang[1], Jing-Ning Zhang[1], Zongping Going[3], Haitao Quan[3], Christopher Jarzynski[2,4], Kihwan Kim[1]

[1]Center for Quantum Information, IIIS, Tsinghua University, Beijing, P. R. China

[2]Department of Physics, University of Maryland, College Park, USA

[3]School of Physics, Peking University, Beijing, P. R. China

[4]Department of Chemistry and Biochemistry, Institute for Physical Science and Technology, University of Maryland, College Park, USA

We report on an experimental test of the non-equilibrium work relation in a quantum system consisting of a single 171Yb+ ion that undergoes decoherence. The non-equilibrium work relation, which relates equilibrium free energy differences to non-equilibrium work distributions [1], has been tested in many classical open systems and recently in isolated quantum systems [2,3]. For open quantum systems, however, the definitions of work and heat are not fully settled [4-5], which hinders experimental verification. Here, the work relation is studied for a specific example of an open quantum system-namely a system which solely undergoes decoherence as a result of contact with its environment. A simple argument is given that the work relation remains valid in the presence of decoherence which is independently verified by experiment using trapped ion. The experimental work represents the first test of the work relation beyond closed quantum systems.

[1] C. Jarzynski, Phys. Rev. Lett. **78**, 2690 (1997).

[2] S. An, et al., Nature Physics, **11**, 193 (2015).

[3] T. B. Batalhão et al, Phys. Rev. Lett. **113**, 140601 (2014).

[4] M. Campisi, P. Hanggi, P. Talkner, Rev. Mod. Phys. **83**, 1653 (2011).

[5] M. Esposito, U. Harbola, S. Mukamel, Rev. Mod. Phys. **81**, 1665 (2009).

# Quantum gas microscope for ytterbium atoms with deep potential in the excited state

Naoki Tambo, Martin Miranda, Izumi Shomura, Takahiro Ogino, Ryotaro Inoue, Mikio Kozuma

Department of Physics, Tokyo Institute of Technology, 2-12-1 O-okayama, Meguro-ku, Tokyo 152-8550, Japan

The highly controllable optical lattice potential provides a fascinating tool for simulating quantum systems [1]. We are using a system of ultracold ytterbium (Yb) atoms trapped in a two-dimensional optical lattice, and have recently implemented a quantum gas microscope (QGM), the high-resolution fluorescence imaging device capable of resolving individual atoms in each site of the optical lattice [2] (a QGM for Yb atoms using different methods has also been developed [3]). Such a QGM will play important roles for the study of strongly correlated system. Evaluating the atom number fluctuation in the lattice sites for the Mott insulator phase, we can estimate the system temperature precisely [4, 5]. In case of studying fermionic atom system, by combining spin-sensitive measurement and QGM we will be able to detect the antiferromagnetic phase directly. The possibility for cooling the temperature of the lattice system with the QGM has also been suggested [6, 7].

The unique feature of our QGM is an imaging process using deep potential in the exited state. During the process of collecting fluorescence from atoms in optical lattice, the atoms should be kept inside the site of the lattice. To achieve this, the atoms are irradiated with cooling light and fluorescence is collected in the experiments with the alkali atoms. In their setups, since fluorescence imaging time is an order of seconds, photoassociation (PA) effects occur and the parity of the atom-number in each site of lattice is detected. In contrary, we create the deep potential in the excited state to prevent atoms being lost due to heating by resonant excitation beam. This method enables us to obtain fluorescence images in very short time (~40μs), and thus gives us the possibility for detecting atom-number in each site directly.

We prepare the laser light having the wavelength of 556nm which causes the PA effect in the inter-combination transition. We take the QGM images in two cases, where atoms in the lattice are irradiated with PA light or not. It is found out that the histograms of photon counts in each site are different between the measurements with and without the radiation of PA light. In this poster we show the details of the experiments described above.

[1] I. Bloch, *et al.*, Rev. Mod. Phys. **80**, 885 (2008).

[2] M. Miranda, *et al.*, Phys. Rev. A **91**, 063414(2015).

[3] R. Yamamoto, *et al.*, New J. Phys. **18**, 023016, (2016)

[4] J. –S. Bernier, *et al.*, Phys. Rev. A, **79**, 061601(R), (2009)

[5] E. Haller, *et al.*, Nature, **11**, 738, (2015)

[6] W. S. Bakr, *et al.*, Nature, **462**, 74, (2009)

[7] J. F. Sherson, *et al.*, Nature, **467**, 68, (2010)

# Spin Hamiltonian, synthetic field, and potential spin squeezing in a strontium optical lattice clock

Tobias Bothwell[1], Shimon Kolkowitz[1], Sarah Bromley[1], Daniel Reed[1], Michael Wall[1], Andrew Koller[1], Xibo Zhang[2], Ana Maria Rey[1], Jun Ye[1]

[1]JILA, National Institute of Standards and Technology and University of Colorado, Boulder, Colorado 80302, USA

[2]International Center for Quantum Materials, Peking University, Beijing 100871, China

Optical lattice clocks (OLC's) represent the most stable and accurate frequency references [1] and offer the prospect of utilizing their metrological precision to explore complex quantum many-body physics [2]. In particular, engineered spin-orbit coupling (SOC) has attracted considerable attention due to the central role of synthetic fields in cold atom systems to simulate exotic phenomena such as the quantum Hall effect and topological edge states [3-5]. Here we report on the implementation of tunable SOC in our OLC based on fermionic Sr atoms. We accomplish this using a highly imbalanced 1D lattice which enables control of tunneling rates while maintaining strong radial confinement. We demonstrate SOC induced splitting of the clock transition due to van Hove singularities in the atomic density of states as well as our ability to precisely select particular quasi-momentum and use coherent Rabi dynamics to map out the eigenstates of our system. On-going work towards studying this SOC system in the presence of interactions will be presented. We also present progress towards the use of a high-finesse optical cavity to generate spin-squeezed atomic states, in order to further improve the performance of our clock by pushing the noise scaling beyond the standard quantum limit.

[1] T. L. Nicholson *et al.*, Nature communications **6**, 6896 (2015).
[2] M. J. Martin *et al.*, Science **341**, 632 (2013).
[3] M. L. Wall *et al.*, Physical Review Letters **116**, 035301 (2016).
[4] M. Mancini *et al.*, Science **349**, 1510 (2015).
[5] B. K. Stuhl *et al.*, Science **349**, 1514 (2015).

# Quantum implementation of unitary coupled cluster ansatz with minimal-basis

Yangchao Shen[1], Xiang Zhang[1], Shuaining Zhang[1], Jing-Ning Zhang[1], Man-Hong Yung[2,1], Kihwan Kim[1]

[1]Center for Quantum Information, Institute for Interdisciplinary Information Sciences, Tsinghua University, Beijing, 100084, P. R. China

[2]Department of Physics, South University of Science and Technology of China, Shenzhen 518055, P. R. China

The lack of an efficient method for preparing molecular ground states is currently the major challenge for simulating molecular physics and quantum chemistry with a quantum computing. In classical computational chemistry, the coupled-cluster ansatz is one of the most commonly used ab initio methods, which is, however, critically limited by its non-unitary nature. Here, we report the first experimental realization of a unitary version of the coupled cluster ansatz with a minimal basis in a quantum device with a trapped ion. We simulate the electronic structure of a molecular ion (HeH+), where the ground-state energy surface curve is probed, energies of excited-states are studied and the bond-dissociation is observed non-perturbatively. Our simulation takes advantages from quantum computation to overcome the intrinsic limitations in classical computation. Our protocol would be the essential method to prepare a molecular ground-state for quantum simulation.

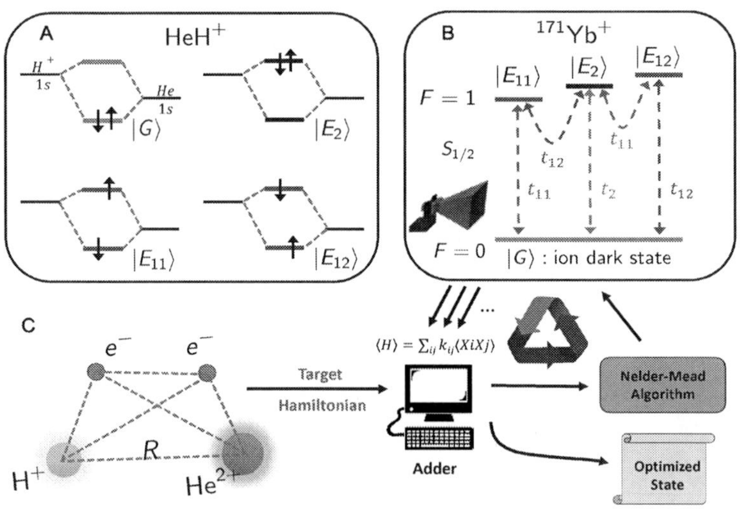


**Figure 1.** Schematic of a trapped $^{171}Yb^{+}$ ion system for the implementation of the unitary coupled-cluster (UCC) method [1].

[1]Yangchao Shen, et al., arXiv:1506.00443 (2015).

# Site-resolved imaging of fermionic quantum many-body systems

Christie S. Chiu[1], Maxwell F. Parsons[1], Anton Mazurenko[1], Geoffrey Ji[1], Daniel Greif[1], Markus Greiner[1]

[1]Department of Physics, Harvard University, 17 Oxford Street, Cambridge, MA 02138, USA

Quantum gas microscopy of ultracold fermionic atoms in an optical lattice opens new perspectives for addressing long-standing open questions on strongly correlated low-temperature phases in the Hubbard model. Here we report on site-resolved imaging of two-component fermionic Mott insulators, metals, and band insulators for a balanced two-component spin mixture of Lithium-6 in a square optical lattice. For strong repulsive interactions we observe Mott insulators with more than 400 atoms and for intermediate interactions we observe a coexistence of phases. From comparison to theory, we find trap-averaged entropies per particle of 1.0 $k_B$ in the Mott insulator and local entropies in the band insulator as low as 0.5 $k_B$.

Using our quantum gas microscope we also report on the observation of antiferromagnetic correlations extending beyond nearest neighbors in our system. The correlations are detected through selectively imaging atoms in either one, the other, or both of the two spin states. From the measured spin correlation function $\langle S_i^z S_{i+d}^z \rangle$ for varying distances $d$ we obtain the spin correlation length, which is found to increase for lower temperatures. By comparing our results to theory we find temperatures as low as $k_B T/t = 0.45(2)$. The ability to measure trap-resolved magnetic correlations will allow us to make use of entropy-redistribution schemes and may provide a way to access the low-temperature phases of the Hubbard model using ultracold atoms.

# Observation of four-body ring-exchange interactions and anyonic fractional statistics in optical lattices

Han-Ning Dai[1,2,3], Bing Yang[1,2,3], Andreas Reingruber[2,5], Hui Sun[1,3], Xiao-Fan Xu[2], Yu-Ao Chen[1,3,4], Zhen-Sheng Yuan[1,2,3,4], Jian-Wei Pan[1,2,3,4]

[1]Hefei National Laboratory for Physical Sciences at Microscale and Department of Modern Physics, University of Science and Technology of China, Hefei, Anhui 230026, China.

[2]Physikalisches Institut, Ruprecht-Karls-Universität Heidelberg,Im Neuenheimer Feld 226, 69120 Heidelberg, Germany

[3]CAS Centre for Excellence and Synergetic Innovation Centre in Quantum Information and Quantum Physics, University of Science and Technology of China, Hefei, Anhui 230026, China

[4]CAS-Alibaba Quantum Computing Laboratory, Shanghai 201315, China

[5]Department of Physics and Research Center OPTIMAS, University of Kaiserslautern,Erwin-Schroedinger-Strasse, Building 46, 67663 Kaiserslautern, Germany

Ring exchange is an elementary interaction for modeling unconventional topological matters which hold promise for efficient quantum information processing [1-3]. We report the observation of four-body ring-exchange interactions and the topological properties of anyonic excitations within an ultracold atom system. A minimum toric code Hamiltonian, in which the ring exchange is the dominant term, was implemented in disconnected four-spin plaquette arrays formed by two orthogonal superlattices. The ring-exchange interactions were resolved from the dynamical evolutions of the spin orders in each plaquette, matching well with the predicted energy gaps between two anyonic excitations of the spin system. A braiding operation was applied to the spins in the plaquettes and an induced phase 1.00(3) $\pi$ in the four-spin state was observed, confirming $\frac{1}{2}$-anynoic statistics. This work represents an essential step towards studying topological matters with ultracold atoms and will contribute to the development of topological quantum computing.

[1] A. Y. Kitaev, Ann. Phys. **303**, 2 (2003).
[2] B. Paredes, I. Bloch, Phys. Rev. A **77**, 023603 (2008).
[3] H. Büchler, M. Hermele, S. Huber, M. P. Fisher, P. Zoller, Phys. Rev. Lett. **95**, 040402 (2005).

# Can theoretical techniques from cold atoms shed light on quantum biology?

David A. W. Hutchinson[1,2], Sue Ann Oh[1], J. A. Provazza[3], M. K. Lee[3], P. Huo[4] and D. F. Coker[3]

[1]Dodd-Walls Centre, Department of Physics, University of Otago, Dunedin, New Zealand
[2]Centre for Quantum Technologies, National University of Singapore, Singapore 117543
[3]Department of Chemistry, Boston University, 590 Commonwealth Avenue, Boston MA 02215 USA
[4]Department of Chemistry, University of Rochester, Rochester, NY 14627-0216 USA

In 2007 long-lived coherences in exciton transport where observed using in the Fenna-Mathews-Olsen (FMO) complex. FMO is a protein complex found in light harvesting bacteria that live in low light environments and is responsible for the transfer of energy from the antenna to the reaction centre in photosynthesis. It does this with near unity efficiency. The implication is that perhaps, in this open system where the active chromophores are coupled to the thermal bath of the surrounding protein scaffold, nature could have utilized quantum effects to enhancing energy transport.

The history of seeking quantum effects in biology is long. Schrödinger's *What is Life?* is the classic starting point, but Frohlich in 1968 also considered the role of long range coherence and energy storage in biological systems.

Here we review some of the evidence for quantum effects in photosynthesis and describe how methods from open quantum systems can be used to model these complexes. We concentrate on our open quantum system modeling results and some thoughts on the role of symmetries in optimizing exciton transport together with the importance of an accurate description of the coupling to the excitation spectrum of the surrounding protein scaffold.

# Cooperative single-photon subradiant states

H. H. Jen[1], M.-S. Chang[2], Y.-C. Chen[2]
[1]Institute of Physics, Academia Sinica, Taipei 11529, Taiwan
[2]Institute of Atomic and Molecular Sciences, Academia Sinica, Taipei 10617, Taiwan

We propose a set of subradiant states which can be prepared and detected in a one-dimensional optical lattice. We find that the decay rates are highly dependent on the spatial phases we manipulate to imprint on the atomic system, which gives systematic investigations of the subradiance in the fluorescence experiments. The time evolution of these states can have long decay time where hundred milliseconds of lifetime is predicted up to one hundred atoms. They can also show decayed Rabi-like oscillations with a beating frequency determined by the difference of cooperative Lamb shift in the subspace. Experimental requirements are also discussed for practical implementation of the subradiant states. Our proposal provides a novel scheme for quantum storage of photons in arrays of two-level atoms through the preparation and detection of subradiant states, which offer opportunities for quantum information processing in optical lattices.

# Observation of polarization rotation spectra for the D2 lines of $^{85}$Rb and $^{87}$Rb atoms

Seung-Chul Yang, Heung-Ryoul Noh
Department of Physics, Chonnam National University, Gwangju 61186, Korea

We present an experimental and theoretical study on the polarization rotation by using two independent lasers of 780 nm in the $D_2$ lines of $^{85}$Rb and $^{87}$Rb atoms. A circularly polarized control beam generates a circular anisotropy in an Rb vapor cell. Then, the refractive indices felt by two orthogonal circular components of the co-propagating linearly polarized probe beam become different from each other and polarization vector of the probe beam rotates. This rotation angle is detected by measuring the difference in the transmissions of the two orthogonal linear components of the probe beam after traversing the cell. In the experiment for $^{85}$Rb ($^{87}$Rb) atoms, the control laser frequency is scanned around the $F_g$=3→$F_e$=2,3,4 ($F_g$=2→$F_e$=1,2,3) transitions and the probe laser frequency is locked to the $F_g$ =3→$F_e$=4 ($F_g$ =2→$F_e$=3) transition. We observe the dependence of the polarization rotation spectra on the control beam power. The experimental results are compared with the results calculated from the full-level density matrix equations, and good agreement between them is found.

# Large Cross-Phase Modulations at the Few-Photon Level

Zi-Yu Liu[1], Yi-Hsin Chen[2], Yen-Chun Chen[1], Hsiang-Yu Lo[1], Pin-Ju Tsai [1],
Ite A. Yu [2], Ying-Cheng Chen[3], Yong-Fan Chen[1]

[1]Department of Physics, National Cheng Kung University, Tainan 70101, Taiwan

[2]Department of Physics, National Tsing Hua University, Hsinchu 30013, Taiwan

[3]Institute of Atomic and Molecular Sciences, Academia Sinica, Taipei 10617, Taiwan

We demonstrate an efficient cross-phase modulation (XPM) based on a closed-loop double-Λ system. The property of the double-Λ medium can be controlled by changing the relative phase of the applied optical fields. This phase-dependent XPM scheme can achieve large phase modulations at low-light intensities and does not require a cavity. With this scheme, we observe a $\pi$-level phase shift of a few-photon pulse induced by another few-photon pulse in cold rubidium atoms. Such novel scheme can work at the few-photon level, and hence it may have potential for applications in all-optical quantum signal processing.

# Spin-Orbit coupling induced Back-action Cooling in Cavity-Optomechanics with Bose-Einstein condensate

Kashif Ammar Yasir, Wu-Ming Liu
Beijing National Laboratory for Condensed Matter Physics, Institute of Physics, Chinese Academy of Sciences, Beijing 100190, China.

Cavity-optomechanics, an exploitation of mechanical effects of light to couple localized optical field with mechanical objects, has played a vital role in quantum-classical interface [1-4]. Meanwhile, spin-orbit (SO) coupling, interaction between spin of a quantum particle and its momentum, has provided foundation to analyze various novel phenomena like spin Hall effect and topological insulator [5]. However, in realistic understanding and implementation of such optomechanical systems, one of the major obstacle or challenge is thermo-mechanical heating of vibrating mirror of the system due to associated nonlinear quantum noises which we intend to solve by the inclusion of spin-orbit coupled hyperfine atomic states. We investigate spin-orbit coupling induced back-action cooling [6] in cavity-optomechanics with spin-orbit coupled Bose-Einstein condensate and a vibrating side mirror [7]. The collective density excitations of hyperfine states - serving as topological oscillators equally coupled to the cavity field - trigger strongly driven atomic back-action. This back-action not only revamps topological properties and low-temperature dynamics of its own but also provides opportunity to cool mechanical mirror to its quantum mechanical ground-state. Further, with spin-orbit coupling, we can superintend dynamic structure factor and squeeze nonlinear quantum noises, like thermo-mechanical and photon shot noise, which enhances optomechanical features of hybrid cavity beyond the previous investigations. Our findings are testable in a realistic setup and provide foundations to manipulate spin-orbit coupled topological states in the field of quantum optics and quantum computation.

[1] M. Aspelmeyer, T. J. Kippenberg, and F. Marquardt, Rev. Mod. Phys. **86**, 1391 (2014).
[2] T. J. Kippenberg, and K. J. Vahala, Science **321**, 1172 (2008).
[3] K. A. Yasir, and W. M. Liu, Sci. Rep. **5**, 10612 (2016).
[4] K. A. Yasir, and W. M. Liu, Sci. Rep. **6**, 22651 (2016).
[5] Y. J. Lin, K. Jimenez-Garcia, and I. B. Spielman, Nature **471**, 83 (2011).
[6] A. D. O'Connell *et al.*, Nature **464**, 697 (2010).
[7] K. A. Yasir, and W. M. Liu, arXiv:1511.01109v1 [quant-ph], (2016).

# Real-space imaging of a topological edge state with ultracold atoms in an amplitude-chirped optical lattice

Martin Leder[1], Christopher Grossert[1], Lukas Sitta[1], Maximilian Genske[2], Achim Rosch[2], Martin Weitz[1]

[1]Institut für Angewandte Physik, Universität Bonn, Wegelerstr. 8, 53115 Bonn, Germany

[2]Institut für Theoretische Physik, Universität zu Köln, Zülpicher Str. 77, 50937 Cologne, Germany

In order to distinguish topological state of matter, as topological insulators, from ordinary matter global measurements are required, revealing topological invariants as the Chern number [1]. At the heart of topological materials are topologically protected edge states that occur at the intersection between regions of different topological order. To study topological states of matter, ultracold atomic gases in optical lattices represent a promising new platform [2], though the observation of edge states has so far been restricted in these systems to the state space imposed by the internal atomic structure [3]. Here we report on the observation of an edge state between two topological distinct phases of an atomic physics system in real space using optical microscopy [4]. An interface between two spatial regions of different topological order is realized in a one-dimensional optical lattice of spatially chirped amplitude. To reach this, a magnetic field gradient causes a spatial variation of the Raman detuning in an atomic rubidium three-level system and a corresponding spatial variation of the coupling between momentum eigenstates. This novel experimental technique realizes a cold atom system described by a Dirac equation with an inhomogeneous mass term closely related to the SSH model [5]. The observed edge state is characterized by measuring the overlap to various initial states, revealing that this topological state has singlet nature in contrast to the other system eigenstates, which occur pairwise. We also determine the size of the energy gap to the adjacent eigenstate doublet. Our findings hold prospects for the spectroscopy of surface states in topological matter and for the quantum simulation of interacting Dirac systems.

[1] M. Z. Hasan and C. L. Kane, Rev. Mod. Phys. **82**, 3045-3067 (2010).

[2] J. Dalibard et al., Rev. Mod. Phys. **83**, 1523–1543 (2011).

[3] B. K. Stuhl et al., Science **349**, 1514-1518 (2015).

[4] M. Leder et al., arXiv:1604.02060 (2016).

[5] W. P. Su, J. R. Schrieffer and A. J. Heeger, Phys. Rev. Lett. **42**, 1698 (1979).

# Reconstruction of the Jaynes-Cummings field state of ionic motion in a harmonic trap

Dingshun Lv[1], Shuoming An[1], Mark Um [1], Junhua Zhang [1], Zhenyu Liu [1], Jing Ning Zhang [1], M.S. Kim [2], Kihwan Kim [1]

[1]Center for Quantum Information, Institute for Interdisciplinary Information Sciences, Tsinghua University, Beijing

[2]QOLS, Blackett Laboratory, Imperial College London, SW7 2AZ, United Kingdom

We experimentally measure the Q function of the vibrational motion for a 171Yb ion interacting with its internal energy states. The scheme for the Q-function is based on highly efficient zero phonon detection using the transitionless adiabatic passage for the the external and internal energy states interaction[1-3]. Under the Jaynes-Cummings dynamics, the Gaussian peak of the initial coherent state has been known to bifurcate and rotate clockwise and counter-clockwise around the origin of phase space[4]. They merge at the so-called revival time at the other side of phase space. The measured Q-function agrees with the theoretical prediction. We confirm the dynamics being coherent by showing that the time reversal brings the state back to the initial peak. At a half revival time, the motional state becomes nearly disentangled from the internal energy states and formes a superposition of two composite states which is confirmed by negativities in the reconstructed Wigner function. The scheme can be applied to other physical setups including circuit-QED and opto-mechanical systems.

[1] M. Berry, Journal of Physics A: Mathematical and Theroretical,**42**, 365303 (2009)

[2] J. Zhang, J. Zhang, X. Zhang, and K. Kim, Phys. Rev.A **89**, 013608 (2014)

[3] Mark. Um. et al, Nature Commun. **7**, 11410 (2016).

[4] J. Gea-Banacloche, Phys. Rev.A **44**, 5913 (1991).

# Combining 1D Nanoscale Waveguides and Cold Atoms

Neil Corzo, Baptiste Gouraud, Aveek Chandra, Julien Laurat

Laboratoire Kastler Brossel, UPMC-Sorbonne Universités, CNRS, ENS-PSL Research University, Collège de France. 4 place Jussieu, Paris, France

Reversible light-matter interfaces are crucial elements in quantum optics and quantum information networks. In particular, the combination of cold atoms and nanoscale waveguides appears as a promising pathway to build light-matter interaction thanks to the tight confinement of light. Recently, our group has developed an interface where light, tightly guided by a subwavelength-diameter optical fiber, strongly interacts with free or trapped atoms near its vicinity. Here, I will describe two of our more recent results using this interface: a) the demonstration of slow light and optical storage at the single-photon level in this all-fibered setting [1]; and b) the observation of a large Bragg reflection off a 1D optical lattice [2].

Our first experimental result using the nanofiber interface is the experimental observation of a slow-light effect and coherent storage in a setting where light is tightly confined in the transverse direction [1]. By interfacing a tapered optical nanofiber with a cold atomic ensemble, electromagnetically induced transparency is observed and light pulses at the single-photon level are stored in and retrieved from the atomic medium with an overall efficiency of 10%. Remarkably, the single-photon signal-to-noise ratio in the retrieved pulse is equal to 20. Collapses and revivals are additionally controlled by an applied magnetic field. The involved decoherence mechanisms are carefully studied. Our result opens up the possibility to implement the seminal Duan-Lukin-Zoller protocol, with the goal of an efficient generation of intrinsically fibered single photons and the remote entanglement of all-fibered ensembles.

In the second experiment, we observe a large Bragg reflection from a 1D lattice made of trapped atoms around the nanofiber. Contrary to previous experimental results with atoms around nanofibers which were obtained with a disordered or incommensurate array of atoms, and only relied on the resulting optical depth of the medium, in our experiment we observe cooperative effects arising from atoms trapped in a close-to-commensurate optical lattice, where Bragg condition is fulfilled close to resonance. The optical lattice potential is generated by pairs of counterpropagating red- and blue-detuned trapping beams inside the nanofiber. By carefully selecting the wavelength of the red-detuned beams, we make the lattice constant almost half of the atomic transition wavelength. Using this close-to-resonant dipole-trap we generate a commensurate array of up to 2000 atoms where we observe a strong Bragg reflection as high as 75% [2]. Furthermore, we observe the effect of the chirality of the tightly focused light on this strong reflection. The lifetime of the Bragg reflection is of the order of milliseconds and is mainly limited by the lifetime of the trap. The ability to control photon transport in 1D waveguides coupled to spin systems would allow for novel quantum network capabilities and many-body effects emerging from long-range interactions.

[1] B. Gouraud *et al.,*Phys. Rev. Lett. **114**, 180503 (2015).

[2] N. Corzo *et al.*, arXiv: 1604.03129 (2016), submitted.

# Nanophotonic Cavity QED with Trapped Neutral Atoms

C. Senko[1], H. Bernien[1], S. Schwartz[1], P. Samutpraphoot[1], T. Dordevic[1], A. Zibrov[1], V. Vuletic[2] , M. Lukin[1]

[1]Harvard University, Department of Physics, Cambridge, MA, USA

[2] MIT, Department of Physics and Research Laboratory of Electronics, Cambridge, MA, USA

Nanophotonics has recently become a toolbox for studying cavity QED in the strong coupling regime. The current technology in trapping and cooling allows us to localize single atoms in the evanescent mode of a photonic crystal cavity. The atoms interact with the light field with high cooperativity due to the small mode volume at sub-wavelength dimensions. This platform enables construction of an efficient atom-photon interface for quantum networks and for the study of quantum many-body physics. We present the ongoing effort towards creating an entangled state of two trapped atoms mediated by cavity photons, building on the previous results on trapping a single atom near the photonic crystal, and the realization of a phase switch [1,2].

[1] J. D. Thompson, T. G. Tiecke, N. P. de Leon, J. Feist, A. V. Akimov, M. Gullans, A. S. Zibrov, V. Vuletić, and M. D. Lukin, Science **340**, 1202 (2013)

[2] T. G. Tiecke, J. D. Thompson, N. P. de Leon, L. R. Liu, V. Vuletic and M. D. Lukin, Nature **508**, 241 (2014)

# Polarization dependence and discrimination of coherence effect in V-type electromagnetically induced transparency of $^{87}$Rb

Hyun-Jong Kang, Heung-Ryoul Noh
Chonnam National University, Gwangju 61186 Korea

We present a theoretical and experimental study of electromagnetically induced transparency (EIT) in V-type systems of $^{87}$Rb atoms [1]. The probe beam frequency is locked to the $F_g=2 \rightarrow F_e=2$ transition of D1 line and the coupling beam frequency is scanned around the $F_g=2 \rightarrow F_e=1,2,3$ transitions of D2 line. We calculate accurate line shapes of V-type EIT spectra using density matrix equations by considering all the magnetic sub-levels and compare them with experimental results. To discriminate the contribution of coherence term, we perform similar calculations by ignoring the coherences between the sublevels in the $5P_{1/2}$ state and those in the $5P_{3/2}$ state. We find that the coherence effect is significant only for the cycling transition line. We also investigate the EIT spectra for the linearly polarized coupling and probe beams in parallel or perpendicular direction.

[1] D. J. Fulton, S. Shepherd, R. R. Mosely, B. D. Sinclair, M. H. Dunn, Phys. Rev. A **52**, 2302 (1995).

# EIT-based Photonic Memory with Near-Unity Storage Efficiency

Ya-Fen Hsiao[1,2], Pin-Ju Tsai[1,3],Hung-Shiue Chen[1], Sheng-Xiang Lin [1],Chih-Chiao Hung[1], Chih-Hsi Lee[1], Yi-Hsin Chen[4], Yong-Fan Chen[5], Ite A. Yu[4] , Ying-Cheng Chen[1]

[1]Institute of Atomic and Molecular Sciences, Academia Sinica, Taipei, Taiwan

[2]Molecular Science and Technology, Taiwan International Graduate Program, Academia Sinica and National Central University, Taipei, Taiwan

[3]Department of Physics, National Taiwan University, Taipei, Taiwan

[4]Department of Physics, National Tsing Hua University, Hsinchu, Taiwan

[5]Department of Physics, National Cheng Kung University, Tainan, Taiwan

Quantum memory is an important component in the long-distance quantum communication based on the quantum repeater protocol. To outperform the direct transmission of photons with quantum repeaters, it is crucial to develop quantum memories with high fidelity, high efficiency and a long storage time. Here, we present our work to achieve a storage efficiency of 96% for a photonic memory based on the electromagnetically induced transparency (EIT) scheme in cold atomic media with an optical depth near 1000. At a storage efficiency of 50%, we also obtain a fractional delay of 1200. At high optical depths, nonlinear optical effects, such as the photon switching and four-wave mixing due to the off-resonant excitation of the EIT control field, may become significant and introduce complications in quantum memory applications. By operating the EIT-based photonic memory with cesium D1 line in addition with Zeeman optical pumping, we show that the photon switching effect, which could limits the storage efficiency at high ODs, can be completely eliminated compared to that of D2 line scheme. We also show that the four-wave mixing effect, which could degrades the fidelity of a quantum memory, can be negligible even with an optical depth of 1000 due to the relative large ground-state hyperfine splitting for cesium and a choice of geometry between the control and probe beam to break the phase matching condition.Our work significantly advances the pursuit of a high-performance optical memory and may find important applications in quantum information science.

# Two-Photon Emission from Coherently Excited Solid Parahydrogen

Hideaki Hara, for the SPAN collaboration
Research Institute for Interdisciplinary Science, Okayama University, Okayama, Japan

Coherence of material interacting with photons induces various interesting phenomena such as spin or photon echo and electromagnetically induced transparency. Superradiance is known as one familiar example of such coherent phenomena. In this case, a single photon emission rate is explosively enhanced due to material coherence. We (SPAN project: SPectroscopy of Atomic Neutrino) have proposed to study extremely small rate processes which emit plural particles involving a neutrino pair with a photon (RENP: Radiative Emission of Neutrino Pair) [1]. Our final goal is to investigate unknown neutrino properties such as absolute masses, mass type (Dirac or Majorana), and CP-violating phases from spectra of RENP. In this research, enhancement of the rare process emitting multi particles is crucial. We have proposed and demonstrated a rate amplification mechanism by using material coherence. We observe coherently amplified two-photon emission signal from the vibrationally excited state of gaseous parahydrogen molecules (p-$H_2$) [2]. This is a QED process without any weak interaction, but it is easier to observe than RENP. While this amplification mechanism is confirmed, realization of high coherence in high density materials is desirable for further study of coherent phenomena. In general, coherence in condensed phases is fragile due to strong interaction with environment. However, decoherence time of vibrational states of solid p-$H_2$, which is estimated from the observed linewidth of the Raman transition [3], is exceptionally longer than those in typical solids. Recently we have implemented two-photon emission from coherently excited solid p-$H_2$. We prepare coherence among the ground and the vibrationally excited states ($\lambda$ = 2410 nm) by the adiabatic Raman method. We inject trigger pulses ($\lambda$ = 4587 nm) for the two-photon emission, then observe coherent two-photon emission signal pulses ($\lambda$ = 5077 nm). We measure the dependence of two-photon emission rate on the timing of the trigger pulses. The emission rate is developed even after the Raman excitation pulses pass away and shows a long tail for hundreds of nanoseconds. This behavior is considered to be due to the long decoherence time of solid p-$H_2$. And it was not observed in our previous experiments using a gaseous p-$H_2$. Furthermore, we measure the dependence of two-photon emission rate on the target temperature and the residual orthohydrogen fraction.

[1] A. Fukumi *et al.*, Prog. Theo. Exp. Phys. **2012**, 04D002 (2012).
[2] Y. Miyamoto *et al.*, Prog. Theo. Exp. Phys. **2015**, 081C01 (2015).
[3] T. Momose *et al.*, J. Mol. Spectrosc. **153**, 760 (1992).

# Multi-sublevel effect on the second-order correlation of a single-atom-cavity system

Jinuk Kim[1], Kyungwon An[1]
[1]Department of Physics and Astronomy, Seoul National University, Seoul 151-747, Korea

Second-order correlation is a well-studied subject in quantum optics. By using the second-order correlation, non-classical effect in cavity QED has been extensively investigated [1,2]. Moreover, it is one of the key indicators for single photon generation [3,4]. In particular, polarization dependence of the second-order correlation function has been studied [5-7]. Analytic solution for a two-level atom in a cavity is obtained [8]. However when Zeeman sub-levels of the atom are considered, the system can be studied only numerically or experimentally. Beside the complexity of calculation due to many levels, the entanglement between Zeeman sub-levels of the atom and the polarization of cavity photons make the situation more complicated but plays an important role in correlation functions. We theoretically investigate several intriguing properties of the second-order correlation functions of a single-atom-cavity system originated from such entanglement. First, a mixture of bunched light beams can have anti-bunching property. Second, sub-Poissonian light can be filtered from mixture of super-Poissonian lights. At last, a classical bound of the second-order correlation function between two orthogonal modes is calculated, and its violation is showed numerically.

[1] G. Rempe, R. J. Thompson, R. J. Brecha, W. D. Lee, and H. J. Kimble, Phys. Rev. Lett. **67**, 1727 (1991).

[2] L. Mielke, G. T. Foster, and L. A. Orozco, Phys. Rev.Lett. **80**, 3948 (1998).

[3] P. Michler *et al.*, Science **290**, 2282 (2000).

[4] C. Kurtsiefer, S. Mayer, P. Zarda, and H. Weinfurter, Phys.Rev. Lett. **85**, 290 (2000).

[5] F. Rohde *et al.*, J. Phys. B: At. Mol. Opt. Phys. **43**, 115401 (2010).

[6] H. J. Lee, T. Jeong, and H. S. Moon, Phys. Rev. A **90**, 033843 (2014).

[7] V. A. Sautenkov, Y. V. Rostovtsev, and M. O. Scully, Phys. Rev. A **72**, 065801 (2005).

[8] H. J. Carmichael, R. J. Brecha, and P. R. Rice, Opt. Commun.**82**, 73 (1991).

# Towards a single-atom register in a fiber micro-cavity for generation of multi-particle entanglement

Francesco Ferri, Claire Lebouteiller, Sébastien Garcia, Jakob Reichel, Romain Long
[1]Laboratoire Kastler Brossel, ENS-PSL/CNRS/UPMC-Sorbonne Universités/CdF, 24 rue Lhomond, 75005 Paris, France

Creating and characterizing multi-particle entanglement in systems of material particles is becoming a major focus of experimental quantum physics. Our approach is to combine methods of optical Cavity Quantum Electro-Dynamics (CQED) with quantum gas microscope techniques developed in the field of optical lattices. We are currently building a new experimental platform aiming at extending the generation of multi-particle entanglement to ensembles of up to around 100 neutral atoms while maintaining analysis at the single-atom level. The first goal of this experiment is to realize a single-atom qubit register in a one-dimensional optical lattice, with single lattice site detection and addressing capabilities, and where each site is maximally and identically coupled to the mode of a high-finesse optical cavity.

At the heart of this experimental system is a fiber-based Fabry-Pérot, developed in our group, which allows the creation of entangled states thanks to the collective interaction of the atoms with the cavity mode. We will present the progress we have made in the production of fiber cavities with controlled birefringence. In addition, these new generation of fiber cavities have a finesse of about 40,000 at two wavelengths an octave apart: 780 nm for probing Rubidium atoms and 1560 nm for trapping the atoms inside the cavity. We have developed a method to measure the overlap between the two intra-cavity standing waves, a key ingredient to get a high and homogeneous coupling between the atoms and the cavity field. We will also describe the combination of the fiber cavity and an "atom microscope" with single-site resolution in a compact setup. Finally, a new scheme, based on an acousto-optically displaced trap, will be presented to perform a fast and efficient transport of cold atoms inside a cavity.

This system will provide an ideal test-bed to investigate different methods for multi-particle entanglement generation and to study their fundamental limits. It will also be possible to implement an experimentally adjustable realization of the Dicke model, which exhibits a quantum phase transition at a critical value of the coupling, in the vicinity of which multi-atom entanglement is predicted.

# Coherent and dynamic beam splitting with quantum memory in a cold atomic ensemble

Kwang-Kyoon Park[1], Tian-Ming Zhao[1], Jong-Chan Lee[1], Young-Tak Chough[2], Yoon-Ho Kim[1]

[1]Department of Physics, Pohang University of Science and Technology, Pohang 37673, Korea

[2]Department of Medical Technology, Gwangju University, Gwangju 61743, Korea

Photonic quantum computation has received much attention over the last decade due to the demonstration of scalability using single-photon sources, projective measurement, and linear optical elements such as beam splitters, phase shifters, and mirrors [1]. In addition to these elements, quantum memories are essential in synchronizing multi-photon events to increase the success probability of linear optical quantum gates [2]. Although recent advances in miniaturized photonic circuits on a silicon chip can reduce the experimental overhead in managing many such linear optical elements [3], it does not actually decrease the number of such elements. Here, we demonstrate a cold-atom based approach to consolidate different linear optical operations and a quantum memory in a single device. Our integration work can curtail the number of required elements in that our single device can work as several optical components and is reconfigurable according to requested operations. By employing Electromagnetically-Induced Transparency based light storage [4], we demonstrate that a weak laser pulse can be coherently and dynamically manipulated in a cold atom ensemble to perform beam splitting, reflection, and phase shifting, in addition to quantum memory operation.

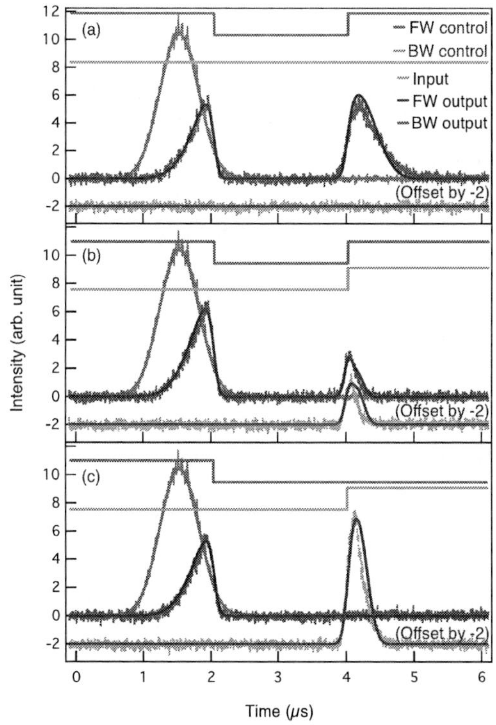


**Figure 1.** Variable beam splitting. (a) All into the FW direction. (b) Both the FW and BW directions. (c) All into the BW direction.

The experimental results for dynamic beam splitting with memory operation are shown in Fig. 1. In Fig. 1(a), only the forward (FW) control is turned on for the FW direction retrieval after the storage. Both the FW and backward (BW) retrievals are demonstrated in Fig. 1(b) when both the controls are applied. Retrieval only into the BW direction is shown in Fig. 1(c).

[1] E. Knill, R. Laflamme, and G. J. Milburn, Nature **409**, 46 (2001).

[2] F. Bussieres, N. Sangouard, M. Afzelius, H. de Riedmatten, C. Simon, and W. Tittel, J. Mod. Opt. **60**, 1519 (2013).

[3] J. Carolan, C. Harrold, C. Sparrow, E. Martín-López, N. J. Russell, J. W. Silverstone, P. J. Shadbolt, N. Matsuda, M. Oguma, M. Itoh, G. D. Marshall, M. G. Thompson, J. C. F. Matthews, T. Hashimoto, J. L. O'Brien, and A. Laing, Science **349**, 711 (2015).

[4] D. F. Phillips, A. Fleischhauer, A. Mair, R. L. Walsworth, and M. D. Lukin, Phys. Rev. Lett. **86**, 783 (2001).

# Non reciprocal devices based on chiral light matter interactions

Adèle Hilico[1], Michael Scheucher[1], Elisa Will[1], Jürgen Volz[1], Arno Rauschenbeutel[1]
[1]Vienna University of Technology, Austria

In order to control the flow of light in an integrated optical environment, one needs to use nanophotonic components which confine the light at the wavelength scale.

This strong confinement leads to an inherent link between the local polarization of the light and its propagation direction - the light obtains a chiral character - and thereby fundamentally alters the physics of light-matter interaction [1].

We employ this effect in order to investigate the realization of novel nonreciprocal optical devices that operate at the single-photon level. For this purpose, we use a single spin-polarized 85Rb atom that is strongly coupled to a novel type of whispering-gallery-mode microresonator - a so-called bottle microresonator [1] - which can be interfaced by two optical nanofibers. These resonators offer the advantage of being fully tuneable and provide very long photon lifetimes in conjunction with near lossless coupling to the nanofibers. Coupling one nanofiber to the bottle microresonator, we studied the on-resonance performance of the system and observed a strong imbalance between the transmissions in forward and reverse direction [2]. This realizes an optical diode whose directional behaviour is controlled by the internal state of a single atom. By interfacing the bottle microresonator with two nanofibers we extended this system to a 4-port device, where photons are nonreciprocally directed from one fibre port to the next. In contrast to the diode such an optical circulator is based on a non-dissipative process which would enable one to employ these devices for quantum information protocols.

[1] C. Junge et al., Phys. Rev. Lett. 110, 213604 (2013).
[2] C. Sayrin et al., Phys. Rev. X 5, 041036 (2015)

# Fiber Integrated Single Photon Source Based on NV Center in Diamond

Stepan Bolshedvorskii[1,2,4], Vadim Vorobyov[1,2,4], Vladimir Soshenko[1,2,3], Javid Javadzade[2,4], Nikolay Lebedev[2,4], Vadim Sorokin[1,2], Andrey Smolyaninov[3], Alexey Akimov[5,1,2]

[1]Lebedev Physical Institute RAS, 53 Leninskij Prospekt, Moscow 119991, Russia

[2]Russian Quantum Center, 100 Novaya Str., Skolkovo, Moscow 143025, Russia

[3]Photonic Nano-Meta Technologies, Str. Nobel b.7, The territory of Skolkovo Innovation Center, Moscow, 143026, Russia

[4]Moscow Institute of Physics and Technology, 9 Institutskiy per., Dolgoprudny, Moscow Region, 141700, Russia

[5]Texas A&M University, College Station, TX 77843, USA

Nowadays handling of the NV center rely on complicated and often low efficient technics. There were several attempts to build a robust optical interface to NV center: by using pulled fibers and AFM setups, photonics crystal fibers and placing nanocrystals inside air holes; by making waveguides within bulk diamond substrate [1,2]. Here we present novel and robust fiber based technology of spin photon interface for the NV center nanodiamonds. Our approach is easy, reproducible and cryogenic-temperature compatible [3]. We carefully studied parasitic fluorescence of the fiber (Fig 1 (c)) and methods to reduce it below the level of NV center emission. With our technic we were able to achieve collection efficiency 1.5% from NV center in one side of the fiber, however, it is limited by the imperfections (transmission of the our fiber line with hand-made connectors from NV center to APD is about 10%). This way we can suppose, that collection efficiency at pulled section of the fiber was about 15% from one side. We expect that using manufactured fiber connectors will give us opportunity to reach mentioned above collection efficiency from single NV center nanodiamonds. Our approach could be further developed using recently demonstrated idea of utilizing fiber Bragg cavity [4]. This way one could use power of cavity quantum electrodynamics within relatively simple technics. This way our approach may become basis for many applications with NV centers in diamond.

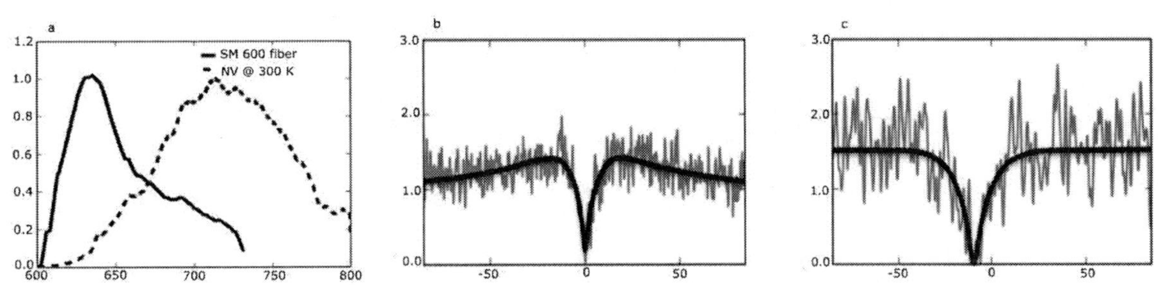


**Figure 1.** (a) Spectra of parasitic fiber fluorescence versus NV fluorescence. (b) Second order autocorrelation function (G2) of the single NV on pulled fiber, collected with objective (c) G2 function collected with tapered fiber.

[1] T. Schroder, et. al., Nano Letters **11**, 198-202 (2011).
[2] M. J. Burek, et. al., Nature Communications **5**, 6718 (2014).
[3] M. Fujiwara, et. al., arxiv: 1601.06188 (2010).
[4] A. W. Schell, et. al., Scientific Reports **5**, 9619 (2015).

# SiV Centres in Microcavities – An Efficient Single Photon Source at Room Temperature

Julia Benedikter[1,2], Elke Neu[3], Christoph Becher[3], Theodor W. Hänsch[1,2], David Hunger[1,2]
[1]Ludwig-Maximilians-Universität, Schellingstraße 4, 80799 München, Germany
[2]Max-Planck-Institut für Quantenoptik, Hans-Kopfermann-Straße 1, 85748 Garching, Germany
[3]Universität des Saarlandes, Campus E2 6, 66123 Saarbrücken, Germany

Single photon sources are an integral part of various quantum information applications. Solid state emitters offer on-demand single photon emission without the need for very involved set-ups [1]. The emission properties, especially the very narrow single phonon line, and stability of the silicon vacancy centre in diamond make it a promising candidate for a single photon source at room temperature. We use fibre-based microcavities [2, 3] to Purcell-enhance and efficiently collect the emission of single SiV centres in nanodiamonds. We operate in the bad emitter regime, where a cavity with a mode volume of a few cubic wavelengths can achieve high effective Purcell factors up to about 20 and near unity collection efficiency.

We report on measurements on narrow-line bright single SiV centres in free space and in an ultra-small mode volume cavity. We observe life time reduction and an enhanced detected photon rate with respect to free space. We thereby demonstrate a narrow-band, bright, and clean single photon source at room temperature. Furthermore, we compare the rate dynamics and discuss an empirical model for shelving and deshelving processes.

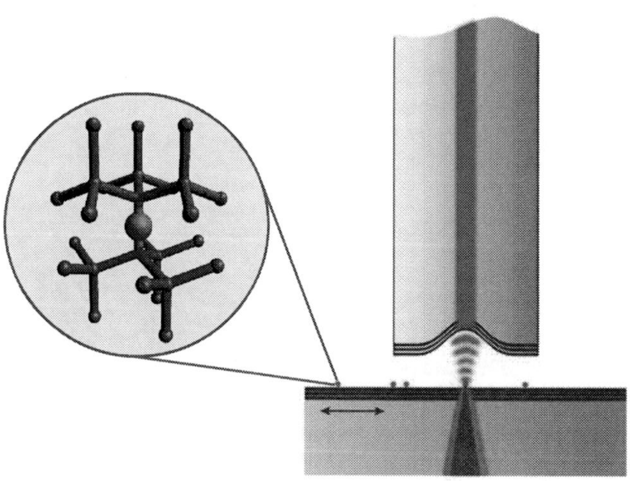

[1] Kaupp, H. et al., arXiv:1606.00167 (2016).
[2] Hunger, D. et al., New J. Phys. **12**, 065038 (2010).
[3] Hunger, D. et al., AIP Advances **2**, 012119 (2012).

# Temporal pure single photons heralded from entangled biphotons generated from cold atomic ensembles

J.F.Chen[1], Peng Qian[1], Zhenjie Gu[1], Rong Cao[1], Rong Wen[1], Z.Y.Ou[2], Weiping Zhang[1],
[1]Department of Physics, East China Normal University, Shanghai, China
[2]Department of Physics, Indiana University-Purdue University, Indianapolis, Indiana, USA

In quantum information processing and quantum network, single-photon-source in Fock state is widely considered to be the most potential candidate as a qubit. It is demonstrated to be available for transferring quantum information and realizing quantum repeater. On one hand, the single photon should be in a temporally pure state in order to carry quantum information efficiently. On the other hand, the time-frequency entanglement is advantageous since it provides a route to manipulate the single-photon-state. Concerning entangled photon pairs and the corresponding heralded single photons, it is commonly believed that, the time-frequency entanglement will damage the temporal purity of the single photons through heralding [1]. Here through observing the Hong-Ou-Mandel interference from independent single photons (Figure 1), we demonstrate that the biphotons with coherence time of 100ns or more, which are generated from spontaneous four-wave mixing process in cold atomic ensembles, are able to maintain the purity and entanglement simultaneously in time domain. When the time response of the detector which registers the trigger photons is shorter than 1/10 of the photon coherence time [2,3], the temporal purity of the single photons is approaching unity.

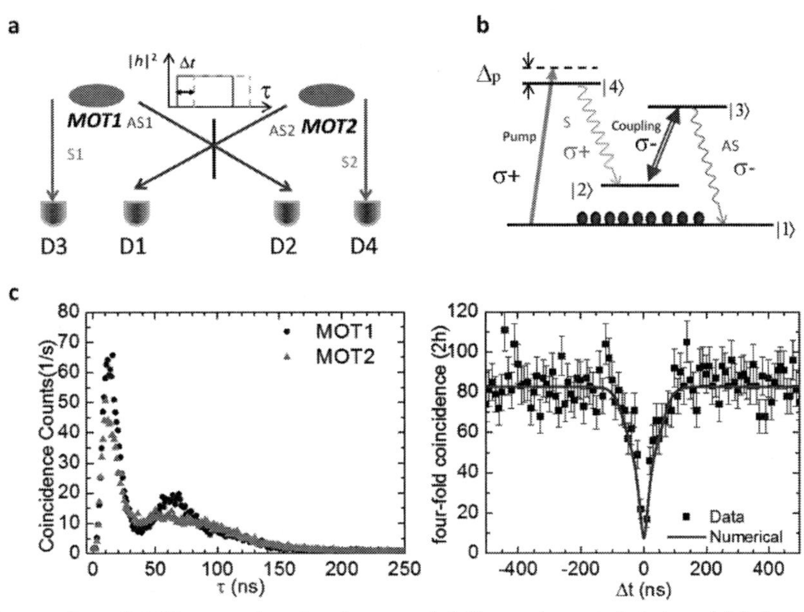


**Figure 1.** (a) Schematics; (b) Energy level scheme; (c) Experimental data of biphoton waveform (red triangles and dark dots in the left graph) and HOM interference (squares with error bars in the right).

[1] A. I. Lvovsky, and M. G. Raymer, Rev. Mod. Phys. **81**, 299 (2009).
[2] S. Du, Phys. Rev. A **92**, 043836 (2015).
[3] P. Qian, *et al.*, Phys. Rev. Lett (accepted)

# Laser-cooled cesium atoms loaded into a hollow-core photonic-crystal fiber with a magic-wavelength optical dipole trap

Taehyun Yoon, Chris Haapamaki, Jeremy Flannery, Michal Bajcsy
Institute for Quantum Computing, University of Waterloo, Waterloo, ON, Canada

A hollow-core photonic-crystal fiber (HCPCF) confining both photons and an atomic ensemble enables strong light-matter interactions which are crucial for implementing optical nonlinearities at single-photon levels. We developed an experimental setup that traps and cools cesium atoms in a magneto optical trap (MOT) and then loads them into a HCPCF (Fig. 1a). The atoms are guided into and inside the fiber by an optical dipole trap at cesium's magic wavelength (935nm). This approach will allow us to perform experiments in the future with the dipole trap continuously on as opposed to the previously reported fiber-loading systems [1] in which the dipole trap was rapidly amplitude-modulated and the atoms probed during the off-times of the trap to avoid the inhomogeneous broadening of the optical transition resulting from the spatially dependent AC Stark shift.

In addition to having a magic wavelength that can be guided by the same HCPCF as its optical transition from the ground state (852nm), cesium can be cooled with polarization-gradient cooling to significantly lower temperatures compared to rubidium, which leads to increased loading efficiency of the atoms into the fiber. Fig. 1b shows the basic signature of atoms loading into the HCPCF. Approximately $10^8$ atoms are collected ~7mm above the fiber tip and cooled to 20μK with polarization gradient cooling. Following this, the cooling laser is shut off and the atomic cloud falls down to the vicinity of the fiber tip where the dipole trap guides the atoms into the fiber core. The optical depth (OD) of the cloud, which is probed during the cloud's fall by a low-power probe laser coupled into the bottom end of the HCPCF, rises sharply as the atoms load into the fiber core. With our simulations of the atoms' motion in the potential created by the diverging dipole beam (Fig. 1c), we estimate that a loading efficiency of ~0.001 should be achievable once the cloud dropping procedure is optimized, leading to a medium with OD of ~500. OD exceeding 2000 should be achievable with further temperature reduction of the initial cloud to below 5μK which would open access to exotic single-photon optical nonlinearities arising in coherently controlled atomic ensembles of large OD [2]. These nonlinearities can be further enhanced by integrating cavities into the fiber, as proposed by [3].

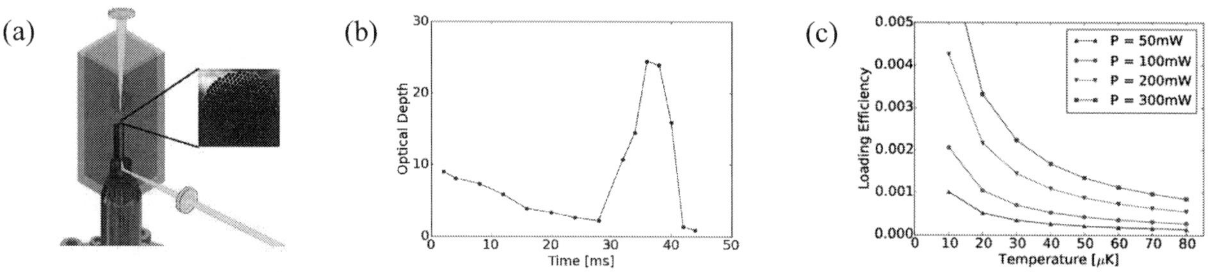


**Figure 1.** (a) Schematic of the mounting of the fiber inside the glass cell and the basic method of coupling light into the fiber. (b) Measured optical depth with the laser shut off at 0ms. (c) Simulated loading efficiency as a function of the MOT temperature for various dipole trap laser power.

[1] M. Bajcsy et al., Phys. Rev. A **83**, 063830 (2011); F. Blatt et al., Opt. Lett. **39**(3), 446 (2014)
[2] D.E. Chang et al., Nat. Photonics **8**, 685 (2014)
[3] C.M. Haapamaki et al., Nanophotonics (2016), Ahead of print

# Generation of a squeezed state and a Schröodinger-cat-like state in a cavity by injecting opposite-phase atomic dipoles

Daeho Yang[1], Junki Kim[1], Moonjoo Lee[2], Young-Tak Chough[3], Kyungwon An[1]

[1]Department of Physics & Astronomy, Seoul National University, Seoul 08826, Korea
[2]Institut für Experimentalphysik, Universtität Innsbruck, Technikerstraβe 25, 6020 Innsbruck, Austria
[3]Department of Medical Technology, Gwangju University, Gwangju 61743, Korea

Generation of a non-classical state has attracted a great deal of interest owing to its unconventional aspects and applications. Especially, squeezed states as well as Schröodinger-cat states have been extensively investigated due to their usefulness in quantum information processing and in quantum metrology. Here, we propose a new method to generate both squeezed states and Schröodinger-cat-like states. Our method of generating squeezed states does not have a limitation in the intracavity quadrature squeezing in contrast to 3dB limitation of the conventional methods. Our method uses continuously injected dipoles with opposite phases represented by $|\Psi_{\pm}\rangle = \alpha|g\rangle \pm \beta|e\rangle$. Opposite-phase dipoles, which can be regarded as atomic squeezed states, traverse the cavity and form a vacuum squeezed state in the cavity by transferring atomic squeezing to the field. When a photon click occurs outside the cavity, the vacuum squeezed state undergoes a quantum jump, and as a result a Schröodinger-cat-like state is generated. The Schröodinger-cat-like state can then be revealed by balanced/unbalanced homodyne detection or atomic state tomography. Generation of squeezed states and cat states is first analytically shown and then quantum trajectory simulation is used to analyze various properties of those states. Based on our scheme, we also propose an experiment within the reach of the current quantum optics technology.

# Experimental study of a coherently pumped cavity-QED microlaser

Junki Kim[1], Daeho Yang[1], Seung-Hoon Oh[1], Younghoon Song[1], Moonjoo Lee[2], Kyungwon An[1]

[1] Department of Physics & Astronomy, Seoul National University, Seoul 151-747, Korea

[2] Institute for Quantum Electronics, ETH Zürich, CH-8093 Zürich, Switzerland

Cooperative interaction between a large number of atoms and a common light field can reveal interesting features such as superradiance and subradiance. A coherently pumped cavity-QED microlaser is a microscopic laser pumped by many atoms interacting with a common cavity field with the same phase. It is predicted to shows distinct intensity properties such as counter-intuitive atomic population dependence and thresholdless intensity growth proportional to the number of atoms. To realize such a laser, we precisely controlled the positions of atoms in the nanometer scale utilizing a nanohole array, an atomic beam aperture with an array of nanoholes with a hole spacing matched with the field wavelength. All of the localized atoms were thus prepared in a superposition state of ground and excited states with the same phase, and therefore would experience an in-phase interaction with a low-loss cavity field. In our experiment, because of remained incoherent effect, we observed a peak in laser intensity with an atomic inversion angle of $0.67\pi$ and a nonlinear growth with a log-log slope of 1.40. The experimental details and the latest data will be presented and discussed.

# Investigation of the enhanced spontaneous emission near an exceptional point in an asymmetric microcavity laser

Younghoon Shin[1], Songky Moon[1], Jinuk Kim[1], Soyun Kim[1], Kyungwon An[1],

[1]Department of Physics and Astronomy, Seoul National University, Korea

When a quantum system is non-Hermitian, *i.e.*, if the system is dissipative or if it has amplifying gain, its eigenmodes are not mutually orthogonal. Instead, they have another biorthogonal basis set which is called adjoint modes. Realistic lasers with atomic and cavity damping are naturally non-Hermitian. In these non-Hermitian lasers, it has been known that each cavity mode has an excess-noise factor or the Petermann factor [1] originating from the non-orthogonality of the modes. Several semiclassical studies have shown that the Petermann factor diverges under a special condition called an exceptional point (EP) [2, 3], where two interacting modes coalesce into one with their differential loss exactly equal to their coupling. However, there have been controversies regarding how atomic spontaneous emission into a mode near an EP is modified. Particularly, Y.-J. Cheng *et al.* showed in their quantum analysis that the total spontaneous emission of atoms is not modified by the Petermann factor when the gain-molecules are uniformly distributed [4]. Here we present a full-quantum analysis of both atomic population and photon rate equations in an asymmetric microcavity. We show that even though the atomic total spontaneous emission is unchanged, the spontaneous emission into a mode can be enhanced by the Petermann factor due to the interference between non-orthogonal modes. Such greatly enhanced spontaneous emission into a mode can be exploited in the development of thresholdless lasers.

[1] K. Petermann, IEEE J. Quantum Electron. **QE-15**, 566 (1979).

[2] M. Berry, J. Mod. Opt. **50**, 63 (2003).

[3] S.-Y. Lee *et al.*, Phys. Rev. A **78**, 015805 (2008).

[4] Y.-J. Cheng and A.E. Siegman, Phys. Rev. A **68**, 043808 (2003).

# Experimental generation of quadruple quantum correlated beams from hot rubidium vapor by cascaded four-wave mixing using spatial multiplexing

Jietai Jing, Leiming Cao, and Jian Qi

State Key Laboratory of Precision Spectroscopy, East China Normal University, Shanghai 200062, China

Multimode quantum states, such as multipartite quantum entanglement or quantum correlations, are important for both fundamental science [1] and the future development of quantum technologies [2]. Here, we theoretically propose and experimentally realize a novel scheme that can generate a large number of strong quantum correlated beams from hot rubidium vapor by cascaded four-wave mixing (FWM) using spatial multiplexing. Our scheme is based on cascaded four-wave mixing (FWM) process in hot rubidium vapor. This system has proven to be very successful for a number of applications in quantum information and precision measurements, such as the tunable delay of EPR entangled states [3], and our recent works about the generation of three quantum correlated beams [4-5], the realization of a SU(1,1) interferometer [6-9] and the realization of a two-mode phase sensitive amplifier [10-11].

Our scheme fully exploits the multi-spatial-mode nature of the FWM process, i.e. spatial multiplexing, and thus integrates multiple FWM processes into a single cell at each stage of the cascaded process. In this way, the number of the generated quantum correlated beams (2n) is exponentially dependent on the number of vapor cells (n), which is substantially different from the linear dependence of our previous schemes [4-5]. In addition, the quantum correlations between the multiple beams also increase as the number of the vapor cell increases. For the case of $n = 2$, we experimentally show that the degree of intensity-difference squeezing between the four quantum correlated beams in our scheme is enhanced to $-8.2 \pm 0.2$ dB from $-5.6 \pm 0.3$/ $-6.5 \pm 0.2$ dB of squeezing obtained with a single FWM process. Our system may find applications in quantum information and precision measurement, such as in the generation of multiple quantum correlated images, the realization of a multiport nonlinear interferometer and the construction of the ultra-large-scale quantum networks using spatial multiplexing.

[1] D. M. Greenberger, M.A. Home, and A. Zeilinger, Phys. Today, **46**, 22, (1993).
[2] S. L. Braunstein, and P. Van Loock, Rev. Mod. Physics, **77**, 77, (2005).
[3] A. M. Marino, R.C. Pooser, V. Boyer, and P.D. Lett, Nature, **457**, 859, (2009).
[4] Z. Qin, L. Cao, H. Wang, A. M. Marino, W. Zhang, and J. Jing, Phys. Rev. Lett, **113**, 023602, (2014).
[5] Z. Qin, L. Cao, and J. Jing, Appl. Phys. Lett, **106**, 211104 (2015).
[6] J. Jing, C. Liu, Z. Zhou, Z.Y. Ou, and W. Zhang, Appl. Phys. Lett, **99**, 011110, (2011).
[7] Jun Xin, Xiaozhou Pan and Jietai Jing, Phys. Rev. Lett, (2016) in press.
[8] F. Hudelist, J. Kong, C. Liu, J. Jing, Z.Y. Ou, and W. Zhang, Nat. Communications, **5**, 3049, (2014)
[9] H. Wang, A. M. Marino, and J. Jing, Appl. Phys. Lett, **107**, 121106 (2015).
[10] Y. Fang, and J. Jing, New Journal of Physics **17**,023027 (2015).
[11] Y. Fang, J. Feng, L. Cao, Y. Wang and J. Jing, Appl. Phys. Lett, **108**, 131106 (2016).

# Quantum dynamics of a two-state system induced by a chirped zero-area pulse

Hangyeol Lee, Yunheung Song, Hyosub Kim, Hanlae Jo, Jaewook Ahn
Department of physics, KAIST, Daejeon 34141, Korea

Complete probability oscillation of a two-state system induced by a chirped zero-area pulse is modeled by stepwise evolutions and experimentally demonstrated [1]. A zero-area pulse is a resonant pulse with zero pulse area and a two-state system driven by a zero-area pulse always results in complete probability return (CPR). However, when our zero-area pulse, a resonant pulse which has a spectral hole at the resonance, is chirped, it induces complete probability oscillation of a two-state system. This unanticipated phenomenon is caused by interplay between adiabatic evolution by the chirped main pulse and Rabi rotation by the spectral hole. In adiabatic basis, total time evolution of the system consists of three stepwise evolutions: two Rabi rotations and an adiabatic evolution (z-rotation) between them. Since the z-rotation in the middle can tune interference between two piecewise Rabi rotations, the two-state system undergoes complete probability oscillation as a function of laser pulse intensity and chirp rate. Experimental demonstration was done by using $5S_{1/2}$ and $5P_{1/2}$ states of cold rubidium atoms as two-state systems. Broadband laser pulses from a Titanium-sapphire laser amplifier were spectrally shaped and chirped simultaneously to realize the chirped zero-area pulses. After the interaction, $5P_{1/2}$ state population was selectively ionized and measured, which showed very good agreement with theoretical expectation.

[1] H. Lee, Y. Song, H. Kim, H. Jo, J. Ahn, Phys. Rev. A **93**, 023423 (2016)

# Shortcuts to adiabatic passage for population transfer in quantum three-level system

Yichao Li[1], Xi chen[1]

[1]Department of Physics, Shanghai University, 200444 Shanghai, People's Republic of China

Speed up adabatic passage has been put forward to transfer state with the applications in quantum information processing. In the large detuning condition, we apply the adiabatic elimination to obtain the effective two-level system by SU(2) symmetry[1], and use the unitary transformation[2] to cancel the supplementary interaction. The pumping and Stokes pulses are modified to achieve the fast population transfer, without additional coupling between level |1> and |3>. This method simplifies the experimental implementation of counter-diabatic driving[3-4] and overcomes the difficulty that phase mismatch between Raman laser and microwave.

[1] R. G. Unanyan, L. P. Yatsenko, K. Bergmann, B. W. Shore, Opt. Commun. **139**, 48 (199 7).

[2] S. Ibáñez, X. Chen, E. Torrontegui, J. G. Muga, and A. Ruschhaupt, Phys. Rev. Lett. **109**, 100403 (2012).

[3] M. V. Berry, J. Phys. A **142**, 365303 (2009).

[4] X. Chen, I. Lizuain, A. Ruschhaupt, D. Guéry-Odelin, and J. G. Muga, Phys. Rev. Lett. **105**, 123003 (2010).

# Quantum-optical switching and entanglement of solid-state qubits in an integrated nanophotonic system

R. E. Evans[1*], A. Sipahigil[1*], D. D. Sukachev[1,2*], M. J. Burek[3], J. Borregaard[1], M. K. Bhaskar[1], C. T. Nguyen[1], J. L. Pacheco[4], H. Atikian[3], R. M. Camacho[4], F. Jelezko[5], E. Bielejec[4], H. Park[1,6], M. Lončar[3], M. D. Lukin[1**]

[1]Department of Physics, Harvard University, Cambridge, MA 02138, USA

[2]Russian Quantum Center, Skolkovo, Moscow 143025, Russia

[3]John A. Paulson School of Engineering and Applied Sciences, Harvard University, Cambridge, MA 02138, USA

[4]Sandia National Laboratories, Albuquerque, NM 87185, USA

[5]Institute for Quantum Optics, University Ulm, 89081 Ulm, Germany

[6]Department of Chemistry and Chemical Biology, Harvard University, Cambridge, MA 02138, USA

* These authors contributed equally to this work.

Efficient interfaces between photons and quantum emitters form the basis for quantum networks and enable nonlinear optical devices operating at the single-photon level. We demonstrate an integrated platform for scalable quantum nanophotonics based on silicon-vacancy (SiV) color centers coupled to nanoscale diamond devices. By placing SiV centers inside diamond photonic crystal cavities, we realize a quantum optical switch controlled by a single color center. We show that the switch can be activated using SiV metastable orbital states and verify optical switching at the single-photon level by using photon correlation measurements. We use Raman transitions to realize a single photon source with tunable frequency and bandwidth in a diamond waveguide. Finally, we create entanglement between two separated SiV centers by detecting indistinguishable Raman photons emitted into a single waveguide. Entanglement is verified using a novel superradiant feature observed in photon correlation measurements.

# Quantum engineering of silicon vacancy centers using diamond cantilevers

Young-Ik Sohn[1,*], Srujan Meesala[1,*], Haig A. Atikian[1], Jeffrey Holzgrafe[1], Mian Zhang[1], Michael J. Burek[1], Jose L. Pacheco[2], John Abraham[2], Edward Bielejec[2], Marko Lončar[1]
[1]John A. Paulson School of Engineering and Applied Sciences, Harvard University, USA
[2]Sandia National Laboratories, USA
* These authors contributed equally to this work.

Silicon vacancy (SiV) color centers in diamond have recently emerged as a promising quantum system in the solid state due to their attractive optical properties [1]. In particular, their electronic structure and spin properties were extensively studied, both theoretically and experimentally [2,3]. Interestingly, ground and excited states of SiV consist of two electronic orbitals separated by approximately 47 GHz and 258 GHz, respectively, and the splitting can be tuned by applying external stress due to its large strain susceptibility [4].

In this work, we perform systematic study of the strain response of single SiVs, and measure the strain susceptibility for both ground and excited state manifolds. To apply stress in a controlled manner, we fabricate cantilevers in single crystal diamond substrate in which we deflect them using capacitive force, as shown in Fig. 1(a). SiVs are implanted at deterministic positions via focused ion beam (Fig. 1(b)), and photoluminescence excitation spectra are taken as the applied stress is varied.

By fitting the experimental data to the group-theory model, we confirm the predicted strain Hamiltonian and extract the fundamental strain susceptibility of ~1 PHz/strain. In combination with favorable optical properties, this high strain-susceptibility can allow coupling of SiVs to mechanical resonators, and realization of phonon-based hybrid quantum systems. Furthermore, we will discuss how the large strain-induced orbital splitting achieved with our MEMS device can be used to improve the phonon-limited spin coherence time of the SiVs.

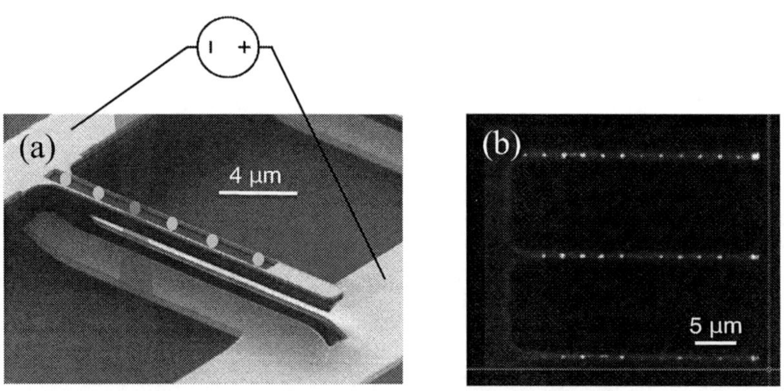


**Figure 1.** (a) SEM of the device. Yellow dots represent deterministic locations where SiVs are implanted via focused ion beam. (b) Scanning fluorescence image of the device.

[1] L. J. Rogers *et al.*, Nature Communications **5**, 4739 (2014).
[2] L. J. Rogers *et al.*, Physical Review B **89**, 235101 (2014).
[3] T. Mueller *et al.*, Nature Communications **5** (2014).
[4] H. Sternschulte, K. Thonke, R. Sauer, and P. C. Münzinger, Physical Review B **50**, 14554 (1994).

# Defect-induced revival of phonon lasing

Hao Lü[1], Jun Qian[1], Yuzhu Wang[1], H. Jing[2]
[1]Key Laboratory for Quantum Optics, Shanghai Institute of Optics and Fine Mechanics, Chinese Academy of Sciences, Shanghai 201800, China
[2]Department of Physics, Henan Normal University, Xinxiang 453007, China

Phonon laser is an acoustic analogy to optical laser and can be realized in a compound cavity setup [1, 2]. Here we theoretically investigate the phonon lasing in a hybid optomechanical system with intrinsic defect that can be described by a two-level system (TLS) [3]. When the TLS damping predominates over the mechanical damping, the photon-phonon-TLS hybrid system reduces to a phonon lasing device. We predict an unexpected defect-induced suppression and revival of the phonon lasing by changing the TLS damping rate, which is analog to loss-induced suppression and revival of lasing in coupled cavities [4]. We contribute the effect to parity-time (PT) phase transition that has been revealed in an open atom-cavity quantum composite [5]. It is found that, in the PT-symmetric regime, energy exchange between the TLS and phonons gives rise to decreasing the mechanical gain and consequently results in the suppression of the phonon laser. On the contrary, in the PT-breaking case, the loss is localized and mechanical gain increases. We also found that the TLS-phonon coupling have remarkable influence on the threshold power of the phonon laser, which shows a saturation feature as increasing the coupling strength. Our results provide a new insight into the role of the intrinsic defect and pave a way to realize phonon amplifiers with loss.

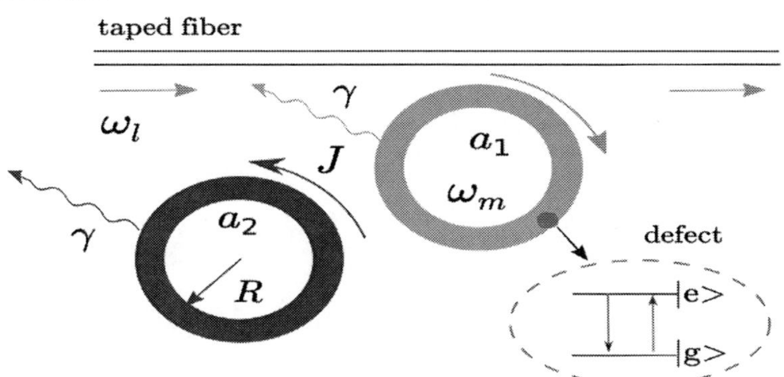


**Figure 1**. Schematic of optomechanical system with compound whispering-gallery resonators.

[1] S. Grudinin, H. Lee, O. Painter, and K. J. Vahala, Phys. Rev. Lett. **104**, 083901 (2010).

[2] H. Jing, S. K. Özdemir, X. -Y. Lü, J. Zhang, L. Yang, and F. Nori, Phys. Rev. Lett. **113**, 053604 (2014).

[3] T. Ramos, V. Sudhir, K. Stannigel, P. Zoller, and T. J. Kippenberg, Phys. Rev. Lett. **110**, 193602 (2013).

[4] B. Peng, Ş. K. Özdemir, S. Rotter, H. Yilmaz, M. Liertzer, F. Monifi, C. M. Bender, F. Nori, and L. Yang, Science **346**, 328 (2014).

[5] Y. Choi, S. Kang, S. Lim, W. Kim, J. -R. Kim, J. -H. Lee, and K. An, Phys. Rev. Lett. **104**, 153601 (2010).

# Optimal shortcut to adiabatic state transfer in Landau-Zener-type systems

Xiaotong Yu[1], Xi Chen[1]

[1]Department of Physics, Shanghai University, 200444 Shanghai, People's Republic of China

Shortcuts to adiabaticity, accelerating adiabatic passage, have been proposed for coherent control in two and three-level quantum systems. However, their physical implementation is sometimes problematic and even impossible, because of the experimental constraint or inflexibility. In this paper, we investigate theoretically the optimal shortcuts to adiabatic state transfer in Landau-Zener-type systems, in which, for example, the detuning is controllable, while the Rabi frequency is considered as a constant, and not changed. By combing the inverse engineering, time-perturbation theory, and optimal control theory, we design the shortcut and further optimize the detuning with respect to various systematic errors in Rabi frequency and detuning, by minimizing the corresponding error sensitivities. Two examples are illustrated, and such dynamics denoted by the trajectories on Bloch sphere are also compared, showing the advantage that the designed shortcuts are more robust than the resonant pi pulse but faster than the conventional adiabatic passages.

[1] A. Ruschhaupt, X. Chen, D. Alonso, and J. G. Muga, New J. Phys. **14**, 093040 (2012).

[2] Xiao-Jing Lu, Xi Chen, A. Ruschhaupt, D. Alonso, S. Guérin, and J. G. Muga, Phys. Rev. A **88**, 033406 (2013).

# Generation of narrowband entangled photon pairs from cold atoms by helicity conservation

Tian-Ming Zhao, Kwang-Kyoon Park, Yong-Sup Ihn, Yoon-Ho, Kim
Department of Physics, Pohang University of Science and Technology, Pohang 37673, Korea

The most important issue in long distance quantum communication is to overcome the losses and decoherence of quantum states through the transmission channels. In order to solve this problem quantum repeater protocol was suggested [1]. A quantum repeater is made up of entanglement swapping, entanglement purification and quantum memory. Recently, quantum memory based on cold atoms has been widely studied. In the cold atom quantum memory system, narrowband entanglement sources are required to fit the narrow linewidth of the atomic transition line. Photons generated via SFWM (spontaneous four-wave mixing) in cold atoms are naturally compatible with the atomic quantum memory. Until now, the generations of narrowband polarized entangled paired photons have been demonstrated in a right-angle scheme [2] or in a small angle scheme with the phase locking [3]. In the right-angle geometry, the pump beam with high power is needed due to the large frequency detuning in the frequency degenerated energy diagram. On the other hand, in the small angle scheme, phase locking makes the experiment to be complicated. In this paper, we demonstrate a new approach to generate polarized entanglement state directly from the SFWM in $^{87}Rb$ atomic ensembles. The simplicity of our scheme, namely the lack of need for phase locking or high power pump, makes this method potentially attractive in the long distance quantum communication.

The experiment setup is the same as that in the previous SFWM experiment [4]. $^{87}Rb$ atoms are trapped in a 2D magneto-optical trap (MOT). In presence of counter-propagation pump and coupling beams, phase-matched backward paired Stokes and anti-Stokes photons are emitted propagating with a small angle with the pump and coupling path. We use pump and coupling beams with orthogonal circular polarizations which are described by $\sigma_p^+$ and $\sigma_c^-$ respectively. Considering the multi Zeeman sublevels, there are two possible configurations for the polarizations of Stokes and anti-Stokes photons satisfying the helicity conservation. If there is no extra magnetic fid, the Zeeman sublevels are frequency degenerate. Thus the two configurations cannot be distinguished, which make us to directly generate polarization entangled states from SFWM in cold atoms. The count rate of this source is 60 pairs per second with a bandwidth of 6MHz. The polarized entanglement is confirmed by quantum state tomography with a fidelity of 89.0%.

[1] L. M. Duan, M. D. Lukin, J. I. Cirac, and P. Zoller, Nature **414**, 413 (2001).

[2] H. Yan, S. C. Zhang, J. F. Chen, M. M. T. Loy, G. K. L. Wong, and S. W. Du, Phys. Rev. Lett. **106**, 033601 (2011).

[3] K. Y. Liao, H. Yan, J. Y. He, S. W. Du, Z. M. Zhang, and S. L. Zhu, Phys. Rev. Lett. **112**, 243602 (2014).

[4] Y. W. Cho, K. K. Park, J. C. Lee, and Y. H. Kim, Journal of the Korean Physical Society **63**, 943 (2013).

# An ion-photon interface for quantum networks and optomechanics

Tracy E. Northup[1], Dario A. Fioretto[1], Konstantin Friebe[1], Moonjoo Lee[1], Florian R. Ong[1], Klemens Schüppert[1], Markus Teller[1], Rainer Blatt[1,2]

[1]Institut für Experimentalphysik, Universität Innsbruck, Technikerstraße 25, 6020 Innsbruck, Austria

[2]Institut für Quantenoptik und Quanteninformation, Österreichische Akademie der Wissenschaften, Technikerstraße 21a, 6020 Innsbruck, Austria

Trapped ions in optical cavities are a promising platform for quantum networks, as they take advantage of tools for coherent manipulation and measurement of ions as well as the light-matter interface provided by the cavity interaction. I will present ongoing work towards a link between two ion-cavity systems: The first is a setup in which we have demonstrated ion-photon entanglement and quantum state mapping, most recently using two calcium ions as a logical quantum bit [1]; the second is a fiber-based cavity with which it should be possible to access the strong coupling regime [2]. A key aspect of this link lies in matching the properties of the single photons produced in the two cavities. I will also discuss the role that an ion in an optical cavity can play as a nonlinear element in quantum optomechanics experiments, drawing on theoretical work that focuses on the preparation of non-Gaussian states [3]. In a new experimental project, an ion and a levitated nanosphere will be coupled to an optical cavity, with the goal of preparing the nanosphere's motion in nonclassical states.

[1] B. Casabone, K. Friebe, B. Brandstätter, K. Schüppert, R. Blatt, T. E. Northup, Phys. Rev. Lett. **114**, 023602 (2015).
[2] B. Brandstätter, A. McClung, K. Schüppert, B. Casabone, K. Friebe, A. Stute, P. O. Schmidt, C. Deutsch, J. Reichel, R. Blatt, and T. E. Northup, Rev. Sci. Instrum. **84**, 123104 (2013).
[3] A. C. Pflanzer, O. Romero-Isart, and J. I. Cirac, Phys. Rev. A **88**, 033804 (2013).

# Shaping a single photon without interacting with it

Denis Sych[1], Valentin Averchenko[1], Gerd Leuchs[1,2]
[1]Max Planck Institute for the Science of Light, Günther-Scharowsky-Straße 1/Bau 24, 91058 Erlangen, Germany
[2]University of Erlangen-Nuremberg, Staudtstraße 7/B2, 91058 Erlangen, Germany

The ability to produce shaped single photons, i.e. photons in a desired spatial-temporal optical mode, is the crucial prerequisite in many modern research areas, particularly, in quantum information science [1].

Methods for photon shaping can be tentatively split into two groups. Methods of the first group shape the photons during the photon generation process by controlling the properties of quantum emitters. The corresponding experimental setups are rather complex, expensive and have limited flexibility. Methods of the second group shape the photons after the photon generation process, and rely either on reshaping or post-selection. They are experimentally simpler, but unavoidable losses that come along with the photon manipulations lead to probabilistic and non-scalable production of shaped photons.

In this work we propose generic single-photon shaping method [2] which is both conditionally lossless, scalable, practical and overcomes the above mentioned limitations. The main idea is schematically illustrated in Fig. 1 and briefly outlined below.

As the main resource the method uses pairs of entangled photons A and B that entangled with respect to the degree of freedom which is subject to shaping. The idler photons B probabilistically pass through a modulator which forms a required single-photon shape. A single photon detector is placed after the modulator and detects idler photons in the basis which is Fourier conjugated to the modulation basis. Clicks of this detector deterministically herald the shaped signal photons (marked green), while the "no-click" cases are discarded (marked white).

The proposed method has several distinctive features. First of all, it is very versatile and can be used to shape single photons with respect to any degree of freedom. Moreover, the method allows for shaping photons with respect to several degrees of freedom simultaneously, e.g. joint shaping of amplitude and polarization. Second, the method does not use any direct manipulation with the signal photon, but only with the idler one. As a consequence of such indirect heralded shaping, shaped photons are produced in a conditionally lossless way. This aspect makes the method scalable for producing many shaped single photons, which is vital for quantum information processing tasks. Last but not least, the method is absolutely feasible to realize in experiment since all the required ingredients are readily accessible.

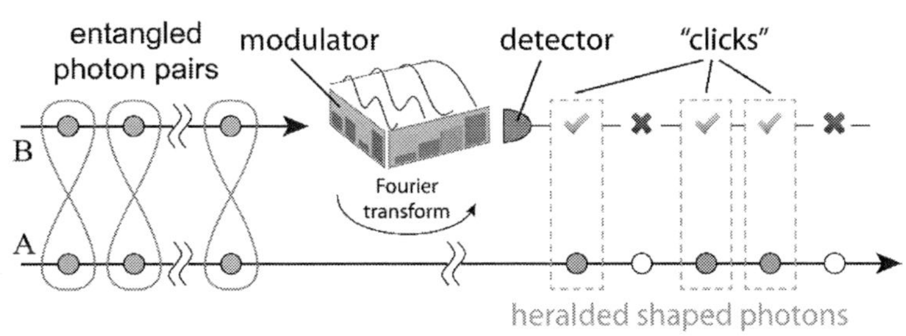


**Figure 1.** Principal scheme of proposed method to produce shaped single photons.

[1] M. Eisaman, J. Fan, A. Migdall, and S. V. Polyakov, Rev. of Sci. Instruments **82**, 071101 (2011).
[2] D. Sych, V. Averchenko, G. Leuchs, arXiv:1605.00023v2 (2016)

# Realization of phonon arithmetic in a quantum system

Mark Um[1], Junhua Zhang[1], Dingshun Lv[1], Yao Lu[1], Shuoming An[1], Jing-Ning Zhang[1], Hyunchul Nha[2,3], M.S. Kim[2,4], Kihwan Kim[1]

[1] Center for Quantum Information, Institute for Interdisciplinary Information Sciences, Tsinghua Univ ersity, P. R. China

[2] Korea Institute for Advanced Study, Seoul, Korea

[3] Texas A&M University at Qatar, Education City, P.O. Box 23874, Doha, Qatar

[4] Blackett Laboratory, QOLS, Physics Department, Imperial College London, Exhibition Road, London SW7 2AZ, UK

We report an the experimental realization of arithmetic conventional arithmetic on a motional mode of a single trapped $^{171}$Yb+ ion trapped in a harmonic potential through the coupling to the internal degre e of freedom, Jaynes-Cummings (JC) interaction. The conventional addition and subtraction are totally different from the quantum operations of creation \hat{a}^{\dagger} and annihilation \hat{a} that have the modification of $\sqrt{n}$ factor due to the symmetric nature of bosons. In our realization, the addition and subtraction do not depend on the number of particles originally in the system and nearly deterministically bring a classical state into a non-classical state. These operations correspond to conventional arithmetic which is commonly used in everyday life but they do n ot come out naturally in quantum mechanics. We implement such operations by applying the scheme of transitionless shortcuts to adiabaticity [1,2] on anti-Jaynes-Cummings transition. We demonstrate these operations on coherent states and superposition states of Fock states in a near deterministic manner, also showing that the technique is an important resource to create non-Gaussian states efficiently. This scheme can be used for the projective measurement of phonon number states and the test of boson sampling problem. This technology enables quantum state engineering and can be applied to many other experimental platforms.

[1] M. V. Berry, J. Phys. A. **42**, 365303 (2009).

[2] J. Zhang, *et al.*, Phys. Rev. A. **89**, 013608 (2014).

# Quantum correlations in the measurement and control of a solid-state mechanical oscillator

V. Sudhir[1], D. J. Wilson[1], S. Fedorov[1], H. Schuetz[1], R. Schilling[1], A. Nunnenkamp[2], T. J. Kippenberg[1]
[1]Institute of Physics, Swiss Federal Institute of Technology (EPFL), Lausanne, Switzerland
[2]Cavendish Laboratory, Cambridge University, UK

We perform a weak continuous measurement of the position fluctuations of a solid-state micro-mechanical oscillator with an imprecision 40dB below that at the standard quantum limit [1]. This is achieved by employing a cavity-enhanced interferometer coupled to the oscillator via a radiation pressure interaction. The quantum back-action due to the measurement, equivalent to several 1000 motional quanta (phonons) of the oscillator, is suppressed by feedback cooling of the oscillator to an occupation of 5 phonons on average.

The act of measurement develops quantum correlations in the radiation field used to perform the measurement. We demonstrate that efficient feedback can be used to "distill" these correlations in an out-of-loop measurement [2]. However, a fundamental tradeoff, relating to the Heisenberg uncertainty principle, limits how well this task can be performed; we study this experimentally and observe the role of a feedback back-action in destroying the quantum correlations due to the measurement. In our case, feedback back-action arises from amplified vacuum noise of the measurement laser, and so constitutes the ultimate limits of linear feedback control.

Finally, we report on two allied results: at cryogenic temperatures, a different manifestation of these quantum correlations – optical squeezing – and its control by measurement-based feedback; and the first observation of the same faint quantum correlations at room temperature [3]. The latter paves the way for optomechanics as a platform for quantum-enhanced metrology at room temperature.

[1] D. Wilson, V. Sudhir, N. Piro, R. Schilling, A. Ghadimi and T. Kippenberg, Nature **524**, 325 (2015)

[2] V. Sudhir, D. Wilson, *et al.*, arXiv:1602.05942 (2016)

[3] H. Schuetz, V. Sudhir, S. Fedorov, R. Schilling, D. Wilson, T. Kippenberg (in preparation)

# Phase shifting a weak coherent beam by a single $^{174}Yb^{+}$ ion

Martin Fischer[1,2], Lucas Alber[1,2], Bharath Srivathsan[1,2], Markus Weber[1,2], Markus Sondermann[1,2], Gerd Leuchs[1,2,3]

[1]Max Planck Institute for the Science of Light, Günther-Scharowsky-Str. 1/ Bldg 24, Erlangen, Germany

[2]University Erlangen-Nürnberg, Department of Physics, Staudtstr. 7/B2, Erlangen, Germany

[3]University of Ottawa, 75 Laurier Avenue East, Ottawa, Canada

Coherent interaction between a single quantum object and a light beam is a key component for many applications in the field of quantum communication and quantum computation. One promising candidate is a photonic phase gate that is controlled by a single atom. We present a setup where a single trapped $^{174}Yb^{+}$ ion is interfaced with a light beam in free space via a 4pi parabolic mirror (4pi-PM). To maximize the phase shift imprinted on the beam, the 4pi-PM converts a radially polarized donut mode to a linearly polarized light field [1] which matches the emission pattern of an atomic $\pi$-transition.

In our experiment, we drive a linearly polarized atomic dipole with a weak coherent beam close to resonance to the $S_{1/2}\leftrightarrow P_{1/2}$ transition in $^{174}Yb^{+}$. In the weak excitation regime, the coherent interaction with the ion causes a phase shift of the transmitted light field depending on the detuning to the atomic resonance. The imprinted phase shift is measured in a common path interferometer via heterodyne detection (see Fig. 1) similar to other free space experiments [2]. In detail, we illuminate the ion with a near-resonant sideband donut mode and an off-resonant carrier donut. After interaction with the trapped ion we record the absolute phase of their beating signals detected on a photomultiplier tube. For retrieving the imprinted phase shift of the near-resonant donut light field we repeat this measurement with the ion pumped into the meta-stable $D_{3/2}$ dark-state manifold. Subtraction of the respective two phase values yields the relative phase shift $\Delta\varphi$ of the near resonant beam.

In first preliminary experiments we achieved a relative phase shift of $\Delta\varphi = (2.2 \pm 0.5)°$, limited by residual ion motion and optical aberrations as well as coupling to the $J_{1/2}\leftrightarrow J_{1/2}$ transition. To boost the phase shift, we are currently working on improving the light-matter coupling efficiency by correcting optical aberrations of the parabolic mirror and by trapping $^{174}Yb^{2+}$ which has a closed two-level linear-dipole transition [3].

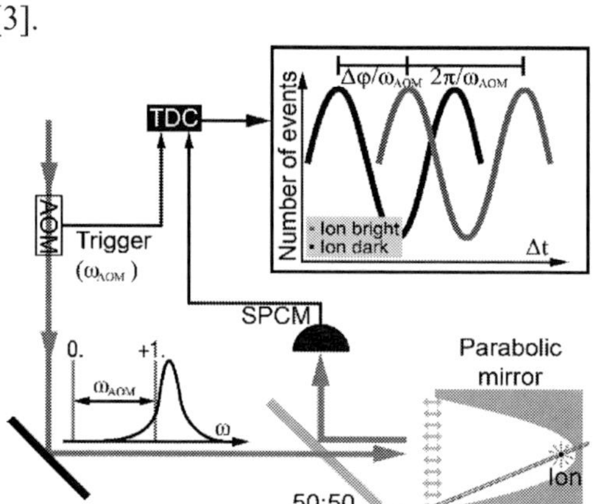


**Figure 1.** Scheme of the setup for measuring the phase shift imprinted by a single trapped $^{174}Yb^{+}$ ion.

[1] M. Sondermann *et al.*, Appl. Phys. B **89**, 489-492 (2007).

[2] M. Pototschnig *et al.*, Phys. Rev. Lett. **107**, 063001 (2011).

[3] S. Heugel *et al.*, J. Phys. B **49**, 015002 (2016).

# Quantum Fisher information as a signature of superradiant quantum phase transition

Guang-Ri Jin
Department of Physics, Beijing Jiaotong University, Beijing 100044, China

The single-mode Dicke model is well known to undergo a quantum phase transition from the so-called normal phase to the superradiant phase (hereinafter called the 'superradiant quantum phase transition'). Normally, quantum phase transitions can be identified by the critical behavior of quantities such as entanglement, quantum fluctuations, and fidelity. Here, we present out recent work about the quantum Fisher information (QFI) [1] and its behavior of both the field mode and the atoms in the ground state of the Dicke Hamiltonian [2]. For a finite but large number of atoms, we show that near the critical atom-field coupling, the QFI of the atomic and the field subsystems can surpass their classical limits, due to the appearance of nonclassical quadrature squeezing [3]. As the coupling increases far beyond the critical point, each subsystem becomes a highly mixed state, which degrades the QFI and hence the ultimate phase sensitivity. In the thermodynamic limit, we present the analytical results of the QFI and their relationship with the reduced variances of the field mode and the atoms. For each subsystem, we find that there is a singularity in the derivative of the QFI at the critical point, a clear signature of the quantum criticality in the Dicke model.

[1] Y. M. Zhang, X. W. Li, W. Yang, and G. R. Jin, Phys. Rev. A 88, 043832 (2013).
[2] T.-L. Wang *et al.*, New J. Phys. **16**, 063039 (2014).
[3] G. R. Jin *et al.*, New J. Phys. **11**, 073049 (2009).

# Radio-frequency dressed detection of atomic clock states

Sindhu Jammi[1], Tadas Pyragius[1], H. M. Florez[1], Mark Bason[1], Thomas Fernholz[1]
[1]School of Physics & Astronomy, University of Nottingham, Nottingham NG7 2RD, United Kingdom

We introduce a new method to dispersively measure populations and population difference of the clock states of alkali atoms. Frequency dependent linear birefringence of the atomic medium allows atom number detection in the two hyperfine states via polarization homodyning. Adiabatically dressing the atomic states with a radio frequency field[1] circumvents low-frequency noise as the atomic signal is modulated and becomes phase-locked to the driving frequency. Measuring the two populations in a single experimental cycle by shining two probe beams lets us correct for atom number fluctuations. Initial experiments with ensembles of $10^8$ $^{87}$Rb atoms allowed us to measure Rabi oscillations with a noise level of only 0.25%, currently limited by photon shot noise. This corresponds to a noise power of approximately 30dB above the atomic shot noise limit, but correct mode matching and direct subtraction of the homodyne signals should enable us to reach quantum noise limited performance and preparation of spin squeezed states via quantum non-demolition measurements[2]. This method can be used for preparation of spin squeezed states and thus enhance atomic clock and interferometer operations.

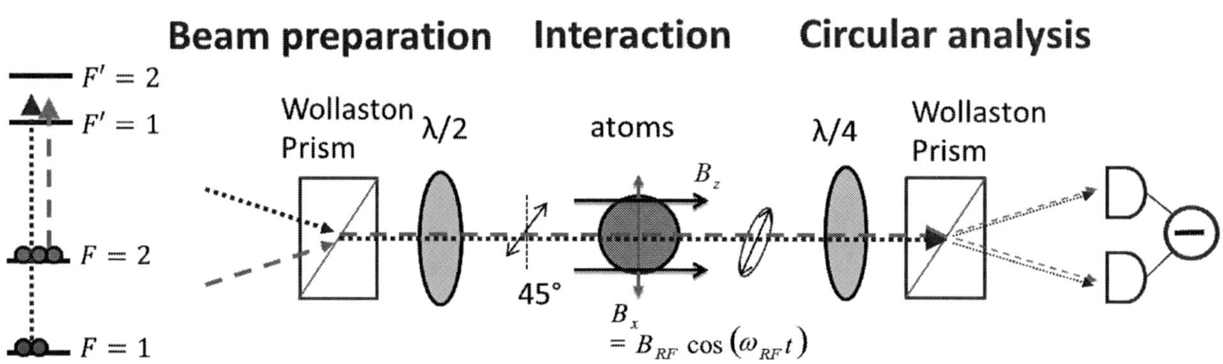


**Figure 1.** Experimental setup. Atoms prepared in a superposition of clock states are adiabatically transformed into radio-frequency dressed states and illuminated with two near-resonant light fields. Measurements of the oscillating, phase-locked linear birefringence allow determination of the two state populations.

[1] I. Lesanovsky et al., Phys. Rev. A **73**, 033619 (2006).

[2] K. Hammerer, A. S. Sorensen and E. S. Polzik, Rev. Mod. Phys. **82**, 1041 (2010).

# Transfer of photon bunching via four-wave mixing process in Rb vapor

Yong Sup Ihn[1], Kwang-Kyoon Park[1], Yosep Kim[1], Yoon-Ho Kim[1]
[1]Department of Physics, POSTECH, Pohang, 151-747 Korea

We report the transfer of photon bunching via four-wave mixing process of collimated blue light generation in rubidium vapor. Atomic coherence induced by the interaction of atoms with coherence light can strongly affect the optical property of atomic medium that gives rise to very efficient nonlinear mixing process. In particular, frequency conversion experiments in alkali atomic vapor have become an attractive issue due to its applicable potential in terms of the realization of coherent sources at blue and IR spectral region [1-3]. Here, based on the second order correlation function, we have investigated that photon bunching effect of pseudo-thermal input field is transferred to emitted photons at 420 nm due to the two-photon coherence.

In a ladder-type rubidium atomic system, a pair of resonant applied fields of 780 nm and 776 nm drives atomic coherence transition from $5S_{1/2}$ to $5D_{5/2}$ that leads amplified spontaneous emission(ASE) on a 5.23 μm transition. A large population transfer into $6P_{3/2}$ contributes to the generation of the blue light at 420 nm. Figure 1(a) shows the experimental set-up. Two pseudo-thermal input fields of 780 and 776 nm are generated by rotating ground disks and co-propagated into the Rb cell. Figure 1(b) represents $g^{(2)}(\tau)$ of 780 nm pseudo-thermal light measured by Hanbury-Brown-Twiss (HBT) interferometer. Here, experimental $g^{(2)}(0)$ value is 2, which is in agreement with the case of thermal light. Based on these pseudo-thermal light generation, we apply two input beams with same or different photon number statistics into a frequency conversion cell, and measure the second order correlation, $g^{(2)}(\tau)$, from the emitted blue light.

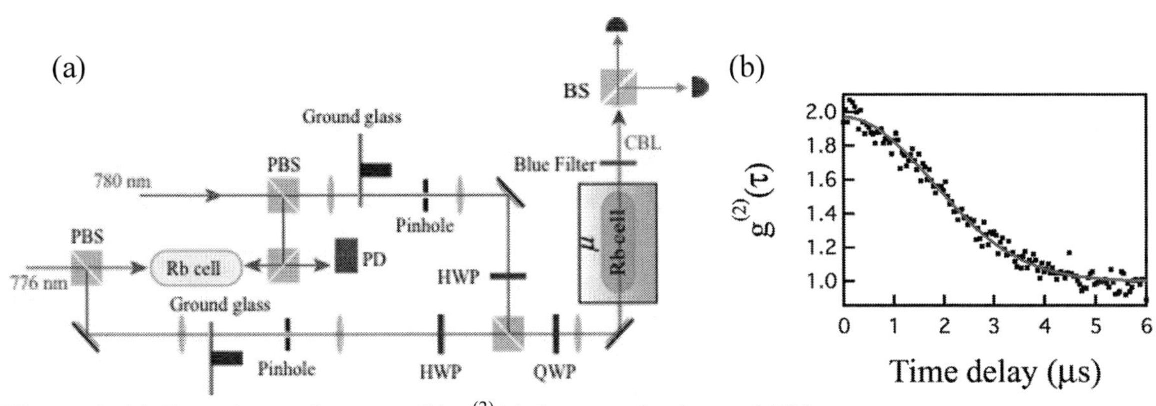


**Figure 1.** (a) Experimental set-up. (b) $g^{(2)}(\tau)$ for pseudo-thermal 780 nm.

[1] A. S. Zibrov *et al*, Phys. Rev. A **65**, 051801(R) (2002).
[2] T. Meijer *et al*, Opt. Lett. **31**, 1002 (2006).
[3] F. E. Beccera *et al*, Phys. Rev. A **78**(1), 013834 (2008).

# Trapping caesium near an optical nanofibre

Samuel Ruddell, Ivan Herrera-Benzaquen, Karen Webb, Maarten Hoogerland
Dodd-Walls Centre for Photonic and Quantum Technologies,
University of Auckland, New Zealand

A sub-wavelength diameter optical nanofibre strongly confines light in the transverse direction, with a significant portion of the guided light propagating in an evanescent field extending from the surface of the nanofibre. This allows for efficient coupling of particles near the surface of the nanofibre to the guided modes of the fibre, and presents an ideal platform for their manipulation and control.

In our experiment, we fabricate optical nanofibres having a radius of 250 nm, and trap neutral caesium atoms near the fibre surface. During fabrication, absolute cleanliness of the optical fibre must be maintained, as any surface contaminants will scatter light and heat the nanofibre. This can cause the nanofibre to melt under ultra-high vacuum conditions, with only a few milliwatts of light passing through, due to the lack of heat dissipation mechanisms in vacuum. In additon, the taper profile must be controlled precisely, as we must ensure that we do not allow any excitation of higher order spatial modes in the fibre, which can also lead to loss.

We have observed single photons emitted from the caesium atoms into the guided modes of the nanofibre, and have used the nanofibre to probe the spatial density of an atomic cloud. Furthermore, we have measured the absorption of a resonant probe laser beam in the fibre by MOT atoms.

By combining two lasers that are red- and blue-detuned from the atomic resonance in the nanofibre, we aim to form a one-dimensional lattice potential for the atoms [1]. We plan to utilise this two-colour trap in combination with an in-fibre cavity, formed by placing a fibre Bragg grating on either side of the nanofibre, to further enhance the coupling of the atoms to the guided modes of the fibre [2].

This poster will present our current progress in setting up this experiment, as well as showing initial results and future plans for the experiment.

[1] E. Vetsch, D. Reitz, G. Sagu´e, R. Schmidt, S. T. Dawkins, and A. Rauschenbeutel, Phys. Rev. Lett. **104**, 203603 (2010).

[2] S Kato, and T Aoki, Phys. Rev. Lett. **115**, 093603 (2015).

# Selective excitation of V-type quantum system by controlled rapid adiabatic passage

Yunheung Song, Han-gyeol Lee, Hanlae Jo, Jaewook Ahn
Department of Physics, KAIST, Daejeon 34141, Korea

Selective complete population transfer (CPT) of multilevel quantum systems is an important problem in coherent control. In a conventional scheme, chirped rapid adiabatic passage (RAP), the excitation selectivity of a 3-level system can be achieved by changing the sign of chirping. In this study, we suggest a new control scheme, controlled chirped RAP [1], which is implemented by using a chirped laser pulse of a spectral hole resonant to an excited level of the system. In this scheme, the laser intensity plays the role of the control knob of excitation selectivity, alternatively, and the sequential application of the pulse can implement the CPT between Raman-coupled states, which can be used as a qubit. We experimentally demonstrated the selective excitation by using the as-shaped femtosecond pulse and $5S_{1/2}$, $5P_{1/2}$, $5P_{3/2}$ levels of atomic rubidium as the 3-level V-type system. We expect that this intensity-dependent selective CPT with a spatial programing of laser beam profile may allow site-resolved sub-wavelength-scale control of atoms and ions arranged in space.

[1] Y. Song, H.-g. Lee, H. Jo, J. Ahn, arXiv: 1604.03699 (2016).

# Atom-atom interactions in an 'Alligator' photonic crystal waveguide

Jonathan Hood[1], Aki Goban[1,2], Ana Ansenjo-Garcia[1], Mingwu Lu[1], Su-Peng Yu[1], Jeff Kimble[1]
[1]California Institute of Technology, USA
[2]JILA, University of Colorado and NIST, Boulder, USA

New opportunities for optical physics emerge from the integration of cold atoms with nanophotonic devices. Due to their small optical loss and tight field confinement, these nanoscale dielectric devices are capable of mediating strong atom-light interactions [1] and open new avenues for quantum transport and quantum many-body phenomena. In particular, coupling atoms to the band edge of a photonic crystal waveguide (PCW) provides a unique platform for generating tunable range coherent atom-atom interactions which are mediated by the guided mode photons. Due to the evanescent nature of the field in the band gap, dissipation into the structure is suppressed exponentially. We have experimentally observed the transition into the bandgap for the first time by shifting the band edge frequency of the PCW relative to the D1 line of atomic cesium with an average of 3 atoms trapped along the PCW [2]. In addition, we have developed a formalism that provides a clear mapping between the transmission spectra and the local Green's function, which allows us to identify signatures of dispersive and dissipative interactions between the atoms [3].

[1] A. Goban, C.-L. Hung, J. D. Hood and S.-P. Yu, Phys. Rev. Lett. **115**, 063601 (2015).
[2] J. D. Hood, A. Goban, A. Asenjo-Garcia, et al. (in press) PNAS (2016).
[3] A. Asenjo-Garcia, J. D. Hood, D. E. Chang, H. J. Kimble, arXiv preprint arXiv:1606.04977 (2016).

# Three-photon correlation of the cavity-QED microlaser

Byoung-moo Ann[1,2], Younghoon Song[1], Junki Kim[1],Daeho Yang[1], Kyungwon An[1]
[1]Department of Physics and Astronomy, Seoul National University, Seoul 147-751, Korea
[2]Max Planck Institute of Quantum Optics, Hans-Kopfermann-Strasse 1, D-85748 Garching, Germany

Measurement of three-photon correlation $g^{(3)}(0,0)$ as well as the higher correlations are of much interest for nonclassical light sources. In usual measurement schemes, a larger number of photon-click events are required as a target order of the photon correlation increases higher. When the operation time of a nonclassical light source such as the cavity-QED microlaser is limited, $g^{(3)}(0,0)$ is thus more difficult to measure than $g^{(2)}(0)$. In this work, we propose a new measurement scheme for photon correlations by utilizing the detector dead time effect. As $g^{(2)}(0)$ is significantly distorted by the dead time even in the two-detector configuration of Hanbury-Brown Twiss type, one can express the observed two-photon correlation $g'^{(2)}(0)$ as a function of the dead time. We found a way to extract the information of three-order correlation from $g'^{(2)}(0)$, and thus we could measure $g^{(3)}(0,0)$ in the two-detector configuration. Our new scheme requires a smaller number of photon-click events [1]. Consequently, we can circumvent the finite operation time problem of the cavity-QED microlaser. Moreover, reducing the number of required photodetectors makes a significant difference in experimental practice. We succeed in measuring $g^{(3)}(0,0)$ of the cavity-QED microlaser by using our new detection scheme. Our experimental setup is shown in the figure below. In parallel, we also develop a theory to describe the relation between $g^{(3)}(0,0)$ and $g^{(2)}(0)$. We eventually confirm that our measurement results are well supported by our theoretical expectation. Our new detection scheme is universal, and thus it can be applied to any types of photon sources.

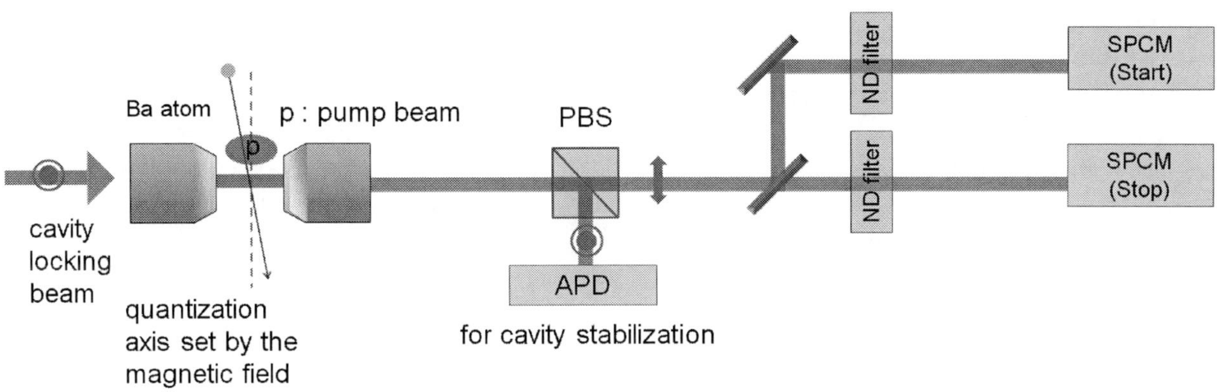


**Figure 1.** Schematic of our experimental setup [2] used in three-photon correlation measurement. Barium atoms are injected into a high-finesse optical cavity after prepared in the exited state. Output field from the optical cavity goes to HBT type measurement system. We can plot $g'^{(2)}(0)$ as a function of the dead time by applying prolonged dead time on photon arrival records.

[1] B. Ann, Y. Song, J. Kim, D. Yang, K. An, Phys. Rev. A **93**, 063816 (2016).

[2] B. Ann, Y. Song, J. Kim, D. Yang, K. An, Phys. Rev. A **92**, 023830 (2015).

# Light spin and quantum walks implied by rotating cylindrical waves with electric field parallel to magnetic field

Hyoung-In Lee[1], Jinsik Mok[2]

[1]Research Institute of Mathematics, Seoul National University, Seoul, Republic of Korea

[2]Department of Industrial and Management Eng., Sunmoon University, Asan, Republic of Korea

A usual wisdom regarding electromagnetic waves is that the E-field, H-field, and the wave-propagation vectors form a right-handed triad in three-dimensional space. However, it was (re-)discovered in [1] for waves stationary in one direction, say, the axial direction in cylindrical coordinates that the electric field (E-field) is parallel to the magnetic field (H-field). Partly because this E-parallel-H waves are existent only for free space, i.e., vacuum, its analytical solution has not maintained further attention of scientists working in the light-matter interactions. Notice that this E-parallel-H waves are describable only in terms of planar coordinates in addition to being temporally sinusoidal. This point has a great potential applicability to two-dimensional electron gas (2DEG) and relevant topological photonics.

We have extended the analytic solution in [1] to rotational waves by taking averages in the azimuthal direction. The resulting rotational waves are found to be established only for even numbers of the azimuthal (angular) index, which is supposed to arise from the coherent combination of two axially counter-propagating waves. We have proved that the even-numbered azimuthal mode indices are completely compatible with energy conservation, when field variables of a wave are summed over all the clock-wise and counter-clock-wise rotational components. The resulting rotational waves turn out to exhibit interesting characteristics of wave energy, light spin, light chirality, and helicity [2]

Although the analytic solution to rotational waves is obtained by solving classical Maxwell's equations, we found that interference effects lead to several quantum-like phenomena. First, our waves exhibit quasi-energy spectrum of the simplest Hubbard-model for bosonic particles. Second, (also related to the first aspect), our waves can be interpreted in terms of one-dimensional quantum walks [3]. From this second viewpoint, the information on the two-dimensional square lattice which has been smeared out during the angular integration turn out to be still hidden underneath our rotational waves. Concerning a dynamics of continuous azimuthal mode index, we will make a reference to our recent work in [4]

[1] K. Uehara, T. Kawai, K. Shimoda, J. Phys. Soc. Jpn. **58**, 3570-3575 (1989).

[2] K. Y. Bliokh, F. Nori, Physics Reports **592**, 1–38 (2015)

[3] H. Obuse, J. K. Asboth, Y. Nishimura, N. Kawakami, Phys. Rev. B **92**, 045424 (2015)

[4] H.-I. Lee and J. Mok, Pacific J. of Mathematics for Industry **8**, 1-16 (2016)

# Heralded single-photon superradiance from a Doppler-broadened ladder-type atomic ensemble

Yoon-Seok Lee, Sang Min Lee, Heonoh Kim, Han Seb Moon
Department of physics, Pusan National University, Busan, Korea

A collective radiation of a single photon in a common spatial mode from an atomic ensemble has been actively used in a wide variety of tasks for the realization of quantum internet, such as generation of photon pair and quantum memory. This directional radiation of a single photon is closely connected with the single-photon superradiance (SPS) [1]. Recently, there has been increasing interest in SPS due to its potential usefulness in the efficient collection of single photons and the enhancement of interaction strength. The SPS has been observed in various physical systems with different ways, such as lifetime shortening, linewidth broadening, and quadratic increment of intensity.

We report on the SPS in a Doppler-broadened ladder-type $^{87}$Rb warm vapor. In the $5S_{1/2}$–$5P_{3/2}$–$5D_{5/2}$ transition of $^{87}$Rb atoms, the two-photon resonance condition can be satisfied in a wide range of velocity groups in the Doppler-broadened atomic ensemble under the counter-propagating configuration of two similar wavelength laser fields, as shown in Fig. 1. Thus most of velocity classes can be coherently excited and collectively radiates the paired photons in a phase-matched direction, and the decay time is significantly reduced. In particular, the detection of a signal photon from excited to intermediate state heralds a collective single excitation of atom in the intermediate state, but it is not known which velocity class is. Consequently the emission of a heralded idler photon is enhanced into the phase-matched direction as a result of a collective radiation of the ensemble. In addition, the probability of detecting a heralded single photon is proportional to the square of the optical density (OD), which is one of the representative consequences from the superradiance. Furthermore, under the condition of high OD, a single-photon superradiant beating is observed [2], for the first time, to the best of our knowledge; this implies the coherent superposition of detection probability amplitudes on heralded idler photon for different velocity classes.

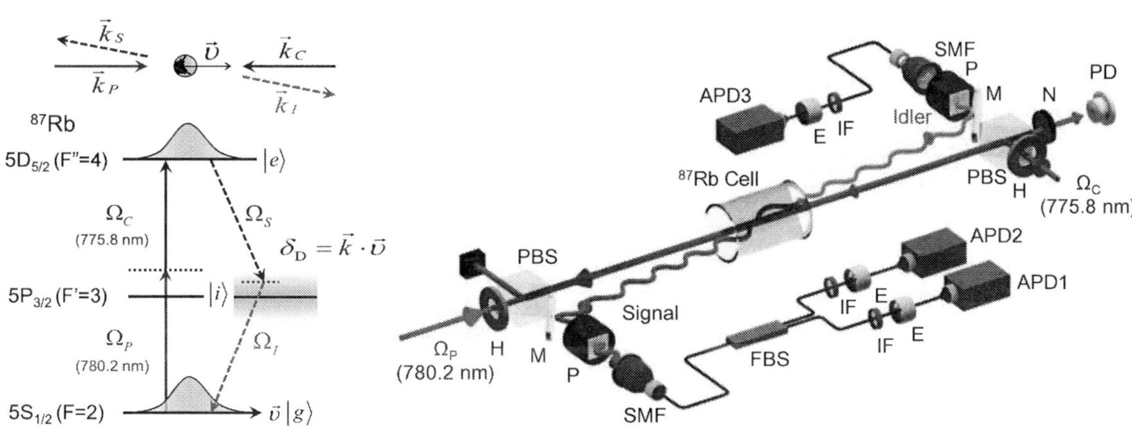


**Figure 1.** Cascade emission of signal and idler photons in a Doppler-broadened three-level ladder-type atomic system and Schematic drawing of experimental set up.

[1] R. H. Dicke, Phys. Rev. **93**, 99 (1954).
[2] M. Gross, J. M. Raimond, and S. Haroche, Phys. Rev. Lett. **40**, 1711 (1978).

# Nondestructive reconstruction of photon number in an optical cavity using trapped ions

Moonjoo Lee[1], Konstantin Friebe[1], Florian R. Ong[1], Dario A. Fioretto[1], Klemens Schüppert[1], Rainer Blatt[1,2], Tracy E. Northup[1]

1. Institut für Experimentalphysik, Universität Innsbruck, Technikerstraße 25, 6020 Innsbruck, Austria
2. Institut für Quantenoptik und Quanteninformation, Österreichische Akademie der Wissenschaften, Technikerstraße 21a, 6020 Innsbruck, Austria

Quantum non-demolition (QND) measurement of cavity photons provides a variety of opportunities in fields ranging from quantum information science to fundamental quantum optics. For example, QND measurement of Schrödinger cat states' number parity corresponds to monitoring of an error syndrome for quantum error correction [1]. Nondestructive measurement of intracavity photons also enables us to probe different squeezing factors between intra- and output cavity photons. In order to carry out such experiments in the optical domain, we must detect optical photons without annihilating them. Here, we report the nondestructive reconstruction of photon number in an optical cavity using trapped ions. We employ a single $^{40}Ca^{+}$ ion that is dispersively coupled to a high-finesse cavity. The detuning between the cavity and the ion is adjusted such that the experiment is performed in the weak-pull regime ($g^2/\Delta \ll \kappa$), where $g$ is the ion-cavity coupling strength, $\Delta$ is the ion-cavity detuning, and $\kappa$ is the cavity decay rate. The cavity is driven by a weak laser field to populate the cavity with mean photon numbers of about unity. Ramsey spectroscopy is performed on the 729 nm qubit transition to identify the shift and the broadening of the atomic energy levels. The shift is due to the ac Stark effect induced by cavity photons, and the broadening is attributed to the photon-number fluctuations of the cavity field. Taking advantage of the maximum-likelihood method, we reconstruct individual cavity photon number components up to four. In combination with currently available mirror-coating technology, QND measurement of optical cavity photons will make it possible to realize quantum feedback control of nonclassical cavity states.

[1] L. Sun *et al.*, Nature **511,** 444 (2014).

# Proposal for spin-nematic squeezing via cavity-assisted Raman transitions

Stuart Masson[1,2], Scott Parkins[1,2], Murray Barrett[3]

[1]Department of Physics, The University of Auckland, New Zealand

[2]Dodd-Walls Centre for Photonic and Quantum Technologies, New Zealand

[3]Center for Quantum Technologies, Singapore

Spin squeezing is a well-known technique to increase the metrological precision of an ensemble of spin-1/2 particles. Working in the SU(2) space $\{S_x, S_y, S_z\}$, squeezing is induced in a component perpendicular to the mean spin direction. Spin-1 particles have more degrees of freedom, and so one may consider squeezing in the SU(2) subspace $\{S_x, Q_{yz}, Q_+\}$ (where $Q_+ = Q_{zz} - Q_{yy}$ and $Q_{ij}$ are elements of the spin-1 nematic tensor). This squeezing is characterized by a parameter $\xi_x$ [1] that quantifies the extra metrological precision over a coherent state for Ramsey spectroscopy. Our proposal uses the $F = 1$ hyperfine level of $^{87}$Rb (Figure (a)), with atom-atom interactions mediated via Raman transitions driven by cavity and laser fields. In the far detuned limit, the internal structure of the $5^2P_{1/2}$ level can be ignored and, due to symmetries in the dipole operator, adiabatic elimination of the excited states leads to a Dicke model Hamiltonian,

$$H = \omega\, a^+ a + \omega_0 J_z + \frac{\lambda_r}{\sqrt{2N}}(a^+ J_- + a J_+) + \frac{\lambda_s}{\sqrt{2N}}(a^+ J_+ + a J_-),$$

where $a\,(a^+)$ is the annihilation (creation) operator for a cavity mode, $J_i$ are collective spin operators for $N$ spin-1 particles, and $\omega$, $\omega_0$, and $\lambda_{r,s}$ are determined by light or Zeeman shifts, or collective Raman transition rates. For sufficiently large Raman detuning, subsequent adiabatic elimination of the cavity mode then yields an effective spin Hamiltonian of the form [2]

$$H = \lambda(J_x^2 + J_y^2) + \gamma J_z.$$

With all atoms initially pumped into the $m_f = 0$ state, this Hamiltonian produces a four-wave mixing process that induces strong spin-nematic squeezing (Figure (b)) on timescales that can be much faster than is possible via collisional spin dynamics in Bose-Einstein condensates [1].

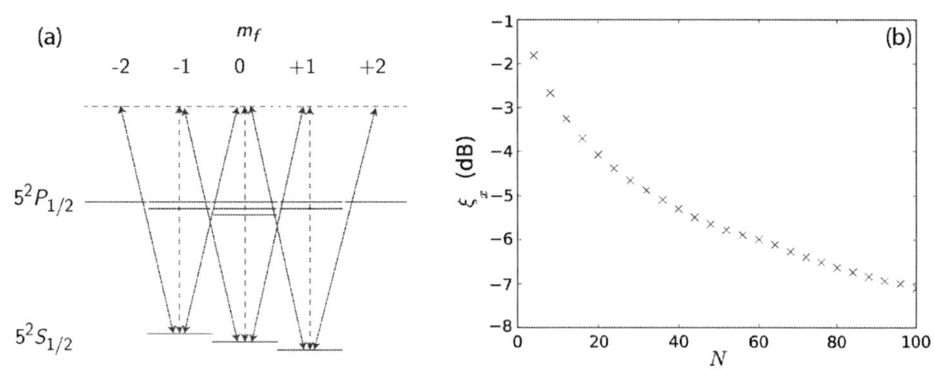


**Figure 1.** (a) Proposed level scheme in Rb$^{87}$ with laser (solid lines) and cavity modes (dotted) driving strongly detuned Raman transitions. (b) Calculated squeezing for different numbers of atoms.

[1] C. D. Hamley et al., Nat. Phys. **8**, 305 (2012)

[2] S. Morrison and A. S. Parkins, Phys. Rev. A **77**, 043810 (2008)

# Spin Squeezing in a Spinor Bose-Einstein Condensate

Xinyao Huang[1], Fengxiao Sun[1], Wei Zhang[2], Qiongyi He[1]

[1]State Key Laboratory of Mesoscopic Physics, School of Physics, Peking University, Beijing 100871, China

[2]Department of Physics, Renmin University of China, Beijing 100872, China

Reduction of quantum noise in one spin component is a valuable tool for enhancing the sensitivity of interferometers and atomic clocks [1]. Spin squeezing can be realized in Bose-Einstein condensate through the use of local nonlinear S-wave scattering interaction between trapped atoms. For the ground state, the obtained planar quantum squeezed state can simultaneously reduce the quantum noise of two orthogonal spin projections below the standard quantum limit, while increasing the noise in a third dimension, which allows the prospect of improved phase measurements at any phase-angle [2]. Spin squeezing can also be realized via spin-orbit coupling (SOC) in BEC. We study spin-squeezing in a spin-1 BEC where all three hyperfine states are coupled by spin-orbit coupling. By deriving an effective spin Hamiltonian in the low excitation limit, we find that the spin squeezing in this system is a direct effect of spin-orbit coupling. In particular, spin-nematic squeezing is enhanced with increasing Rabi frequency and two-photon detuning, in clear contrast to the behavior of spin squeezing in spin-1/2 systems. In addition, spin-nemtatic squeezing can be interpreted as the two-mode entanglement (squeezing). We further investigate the effects of harmonic trap and interaction with realistic experimental parameters. Finally, we discuss possible experimental signatures to observe such novel type of spin squeezing [3].

[1] Kitagawa M, Ueda M, Phys. Rev. A **47**, 5138-5143 (1993).

[2] Xinyao Huang et al, Acta Phys. Sin **64**, 160304 (2015).

[3] Xinyao Huang et al, in preparation (2016).

# High-efficiency wavelength conversion based on coherent optical memory

Pin-Ju Tsai[1,2], Ya-Fen Hsiao[1,3], Hung-Shiue Chen[1], Yong-Fang Chen[4], Ite A. Yu[5] and Ying-Cheng Chen[1,5]

[1]Institute of Atomic and Molecular Sciences, Academia Sinica, Taipei 10617, Taiwan
[2]Department of Physics, National Taiwan University, Taipei 10617, Taiwan
[3]Molecular Science Technology, Taiwan International Graduate Program, Academia Sinica and National Central University, Taiwan
[4]Department of Physics, National Cheng Kung University, Tainan 70101, Taiwan
[5]Department of Physics, National Tsing Hua University, Hsinchu 30043, Taiwan

We realized a high-efficiency (> 90%) wavelength conversion from 895 nm to 852 nm based on coherent optical memory. The coherent optical memory is based on electromagnetically-induced transparency (EIT) scheme in cold atoms. The wavelength conversion contains two procedures. By adiabatic turning off the EIT coupling field which drives cesium $D_1$ line at 895 nm, a weak probe pulse which also drives 895 nm is firstly converted to the collective atomic coherence and stored inside the medium. By turning on another coupling field which drives cesium $D_2$ line at 852 nm, the collective atomic coherence is retrieved as a light pulse at 852 nm. Due to the adiabatic nature, the wavelength conversion process is nearly perfect. The overall conversion efficiency is limited by the storage efficiency of the first procedure, which depends on the optical depth and the ground-state decoherence rate. This wavelength conversion scheme is much simpler to that based on the four-wave mixing (FWM) scheme[1], in which the optimum detuning for the driving field is dependent on the optical depth. Additionally, this wavelength conversion scheme has the potential to generate a frequency-entangled single-photon state for the application in quantum information science.

**References**

[1] C-K Chiu, Y-H Chen, Y-C Chen, Ite A. Yu, Y-C Chen, and Y-F Chen, Phys. Rev. A, 89, 023839(2014).

[2] M. Fleischhauer and M. D. Lukin, Phys. Rev. Lett, 84, 5094(2000).

[3] M.G. Payne and L. Deng , Phys. Rev. Lett, 91, 123602(2003).

[4] D. Viscor, V. Ahufinger, J. Mompart, A. Zavatta, G. C. La Rocca,M. Artoni, Phys. Rev. A, 86, 503827(2012).

# Detection of Bell correlations in a spin-squeezed Bose-Einstein condensate

R. Schmied[1], J.-D. Bancal[1], B. Allard[1], Matteo Fadel[1],
V. Scarani[2], P. Treutlein[1], N. Sangouard[1]
[1]Department of Physics, University of Basel
[2]Centre for Quantum Technologies, National University of Singapore

The results of measurements by different observers can show correlations stronger than any local deterministic theory allows. These Bell correlations can be confirmed by violating a Bell inequality, for which quantum entanglement is necessary but not sufficient.

A family of Bell inequalities has recently been proposed that requires only single- and two-particle measurements even for many-body systems [1]. Based on this work, we present the first experimental observation of Bell correlations in squeezed states of a two-mode Bose-Einstein condensate (BEC).

The system consists of a two-component 87Rb BEC of a few hundred atoms created on an atom-chip [2]. A state-selective potential gives rise to nonlinear one-axis twisting dynamics [3], which we use to prepare spin-squeezed states of the condensate, with up to 7dB of squeezing according to the Wineland criterion.

We measure the first and second moments of the collective spin operator along several axes using a combination of well-controlled Rabi rotations and projective measurements. These measurement results demonstrate the presence of Bell correlations in our system by more than three standard deviations [4].

[1] J. Tura, R. Augusiak, A. B. Sainz, T. Vértesi, M. Lewenstein, A. Acin, Science **344**, 1256 (2014).
[2] M.F. Riedel, P. Böhi, Y. Li, T.W. Hänsch, A. Sinatra, P. Treutlein, Nature **464**, 1170 (2010).
[3] M. Kitagawa, M. Ueda, Phys. Rev. A **47** 5138-5143 (1993).
[4] R. Schmied, J.-D. Bancal, B. Allard, M. Fadel, V. Scarani, P. Treutlein, N. Sangouard, Science **352**, 441 (2016).

# Experimental measurement of correlation functions in a trapped ion

Shuaining Zhang, Yangchao Shen, Yao Lu, Kuan Zhang, Jing-Ning Zhang, Kihwan Kim
Center for Quantum Information, IIIS, Tsinghua University, Beijing, P. R. China

J. S. Pedernales, Lucas Lamata, Enrique Solano,
Department of Physical Chemistry, University of the Basque Country UPV/EHU, Bilbao, Spain

J. Casanova,
Institut für Theoretische Physik, Universität Ulm, Ulm, Germany

We implement an algorithm to measure the time-correlation functions of spins and phonons of a system that evolves under a Hamiltonian of Jaynes-Cumming model or Dirac equation with a trapped 171Yb+ ion. The algorithm proposed in Ref. [1] requires only one ancillary qubit, where conditional gates are performed to the system depending on the state of the qubit. We measure the spin-spin time-correlations in spin system and the spin-phonon time-correlation in bosonic field. According to the linear response theory, the time-correlations can be used to characterize relevant physical magnitude such as magnetic susceptibility. This scheme can be extended to a system including n-time spins and phonons correlations.

[1] J. S. Pedernales, et al., Phys. Rev. Lett. 113, 020505 (2014).

This work was supported by the National Basic Research Program of China under Grants No. 2011CBA00300 (No. 2011CBA00301), the National Natural Science Foundation of China 11374178, 11574002 and 11504197.

# Coherent-state discrimination with non-destructive measurements

Rui Han[1,2], Janos Bergou[3], Gerd Leuchs[1,2]

[1]Max Planck Institute for the Science of Light - 91058 Erlangen, Germany, EU

[2]Institute of Optics, Information and Photonics, University of Erlangen-Nürnberg - 91058 Erlangen, Germany, EU

[3]Department of Physics and Astronomy, Hunter College of the City, University of New York - 10065 New York, NY, USA

The impossibility to discriminate perfectly between two non-orthogonal states is at the heart of the communication of quantum information, and the security of quantum key distribution relies on this property. At the same time, this poses the problem and the challenge to find the best possible discrimination procedure. We explore a novel scheme for discriminating coherent light states, with significant overlaps, by non-destructive measurements on the light states through measuring atomic states that is entangled to the coherent states via dipole coupling. In this way, the problem of measuring and discriminating coherent light state is shifted to finding the optimal light-atom interaction and the appropriate atomic measurements. We demonstrate that, for discriminating binary coherent state, the error probability of this scheme can be extremely close to the Helstrom bound with optimized light-atom dipole coupling and simple projective measurements on the atomic qubit ancilla. Moreover, because the measurements are non-destructive on the light states, information failing to be detected by the measurements owning to experimental imperfection or limitations can be extracted through sequential measurements on the post-measurements light states.

# Toric-boson model and finite temperature quantum memory

Yan-Bin Yang[1], Alioscia Hamma[1]
[1]Center for Quantum Information, IIIS, Tsinghua University, Beijing 100084, China

Toric-boson model is the toric code with an additional bosonic field that couples to the defects. The interaction with the bosonic field induces a long-range attractive force between the end points of open strings [1]. We study the two-dimensional model in the presence of a generic environment at finite temperature. To solve the Hamiltonian with logarithmic attraction between anyons, we consider one type of excitation and use the mean field approximation to calculate the partition function and obtain the self-consistent equation. Then we get the finite critical temperature proportional to coupling with phonon in thermodynamic limit which corresponds to the confinement-deconfinement transition. So this means we can have some quantum memory at finite temperature [3]. Furthermore, we consider other forms of interaction in different dimensions and find that the logarithmic interaction is special to yield the finite transition temperature in the criterion of confinement of defects. We also try to calculate the topological entanglement entropy at finite temperature exactly as the toric code model [2].

[1] A. Hamma, C. Castelnovo, C. Chamon, Phys. Rev. B **79**, 245122 (2009).
[2] C. Castelnovo, C. Chamon, Phys. Rev. B **76**, 184442 (2007).
[3] B J. Brown, D. Loss, J K. Pachos, et al, arXiv:1411.6643 (2014).

# Towards storage of quantum dot single photons in a rubidium quantum memory

Lucas Béguin[1], Janik Wolters[1], Jan-Philipp Jahn[1], Mathieu Munsch[1], Andrew Horsley[1], Fei Ding[2], Armando Rastelli[3], Oliver G. Schmidt[2], Richard J. Warburton[1], Philipp Treutlein[1]

[1]University of Basel, Switzerland

[2]IFW, Dresden, Germany

[3]Johannes Kepler University, Linz, Austria

Single photon sources and quantum memories are essential resources to implement quantum networks for communication, computation and simulation [1]. We investigate a hybrid system combining a semiconductor quantum dot (QD) as a deterministic source of single photons with a frequency-matched memory made of an ensemble of $^{87}$Rb atoms. We have characterized a promising sample of droplet GaAs/AlGaAs QDs, which emit lifetime-limited single photons at 780 nm, the wavelength of the Rb D2 line [2]. Fine-tuning of the QDs emission is achieved using a PZT to induce biaxial strain in the sample. The scanning range covers all the hyperfine structure of the D2 line, as the spectroscopy of a room temperature Rb vapor shows in Figure 1. Independently, our new light-atom interface allows us to prepare cold and dense ensembles of $^{87}$Rb atoms (OD > 2000). By combining the two systems we are now developing an EIT-based memory that predicts storage-and-retrieval efficiency exceeding 30% [3]. Such a memory will form the basis for experiments on hybrid entanglement and quantum networks.

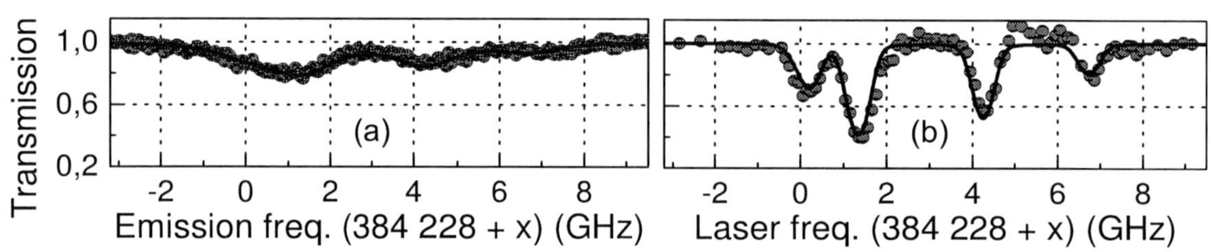


**Figure 1. Spectroscopy of a Rb vapor with QD photons.** (a) The QD is excited non-resonantly and the emission frequency is tuned via strain. (b) The QD is excited resonantly in the coherent scattering regime. The frequency of the driving laser is scanned across the Rb D2 line while keeping the QD transition at resonance.

[1] N. Sangouard, C. Simon, J. Minar, H. Zbinden, H. de Riedmatten, N. Gisin, Physical Review A **76**, 050301(R) (2007).

[2] J-P. Jahn, M. Munsch, L. Béguin, A.V. Kuhlmann, M. Renggli, Y. Huo, F. Ding, R. Trotta, M. Reindl, O.G. Schmidt, A. Rastelli, P. Treutlein, and R.J. Warburton, Physical Review B **92**, 245439 (2015).

[3] M.T. Rakher, R.J. Warburton, and P. Treutlein. Physical Review A **88**, 053834 (2013).

# Quantum information approach to Bose-Einstein condensation of composite bosons

Su-Yong Lee[1], Jayne Thompson[2], Sadegh Raeisi[3], Pawel Kurzynski[4], Dagomir Kaszlikowski[2]

[1]School of Computational Sciences, Korea Institute for Advanced Study, Korea

[2]Centre for Quantum Technologies, National University of Singapore, Singapore

[3]Institute for Theoretical Physics II, Friedrich-Alexander-Universitat Erlangen-Nurnberg, Germany

[4]Faculty of Physics, Adam Mickiewicz University, Poznan, Poland

We consider composite bosons (cobosons) comprised of two elementary particles, fermions or bosons, in an entangled state. First, we show that the effective number of cobosons implies the level of correlation between the two constituent particles. For the maximum level of correlation, the effective number of cobosons is the same as the total number of cobosons, which can exhibit the original Bose- Einstein condensation (BEC). In this context, we study a model of BEC for indistinguishable cobosons with a controllable parameter, i.e., entanglement between the two constituent particles [1].

[1] S.-Y. Lee, J. Thompson, S. Raeisi, P. Kurzynski, and D. Kaszlikowski, New Journal of Physics **17**, 113015 (2015).

# Waveguide based photonic chip for manipulating single atom array

Xiao Li,[1,2] Min Ke,[1,3] Jin Wang,[1,3] Mingsheng Zhan[1,3]

[1]State Key Laboratory of Magnetic Resonance and Atomic and Molecular Physics, Wuhan Institute of Physics and Mathematics, Chinese Academy of Sciences - Wuhan National Laboratory for Optoelectronics, Wuhan 430071, China

[2]School of Optical and Electronic Information, Huazhong University of Science and Technology, Wuhan 430074, China

[3]Center for Cold Atom Physics, Chinese Academy of Sciences, Wuhan 430071, China

Trapped single neutral atoms as promising candidates for storing and processing quantum information have attracted tremendous interest [1]. But the compatibility of scalability and individually addressing is still a challenge. In order to solve this problem, we propose a waveguide based photonic chip which is capable of trapping, coherently manipulating, detecting and individually addressing an array of single neutral atoms [2]. This chip consists of an array of 7 independent functional units spaced by a few micrometers, each of which is comprised of one silica-on-silica optical waveguide and one phase Fresnel microlens etched in the middle of waveguide output interface. The waist size of the light at 830 nm focused by the microlens is about 1.5 μm measured in the experiment, which is in collisional blockade region. Two-photon Rydberg excitation scheme in our chip is also illustrated. This photonic quantum chip is scalable and capable of individually addressing single neutral atoms, which will play an important role in the future quantum devices.

[1] M. Saffman, T. G. Walker, and K. Mlmer, Rev. Mod. Phys. **82**, 2313(2010).

[2] M. Ke, F. Zhou, X. Li, J. Wang, and M. S. Zhan, Opt. Express **24**, 9157(2016).

# An Atomtronics Flux Qubit: A ring lattice of Bose-Einstein condensates interrupted by three weak links

D. Aghamalyan[2], N.T. Nguyen[2], F. Auksztol[2], K. S. Gan[2], M. Martinez Valado[2], P. C. Condylis[2], L.-C. Kwek[2,3], R. Dumke[2], L. Amico[2,4,5]

[1]Institut de Physique de Rennes, UMR 6251 du CNRS and Universite de Rennes 1, 35042 Rennes Cedex, France

[2]Centre for Quantum Technologies, National University of Singapore, Singapore 117543

[3]National Institute of Education and Institute of Advanced Studies, Nanyang Technological University, 1 Nanyang Walk, Singapore 637616

[4]CNR-MATIS-IMM & Dipartimento di Fisica e Astronomia, Via S. Sofia 64, 95127 Catania, Italy

[5]Laboratori Nazionali del Sud, INFN, via S. Sofia 62, 95123 Catania, Italy

We study a physical system consisting of a Bose-Einstein condensate confined to a ring shaped lattice potential interrupted by three weak links. The system is assumed to be driven by an effective flux piercing the ring lattice. By employing path integral techniques, we explore the effective quantum dynamics of the system in a pure quantum phase dynamics regime. Moreover, the effects of the density's quantum fluctuations are studied through exact diagonalization analysis of the spectroscopy of the Bose-Hubbard model. We demonstrate that a clear two-level system emerges by tuning the magnetic flux at degeneracy [1].

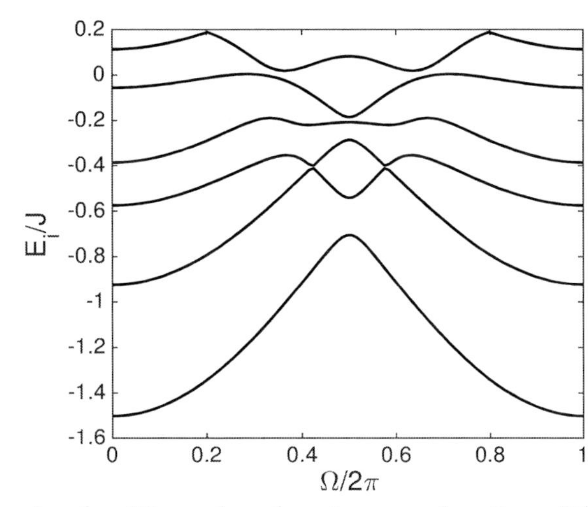


**Figure 1.** Six first energy levels of the reduced system as a function of the dimensionless external flux $\Omega$. The first two energy levels form a two level Qubit system.

[1] D. Aghamalyan, N.T. Nguyen, F. Auksztol, K. S. Gan, M. Martinez Valado, P. C. Condylis, L.-C. Kwek, R. Dumke, and L. Amico, arxiv: 1512.08376 [quant-ph] (2016).

# Manifesting nonclassicality beyond Gaussian states by observing a single marginal distribution

Jiyong Park[1], Yao Lu[2], Jaehak Lee[1], Yangchao Shen[2], Kuan Zhang[2], Shuaining Zhang[2], M. Suhail Zubairy[3], Kihwan Kim[2], Hyunchul Nha[1,4]

[1]Department of Physics, Texas A&M University at Qatar, Education City, P.O.Box 23874, Doha, Qatar

[2]Center for Quantum Information, Institute for Interdisciplinary Information Sciences, Tsinghua University, Beijing 100084, Peoples Republic of China

[3]Department of Physics and Institute of Quantum Studies, Texas A&M University, College Station, TX 77843, USA

[4]School of Computational Sciences, Korea Institute for Advanced Study, Seoul 130-722, Korea

We theoretically propose and experimentally demonstrate a method to manifest nonclassicality by observing a single marginal distribution alone, which may provide a novel insight into nonclassicality with practical applicability to various quantum systems. Our approach depicted in Figure 1 detects a broad class of nonclassical states both Gaussian and non-Gaussian, and also yields a lower bound for entanglement potential—a measure of entanglement generated using a nonclassical state with a beam splitter setting. Remarkably, our method works regardless of measurement axis for all non-Gaussian states in finite-dimensional Fock space of any size, which also extends to infinite-dimensional states of experimental relevance for CV quantum informatics. We experimentally illustrate the power of our criterion for motional states of a trapped ion confirming their nonclassicality in a measurement-axis independent manner. We also address an extension of our approach combined with phase-shift operations, which leads to a stronger test of nonclassicality, i.e. detection of genuine non-Gaussianity under a CV measurement in contrast to the previously suggested methods.

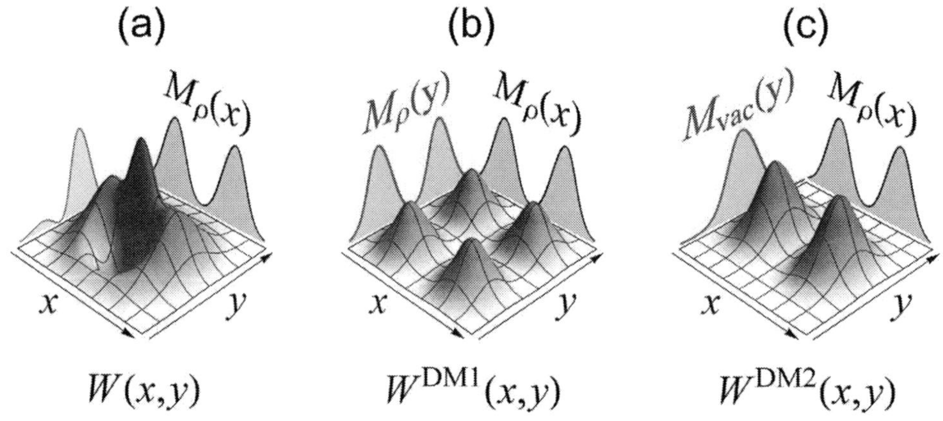


**Figure 1.** Illustration of our approach for nonclassicality tests. (a) original Wigner function $W(x,y)$ with its marginal distribution $M_\rho(x)$ in the backdrop. (b) a fictitious Wigner function $W^{DM1}(x,y)=M_\rho(x)M_\rho(y)$ where the same marginal distribution is replicated along the orthogonal axis (red solid curve). (c) a fictitious Wigner function $W^{DM2}(x,y)=M_\rho(x)M_{vac}(y)$ where the vacuum state distribution is employed along the orthogonal axis (red solid curve). If the resulting $W^{DM1}(x,y)$ and $W^{DM2}(x,y)$ do not represent any physical states, it confirms the nonclassicality of the given state.

# Higher-order uncertainty relations for SU(2) symmetric quantum states

U. Seyfarth[1], A. B. Klimov[3], L. L. Sánchez-Soto[1,2,4], G. Leuchs[1,2]

[1]Max Planck Institute for the Science of Light, 91058 Erlangen, Germany

[2]Institut für Optik, Information und Photonik, Universität Erlangen-Nürnberg, 91058 Erlangen, Germany

[3]Departamento de Física, Universidad de Guadalajara, 44420 Guadalajara, Jalisco, Mexico

[4]Departamento de Óptica, Facultad de Física, Universidad Complutense, 28040 Madrid, Spain

In the past decades experimental techniques in the area of quantum optics have undergone a considerable progress in the ability to control and measure quantum systems very precisely, which asks for pushing the theoretical basis for future steps. Quantum uncertainty relations give limitations in the accuracy of quantum states.

In this contribution we focus on uncertainty relations for SU(2) symmetric states, where we derive a recipe to generate the higher-order uncertainty relations for SU(2) symmetric states. Similar uncertainty relations can be derived for shrinked operators which are useful in the context of different applications.

# Characterization and optimization of the quantum operation of two qubit system using 3D transmon qubits

Taewan Noh[1], Gwan Yeol Park [1,2] , Dong Gwang Ha[1], Woon Song[1], Soon Gul Lee[2], Yonuk Chong[1,3*]

[1]Korea Research Institute of Standards and Science, Daejeon 34113, Korea

[2]Korea University Sejong campus, Sejong 30019, Korea

[3]University of Science and Technology, Daejeon 34113, Korea

*yonuk@kriss.re.kr

Precise construction, manipulation and characterization of aimed quantum states with reliable fidelities are crucial elements in developing a system for quantum information process. In order to characterize and optimize the quantum operations of two qubit system, we studied all microwave entangling scheme in two fixed-frequency superconducting transmon qubits. Here we report characterization and fidelity evaluation of the quantum operations on our two qubit system. We characterized the resulting states by using two qubit state tomography. To reduce and characterize the error rates caused by SPAM (state preparation and measurement), we implemented DRAG (derivative removal via adiabatic gating) scheme, cosine-shaped pulse complemented by the derivative on the quadrature with a calibrated weight. We also present our results of randomized benchmarking of quantum gate operations in our system.

# Experimental simulation of quantum universal-not gate on parallel state

Kuan Zhang[1], Jiajun Ma[1], Xiang Zhang[1], Jayne Thompson[2],
Vlatko Vedral[2], Mile Gu[3], Kihwan Kim[1]
[1]Center for Quantum Information, IIIS, Tsinghua University, Beijing, P. R. China
[2]Centre for Quantum Technologies, National University of Singapore, Singapore
[3]Centre for Quantum Technologies, Nanyang Technology University, Singapore

A fundamental discrepancy between classical and quantum information processing is the availability of the universal-NOT (U-NOT) gate. While flipping the value of a single bit is classically trivial, the quantum analogue that flips all measurement statistics on a qubit is strictly impossible. Here we show how this discrepancy can be understood operationally using the concept of classical and quantum correlations. When the UNOT gate applies to one of a pair of correlated qubits, it changes the amount of quantum correlations, but not the classical correlations. We demonstrate the result through an experimental simulation - by harnessing ion trap technology that allows simulation of unphysical operations on qubits embedded in a multi-level trapped ion. Our work simultaneously offers a new perspective in understanding why the U-NOT gate is unphysical, while opening potentially new applications for the experimental simulation of unphysical operations.

[1] H. Ollivier and W. H. Zurek, Phys. Rev. Lett. **88**, 017901 (2000).
[2] N. Gisin and S. Popescu, Phys. Rev. Lett. **83**, 432 (1999).

# A loophole-free test of Bell's inequality with atoms entangled over a distance of 400m

Daniel Burchardt[1], Kai Redeker[1], Robert Garthoff[1], Norbert Ortegel[1], Markus Rau[1], Wenjamin Rosenfeld[1,2], Harald Weinfurter[1,2]

[1]Fakultät für Physik, Ludwig-Maximilians-Universität München, D-80799 München, Germany

[2]Max-Planck-Institut für Quantenoptik, D-85748 Garching, Germany

Bell's inequality allows to test experimentally whether nature can be described in a local-realistic way by measuring correlations between two particles in an entangled state for a set of four measurement directions. In a conclusive test, the detection efficiency of the particles has to be high enough and the measurements on the two sides have to be space-like separated in order to avoid loopholes.

Here we present our results for a conclusive Bell test using heralded entanglement of two neutral Rb-atoms over a distance of 400m. Therefore, each atom is first entangled with a photon. The photons are coupled into single-mode fibers and interfere at a beam splitter. The projection of the photons onto a Bell state then projects the atoms onto an entangled atom-atom state [1]. To close the locality loophole the measurements on both sides, including the random choice of the measurement basis, have to be performed in a time shorter than 1.33μs. The basis is chosen independently on each side by a fast quantum random number generator. The atomic state readout itself is based on a state-selective ionization and subsequent detection of the ionization fragments by channel-electron-multipliers (CEM) [2]. Altogether the time for the state readout is below 1μs and the locality condition is fulfilled. Evaluation of 5000 events for each of the two prepared atom-atom states shows a significant violation of Bell's inequality with S-parameters of 2.237 ± 0.047 and 2.202 ± 0.047. Reliable violations of Bell's inequality form the basis for certified generation of random numbers or device-independent quantum key distribution.

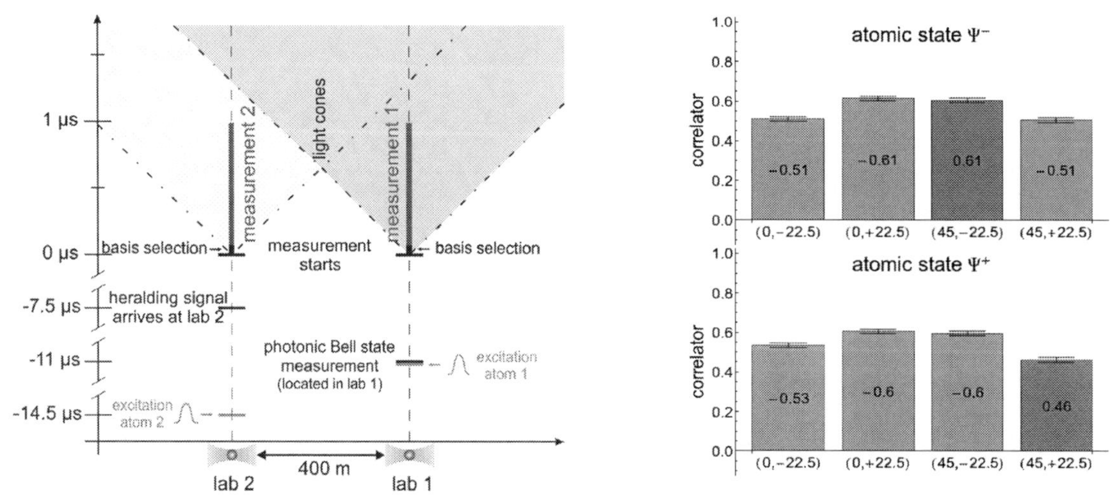


**Figure 1.** Left: Space-time diagram of the Bell experiment. After a successful preparation measurements on the two particles are performed within 1μs, thus fulfilling the criterion of space-like separation. Right: Measured correlators for the four sets of measurement directions for both states.

[1] J. Hofmann et. al., Science **337**, 72 (2012).

[2] F. Henkel et. al., Phys. Rev. Lett. **105**, 253001 (2010).

# Detection of genuine tripartite entanglement and steering in hybrid optomechanics

Y. Xiang[1,3], **F. X. Sun**[1], M. Wang[1,2], Q. H. Gong[1,2,3], Q. Y. He[1,2,3]

[1]State Key Laboratory of Mesoscopic Physics, School of Physics, Peking University, Beijing 100871, China

[2]Collaborative Innovation Center of Quantum Matter, Beijing 100871, China

[3]Collaborative Innovation Center of Extreme Optics, Shanxi University, Taiyuan, Shanxi 030006, China

Multipartite quantum entanglement is a key resource for ensuring security in quantum network. We show that by using a unified parameter in terms of reduced noise variances one can determine different types of tripartite entanglement of a given state generated in a hybrid optomechanical system, where an atomic ensemble is located inside a single-mode cavity with a movable mirror, with different thresholds for each type [1]. In particular, the special quantum states which allow both entanglement and steering genuinely shared among atom-light-mirror modes can be observed, even though there is no direct interaction between the mirror and the atomic ensemble. We further show the robustness against mechanical thermal noise and damping, the relaxation time of atomic ensemble, as well as the effect of gain factors involved in the criteria. Our analysis provides an experimentally achievable method to determine the type of tripartite quantum correlation in a way.

[1] Y. Xiang, F. X. Sun, M. Wang, Q. H. Gong, Q. Y. He. Optics Express **23**, 30104-30117 (2015).

# Experimental preparation of high NOON states for phonons

Junhua Zhang[1], Mark Um[1], Dingshun Lv[1], Jingning Zhang[1], Luming Duan[1], Kihwan Kim[1]
[1]Center for Quantum Information, IIIS, Tsinghua University, Beijing, P. R. China

Multi-party entangled state, in particular, NOON state has brought great interests because it allows the precision measurement to the ultimate limit as the number of particles increases. However, experimental preparation of the NOON state with sufficiently high N remains as a challenge. Here we develop a deterministic method to generate the NOON state of arbitrary phonon numbers and experimentally create the states up to $N = 9$ phonons in two radial modes of a single trapped $^{171}Yb^{+}$ ion. We clearly observe the fidelity of the NOON state over classical limit by measuring the contrast of the characteristic phase oscillations and the populations through the phonon projective measurement of two motional modes. We also observe the Heisenberg scaling of phase sensitivity in the NOON states through quantum Fisher information. Our scheme is generic and directly applicable to cavity QED or circuit QED systems and optomechanical systems.

# Arbitrary two-qubits controlled-unitary quantum gates on a reconfigurable silicon photonic circuits for quantum information processing

Jianguo Huang[1, 3], Jiuhui Wu[1], Xiaoqi Zhou[2], Aiqun Liu[3]

[1]School of Mechanical Engineering, Xi'an Jiaotong University, Xian 710049, China

[2]School of Physics and Engineering, Sun Yat-sen University, Guangzhou 510275, China

[3]School of Electrical & Electronic Engineering, Nanyang Technological University, Singapore

We demonstrate the arbitrary two-qubits controlled unitary quantum gates on a chip-scale silicon photonic circuits by extending the Hilbert space to simplify adding control qubits. It has the advantage of greatly reducing size of the circuits without decomposition of a functional quantum circuits into plenty elementary logic gate. Most important of all, it enables the ability to add control to the unitary operations even though the operations is unknown. By utilizing this approach, we demonstrated a series of controlled-unitary (CU) gates, including CNOT, C-Hadamard (CH), CPhase (CZ) and CSIGN by designing the proper operations. The schematic of the proposed CU gates is shown in the Figure 1. The phase shifter α can control the pumping of the two ring source. The correlated photon pairs generated by the nonlinear spontaneous four-wave mixing effect is enhanced by the ring resonator cavity. The signal and idler photons are then separated into different channels by tuning the signal and idler ring resonance wavelength. The top idler (signal) photons is operated by A1 (A2) unitary function, which is controlled by applying voltage to the phase shifter. It's the same operation to the bottom signal and idler photons. After the operation, the A1 (A2) and B1 (B2) channel are reconfigured to form C1 (C2) and C1′ (C2′) [1]. The total operation can be understood as $U=A_1 \otimes A_2 \pm B_1 \otimes B_2$ depending on the measurement [2]. The on chip qubit state is finally obtained by quantum state tomography. In conclusion, we have demonstrated the arbitrary two-qubits controlled unitary quantum gates on a chip-scale silicon photonic circuits. By combing the advantage of silicon fabrication technology and new quantum algorithms, the new approach will provide a novel way for the on chip quantum information processing.

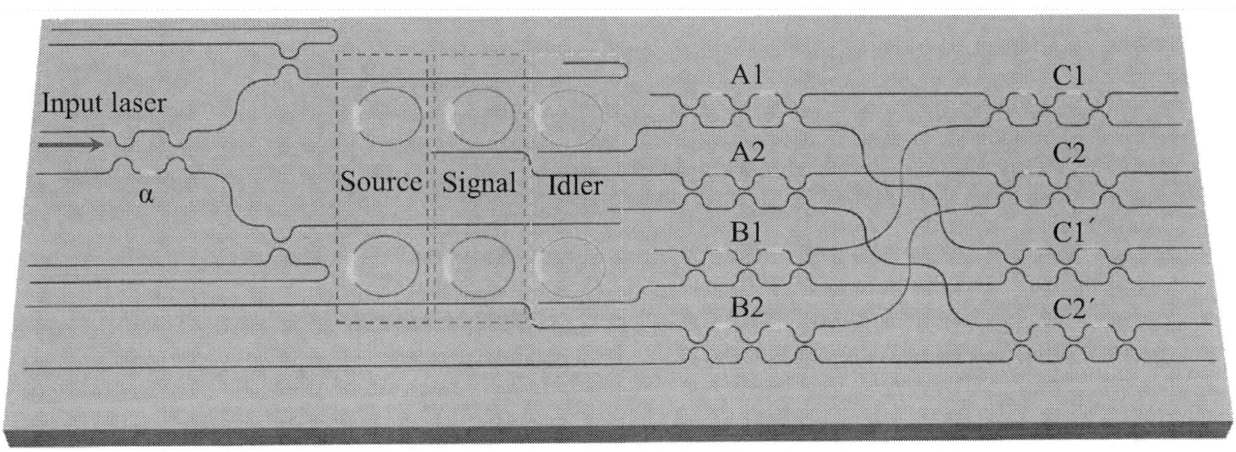


**Figure 1.** The schematic of the design and operation of the devices.

[1] Zhou, Xiao-Qi, et al. Nature communications **2**, 413 (2011).

[2] Zhou, Xiao-Qi, et al. Nature Photonics **7**, 223-228 (2013).

# Unified view of quantum correlations and quantum coherence

Kok Chuan Tan, Hyukjoon Kwon, Chae-Yeun Park, Hyunseok Jeong
Center for Macroscopic Quantum Control, Department of Physics and Astronomy,
Seoul National University, Seoul, 151-742, Korea

We present arguments that quantum correlations may be viewed as a type of measurement of the amount of quantum coherence present in the system. In order to do this, we introduce a new type of quantum coherence called Correlated Coherence and show that coherence of this type is able to define the same set of quantum correlations as quantum discord (both the symmetrized and the non-symmetrized versions) as well as entanglement – two of the most well studied forms of quantum correlations today. We also demonstrate that it is possible to construct from the ground up, using the resource theory of coherence [1], a reasonable measure of entanglement which provides a direct connection between these 2 theories.

[1] T. Baumgratz, M. Cramer, and M.B. Plenio, Phys. Rev. Lett. **113**, 140401 (2014).

# Site-specific loading and control of single atoms in a 1D optical lattice

Hyok Sang Han, Hyun Gyung Lee, Seokchan Yoon, D. Cho
Department of physics, Korea University, Seoul 136-713, Korea

Initialization and selective qubit-control are key elements for quantum information processing (QIP). Although an optical lattice has been considered as a promising platform for QIP, initialization of a quantum registry in a lattice proved to be difficult. The main difficulty is from inhomogeneous broadening of an RF transition due to differential ac Stark shift from the lattice field itself.

We report our experiment to load and manipulate a single atom at a specific site of a 1D lattice in a programmable manner. We also developed a fluorescence-imaging technique to measure the number of atoms in the lattice, which is essential for loading one and only one atom with high confidence level. For the programmable loading and selective operation, we distinguished individual sites by spatial dependent Zeeman shift produced by a magnetic field gradient. Although this method has already been proposed and demonstrated in other experiments, the above-mentioned inhomogeneous broadening limited resolution. We use "magic polarization" to eliminate the Stark-induced broadening to achieve a narrow linewidth [1]. Each site is resolved in an RF scan and it enables high-fidelity qubit control in a site-specific manner. Instead of employing a complicated imaging system, manipulation of each qubit is achieved by a spectroscopic method. With the capability we plan to demonstrate a lattice-based single-atom CNOT gate, in a way similar to that of an ion-trap gate [2].

[1] Huidong Kim, Hyok Sang Han, and D. Cho, Phys. Rev. Lett. **111**, 243004 (2013).

[2] C. Monroe, D. M. Meekhof, B. E. King, W. M. Itano, and D. J. Wineland, Phys. Rev. Lett. **75**, 4714 (1995).

# Atom-by-atom assembly of cold quantum matter

Hannes Bernien[1], Manuel Endres[1,2], Alexander Keesling[1], Harry Levine[1], Eric Anschuetz[1], Crystal Senko[1], Markus Greiner[1], Vladan Vuletic[3], Mikhail Lukin[1]

[1]Harvard University, Cambridge, USA

[2]California Institute of Technology, Pasadena, USA

[3]Massachusetts Institute of Technology, Cambridge, USA

The realization of large-scale fully controllable quantum systems is an exciting frontier in modern physical science. We use atom-by-atom assembly to implement a novel platform for the deterministic preparation of regular arrays of individually controlled cold atoms. A measurement and feedback procedure eliminates the entropy associated with the probabilistic trap occupation and results in defect-free arrays of over 50 atoms in less than 400 ms. The technique is based on fast, real-time control of 100 optical tweezers, which we use to arrange atoms in desired geometric patterns and to maintain the arrays by replacing lost atoms from a reservoir. This bottom-up approach enables controlled engineering of scalable many-body systems for quantum information processing, quantum simulations, and precision measurements.

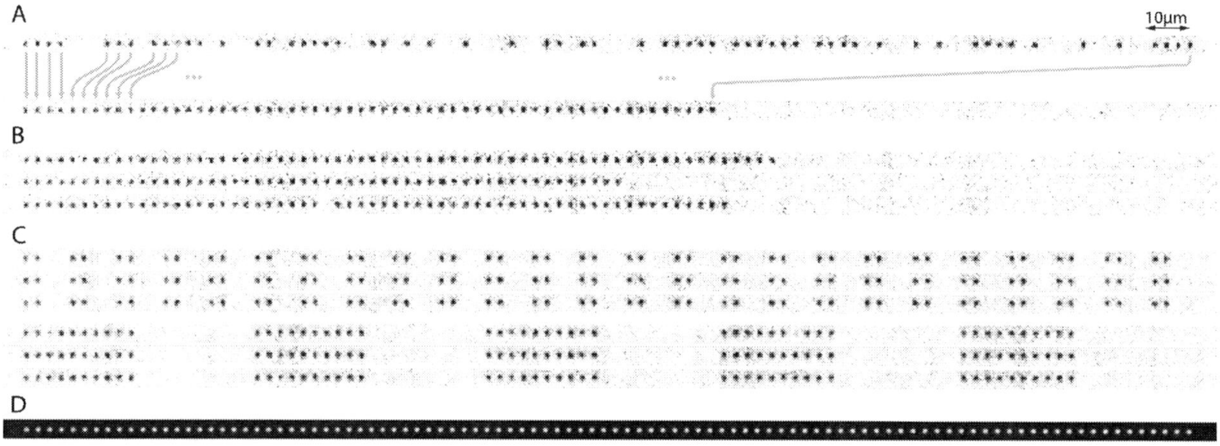


**Assembly of regular atom arrays.** (**A**) Single-shot, single atom fluorescence images before (top) and after (bottom) rearrangement. Defects are identified and the loaded traps are rearranged, indicated by arrows for a few selected atoms. (**B**) Two instances of successfully rearranged arrays (first two pictures), and one instances (last picture) where a defect is visible after rearrangement. (**C**) The final arrangement of atoms is flexible, and we generate, e.g., clusters of two (top) or ten (bottom) atoms. Non-periodic arrangements and adjustable lattice spacings are also possible. (**D**) High-resolution CCD image of trap array. Our default configuration for loading atoms consists of an array of 100 optical tweezers with separation of 2.6 μm between the focused beams.

# Coupling ultracold atoms to a superconducting microwave resonator

Helge Hattermann[1], Li Yuan Ley[1], Daniel Bothner[1,2], Benedikt Ferdinand[1], Lőrinc Sárkány[1], Reinhold Kleiner[1], Dieter Koelle[1], József Fortágh[1]

[1]CQ Center for Collective Quantum Phenomena, Physikalisches Institut Tübingen, Auf der Morgenstelle 14, 72076 Tübingen, Germany

[2] Present Address: Kavli Institute of Nanoscience, Delft University of Technology, PO Box 5046, 2600 GA Delft, The Netherlands

Superconducting microwave circuits in the quantum regime are one of the most promising candidates for quantum information processing. However, while they can process information on very short timescales, they lack the ability to store information for extended periods of time. Hybrid quantum systems, in which quantum information is processed by superconducting circuits and stored in a trapped ensemble of cold atoms, have therefore been proposed as a possible alternative. We demonstrate loading of ultracold Rubidium atoms to a superconducting coplanar microwave cavity and show that the coherence of atomic hyperfine superposition states is preserved on the order of seconds [1]. We show that non-resonant microwave dressing can be used to render atomic qubit states insensitive to magnetic field fluctuations. The differential Zeeman shift of the clock transition is suppressed up to second order [2]. As a next step towards the realization of cold atom and superconductor hybrid quantum systems, we demonstrate the coupling between magnetically trapped ultracold $^{87}$Rb atoms and a driven coplanar waveguide resonator, which is near-resonant to the atomic hyperfine transition.

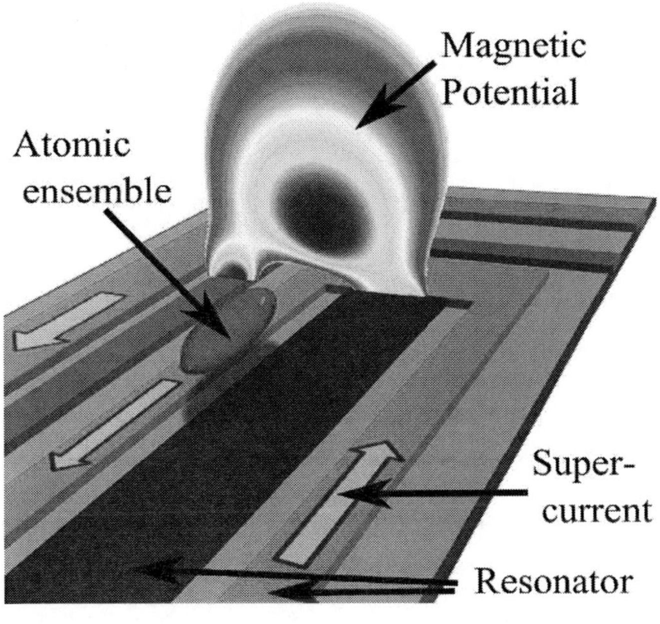


**Figure 1.** Schematic trapping of ultracold atoms in a superconducting coplanar waveguide resonator and simulated magnetic potential.

[1] S. Bernon et al., Nat. Commun. **4**, 2380 (2013).
[2] L. Sárkány et al., Phys. Rev. A **90**, 053416 (2014).

# Simulation of the small number of qubits in 1D array using MPS method

Donggeun Lee[1], Kyunghyun Baek[1], Wonmin Son[1,2]
[1]Department of Physics, Sogang University, Seoul, 121-742, Korea
[2]University of Oxford, Department of Physics, Parks Road, Oxford OX1 3PU, United Kingdom

Description of many-body quantum system has been known as an intractable problem in physics. Within the framework of quantum information, we have researched the physical properties of the many-body system by using the analytical methods [1, 2, 3] and recently embark on the numerical methods beyond exactly solvable models. Matrix Product State (MPS) method for one-dimensional system is one of the most popular and powerful numerical methods in tensor network theory. In this method, the exponential order of the quantum system O(exp(N)) can be reduced to the linear order O(N), which results in the efficient simulation of the many-body quantum system. By using the MPS method [4, 5], we program the algorithms for searching the many-body ground state pertaining to the class of its decomposed matrices and the variational method. Then, we test this programming for the s=1/2 Heisenberg model and analyze its error by calculating its ground energy as shown in Figure 1. Furthermore, we study the s=1/2 isotropic Heisenberg chain with nearest and next-nearest exchange at the Majumdar-Ghosh point within the open and periodic boundary conditions.

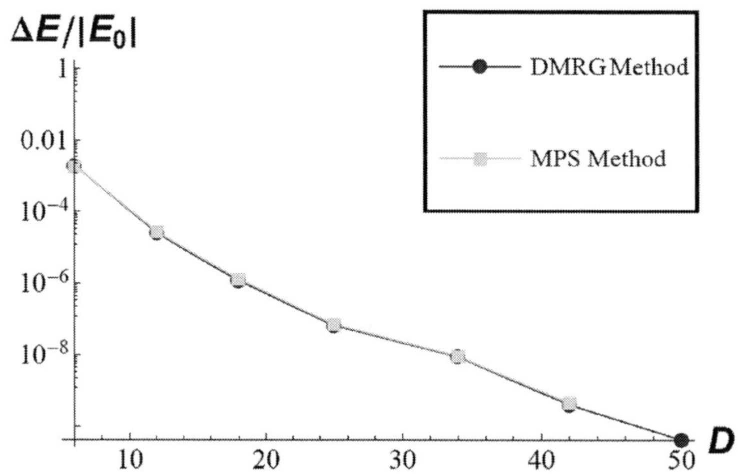


**Figure 1.** Relative error of the ground energy for N=28 is a function of the bond dimension $D$. The exact energy $E_0$ is obtained from the DMRG method with D=58 for the open boundary condition.

[1] W. Son, L. Amico, V. Vedral, Quantum Inf. Process **11**, 1961 (2012).
[2] W. Son, L. Amico, R. Fazio, A. Hamma, S. Pascazio, V. Vedral, Europhys. Lett. **95**, 50001 (2011).
[3] W. Son, L. Amico, F. Plastina, V. Vedral, Phys. Rev. A **79**, 022302 (2009).
[4] U. Schollwöck, Ann. Phys. **326**, 96 (2011).
[5] R. Orús, Ann. Phys. **349**, 117 (2014).

# Building an ion trap for quantum information processing

Rafael R de Paiva[1], Daniel V. Magalhães[1], Vanderlei S. Bagnato[1]
[1]Instituto de Física de São Carlos, Universidade de São Paulo, Brasil

Quantum information processing (QIP) uses qubits and pure quantum phenomena (entanglement, discord, etc) to efficiently simulate complex physical systems in order to understand them better. Ions traps are among the systems that are more suitable for QIP. Their characteristics were widely tested and they meet all the criteria necessary for quantum information processing. In this presentation, we will show a compact and low-cost experimental setup for QIP. The system comprises of home-made laser systems, compact vacuum setup with good optical accesses and a linear Paul trap based on the Innsbruck design. It will also be presented the futures experiments and ongoing collaborations. We acknowledge the support from FAPESP for the realization of this work.

# Optical trapping of ions

Christoph Fischer[1], Matthew Grau[1], Oliver Wipfli[1], Matteo Marinelli[1] and Jonathan P. Home[1]
[1]Institute for Quantum Electronics, ETH Zürich, CH-8093, Zurich, Switzerland

One of the major challenges of simulating quantum systems using ions is the scalability of the trap structure. Linear Paul traps can confine ions only in one-dimensional strings and micro-structured potentials generated by two-dimensional arrays of radio-frequency (rf) electrodes are difficult to fabricate [1].

Recent experiments have shown loading of a single ion into a far-detuned optical dipole trap formed by two counter-propagating beams without the micromotion typically associated with ion traps [2]. Based on these results we are planning a new experiment for trapping magnesium ions ($^{25}Mg^{+}$) in an optical lattice formed by a high-finesse optical cavity at a wavelength of $\lambda = 532$ nm and supported by additional static electric quadrupole fields. Numerical simulations indicate that this trap geometry might be able to trap up to two layers of two-dimensional Coulomb crystals and configurations with up to three optical cavities have the potential of confining two- and three-dimensional arrays of ions.

We will present the recent status of the experiment along with numerical simulations of different trap configurations. Furthermore, we will outline the theoretical considerations which go into the system design.

[1] M. Kumph, M. Brownnutt and R. Blatt, New. J. Phys. **31**, 073043 (2011).

[2] T. Huber, A. Lambrecht, J. Schmidt, L. Karpa and T. Schätz, Nat. Comm. **5**, 5587 (2014).

# Cooling and Coherent Manipulation of Small Coulomb Crystals in a Penning Trap

[1]P. Hrmo, [1]M. Joshi, [1]V. Jarlaud, [2]G. Stutter, [1]J.F. Goodwin, [1]D.M. Segal, [1]R.C. Thompson

[1]Quantum Optics and Laser Science, Blackett Laboratory, Imperial College London, Prince Consort Road, London SW7 2AZ, United Kingdom

[2]Department of Physics and Astronomy, Aarhus University, DK-8000 Aarhus C, Denmark.

We have recently demonstrated the laser cooling of the axial motion of a single $^{40}Ca^{+}$ ion to the motional ground state in a Penning trap using resolved-sideband laser cooling on the $S_{1/2} \leftrightarrow D_{5/2}$ electric quadrupole transition.[1] We report on the extension of this technique to small ion Coulomb crystals (ICC) made of two or three ions. Efficient cooling of the axial motion is achieved outside the Lamb-Dicke regime on a two ion chain along the magnetic field axis as well as on two and three-ion planar crystals. The ground state occupation of both the centre of mass mode and breathing mode of the two ion chain is measured to be 92±2%, while the two-ion planar crystal has a 98±2% for its centre of mass and tilt modes. Preliminary results for three ions in a radial plane are also presented. We also measure typical heating rates of the order of a phonon per second for most of the modes.

[1] J.F. Goodwin, G. Stutter, R.C. Thompson, and D.M. Segal, Physical Review Letters **116**, 143002 (2016).

# Quantum information processing with mixed-species ion crystals

Hsiang-Yu Lo, Matteo Marinelli, Vlad Negnevitsky, Christa Flühmann, Robin Oswald, Ludwig de Clercq, Daniel Kienzler, Jonathan P. Home

Institute for Quantum Electronics, ETH Zürich, Otto-Stern-Weg 1, 8093 Zürich, Switzerland

Trapped ions are one of the most promising technologies for realising quantum information processing and quantum simulation. In our work, we use beryllium and calcium ions to implement quantum information processing. The advantage of using two species of ion is that we can individually manipulate each ion species with a wide range of light fields without disturbing the internal state of the other.

We demonstrate high-fidelity quantum single- and two-qubit gates using calcium and beryllium ions, respectively, as well as mixed-species entangling gates using a calcium and a beryllium ion. Additionally, we will present parallel quantum gate operations implemented by transporting ions through stationary laser beams. The methods demonstrated here are key ingredients for realising hybrid universal quantum computation. Using mixed-species ion crystals accompanied with transport techniques has advantages in scalable quantum information processing [1], and should also enable the investigation of new avenues in quantum simulation and quantum state control.

[1] J. P. Home, D. Hanneke, J. D. Jost, J. M. Amini, D. Leibfried, and D. Wineland, Science **325**, 1227 (2009).

# Coherent Manipulation of a Single Sideband Cooled Ion in a Penning Trap

V. Jarlaud[1], P. Hrmo[1], M. K. Joshi[1], G. Stutter[1,2], J. F. Goodwin[1] and R. C. Thompson[1]

[1]Quantum Optics and Laser Science, Blackett Laboratory, Imperial College London, Prince Consort Road, London SW7 2AZ, United Kingdom

[2]Department of Physics and Astronomy, Aarhus University, DK-8000 Aarhus C, Denmark

We demonstrate coherent manipulation of a single $^{40}Ca^{+}$ ion in a Penning trap. The axial motion of the ion is initially cooled to the ground state with a high occupation probability using laser Doppler cooling and resolved sideband cooling on the electric quadrupole transition $S_{1/2} \leftrightarrow D_{5/2}$. The observation of Rabi oscillations demonstrates the coherence of the system with a coherence time estimated to be 1400 ± 100 μs while Ramsey experiments suggest 1280 ± 30 μs. The coherence time can be significantly increased by implementing the Uhrig dynamical decoupling (UDD) scheme [1]. A coherence time of 9200 ± 130 μs was measured with a UDD sequence of three π pulses. We also measure a heating rate of 0.3(2) phonon/s, the lowest reported to date in an ion trap [2].

[1] G. S. Uhrig, Physical Review Letters 98, 100504 (2007).

[2] J. F. Goodwin *et al*, Physical Review Letters 116, 143002 (2016).

# Nonlinear interactions between normal modes of motion of trapped ions

Shiqian Ding[1], Gleb Maslennikov[2], Roland Hablutzel[1], Jaren Gan[1], Dzmitry Matsukevich[1,2]
[1]Centre for Quantum Technologies, National University of Singapore, Singapore
[2]Departmnt of Physics, National University of Singapore, Singapore

We experimentally investigate nonlinear coupling between modes of motion in the system of trapped ions induced by anharmoincity of the Coulomb interaction between the ions. We demonstrate coupling between normal modes of motion in the crystal of two trapped $^{171}$Yb ions equivalent to the degenerate parametric oscillator on the level of single quanta and exploit it to measure parity and Wigner functions of ions motional states [1]. We also observe similar coupling between three modes of motion in the three ion crystal, analogous to the non degenerate parametric oscillator and apply it to simulate the process of non-degenerate parametric down conversion in a regime of depleted pump. We demonstrate deviation from the thermal statistic for the `signal' and `idler' modes and discuss its relation with a simple model of Hawking radiation. We also present experimental results on simulation of Jaynes-Cummings model using this Hamiltonian.

[1] S. Ding, G. Maslennikov, R. Hablutzel, H. Loh, D.Matsukevich, arXiv:1512. 01670

# Constructing a microwave quantum-computer demonstrator device

S. Weidt,[1] J. Randall,[1,2] S. C. Webster,[1] K. Lake,[1] A. E. Webb,[1] I. Cohen,[3] T. Navickas,[1] B. Lekitsch,[1] A. Retzker,[3] N. Johnson[1], R. LeBrun-Ricalens[1] and W. K. Hensinger[1]

*[1]Department of Physics and Astronomy, University of Sussex, Brighton, BN1 9QH, UK*
*[2]QOLS, Blackett Laboratory, Imperial College London, London, SW7 2BW, UK*
*[3]Racah Institute of Physics, The Hebrew University of Jerusalem, Jerusalem 91904, Givat Ram, Israel*

Trapped ions are a promising tool for building a large-scale quantum computer. The number of required radiation fields for the realisation of quantum gates in any proposed ion-based architecture scales with the number of ions inside the quantum computer, posing a major challenge when imagining a device with millions of qubits. Here we present a fundamentally different approach, where this scaling entirely vanishes. The method is based on individually controlled voltages applied to each logic gate location to facilitate the actual gate operation analogous to a traditional transistor architecture within a classical computer processor. We have demonstrated the key principle of this approach by implementing a two-qubit quantum gate based on long-wavelength radiation where we generate a maximally entangled two-qubit state with fidelity 0.985(12) [1]. We will then discuss the construction of a microwave trapped-ion quantum computer demonstrator device [2] as part of the UK Quantum Technology Hub on Networked Quantum Information Technologies (NQIT).

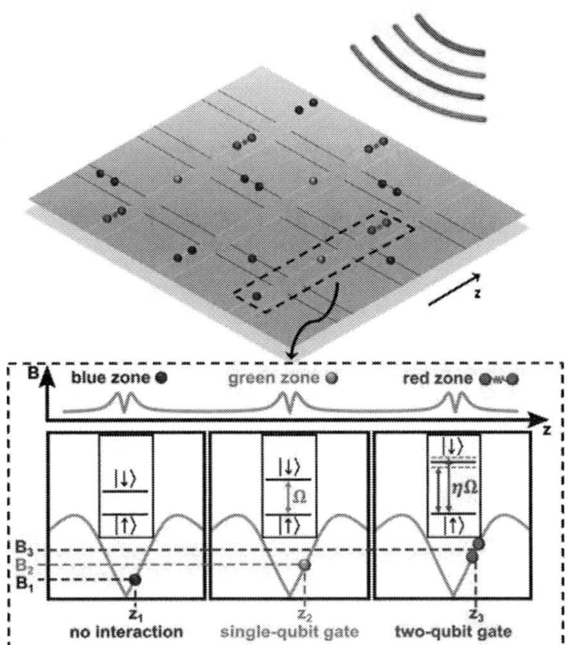


**Figure 1.** Ions are confined in a two-dimensional X-junction surface trap architecture. Local DC electrodes are used to shift the centre of the trapping potentials in the magnetic field gradient in order to tune a particular zone in resonance with a particular set of microwave and RF fields (illustrated in the dashed box). The ion displacements in the green (red) zones tune the respective ions into resonance with the global fields to realize single- (two-) qubit gates while no shift is applied to the blue zones, making all globally applied fields off-resonant for ions located in these zones.

[1] Trapped-ion quantum logic with global radiation fields, S. Weidt, J. Randall, S. C. Webster, K. Lake, A. E. Webb, I. Cohen, T. Navickas, B. Lekitsch, A. Retzker, and W. K. Hensinger, arXiv:1603.03384 [quant-ph] (2016)
[2] Blueprint for a microwave ion trap quantum computer, B. Lekitsch, S. Weidt, A.G. Fowler, K. Mølmer, S.J. Devitt, Ch. Wunderlich, and W.K. Hensinger, arXiv:1508.00420 [quant-ph] (2015)

# Towards portable ion traps for highly sensitive magnetometer

H. M. Bostock,[1] E. R. Potter,[1] A. H. Nizamani,[1] A. E. Webb,[1] J. Randall,[1] S. Weidt,[1] B. Lekitsch,[1] A. Rodriguez Blanco,[1] A. Grounds[1] and W. K. Hensinger[1]

[1]*Department of Physics and Astronomy, University of Sussex, Brighton, BN1 9QH, UK*

Highly sensitive magnetometers in the DC - GHz frequency range have uses such as detecting explosives and drugs and can be used as a biological scanner. Single trapped ytterbium ions have already successfully been used to sense magnetic fields with sensitivities of 4pTH^(-1/2) for AC magnetic fields [1]. We will present progress towards making a portable, highly sensitive device based on said method. For quantum sensing of magnetic fields with trapped ions, increasing the number of trapped ions and coherence time improve sensitivity. We will describe a design capable of trapping more than 100 ions along with progress towards miniaturized ion trap sub-systems technologies that are required for a portable device.

[1] Ultrasensitive magnetometer using a single atom, I. Baumgart, J.-M. Cai, A. Retzker, M. B. Plenio and Ch. Wunderlich, arXiv:1411.7893 [quant-ph] (2014)

# Frequency measurement of the $^1S_0 \rightarrow {}^3P_1$ transition of $Al^+$ via quantum logic spectroscopy with $Ca^+$

Michael Guggemos[1,2], Daniel Heinrich[1,2], Óscar-Andrey Herrera-Sancho[2,3,4], Yves Colombe[2], Rainer Blatt[1,2], and Christian F. Roos[1,2]

[1]Institut für Quantenoptik und Quanteninformation, Österreichische Akademie der Wissenschaften, Technikerstr. 21a, 6020 Innsbruck, Austria

[2]Institut für Experimentalphysik, Universität Innsbruck, Technikerstr. 25, 6020 Innsbruck, Austria

[3]Escuela de Física, Facultad de Ciencias, Universidad de Costa Rica, 2060 San Pedro, San José, Costa Rica

[4]Centro de Investigación en Ciencia e Ingeniería de Materiales, Universidad de Costa Rica, 2060 San Pedro, San José, Costa Rica

Over the years, the physical quantity frequency has always had a special status in physics primarily because accurate knowledge of emission lines shaped by quantum state transitions in atoms is crucial in the realization of fundamental physics tests with the development of better clocks and measurement standards [1, 2]. Recently with the inclusion of essential ingredients from quantum information science it was possible to perform high precision spectroscopy in the otherwise difficult-to-reach transitions of $^{27}Al^+$ ions by means of Quantum Logic Spectroscopy (QLS) [3]. Briefly, QLS uses an auxiliary "logic" ion, which is stored in a miniature linear radio-frequency trap together with a spectroscopic ion. The "logic" ion is then used to cool down the initially hot "spectroscopic" ion via the Coulomb interaction and, here comes a key ingredient taken from quantum-information processing, additionally allows the preparation and detection of the internal state of the "spectroscopic" ion with high efficiency.

In our experimental demonstration of QLS, a single $^{40}Ca^+$ ion ("logic" ion) and a single $^{27}Al^+$ ion ("spectroscopic" ion) are laser-ablation loaded and simultaneously trapped in a linear Paul trap in order to perform optical spectroscopy on the $Al^+$ ion. We observe long sympathetic cooling times before crystallization occurs. Therefore, we develop a new technique for detecting the loading of an initially hot ion ("spectroscopic" ion) with energy in the electron-volt range by monitoring the motional state of a Doppler-cooled ion ("logic" ion) already present in the trap [4]. When the two-ion crystal of $Ca^+$ and $Al^+$ is cooled near its motional ground state by continuously laser-cooling the $Ca^+$ ion, we interrogate the "spectroscopic" ion and map its internal state onto the $Ca^+$ ion via the common vibrational state. By employing QLS, we have measured the absolute frequency of the $|3s^2\ {}^1S_0, F{=}5/2, m_F = \pm 5/2\rangle \rightarrow |3s3p\ {}^3P_1, F{=}7/2, m_{F'} = \pm 7/2\rangle$ in $^{27}Al^+$ to be 1 122 842 857 335 kHz by using the $^{40}Ca^+$ $|4s^2\ S_{1/2}\rangle \rightarrow |3d^2\ D_{5/2}\rangle$ electric-quadruple transition frequency as a reference.

[1] Chou C.W., Hume D.B., Koelemeij J.C.J., Wineland D.J., and Rosenband T. PRL **104**, 070802 (2010).
[2] Chou C.W., Hume D.B., Rosenband T., and Wineland D.J. Science **329**, 1630 (2010).
[3] P.O. Schmidt, Rosenband T., Langer C., Itano W.M., Bergquist J. C. and Wineland D. J. Science **309**, 749 (2005).
[4] M. Guggemos, D. Heinrich, O.A. Herrera-Sancho, R. Blatt and C.F. Roos, New J. Phys. **17**, 103001 (2015).
[5] M. Guggemos *et al.*, to be published.

# Towards high precision spectroscopy of sympathetically cooled $H_2^+$

Johannes Heinrich[1], Nicolas Sillitoe[1], Thomas Louvradoux[1],
Albane Douillet[1,2], Jean-Philippe Karr[1,2], Laurent Hilico[1,2]
[1]Laboratoire Kastler Brossel, UPMC-Sorbonne Universités,
CNRS , EMS-PSL Research University, Collège de France
Case 74, 4 place Jussieu, F-75005 Paris, France
[2]Université d'Evry-Val d'Essone
Boulevard François Mitterand, 91025 Evry Cedex, France

The development of an experiment to capture and cool $H_2^+$ ions in order to perform high precision spectroscopy is discussed. There are two main motivations for this project: first, to perform a test of quantum-electrodynamics in molecules [1], and second, to measure the proton to electron mass ratio, one of the fundamental constants in the CODATA adjustment [2]. Results with single-photon spectroscopy obtained by other groups [3,4] show the feasibility of this method of direct optical determination of $m_p/m_e$. Moreover we plan to probe a Doppler-free two photon transition to improve the accuracy of the frequency measurement.

Since $H_2^+$ can not be directly laser-cooled we are working on the implementation of sympathetic cooling with $Be^+$. Currently a linear Paul trap is realized, designed to accommodate the two different ion species, a Fabry-Perot cavity for Doppler-free two photon spectroscopy, as well as lasers for in situ creation of $H_2^+$, cooling of $Be^+$ and photodissociation of the excited hydrogen molecules.

To cool the $Be^+$ ions a 626nm DBR-based UV laser source is designed and constructed. A master-slave configuration provides enough power for frequency doubling in a bow tie cavity.

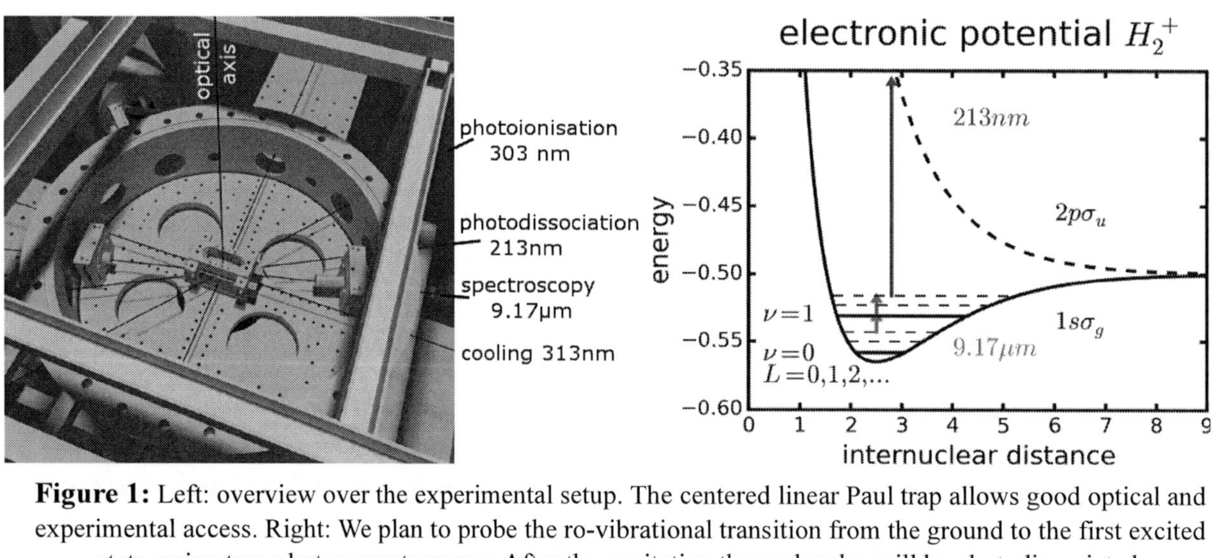


**Figure 1:** Left: overview over the experimental setup. The centered linear Paul trap allows good optical and experimental access. Right: We plan to probe the ro-vibrational transition from the ground to the first excited state, using two-photon spectroscopy. After the excitation the molecules will be photodissociated.

[1] V. I. Korobov, L. Hilico and J-Ph. Karr, Physical Review A 89, 032511 (2014).

[2] P. J. Mohr, D. B. Newell and B. N. Taylor, arXiv: 1507.07956v1 (2015).

[3] U. Bressel et al., Physical Review Letters 108, 183003 (2012).

[4] J. Biesheuvel et al. Nature Communications 7, 10385 (2016).

# Fidelity measurement for two motional states with the phonon beam splitter

Shuoming An[1], Dingshun Lv[1], Zhenyu Liu[1], Kihwan Kim[1]
[1]Center for Quantum Information, Institute for Interdisciplinary Information Sciences, Tsinghua University, Beijing 100084, People's Republic of China

The fast and reliable measure measurement of the quantum state is important task for the quantum information. Here we present the fidelity measurement between two motional state with the phonon beam splitter [1]. After preparing the two input states, we use the laser induced beam splitter operation to mix them. Then a phonon number dependent ac Stark shift combined with a Ramsey type measurement is used to take out the parity of one of the output mode, which is the fidelity between the two input modes [2]. When one of the input mode is a specific number state, we can get the phonon distribution of the other mode. During the whole process the traditional sideband type measurement is not involved.

[1] M. S. Kim, J. Lee and W. J. Munro, Phys. Rev. A **66**, 030301(R) (2002).
[2] L. G. Lutterbach and L. Davidovich, Phys. Rev. Lett. **78**, 2547 (1997).

# Sympathetic cooling of $^{171}Yb^+$ qubit ions trapped in a chip using different Yb isotope ions

Yeong-Dae Kwon[1], Jun Sik Ahn[2], Seokjun Hong[1], Minjae Lee[1], Yunjae Park[1], Dongil Dan Cho[1], Taehyun Kim[2]

[1]ASRI/ISRC and Department of Electrical and Computer Engineering, Seoul National University, Republic of Korea

[2]Quantum Tech. Lab., SK Telecom, Republic of Korea

To develop ion-trap-based large-scale quantum computing devices [1], the motional states of qubit ions need to be kept close to the ground state in the presence of constant heating caused by the electrical noise from the electrodes. Sympathetic cooling is a natural solution for this problem [2], but trapping two different species of ions generally requires two separate sets of optical components such as lenses, cameras, and ionization/cooling/repumping lasers for each ion type. Such a "double-setup" increases the complexity and the cost of the ion trap setup that incorporates sympathetic cooling. However, since isotopes have energy levels close to each other, if isotope ions of the same species are used for sympathetic cooling, the optical components prepared for the qubit ions can also be used for the cooling ions, vastly simplifying the experimental design. In this paper, we demonstrate sympathetic cooling of $^{171}Yb^+$ qubit ions using Doppler-cooled $^{170}Yb^+$ ions. $^{170}Yb^+$ was found to have the most favorable energy spectrum among various stable isotopes of Ytterbium for our purpose. To prevent the cooling beam from exciting the qubit ions and destroying their coherence, the beam was focused using our high-NA imaging optics to impinge only on the cooling ions, minimizing the exposure on the qubit ions. We found that such tightly focused cooling beam showed some interesting cooling characteristics that are different from those of normal beams. With application of dynamic decoupling scheme, more than 10 seconds of qubit coherence time was observed with continuous cooling of ion strings in our surface trap, which is roughly an order of magnitude improvement from our previous measurement without cooling. With our next generation of ion trap chips that has lower heating rates, we expect to further increase the coherence time of our qubit ions.

[1] D. Kielpinski, C. Monroe, D. Wineland, Nature **417**, 709 (2002).

[2] G.-D. Lin *et al.*, EPL **86**, 6 (2009).

# Single Trapped $^{40}$Ca-Ion as Wave-Front Sensor

Simon Ragg[1], Frieder Lindenfelser[1], Jonathan Home[1]
[1]Institute for Quantum Electronics, ETH Zurich, Otto-Stern-Weg 1, 8093 Zurich, Switzerland

A major current goal in the control of experimental quantum systems is quantum information processing. Atomic ions trapped in radio frequency traps are a leading candidate for realizing a quantum information processor. The traditional scheme for cooling and detecting $^{40}$Ca-ions relies on a dipole transition in the ultraviolet. An alternative approach is to use the dipole forbidden transition at 729 nm for cooling [1,2] providing background free state detection and removing the need for ultra-violet lasers for controlling the ion. I will describe recent work in which we have demonstrated trapping, cooling and fluorescence detection with 729 nm and 854 nm light in a novel "Photonic-Crystal Fibre" ion trap [3], allowing us to tune continuously between the weak binding (Doppler cooling) and strong-binding (sideband cooling) limits. We have performed a thorough investigation of the cooling rates and minimum temperatures, taking into account the multi-level nature of the ion. To implement cooling in the weak binding limit a high 729 nm laser intensity is required. In order to achieve this we attempt to produce a diffraction limited spot. In order to optimize performance, we are investigating the use of a MEMS mirror array and the possibility to optimize the spot size using the signal from the ion to obtain information about the beam. Theoretical work indicates that it should be possible to obtain the full complex amplitude using a phase retrieval algorithm, where the obtained phase provides us information about optical aberrations which can be used to suppress these.

[1] R. J. Hendricks, J. L. Sørensen, C. Champenois, M. Knoop, and M. Drewsen, Phys. Rev. A **77**, 021401 (2008).
[2] Z. Fei, X. Yi, C. Liang, W. Wei, W. Hao-Yu, and F. Mang, Chin. Phys. Lett. **30**,033701 (2013).
[3] F. Lindenfelser, B. Keitch, D. Kienzler, D. Bykov, P. Uebel, M. Schmidt, P. S. J. Russell, and J. Home, Rev. Sci. Instrum. **86**, 033107 (2015).

# Microfabrication of Ion-Trap Chips with Metal-Coated Dielectric Pillars to Reduce Heating of Trapped Ions

Seokjun Hong[1], Yeong-Dae Kwon[1], Minjae Lee[1], Yunjae Park[1], Changhyun Jung[1], Jun Sik Ahn[2], Taehyun Kim[2], and Dong-il "Dan" Cho[1,*]

[1]ASRI/ISRC and Department of Electrical and Computer Engineering, Seoul National University, Seoul, 151-744, Republic of Korea

[2]Quantum Tech. Lab., SK Telecom, Seongnam-si, Gyeonggi-do, 463-784, Republic of Korea

** Corresponding author (E-mail: dicho@snu.ac.kr)*

One of the major goals for ion-trap technologies is the reduction of ion heating. Ion heating can induce an unintended increase of phonon numbers, which in turn, results in lower fidelity of two-qubit operations that utilize the phonon states. The ion heating phenomenon is called anomalous, because the dominant source of the heating cannot be explained by well-known noise sources such as thermal electronic noise and external electronic noise. Previous research has shown that the electric-field fluctuation induced by the built-up charges on dielectric surfaces is one of the factors that induce ion heating [1]. To reduce the effect of built-up charges, the area of dielectric surfaces that are exposed to trapped ions should be minimized. Stick *et al.* proposed an ion-trap chip with an electrode overhang structure [2], and Niedermayr *et al.* proposed an ion-trap chip with a deep trench between electrodes [3]. However, these structures cannot entirely eliminate exposing dielectric surfaces from trapped ions. In this paper, a microfabricated ion-trap chip with metal-coated dielectric pillars is proposed (Figure 1). In the proposed ion-trap structure, the electrode metal layer and the ground metal layer are extended to cover the sidewalls of the dielectric pillars. The two layers are electrically isolated by the overhang structures of dielectric pillars. The overhang structures can be fabricated through a novel sacrificial process. Preliminary fabrication results of the proposed ion-trap chip are shown in Figure 2. The sidewalls of the dielectric pillars are coated with the metal layers. The RF electrodes, DC electrodes, inner DC rails, and ground plane are electrically isolated from each other. The proposed ion-trap chip is expected to reduce the heating phenomenon of trapped ions, which can lead to a higher fidelity of qubit operations.

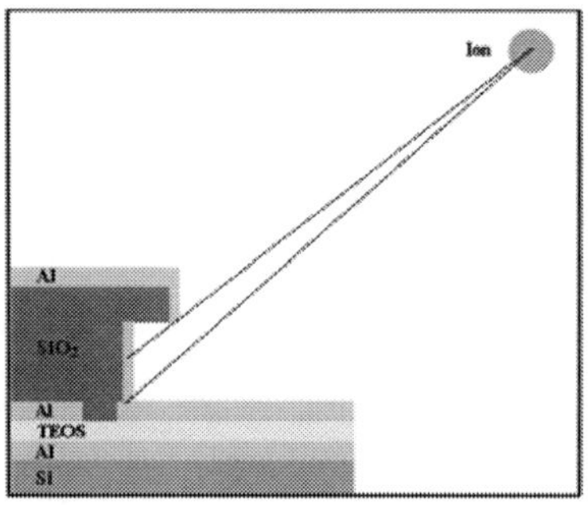


**Figure 1.** Schematic of the proposed ion-trap chip structure

**Figure 2.** Fabrication results of the proposed ion-trap chip (left: cross-sectional view, right: top view)

[1] M. Harlander *et al.*, New Journal of Physics **12**, 093035 (2010).

[2] D. Allcock *et al.*, Applied Physics B **107**, 913-919 (2012).

[3] M. Niedermayr *et al.*, New Journal of Physics **16**, 113068 (2014).

# Quantum-state control of ions in surface-electrode ion-trap systems

Chi Zhang[1], Joseba Alonso[1], Florian Leupold[1], Maciej Malinowski[1], Roland Matt[1], and Jonathan Home[1]

[1]Institute for Quantum Electronics, ETH Zürich, Otto-Stern-Weg 1, 8039 Zürich, Switzerland

Surface-electrode ion-traps have been recognized as a promising platform to scale up trapped-ion systems for quantum computation and simulation. To make use of the quantum properties of trapped-ions, it is essential to control both the ions' motional and electronic states. In our surface-electrode ion-trap system, we have implemented nanosecond electronic switches inside the 4 Kelvin vacuum system to realize quasi-instantaneous “bang-bang” control of the motional states of trapped-ion oscillators. By switching between spatially displaced trapping potentials, large coherent states with up to 10,000 quanta of energy have been demonstrated [1]. With a similar scheme, we are working towards squeezing the ion's motional state as well as fast ion-transportation [2]. In the same apparatus, we have recently developed the capability to perform universal logical gates on qutrits formed by three of the electronic energy levels of our ions. While benchmarking control, we have performed the Klyachko-Can-Binicioğlu-Schumovsky (KCBS) contextuality tests [3] with a single qutrit, demonstrating the violation of non-contextual hidden-variable inequalities as predicted by quantum mechanics.

In order to move towards a more scalable system, we are investigating the use of industrial standard complementary metal-oxide semiconductor (CMOS) technology [4], following recent work at Lincoln labs [5]. I will present our work on trap design which aims to take advantage of the increased complexity available through such fabrication methods, facilitating the integration of current carrying wires for microwave qubit manipulation [6] and junctions between different ion trap regions [7].

[1] J. Alonso, F. M. Leupold, Z. U. Soler, M. Fadel, M. Marinelli, B. C. Keitch, V. Negnevitsky, and J. P. Home, Nature Communications **7**, 11243 (2016).

[2] J. Alonso, F. M. Leupold, B. C. Keitch, and J. P. Home, New Journal of Physics **15**, 023001 (2013).

[3] A. A. Klyachko, M. A. Can, S. Binicioğlu, and A. S. Shumovsky, Physical Review Letters, **101**, 020403 (2008).

[4] C. Zhang. CMOS fabrication for scalable trapped-ion quantum information processing (Master thesis). ETH Zurich (2015).

[5] K. K. Mehta, A. M. Eltony, C. D. Bruzewicz, I. L. Chuang, R. J. Ram, J. M. Sage, and J. Chiaverini, Applied Physics Letters, **105**, 044103 (2014).

[6] C. Ospelkaus, U. Warring, Y. Colombe, K.R. Brown, J. M. Amini, D. Leibfried, and D. J. Wineland, Nature, **476**, 181-184 (2011).

[7] R. B. Blakestad, C. Ospelkaus, A. P. VanDevender, J. M. Amini, J. Britton, D. Leibfried, and D. J. Wineland, Physical Review Letters, **102**, 153002 (2009).

# Long coherence time of single quantum bit in a hybrid ion trap

Ye Wang[1], Mark Um[1], Junhua Zhang[1], Shuoming An[1], Ming Lyu[1], Lu-Ming Duan[1], Dahyun Yum[1], Kihwan Kim[1]

[1]Center for Quantum Information, Institute for Interdisciplinary Information Sciences, Tsinghua University, Beijing, 100084, P. R. China

For a single qubit, the longest reported coherence time is about tens of seconds [1], which is realized in trapped ion system and mainly limited by the heating of the ion. We increase the coherence time of an ion qubit to around 10 minutes in a hybrid ion trap system. Our hybrid ion trap has 171Yb+ as the memory ion qubit and 138Ba+ as the cooling ion. The qubit ion is kept near motional ground state through sympathetic cooling with cooling ion. We apply the dynamical decoupling scheme. The coherence time in our system is mainly limited by the gate fidelity for the dynamical decoupling pulses and the low frequency noise.

[1] C. Langer, et al., Phys. Rev. Lett. 95, 060502 (2005); T. P. Harty, et al., Phys. Rev. Lett. 113, 220501 (2014).

# Towards a new class of trapped ion experiments with ion rings

Hao-Kun Li[1], Erik Urban[2], Crystal Noel[2], Alexander Chuang[2], Yang Xia[1], Anthony Ransford[2], Boerge Hemmerling[2], Yuan Wang[1,3], Tongcang Li[1], Hartmut Häffner[2,3], Xiang Zhang[1,3]

[1]NSF Nanoscale Science and Engineering Center, 3112 Etcheverry Hall, University of California, Berkeley, California 94720, USA
[2]Department of Physics, University of California, Berkeley, California 94720, USA
[3]Materials Sciences Division, Lawrence Berkeley National Laboratory, 1 Cyclotron Road, Berkeley, CA 94720, USA

We present the design and implementation of a novel surface ion trap allowing for trapping of ions in a ring configuration. A ring topology introduces a number of new features that are not present in conventional, linear ion traps, such as extremely low trap frequencies, periodic boundary conditions, translational symmetry and the ability to rotate. These new features open the possibility to conduct a new class of experiments previously inaccessible to ion traps. Towards this goal, we trap up to 20 ions 400 μm above the plane of a trap surface in a ring geometry with a diameter of 90 μm. The large trapping height relative to the ion-ion spacing gives us the the ability to control electric fields at the trapping point with only a few compensation electrodes. This control allows us to pin the ions either into a localized crystal on one side of the ring potential or to fully delocalize them over the full extent of the ring, demonstrating any symmetry breaking in our ring is on energy scales below the temperature of the laser cooled ions. We present studies of the trap frequencies in a pinned configuration and its transition to a depinned, rotating state.

# Geometric quantum computation with a tripod system of Strontium

Frederic Leroux[1], Riadh Rebi[1] , Kanhaiya Pandey[1], David Wilkowski[1,2]
[1]Center for Quantum Technologies, National University of Singapore, Singapore 117543
[2]MajuLab, CNRS-UNS-NUS-NTU Interbnational Joint Research Unit (UMI 3654), Singapore

A four-level tripod system between two hyperfine states $^1S_0$ and $^3P_1$ of ultracold $^{87}$Sr[1] is driven continuously by three resonant lasers. We perform operations of quantum computation on a single geometrical qubit composed of the two dark states[2] issued from the light-matter interaction by tuning the phases of two lasers. I will present results about the non-Abelian character of the geometrical gates and the time evolution of these dark states.

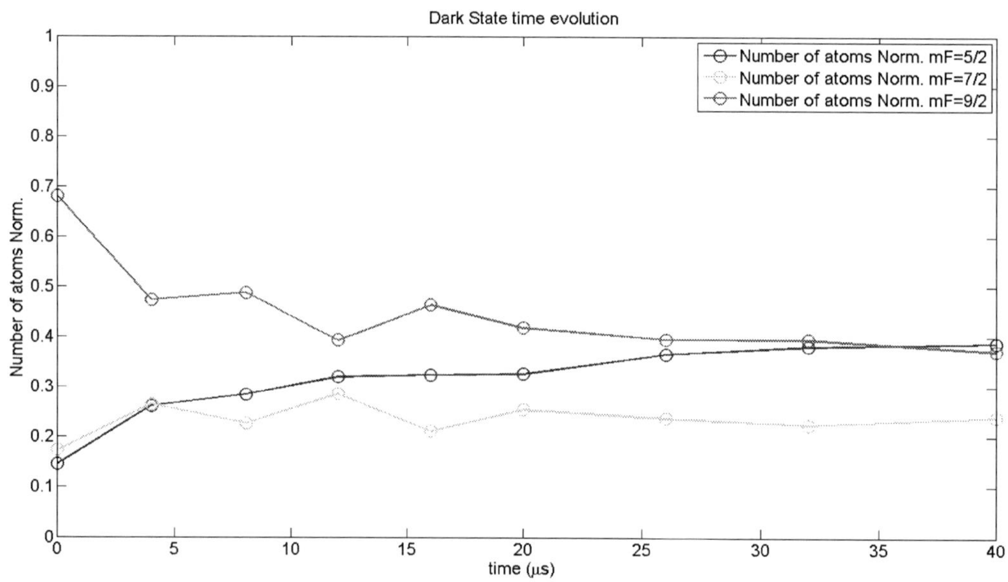


**Figure 1.** Dark state time evolution represented as population change for $^1S_0$ $m_F$=5/2,7/2,9/2

[1] A high flux source of cold strontium atoms, Yang, T et al. Eur. Phys. J. D (2015) 69: 226
[2] Non-Abelian Gauge Potentials for Ultracold Atoms with Degenerate Dark States, Ruseckas, PRL 95 (2005)

# Author Index

ibs 기초과학연구원 Institute for Basic Science

GIST 광주과학기술원 Gwangju Institute of Science and Technology

# Center for Relativistic Laser Science

## - Exploration of superintense laser-matter interactions -

- Exploration of the interactions between ultra-intense laser field and matter in the relativistic regime by using femtosecond petawatt lasers
- Investigation of ultrafast (attosecond) atomic/molecular processes

## Research topics

- **High-power femtosecond lasers**
- **Relativistic laser-plasma interactions**
- **Laboratory astrophysics**
- **High-energy density physics**
- **Attosecond science**

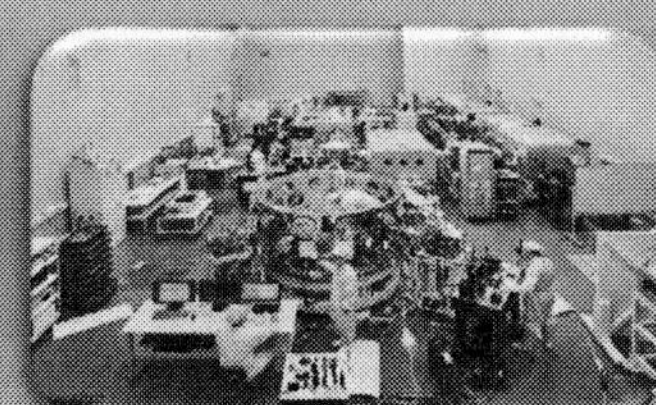

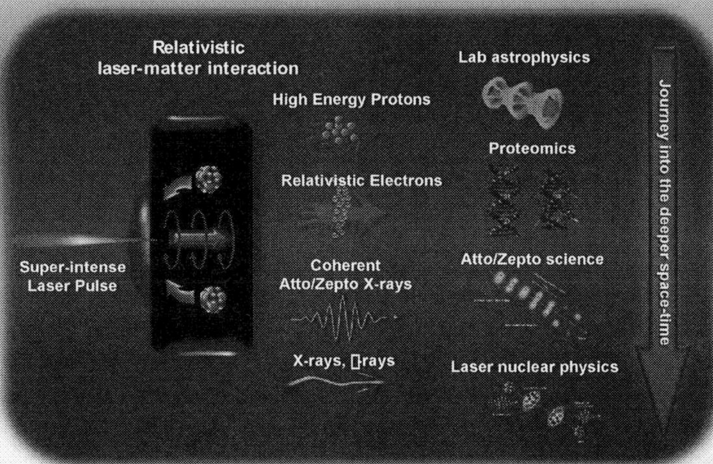


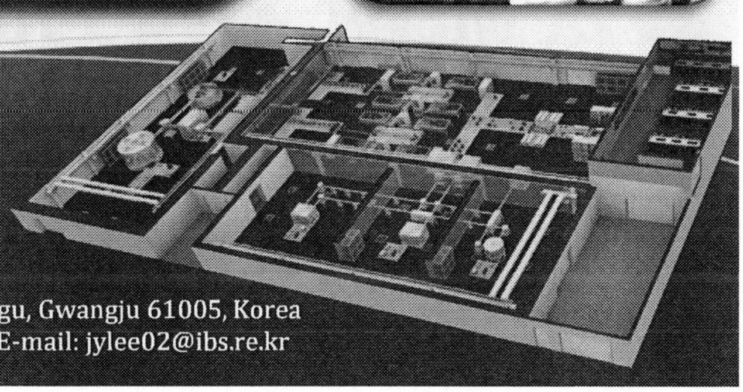


Center for Relativistic Laser Science (CoReLS)
Institute for Basic Science (IBS)
Addr.: UQBF, GIST, 123 Cheomdangwagi-ro, Buk-gu, Gwangju 61005, Korea
Tel.: +82-62-715-4703; Fax: +82-62-715-4705; E-mail: jylee02@ibs.re.kr

# Program Table

| July 24 (Sun) | July 25 (Mon) | July 26 (Tue) | July 27 (Wed) | July 28 (Thu) | July 29 (Fri) |
|---|---|---|---|---|---|
| | 8:30 - 8:45<br>**Welcome Remarks**<br>**W. Jhe**<br><br>8:45 - 10:15<br>**Plenary Session**<br>**E. Hinds**<br>45min) Harald Hess<br>45min) J. Saunders | 8:30 - 10:15<br>**Quantum optics and cavity QED I**<br>**X. Chen**<br>45min) Misha Lukin<br>30min) Stephan Ritter<br>30min) Jakob Reichel | 8:30 - 10:45<br>**Rydberg and artificial atoms**<br>**S. Whitlock**<br>45min) Tilman Pfau<br>30min) R. Schoelkopf<br>30min) Wu-Ming Liu<br>30min) Ana Maria Rey | 8:30 - 10:15<br>**Quantum computation & quantum simulation III**<br>**P. Naidon**<br>45min) Markus Greiner<br>30min) Stefan Kuhr<br>30min) Selim Jochim | 8:30 - 10:15<br>**Cold molecules**<br>**J. Ye**<br>45min) David DeMille<br>30min) Michael Tarbutt<br>30min) Simon Cornish |
| | 10:15 - 10:40<br>**Coffee Break** | 10:15 - 10:40<br>**Coffee Break** | 10:45 - 11:15<br>**Coffee Break** | 10:15 - 10:40<br>**Coffee Break** | 10:15 - 10:40<br>**Coffee Break** |
| | 10:40 - 12:25<br>**Atomic clocks**<br>**T. Udem**<br>45min) Jun Ye<br>30min) H. Katori<br>30min) Mariana Safronova | 10:40 - 12:25<br>**Fundamental tests and precision meas.**<br>**B. Das**<br>45min) G. Gabrielse<br>30min) Dmitry Budker<br>30min) H. Haeffner | 11:15 - 11:45<br>**Quantum stat. phys. & matter-wave inf.**<br>**D. Meschede**<br>30min) P. Bouyer<br><br>11:45 - 12:10<br>**D. Meschede**<br>**IUPAP C-15 Young Scientist Prize winner**<br>25min) Yuao Chen | 10:40 - 12:25<br>**Quantum gases I**<br>**J. Walraven**<br>45min) C. Salomon<br>30min) C. Naegerl<br>30min) M. Zwierlein | 10:40 - 12:20<br>**Hot Topics 2**<br>**G. Jo**<br>25min) R. Moszynski<br>25min) Jing Zhang<br>25min) Jean-Michel Raimond<br>25min) Carsten Klempt |
| | 12:25 - 14:50<br>**Lunch & Poster Session 1**<br>**The LaVille** | 12:25 - 14:50<br>**Lunch & Poster Session 2**<br>**The LaVille** | 12:30 -<br>**Lunch box & Excursion** | 12:25 - 14:50<br>**Lunch & Poster Session 3**<br>**Meet the editors (2F)**<br>**The LaVille** | 12:20 - 13:00<br>**D. Cho**<br>**Closing & B-Boy Performance** |
| | 14:50 - 16:05<br>**Quantum computation & quantum simulation I**<br>**C. Monroe**<br>45min) Randy Hulet<br>30min) F. Ferlaino | 14:50 - 16:05<br>**Intense fields and ultrafast science**<br>**M. Lewenstein**<br>45min) C. H. Nam<br>30min) R. Kienberger | | 14:50 - 16:20<br>**Quantum gases II**<br>**Y. Shin**<br>30min) Chris Vale<br>30min) V. Bagnato<br>30min) A. Hemmerich | |
| | 16:05 - 16:30<br>**Coffee Break** | 16:05 - 16:30<br>**Coffee Break** | | 16:20 - 16:45<br>**Coffee Break** | |
| 17:00 - 19:00<br>**Registration Open**<br>**COEX Auditorium** | 16:30 - 18:30<br>**Quantum computation & quantum simulation II**<br>**A. Derevianko**<br>45min) Rainer Blatt<br>45min) Chris Monroe<br>30min) Akira Furusawa | 16:30 - 18:35<br>**Hot Topics 1**<br>**J. Ahn**<br>25min) Luca Argenti<br>25min) Andy Kung<br>25min) Trey Porto<br>25min) Giacomo Roati<br>25min) S. Gupta | | 16:45 - 18:30<br>**Quantum optics and cavity QED II**<br>**T. Northup**<br>45min) Vladan Vuletic<br>30min) Jian-Wei Pan<br>30min) Yoon Ho Kim | |
| 18:00 - 21:00<br>**Welcome Reception**<br>**The LaVille** | | 21:00 - 22:00<br>**Mini Concert**<br>String Quintet<br>**The LaVille** | | 18:40 - 21:30<br>**Drink &Banquet**<br>**Harmony Ballroom**<br>**COEX Intercontinental Hotel** | |

Map of ICAP2016

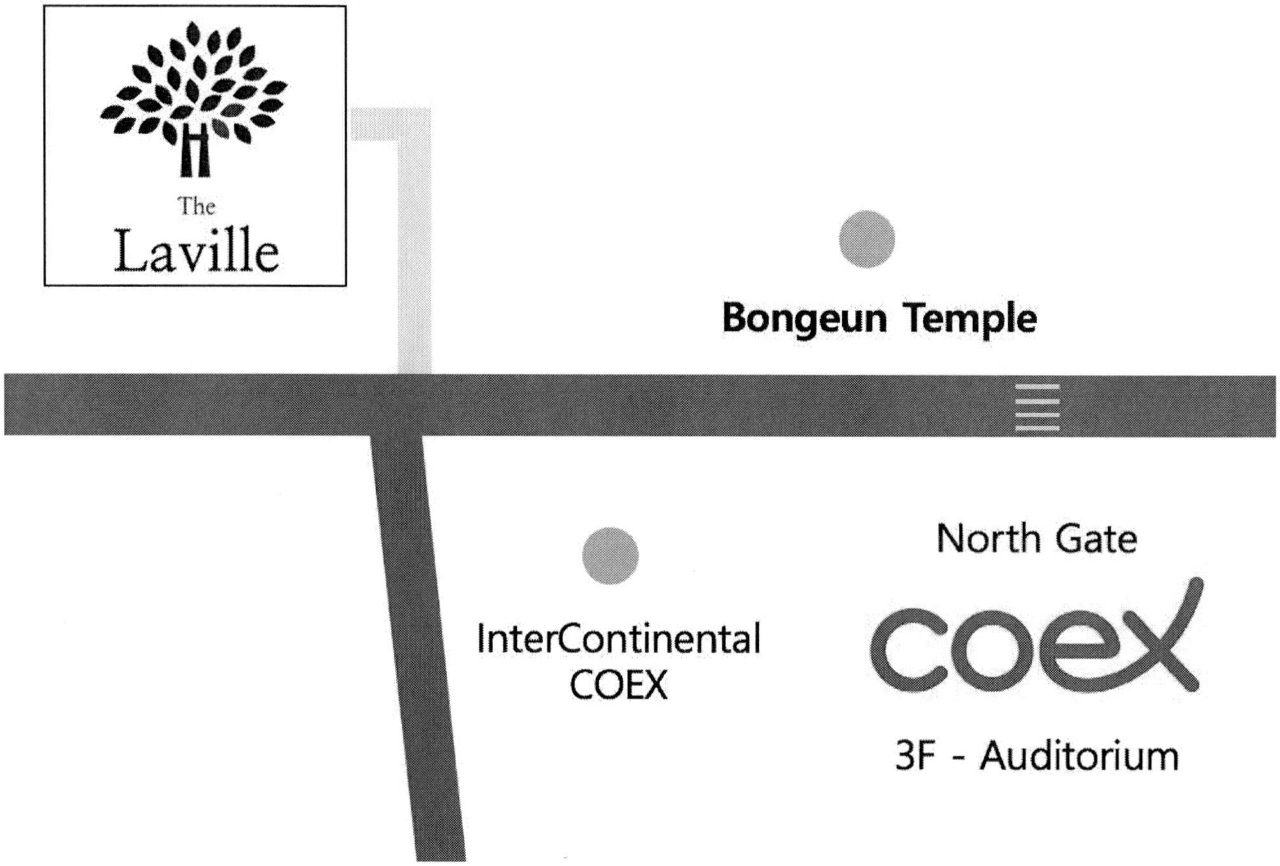

The host agency:

KPS 한국물리학회 The Korean Physical Society 원자분자물리분과 (Div. Atom. Mol. Phys.)

IAP INSTITUTE OF APPLIED PHYSICS SEOUL NATIONAL UNIVERSITY

The lead agency:

IAP INSTITUTE OF APPLIED PHYSICS SEOUL NATIONAL UNIVERSITY

고려대학교 기초과학연구원

KAERI Korea Atomic Energy Research Institute

KPS 한국물리학회 The Korean Physical Society

This work was supported by the Korean Federation of Science and Technology Societies Grant funded by the Korean Government.